Essentials of Aviation Management: *A Guide for Aviation Service Businesses*

Ninth Edition

John H. Mott
Thomas Q. Carney

Cover image © NBAA. Used with permission.

Kendall Hunt
publishing company

www.kendallhunt.com
Send all inquiries to:
4050 Westmark Drive
Dubuque, IA 52004-1840

Copyright © 1977, 1981, 1985, 1990, 1995, 2003, 2010, 2015, 2021 by Kendall/Hunt Publishing Company

PAK ISBN: 978-1-7924-6762-2
Textbook ISBN: 978-1-5249-8901-9

All rights reserved. No part of this publication may be reproduced, stored in a retrieval system, or transmitted, in any form or by any means, electronic, mechanical, photocopying, recording, or otherwise, without the prior written permission of the copyright owner.

Printed in the United States of America

To our families

Contents

List of Figures, xiii
Preface, xv
About the Authors, xvi

Chapter 1 ✈ The Role of the General Aviation Service Center or "Fixed-Base Operator" in the National Aviation System, 1

Objectives, 1
Introduction, 1
   Terminology, 1
   Scope of Book, 2
Aviation's Early Economic History, 2
   Aviation Pioneers and Economic Milestones, 2
   The Jet Engine: New Horizons, 3
   Rotorcraft: New Functions, 3
   Tiltrotor, 4
   The Full Circle: Ultralights and Homebuilts, 4
Market Changes of Recent Decades and Their Implications for Aviation, 4
   1990s Onward—New Wealth, 4
   Security and Terrorism, 4
   The World Wide Web, 4
   Globalization of the Market, 5
   Airline Deregulation, 5
   General Aviation Revitalization Act of 1994 (GARA), 5
   The "Revenue Diversion" Issue, 6
   Threat of Airport Closures, 6
   Airport Improvement Funding, 7
   Fractional Aircraft Ownership, 7
   NASA's Involvement in General Aviation, 7
   Transportation Security Administration, 7
   Revolution in Avionics Technology, 7

Components of the Modern Aviation Industry, 8
   Summary, 8
   Pilots, 8
   Other Aviation Support Personnel, 8
   The Airport System, 9
   The Air Navigation System, 9
   Aviation Manufacturers, 10
   Scheduled Air Carriers, 10
   General Aviation, 10
   Aviation Industry Groups, 11
Aviation Regulation, 14
   Introduction, 14
   Federal Aviation Administration (FAA), 14
   National Transport Safety Board (NTSB), 15
   Transportation Security Administration (TSA), 15
Fixed-Base Operators, 15
   Fixed-Base Operators: Their Role, 15
   Scale and Prospects of the Industry, 16
   Airport Management, 16
FBO Industry Trends and Issues, 16
   Maturity and Professionalism, 16
   Public Awareness, 17
   Technical Issues, 18
Conclusion, 18
Summary, 19
Discussion Topics, 19
Endnotes, 19

Chapter 2 ✈ Management Functions, 21

Objectives, 21
Introduction, 21
   The Four Traditional Functions of Management, 21
Common Managerial Errors and How to Address Them, 22

v

1. Failure to Anticipate Industry Trends, 22
2. Lack of Priorities, 23
3. Indecisiveness and Lack of Systems, 23
4. Poor Time Management, 23
5. Poor Communication Skills, 23
6. Lack of Personal Accountability and Ethics, 23
7. Failure to Develop, Train, and Acknowledge People, 23
8. Failure to Support Company Policy in Public, 24
9. Failure to Acknowledge and Accommodate People's Workstyles, 24
10. Failure to Focus on Profit, 24
11. Failure to Recognize the Needs People Fulfill by Working, 24
12. Failure to Establish and Adhere to Standards, 24

Planning and Organizing, 24
The Need for a Business Plan, 24
Business Plan Outline, 25

Directing, Coordinating, and Controlling, 29
Managing, Versus Doing, 29
Style of Problem-Solving and Delegation, 30
Leadership Styles, 30
Objectives of Delegation, 31
Managerial Control, 31
Choosing the Areas to Delegate, 32
Dos and Don'ts of Delegation, 32

The Decision-Making Process, 32
Decision-Making Tools, 34
Decide Something!, 35

Time Management, 35
Sources of Time-Management Problems, 35
Time Management Strategies, 35

Business Ethics, 37
Summary, 38
Discussion Topics, 38
Endnotes, 39

Chapter 3 ✈ Marketing, 41

Objectives, 41
Introduction, 41
The Need for Marketing in Aviation, 41
Marketing Orientation, 42
Definition of Marketing, 42
Natural Segmentation of Aviation Markets, 43
Market Research, 43
The Nature of Aviation Market Research, 44

National Trends That May Affect General Aviation, 44
Introduction, 44
Terrorism and Its Aftermath, 44
Fractional Aircraft Ownership, 44
Encouraging New Pilots, 46
Sport Pilot Certificate, 46
Airport Closures, 46
Technology, 46
Tort Reform, 47

Sources of Aviation Forecasts and Market Projections, 47
Understanding the Local Aviation Market, 53
Forecasting Techniques and the Individual General Aviation Business, 53
Customer Needs and Identification of New Prospects, 55

Product and Service Definition, 56
Total Product, 56
Product Classification, 56
The Competition, 57
Market Niche, 57
Price, 57
Elasticity, 59
Pricing Policy, 60
Place, 61
Distribution Systems, 61
Promotion, 62
Sales, 65
Collecting, 66
Marketing Controls, 66
Marketing Plan, 66
Contribution Analysis, 66
Performance Evaluation, 66
Quality Control, 67
Budgeting, 67
Information Systems, 67
Integrated Marketing, 67

Marketing the Airport, 67
Summary, 68
Discussion Topics, 68
Endnotes, 69

Chapter 4 ✈ Profits, Cash Flow, and Financing, 71

Objectives, 71
Introduction, 71
Historic Context, 71
Profit and Cash Flow Today, 71

Definitions of Profit, 72
    Reward for Effort, 72
    Reward for Risk, 72
    Return on Investment, 72
    Profit to Sales Ratio, 72
Profit Objectives, 72
    Profit Maximization, 72
    Satisfactory Profit, 73
    Non-Monetary Profit, 73
    Hobby/Business, 73
    Social Responsibility, 73
Profit Levels, 73
    FBO Reports, 73
    What Is Your Profit?, 74
Realizing Profit, 74
    Profit Orientation, 74
    Cost Control, 74
    Planning, 74
    Marketing Orientation, 75
    Information System Design, 75
    Records, 75
    Depreciation Practices, 75
    Inventory, 75
    Bad Debts, 75
    Managerial Decisions, 75
    Cash Flow, 75
Setting Your Cash Flow and Profit Goals, 76
Part 1—Planning for Positive Cash Flow, 76
    Forecasting Sales and Revenues, 76
    Forecasting Expenses, 76
    Month-by-Month Cash Flow Analysis, 76
    Improving the Cash Position, 77
Part 2—Planning for Profits, 78
    Profit Objectives, 78
    Break-Even Analysis, 78
    Profitability Variations among Product Lines, 79
    Profit Centers, 80
    Profit and Loss Statement/Income Statement, 80
    Balance Sheet, 81
Part 3—Budgeting, 81
    Introduction, 81
    Aviation Budget Development, 81
    Steps in Budget Development, 81
    Budget Operation and Control, 91
Part 4—Other Considerations, 91
    Tax Planning, 91
    Competition, 92
    Retained Earnings, 92
    New Revenue Sources, 92

Financing, 93
    Types of Money, 93
    Sources of Money, 93
Credit Management, 94
    Nature and Reason for Credit, 94
    Creating a Credit Policy, 94
    Functions of Credit Management, 94
    Credit Process, 94
    Terms and Definitions, 99
    Cash or Credit Card?, 99
Summary, 99
Discussion Topics, 100
Endnotes, 100

Chapter 5 ✈ Human Resources, 101

Objectives, 101
Pipeline Concept, 101
    Control Points, 101
    Scope of Chapter 5, 102
Aviation Industry Trends, 103
    Labor Market Trends in Aviation, 103
    Industry Maturity and Professionalism, 106
Identifying Human Resource Needs, 106
    The Human Resources Component of the Business Plan, 106
    Permanent or Temporary Needs, 107
    Skills Required, 107
    Job Descriptions and Specifications, 107
    Special Activities, 107
Laws and Regulations, 109
    Introduction, 109
    Employment Discrimination, 111
    Workplace Safety, 112
    Workers Compensation: An Overview, 113
    Comparable Worth, 113
    Payroll Taxes and Deductions, 114
    Employee Access to Records and Fair Information Practices, 114
Recruiting Qualified Candidates, 114
    Industry Contacts, 114
    Recruiters, 114
    Employment Agencies, 115
    Colleges and Trade Schools, 115
    Advertising, 115
Selecting Employees, 116
    Preliminary Screening, 116
    The Application Form, 116
    The Interview, 116
    Testing and Investigation, 123

Background and References, 123
The Physical Examination, 125
The Job Offer, 125
Orientation and Training, 127
New Employees, 127
Training 127
Communicating, 129
Basic Elements, 129
Barriers to Effective Communication, 130
Verbal and Non-Verbal Communications, 130
Improving Communications, 131
Motivating, 132
Array of Needs, 132
Needs Satisfied by Working, 133
Creating a Motivating Environment, 133
Leadership, 134
Evaluating Employees, 134
Promoting Employees, 139
Compensation Systems, 139
Job Evaluation, 139
Fringe Benefits, 139
Administration of the Total Compensation Plan, 140
Disciplinary Problems, 140
Conflict Resolution, 140
Administering Discipline, 140
The "Red-Hot-Stove" Rule, 141
The Troubled Worker, 141
Separation, 142
Personnel Policy Manual, 142
Manual Style, 144
Personnel Records, 144
Employee Organizations, 144
Impact on Management, 144
Summary, 145
Discussion Topics, 146
Endnotes, 146

Chapter 6 ✦ Organization and Administration, 147

Objectives, 147
Introduction, 147
Goals and Objectives, 147
General, 148
Financial Practices, 148
Personnel, 148
Physical Assets, 148
Selecting Information, 148
Routine-ize the Routine, 149

A World of Change, 149
Legal Structure, 149
Sole Proprietorship, 149
Partnership, 150
Corporation, 151
Principles of Internal Organization, 151
The Rational Model vs. Some New Approaches, 152
Explicit Corporate Philosophy, 152
Organizational Culture, 152
Specialization and Job Rotation, 153
Decentralization and Decisions by Consensus, 153
Management by Walking Around, 154
Management by Results, 154
Span of Control, 154
Effective Work Groups and Teams, 155
Staff Support, 155
Human Factors, 156
New Approaches to Organization, 157
Communications Technology and Applications for Internal Organization, 157
Internal Structure Design, 158
Formal Internal Structure, 158
Functional or Matrix Management, 158
Line Organization, 159
Line and Staff Organization, 159
Informal Internal Structures, 159
Quality Circles, 161
Task Forces, 161
Social Structures, 162
Other Networks, 162
External Pressures on Choices of Structure, 162
Industry Norms, 162
Government and Regulatory Influences on Organizational Structure, 162
Practical Applications Guidelines, 162
Problems, 163
The Organization Manual, 164
Manual Outline, 164
Summary, 164
Endnotes, 164
Discussion Topics, 165

Chapter 7 ✦ Management Information Systems, 167

Objectives, 167
Introduction, 167

Industry Changes—Franchises
and Computerization, 167
System Purposes, 168
    Financial Survival, 168
    Performance Monitoring and
      Improvement, 169
    Special Reports, 169
    Taxes and Legal Obligations, 170
    Checks on Reality, 170
System Processes, 170
    Integrated Flow, 170
    Requirements of an Effective System, 172
    First Steps, 172
Aviation Management Information Systems, 172
    Human Resources, 172
    Financial, 172
    Material, 172
    Aviation Operations, 176
    Legal and Tax Information, 176
    Market Information, 176
    Technological Information, 176
Analyzing Business Activity, 177
    General Procedure, 177
    Analyzing the Business as a Whole, 177
    Management Audit, 179
Analyzing Departmental Activity, 179
    Introduction, 179
    Net Income for the Period—Profit
      or Loss, 180
    Taking Action, 183
Records and Record-Keeping, 183
    Records Design, 184
    Correspondence, 184
    Records Management, 185
    Communications, 189
    Duplicating Information, 190
Aviation Accounting, 190
    Accounting Flow, 190
    Source Documents, 190
    Journals, 191
    General Ledger, 193
    Financial Statements, 193
    Accounting Activity Flow Chart, 193
    Accounting System, 193
    Profit Center Accounting, 194
Information System Tools, 196
    Introduction, 196
    Desktop Computers, 197
    Selecting the Right Functionality, 197
    Computer Selection Considerations, 198

    FBO Management Software Choices, 198
    Computer Service Bureaus, 200
    Other Devices, 200
Business Security, 200
    Confidentiality and Control of Information, 200
    Types of Losses, 200
    Methods of Combating Losses, 201
Summary, 201
Discussion Topics, 201
Endnote, 202

Chapter 8 ✈ Operations: Flight Line and Front Desk, 203

Objectives, 203
Introduction, 203
Customer Service, 203
Staying Up to Date, 204
    Technological Change, 204
    Facility Appearance, 204
Flight Line, 204
    Line Layout, 205
    Line Operations, 206
    Line Administration, 207
    Training Line Personnel, 211
    Service Array and Profitability, 212
Fueling, 212
    Trends in Use of Alternative Fuels, 213
    Self-Fueling, 213
Front Desk, 213
    Procedures, 214
    Staying Close to the Customer, 215
    Related Services, 215
    Flight Planning and Services, 215
    FAA Weather Services, 215
    Weather Information Systems, 216
    Other Pilot and Passenger Services, 216
Summary, 217
Discussion Topics, 217
Endnotes, 217

Chapter 9 ✈ Flight Operations, 219

Objectives, 219
Introduction, 219
    Types of Flights, 219
    Market Trends, 220
    System Issues Affecting Flight Operations, 220
    Choosing What Services to Offer, 226
    Organization, 226

Air Transportation, 226
    Benefits, 226
    Charter and Air Taxi, 227
    Aircraft Rental, 228
    Fractional Aircraft, 229
    Aircrew and Ferry Services, 229
    Air Cargo, 229
    Air Ambulance/Medical Evacuation, 230
Other Commercial Flight Operations, 231
    Aerial Patrol, 231
    Aerial Application, 231
    Aerial Advertising, 234
    Fish Spotting, 234
    Aerial Photography, 234
Flight Instruction, 234
    The Changing Market, 234
    Training Programs, 234
    Instruction Administration, 236
    Flight Instructors, 237
    Instructor Training, 238
    Freelance Instructors, 238
    Simulator/FTD Usage, 238
Sport and Recreational Flyers, 239
    Gliders and Sailplanes, 239
    Parachuting, 239
    Ultralights, 239
    Experimental and Home-Built Aircraft, 239
    Balloons, 239
    Rotorcraft, 240
    Sight-Seeing, 240
Aircraft Sales, 241
    New Aircraft, 241
    Used Aircraft, 241
    Brokerage, 241
    Demonstration Flights, 241
Flight Operations Manual, 242
Summary, 242
Discussion Topics, 242
Endnotes, 243

Chapter 10 ✈ Aviation Maintenance, 245

Objectives, 245
Introduction, 245
    Goals of the Maintenance Shop, 245
    Changing Issues, 246
Maintenance Activity, 247
    Overview, 247
    Organization, 248
    Certification, 249

Personnel, 252
    Qualifications, 252
    Training, 253
    Certification, 254
    Capabilities and Limitations, 254
    Inspection Authorization, 254
    Repairmen, 254
Facilities and Equipment, 255
    Overview, 255
    Managerial Concerns, 255
Parts and Supplies, 256
    Inventory Control, 256
Quality Control, 259
    Training, 260
    Checklists, 260
    Inspection, 260
    Recognition, 260
    Balance, 260
Competition, 264
    Nonexclusive Rights, 264
    Referrals, 264
    Outsourcing, 264
    "Through-the-Fence" Operations, 264
    Tailgate, Shade Tree, and Gypsy Mechanics, 265
    Corporate and Other Self-Maintenance, 265
Administration, 265
    Flat-Rate Pricing, 266
    Computer-Assisted Maintenance, 272
    Profitability, 274
    Information, 274
Analysis, 274
    Control, 274
    Techniques, 274
Professional Maintenance Organizations, 275
Avionics Repair Stations, 275
Summary, 277
Discussion Topics, 277
Endnotes, 277

Chapter 11 ✈ Safety, Security, and Liability, 279

Objectives, 279
Introduction, 279
    The Need for Risk Management Procedures, 279
    Interaction of Safety, Security, and Liability, 280
Risk Exposure, 280
    Normal Business Exposure, 280
    Aviation Risk Exposure, 280

Risk Reduction, 281
    Normal Risk Reduction, 281
    Aviation Risk Reduction, 283
    Aviation Safety and Security Regulations
        and Guidelines, 284
    Airport Risk Audit, 284
    Procedures Manual, 284
    Documentation, 285
    Inclement Weather, 285
Risk Transfer, 285
    Principles of Insurance, 285
    Insurance Regulations, 285
    The U.S. Insurance Market, 285
    Normal Business Insurance, 286
    Special Aviation Coverages, 289
    Aviation Tenant-Landlord Agreement, 291
    Selection of Aviation Insurance, 291
    Selection of Aviation Insurer, 292
Accident Policy and Procedures, 292
    Federal Reporting Requirements, 292
    State and Local Reporting Requirements, 294
    Aircraft Rescue and Firefighting (ARFF)
        Procedures, 294
    Search and Rescue, 295
Aviation Security, 295
    Flight Security, 295
Summary, 297
Discussion Topics, 297
Endnotes, 297

Chapter 12 ✈ Physical Facilities, 299

Objectives, 299
Introduction, 299
The Four Levels of Airport Service Business Involvement in Physical Facilities, 299
The National Airport Hierarchy, 300
    The Airport System, 300
    The Airspace System, 302
    Public Airport Organizational Structure, 302
The Airport's Wider Environment, 302
    Overview, 302
    How FAA Handles Aviation Noise, 303
    FAR Parts 150 and 161, 303

Facilities on the Airport, 304
    Introduction, 304
    The Airport Master Plan and ALP, 304
    Participation by Airport Businesses in
        Airport Policy and Planning, 305
    Private Airports, 305
    Airport Revenue Planning, 306
The FBO's Own Facilities, 306
    Data Collection, 306
    Planning a New Airport, 307
    Facility Expansion, 307
    Preventive Maintenance, 307
    Scheduled Replacement of Plant
        and Equipment, 307
    Zoning and Other Local Controls, 308
    Environmental Compliance, 308
    Leases, 308
    Competition with Other FBOs, 319
Threats to the General Aviation Physical System, 320
    Lack of Appreciation of GA's Role, 320
    Airspace Restrictions, 320
    Noise and Operating Restrictions, 321
    Airport Closures, 321
    Disappearance of Private Airports
        and Back Country Strips, 321
    Siting New and Expanded Facilities, 322
    Lack of Suitable Land Use Controls, 322
    Bird Strike Threat, 323
    Revenue Diversion, 323
    Inadequate General Aviation Funding, 323
New Initiatives and Opportunities, 325
    Noise Abatement Programs, 325
    Continuous Noise Abatement Strategy, 325
    Economic Benefits Studies, 325
    Airport Privatization Program, 326
    NASA Small Aircraft Program, 327
    Airport and Aviation System Planning, 327
Summary, 327
Endnotes, 327
Discussion Topics, 328

Index, 331

List of Figures

1.1	General Aviation Aircraft Shipments, 10	4.1	Monthly Cash Flow Chart, 77
1.2	U.S. Civil Aircraft Fleet, 2010 to 2040, 11	4.2	Break-Even Chart, 78
1.3	Thousands of Aviation Hours Flown, FY 2010 to 2040, 12	4.3	Income Statement, 80
1.4	A Taxonomy of General Aviation Flying, 13	4.4	Balance Sheet, 82
1.5	Expanded Market for Flying, 17	4.5	Annual Sales Budget Worksheet, 85
2.1	Business Plan Outline, 26	4.6	Monthly Budget Report, 86
2.2	Managing Versus Doing, 29	4.7	Combined Annual and Monthly Purchases Budget Worksheet, 88
2.3	Dos and Don'ts of Delegation, 33	4.8	Expense Budget Worksheet, 89
2.4	Practical Decision Tools for the Aviation Manager, 34	4.9	Budget Worksheet Used to Formulate Complete Budget, 90
2.5	Meeting Management, 37	4.10	Cash Budget Worksheet, 92
3.1	Typical Marketing Department, 42	4.11	Sample Credit Policy, 95
3.2	Sport Pilot Certificate, 47	4.12	Credit Application Form Available to Piper Aircraft Dealers, 96
3.3	An Alternative View of GARA, 48	4.13	Credit Check Form, 98
3.4	Active GA Fleet, 2020, 49	5.1	Human Resource Pipeline, 102
3.5	GA Flight Hours, 2020, 49	5.2	Tasks Required in FBO Positions, 108
3.6	FAA's Assessment of General Aviation Trends Through 2040, 50	5.3	Skills Required of Various FBO Positions, 109
3.7	Aircraft Utilization Assumptions through 2040, 50	5.4	Training Required for Aviation Maintenance Positions, 110
3.8	Active Pilots by Type of Certificate, 51	5.5	Job Description Format, 117
3.9	Factors Affecting Future GA Trends, 51	5.6	Screening Guide to Assist in Hiring Review, 118
3.10	Example of Data Provided in FAA Terminal Area Forecasts, 52	5.7	Application for Flight Instructor Position, 119
3.11	General Aviation Publications, 54	5.8	Appropriate and Inappropriate Questions During Interactions with Candidates, 124
3.12	Industry Market Research and Sales Planning Process, 54	5.9	Twelve Steps to Assessing a Job Candidate's Suitability, 125
3.13	Shape of Typical Fuel Cost Curves when Average Variable Cost Is Assumed Constant, 58	5.10	Telephone Reference Check Guide, 126
3.14	Demand Curve for Aviation Fuel, 59	5.11	Model of the Communication Process, 129
3.15	Supply Curve for Aviation Fuel, 60	5.12	Maslow's Hierarchy of Needs: Two Views, 132
3.16	Equilibrium of Supply and Demand for Aviation Fuel, 60	5.13	Sample Performance Appraisal, 135
3.17	Four Basic Channels of Distribution, 62	5.14	21 Steps to Resolving Disputes, 141
3.18	Promotion Planning Chart for a General Aviation Organization, 63		

5.15 Exit Interview Form to Be Used with the Departure of Each Employee, 143
5.16 Personnel Forms and Records, 145
6.1 Key Characteristics of a Group, 155
6.2 Functional or Matrix Organization, 158
6.3 Line Organization, 159
6.4 Line-Staff Organization, 160
7.1 Fuel Sales by Type by Customer, 171
7.2 Human Resource Data Needs, 173
7.3 Cash Data Needs of an Organization, 173
7.4 Summary of Business Assets, 173
7.5 Sample Physical Inventory Worksheet, 174
7.6 Aviation Operations Data, 176
7.7 General Ledger Standard Income Report, 181
7.8 Management Analysis and Action, 183
7.9 Information Cycle, 184
7.10 Information System Activity Flow Chart, 191
7.11 Sample Computerized Invoice, 192
7.12 Contribution Margin as Graphically Illustrated in a Break-Even Chart, 196
8.1 Sample Line Service Request Form, 208
8.2 Sample Line Invoice Form, 209
8.3 Sample Line Invoice Form, 210
9.1 NBAA Noise Abatement Techniques, 223
9.2 AOPA Guidelines for Noise Abatement, 224
10.1 Typical Organization of a Maintenance Department, 249
10.2 Federal Aviation Administration Application for Repair Station Certificate, 250
10.3 Parts Sales Summary, 257
10.4 Piper Aircraft Inventory Card, 258
10.5 Basic EOQ Graph, 259
10.6 Aircraft Maintenance Inspection Checklist, 261
10.7 Shop Order, 267
10.8 Shop Order Screen, 268
10.9 Internal Shop Order Form, 271
10.10 Sample Page From an Aviation Flat-Rate Manual, 272
10.11 Maintenance Warning Report, 273
10.12 Individual and Shop Productivity, 275
10.13 Break-Even Chart Used for Controlling Overall Service Shop Productivity, 276
10.14 Key Ratios and Normally Accepted Standards for Maintenance Departments, 276
11.1 Fire Inspection Checklist, 282
11.2 Sample Security Mission Statement, 296
12.1 National Hierarchy of NPIAS Airports, 300
12.2 Distribution of Activity, 301
12.3 Grant Conditions Under Airport Improvement Program, 306
12.4 NPIAS Development Cost Estimates by Airport Category, 324
12.5 Total NPIAS Cost ($43.6 Billion) Estimates by Type of Development, 324

Preface

Since *Essentials of Aviation Management* was first published in 1977, much has changed within the U.S. general aviation industry. In some sense, it is an industry that is embattled and one that faces challenges on multiple fronts. Due to the increasingly prohibitive costs inherent in aircraft operation, annual shipments of new aircraft have declined dramatically from their post-war peak of over 17,000 in 1978, and aircraft ownership has decreased as well. Not even the passage of the General Aviation Revitalization Act in 1994 could cause the level of general aviation aircraft sales to return to anywhere close to the numbers found in the 1970s. To be sure, certain niche areas within the overall category of general aviation, such as experimental and light sport aircraft, and higher-end business jets and the fractional ownership subindustry, have seen increases in recent years. But general aviation aircraft ownership has never quite made the jump from luxury to legitimacy, despite diligent efforts of such organizations as AOPA, NBAA, and others, and this image problem continues to have a widespread effect on the industry.

While results remain to be seen, the Covid-19 pandemic of 2020–21 may actually present a bright spot for general aviation, however, as individuals choose to travel in smaller groups and utilize smaller aircraft, moving away from crowded airport terminals and airliners that operate with what had been, pre-pandemic, the highest industry load factors on record.

With the prevalent lack of understanding of general aviation by much of the public and politicians, funding for general aviation airports is increasingly threatened. Few other businesses are so dependent for their physical facilities on landlords who are rarely exclusively committed to the GA cause. With the U.S. now losing scores of general aviation airports per year due to funding issues and repurposing, it is imperative that aviation businesses play an active role in ensuring a positive business climate for their home fields and for other airports as well.

With this decline in deliveries, ownership, and facilities has come a similar decline in general aviation operations, pilot certification, and, more importantly for the readers of this book, sales at fixed base operators (FBOs). The number of FBOs operating in the U.S. in the early 1980s was estimated to be over 10,000, whereas there are roughly a third of that number operating today. The trend in the FBO business is one of consolidation, with many smaller FBOs being acquired by larger chains. Still, there remains a place for smaller FBOs at smaller general aviation airports. Regardless of size, however, it is clear that, because of the challenges faced by the industry, FBOs must be well managed to survive.

Information technology may be considered the great equalizer of the late 20th and early 21st centuries, just as Samuel Colt's revolver was viewed on the American frontier in the late 19th century. General aviation customers with Internet access now have virtually unlimited ability to research airports and FBOs before patronizing them. It is easy to determine pricing on the industry's number one commodity, fuel; prior to purchasing it; thus, value-added and non-commoditized services provided by FBOs will continue to increase in importance. Managers will want to take advantage of the wealth of operational data afforded them by their in-house information systems as well, allowing them to analyze profit centers and potential areas for cost savings as never before.

Globalization is a major force that affects the aviation industry as a whole, with increased levels of international business, foreign travel, and foreign students in the U.S., and the trend of U.S. companies to move manufacturing jobs abroad as well as that of foreign companies to establish factories in the United States. We in aviation are now players in a global marketplace, whether or not we desire to be, and we must adjust our perceptions of our place and function in the marketplace accordingly.

This book seeks to provide the analytical tools that will assist the manager of the aviation service business who wishes to be competitive in today's aviation environment capitalize on the challenges and opportunities that we have briefly mentioned in this preface. It is intended to serve both undergraduate and graduate academic audiences, as well as to be a practical reference for the individual in industry. Additional trade and industry materials, many of which are available on the Internet from a host of professional organizations, may be used to supplement the text. The appendices from previous editions are now located on the companion website for greater utility, and all URLs have been updated. New to this edition are end of chapter questions, PowerPoint slides, and test bank questions with answers for instructors. Numerous suggestions regarding format and content from both academic instructors and industry leaders have been incorporated as well.

The thoughts and ideas presented within the book are synthesized from both the academic and practical experiences of the authors as they have been involved in their respective aviation careers.

We hope that the ninth edition of *Essentials of Aviation Management* will continue to be valuable to both students of and practitioners in the general aviation industry.

John H. Mott
Thomas Q. Carney

About the Authors

John H. Mott

Dr. John H. Mott is an Associate Professor in the School of Aviation & Transportation Technology at Purdue University. A summa cum laude graduate of the University of Alabama with bachelor's and master's Degrees in Electrical Engineering, and of Purdue University with a Ph.D. in Civil Engineering, he also possesses an FAA Commercial Pilot Certificate with Instrument and Multiengine ratings, an FAA Flight Instructor Certificate with Airplane Single-and Multiengine and Instrument ratings, and an FAA Ground Instructor Certificate with Advanced and Instrument ratings. He holds type ratings in the Beech King Air 300, Hawker HS-125, and Canadair Challenger 600. He has worked as a flight instructor, a charter pilot, chief pilot, and director of training for a FAR 135 operator, an airline pilot flying the SA-226TC Metroliner, and a corporate pilot in many different models of aircraft.

Dr. Mott serves as the Director of the Advanced Aviation Analytics Center of Research Excellence (A3IR-CORE) and was the founding editor of the *Journal of Aviation Technology & Engineering*, where he served as executive editor through 2018. Dr. Mott's research is focused primarily on the aggregation of distributed data related to various operational aspects of transportation systems, the analysis of that data using deterministic and stochastic mathematical modeling, and the development of related tools to facilitate improvements to the safety and efficiency of those systems. While this research is inherently interdisciplinary and crosses boundaries between the school's focus areas of safety, quality, and sustainability, Dr. Mott's principal focal areas are those of quality and safety. The related challenge is the issue of employing data and decision science to improve the quality and safety of the services provided within the transportation industry, which aligns with the restoration and improvement of urban infrastructure, one of the National Academy of Engineering's 14 Grand Challenges. The collection and analysis of data from transportation systems facilitates improved management decisions regarding those systems and ultimately improves the quality of life of those who utilize transportation services.

Prior to joining Purdue, Dr. Mott was Engineering Manager for a firm in Birmingham, Alabama, and also operated an electronics design and manufacturing business for many years. He is a Senior Member of the Institute of Electrical and Electronics Engineers (IEEE) and the Chair of the Signal Processing and Communications Society Chapters of the Central Indiana Section, a member of the Board of Trustees of the University Aviation Association (UAA), and a member of the Standing Committee on Airfield Capacity and Delay of the Transportation Research Board of the National Academies of Science, Engineering, and Medicine. He is also a member of the Institute of Transportation Engineers (ITE).

Dr. Thomas Carney

Dr. Thomas Carney is a Professor Emeritus of Aviation & Transportation Technology and former department head of the Department of Aviation Technology at Purdue University, where he taught from 1972 through 2018. Dr. Carney has more than 53 years of experience as a pilot, with more than 11,000 flight hours, and holds the Airline Transport Pilot certificate with multiengine, Mitsubishi Diamond, and Beechjet type ratings, in addition to the Certified Flight Instructor certificate with airplane single- and multiengine, and instrument airplane ratings. He was the 1996 recipient of the James G. Dwyer award for outstanding undergraduate teaching in the School of Technology and is a member of the inaugural group of 225 faculty members included in the Purdue University Book of Great Teachers. In 2002, he was awarded the William A. Wheatley award by the University Aviation Association. Dr. Carney was the recipient of the 2003 Outstanding Alumnus Award from the Earth and Atmospheric Sciences Department at Purdue, and in 2012 he received the Distinguished Science Alumnus award from the Purdue College of Science. In 2004, he was named "Member of the Year" by the Council on Aviation Accreditation, and in 2006, he received the President's Award from the University Aviation Association, for long-term leadership and service. In 2008, he received a recognition award for "leadership and guidance…provided to future airline pilots", from the Airline Pilots Association. Dr. Carney received the Aviation Accreditation Board International (AABI) Chairperson's Award, in July 2017, and in April 2018, he was presented with the FAA Wright Brothers' Master Pilot Award.

In addition to his flight background, Dr. Carney holds M.S. and Ph.D. degrees in Atmospheric Science, and prior to his retirement maintained a courtesy faculty appointment in the Purdue Department of Earth and Atmospheric Sciences. His primary research areas of interest in aviation and atmospheric science include aviation meteorology and the impact of weather on aviation operations, synoptic-scale dynamics and energetics and the interactions between synoptic- and mesoscale motion fields, aviation leadership and management, and effective methods for recruiting and preparing the next generation of aviation professionals. He is the author, or co-author, of more than 100 publications and presentations, and is a current or former member of a number of national and international organizations and committees relating to collegiate flight education and accreditation, aviation meteorology, corporate flight operations, and the ICAO Next Generation of Aviation Professionals Task Force.

Dr. Carney formerly served as the senior editor of the Collegiate Aviation Review and is a member of the editorial board of the Journal of Aviation/Aerospace Education and Research. He currently serves as the AABI Vice President of Accreditation and chairman of the Accreditation Committee and is a past chairman of the Certified Aviation Manager Governing Board (CAMGB) of the National Business Aviation Association (NBAA). He also served as the founding chairman of the NGAP (Next Generation of Aviation Professionals) Task Force of the International Civil Aviation Organization (ICAO). Dr. Carney currently serves as chairman of the Friends of Mammoth Cave National Park Board of Directors and as expert witness in litigation involving flight operations and aviation meteorology.

The Role of the General Aviation Service Center or "Fixed-Base Operator" in the National Aviation System

OBJECTIVES

› Explain the term "FBO," and the functions such businesses fulfill in the general aviation community.

› Discuss the historical development of powered aircraft and why some believe we have come "full circle."

› Describe some of the aviation trends and issues relative to the fixed-base operator/airport service business.

› Increase the reader's awareness of external and non-aviation issues facing general aviation and impacting its future.

Introduction

Terminology

This book is a small business text for the management of United States general aviation service centers or Fixed-base Operators (FBOs), which are the service stations of the general aviation system. It addresses two audiences: the aviation student who is seeking a basic understanding of FBO management, and the fixed-base operator owner, manager, and staff who are practitioners in the field with varying levels of formal business training.

Early aviation consisted mainly of acrobatic pilots, who traveled across the country conducting air shows. These pilots, who used fields as landing strips, were typically accompanied by teams of independent mechanics and flight instructors. As aviation grew, so did the need for the establishment of airports. Many traveling mechanics and flight instructors chose to locate their businesses at "fixed" locations on these airports. In order to differentiate between those who were operating at fixed locations and those acting in a roving capacity, the Civil Aeronautics Authority (CAA), predecessor of the Federal Aviation Administration (FAA), coined the term Fixed-base Operator (FBO).

"Fixing" these general aviation centers, whose services grew to include charter flights and refueling, played a major role in the development of general aviation. One of the biggest benefits of this trend was

1

its positive impact on safety. The CAA could impose more stringent documentation and inspection standards on operators at fixed locations. As a result, even the most seasoned businessmen were surprised by the amount of paperwork that was required to run an FBO—a trend that continues today. Reed Fuller of Ascent Aviation often jokes, "Planes don't fly on fuel. Planes fly on paperwork."

In an effort to reduce the number of roving operators, who were harder to control, the CAA allowed FBOs a greater degree of operational leniency. Many of these regulatory differences are still evident today. For example, private flight instructors must abide by Part 61 of the Federal Aviation Regulations in terms of flight hour requirements for their students. These requirements exceed those found in Part 141, which governs fully-qualified flight schools.

"FBO" is an elusive term that is now applied to almost any general aviation business existing on an airport. Over the years, there has been considerable industry discussion about the idea of changing the name "FBO" to some other term more meaningful to the non-aviation public. "General aviation service center" or "aviation service business" are both probably more descriptive. However, old habits die hard, and within the industry, "FBO" is well understood; therefore, in this book we use all of these terms interchangeably, and the reader should assume they are synonymous. Note that, for the purposes of the book, "FBO" does not necessarily mean a full-service organization.

Scope of Book

This book places the aviation service business in context as part of the national aviation system. It reviews, from an aviation perspective, current small-business practice and theory in areas such as business planning, marketing, financing, human resources, administration and information systems. The principal areas of general aviation center activity, such as flight operations, flight instruction, front desk, and aviation maintenance, are explored. The regulatory context of each area is summarized in its corresponding chapter. Physical facility planning for general aviation center areas and other parts of an airport is discussed. Finally, the book examines future trends likely to affect the general aviation industry.

Various appendices (provided at this book's website), chapter notes, and chapter questions provide additional technical data and aviation website references for further reading.

Aviation's Early Economic History

Aviation Pioneers and Economic Milestones

Historically, aviation development in the U.S. was initially led by the sake of technological innovation. However, within a couple of decades, aviation had grown as a key element of the rapidly evolving national transportation system, and thus, of the economy. The first part of the 20th century saw an extraordinary rate of aeronautical experimentation and development that changed how not only the U.S., but also the whole world, interacts and does business. Technological advances in pursuit of new records, and feats of endurance in aviation and space have quickly led to commercial applications. This history is well documented in numerous texts.[1] What follows is a very brief overview.

When human beings first began to build flying machines, particularly when they began to build powered craft, their primary preoccupation was to understand and master aerodynamics; potential markets for the technology were merely a secondary consideration. Orville Wright, even ten years after Kitty Hawk, felt there was almost no chance of successfully completing a transatlantic flight.[2]

Yet some, even in the very early days, understood the economic and transportation role of aviation. Experimenters such as Samuel Langley succeeded in getting financial backing from people interested in the large-scale practical uses of aircraft.[3] With Lindbergh's solo transatlantic flight in 1927, and the huge public acclaim that accompanied this feat, the perception of the market potential of aviation began to change. From around this time, passenger flight became a market reality.

For airmail, it changed much earlier. The first U.S. airmail service was conducted during the week of September 23, 1911 by Earle L. Ovington, who became the first air mail carrier. During this seven-day period, over 37,000 pieces of mail were carried between Long Island, NY and Minneola, NY. It did not take the Post Office Department long to recognize the benefits of developing the airplane into a practicable means of aerial transportation. Airmail represented the government's first, non-military interest in aviation.

By the mid-1920s, especially in the United States, the capabilities of heavier-than-air craft were apparent. Passenger service began to be popular in the 1920s and 1930s. Between 1911 and 1927, the post office handled airmail in its own aircraft. However, the

Airmail Act of 1925 enabled passenger air carriers to contract for mail service.

By early 1926, twelve contract airmail routes had been awarded. In 1927, the Boeing Aircraft Company received a bid award (later transferred to Boeing Air Transport) and in the first two years of operation carried about 6,000 passengers as well as 1,200 tons of airmail. The wider acceptance of aircraft as passenger carriers did not occur until the 1930s after several legislative changes designed to strengthen national airmail routes and encourage financially viable carriers. This was a stormy period in the development of commercial aviation; however, by the end of 1936 the air carriers were finally making more revenue from passengers than from mail.

Some earlier acceptance of air travel had resulted from nineteenth and twentieth century European lighter-than-air flying machines. Henri Giffard built the first controllable dirigible as early as 1852, and in 1909 Count Zeppelin started the world's first airline, called DERLAG.

During World War I, the Germans used dirigibles extensively for bombing. In the 1920s and 1930s several other nations became active in the development of passenger and military airships. The passenger airship era ended after a series of crashes, punctuated by the infamous explosion of the Hindenburg in 1937. Since then, airships have been used mostly for war or marketing purposes.

The Jet Engine: New Horizons

World War II military necessity gave the jet engine its impetus, although experimenters in this area had been working on it for decades.[4,5] The commercial jet's greater speed and range opened up significant new horizons for aviation, especially in larger countries such as the United States and Canada. During World War II, jet technology became refined, and after the war commercial jet applications began. The implications of jet aircraft travel pertained not only to speed but also to:

> Range of flight without refueling;
> Higher noise levels (with a few exceptions);
> Runway length requirements;
> Fuel type requirements;
> Fuel consumption; and
> Traffic mix in the flight pattern.

The subsonic *(slower than the speed of sound)* jet is now standard for most air transportation. Decreasing transatlantic or transcontinental travel time to about six hours, and thus eliminating refueling stops was a greatly significant achievement for the flying public.

Supersonic *(faster than the speed of sound)* transports first entered commercial service in 1975. The earliest supersonic transport, the Tupolev TU-144, was introduced by Aeroflot in late 1975 and operated until 1978.[6] The later and more successful Concorde, manufactured by a joint venture of Aerospatiale and British Aircraft Corporation, flew regular transatlantic flights from London and Paris to Washington, DC beginning in 1976. Due to economic factors, as well as a July 25, 2000 crash in which 113 people were killed, the Concorde was taken out of service permanently.

Hypersonic *(five times or more faster than the speed of sound)* transports may become a reality in the near future through technology, such as the scramjet, currently in development. The primary difference between a scramjet engine and a rocket is that a scramjet uses air from the atmosphere to burn its fuel, rather than carrying an oxidizing substance on board. Top speed projections for a scramjet engine range between Mach 12 and Mach 24, which is over seven times faster than the SR-71 Blackbird. On May 22, 2009 a joint research program between the Defense Science and Technology Organization of Australia and the U.S. Air Force successfully conducted the first test flight of a hypersonic aircraft at the Woomera test range in Australia.[7]

Rotorcraft: New Functions

The concept of a flying machine that could hover, land almost anywhere, and perform operations without even needing to land was a dream for many aircraft inventors over the years. Its development has evolved to the design and use of very sophisticated machines with range, speed, and load capacity close to midsized, fixed-wing aircraft. Because of their ability to operate from a site as small as a rooftop, increasing numbers of corporations have turned in the past few decades to the use of rotorcraft. Their use has been of major importance in the development of offshore oil resources and servicing of rigs, in forest-fire fighting, in pipeline laying and patrol, in construction of tall buildings, and in search and rescue. They have also been important in activities such as logging, which otherwise requires the construction of roads for access. Rotorcraft, though not a major portion of the total aviation fleet, do comprise a rapidly growing one.

With the downsizing of the military in the late 1980s and early 1990s, the supply of helicopter-rated pilots available to meet the growing need by industry became an issue.

Tiltrotor

A tiltrotor offers the vertical takeoff and landing capability of a helicopter, with the cruising speed, altitude, and range of a fixed-wing aircraft. It represents a totally new approach to executive transportation, search and rescue, and law enforcement applications. The AW609 Tiltrotor, manufactured by AgustaWestland NV, is the world's first civilian tiltrotor. It combines the speed, altitude, and comfort of a turboprop with the vertical takeoff and landing capabilities of a helicopter. Long delayed by testing difficulties, certification of the AW609 is slated for 2021.[8]

The Full Circle: Ultralights and Homebuilts

It has been said that if the Wright Brothers had dacron fibers and aluminum tubing at their disposal, they would have invented the ultralight. Certainly the appearance and handling of the first generation of ultralights (such as B1-RD and Kasperwing) were very similar to the aircraft in which the Wrights first undertook powered flight.

There has been a tremendous surge of interest in recent years not only in ultralights, but also in related heavier aircraft including the new category of light-sport aircraft approved by the FAA in 2004. Interest has grown in a great array of experimental and homebuilt aircraft. The FAA recognized this interest as early as 1987 with the issuance of its Recreational Pilot certificate; however, even that certificate proved too restrictive due in part to its requirement for the pilot to stay within 50 miles of home base. It was not until FAA's Sport Pilot certificate was created in 2004 that the licensing barrier to easy sport flying was really eliminated. Projections by the FAA show significant growth in the number of Sport pilots over the next 15 years.[9]

Market Changes of Recent Decades and Their Implications for Aviation

1990s Onward—New Wealth

Owing to the booming U.S. economy from the late 1980s through the latter part of 2008, a new generation of younger wealthy people with significant discretionary income that might be used in aviation has emerged. Some of these people turned to flying as an outlet for their energies and their wealth, and as a business tool, giving a boost to an industry that had seen shrinkage in many areas for over a decade since the late 1970s.

Security and Terrorism

In the era of global terrorism, security repercussions imposed by the Transportation Security Administration (TSA) sometimes border on oppressive. Especially in aviation, there has traditionally been a tradeoff between security and privacy or profitability. In general aviation, the TSA Security Directive 1542–04–8F (SD-08F), is a prime example of this delicate balance. Billed by the TSA as an essential security measure, SD-08F expanded airport badging requirements to include private aircraft owners, aviation maintenance providers, FBO employees, flight instructors and students, and various other general aviation stakeholders. General aviation advocacy groups claimed that SD-08F would adversely impact the profitability of operators and limit pilots' access to aircraft or airports. The TSA replaced SD-08F with SD-08G, which removed the badging requirement for transient pilots.[10]

The World Wide Web

In the larger environment in which U.S. aviation businesses operate, the single most pervasive external change in the last couple of decades has been the advent of the Internet, which has changed how virtually all companies operate. The Internet allows the access and exchange of information on a scale never before possible.

The tech-savvy aviation consumers of today use the Internet to find and compare companies before making a purchase. For this reason, aviation businesses must, at the very least, have a static web presence. To stay competitive in the modern online marketplace, however, FBOs will need database-driven websites. A database-driven website is different from a static website in that its webpages intelligently and automatically change viewable contents based on the user's identity and inputted information.

To increase the quality and quantity of web traffic to their websites from search engines, businesses must routinely conduct a search engine optimization

(SEO). On average, this process takes six months and can cost in the range of one to three thousand dollars. Many small business owners who are unaware of the importance of performing an SEO may opt to have this process completed by inexperienced staff members or, even worse, choose not to do it at all.

Globalization of the Market

The world is moving rapidly to a global economy and has been doing so for some time. Despite the economic slowdown of 2008–2010, globalization of the economy continues, in part because of the global computerization of financial markets. At the macro level, the gross annual output of a number of "transnational" corporations is larger than the gross output of many nations. Vastly increased global flows include commodities such as money, manufacturing production, imports, information, cultures, and workers. For aviation, the globalization trend means several things: more international business flights in corporate aircraft and more FBO chains looking at establishing overseas operations.

Airline Deregulation

The Airline Deregulation Act of 1978 was a major economic milestone in the history of U.S. aviation. By unshackling airline services to be a virtually free-market operation, it eliminated the industry's protected status and most of its subsidized role as an "infant industry." Deregulation permitted the scheduled airlines to abandon unprofitable small destinations they were formerly required to serve. New, competitive subsidy processes for these small points were established under the Essential Air Service provisions of the Act. Many commuter carriers were successful in bidding for this small community service while receiving federal subsidies to assist them.

At the same time, rapid fuel cost increases, fare increases, intensive competition and, in some cases, overambitious expansion resulted in major stresses for many of the larger carriers. This has led to numerous mergers and several bankruptcies. As a result, other new markets have opened up for commuters and former air-taxi operators. In the decades since deregulation, fewer and fewer scheduled service points in the United States continue to have nonstop service. This is primarily due to the airlines' development of hub-and-spoke route systems.

The economic impact of the deregulation act seems likely to continue and to result in new opportunity for not only scheduled commuters, but also for the traditional FBO charter and air taxi services.

One of the success stories resulting from the Airline Deregulation Act has been Southwest Airlines. With a "no frills" service and marketing strategy, the company has steadily increased its market area. Beginning as a regionally based carrier, operating primarily in the southwest, it has now expanded its routes to include over half the U.S.

A unique management style, designed to be both customer- and employee-oriented, has led to its success. Reducing operating costs by offering a snack only, eliminating boarding passes, and providing quick turn-around times at the gate were all factors. Many established carriers initially believed Southwest would not survive; however, many of those same airlines are now trying to duplicate the successful strategies initiated by Southwest. The point-to-point, "no-frills" business model is emulated internationally by carriers such as Europe's EasyJet and Ryanair, Indonesia's Lion Air, Canada's WestJet, Mexico's Volaris, and Malaysia's AirAsia.[11]

The formation of several new carriers since deregulation, such as JetBlue and Spirit, has maintained downward pressure on airfares. Stiff competition has also led to the demise or merger of several airlines. The airline industry continues to be dynamic and volatile due in part to deregulation, yet this situation continues to create opportunities for the general aviation industry.

Market issues for airport service businesses are discussed more fully in Chapter 3, Marketing.

General Aviation Revitalization Act of 1994 (GARA)

One of the major milestones of recent decades for general aviation was GARA. The general aviation aircraft production industry had been crippled by a difficult economy and a lack of protection against insurance claims and lawsuits. Product liability insurance was becoming an ever-increasing proportion of the cost of each new aircraft, inflating prices and resulting in lower sales. Thus, the insurance costs of aircraft production had to be spread over an ever-falling volume of aircraft, resulting even higher unit costs and even larger proportions of unit costs being attributed to insurance. New aircraft production in the U.S. reached a peak in 1978 and decreased drastically in the years immediately thereafter. The result of this downward spiral was that the nation's general aviation aircraft producers were hardly able to stay afloat. The industry lobbied for product liability time limits, and legislation was finally passed in 1994. A news article three years later reported as follows:

"On August 17, 1994, President Clinton signed into law the General Aviation Revitalization Act (GARA) in order to breathe life into an industry that had lost over 100,000 jobs and experienced a 95 percent decline in production. Today, over three and a half years (sic) after the bill was signed into law, it is clear that GARA has been a tremendous success.

"Thanks to GARA the general aviation industry is in better shape today than it has been at any time in well over a decade. Employment at general aviation manufacturing companies has increased every single year since enactment of the statute of repose. Overall employment at general aviation manufacturing companies has increased over 46 percent since the legislation was signed into law.

"The production of general aviation aircraft in the United States has also increased every single year that GARA has been in existence. The total increase in general aviation production since enactment is over 69 percent. Production of single engine piston-powered aircraft, the type normally associated with the entry-level market, has increased over 103 percent.

"Without GARA many of the positive things that are now happening in the general aviation industry would never have taken place. For example, the world's two largest producers of piston-powered aircraft—Cessna Aircraft Company and The New Piper Aircraft, Inc.—would not be building piston-powered airplanes. Exciting new computer based training programs for the student pilots would not have been developed. And GA Team 2000, the largest "learn how to fly program" in civil aviation history, would not have been launched.

"Clearly, the general aviation industry is in a much different position now than it was before passage of GARA—it has been revitalized. Companies have once again begun investing in general aviation, the industry is growing, and its future outlook is bright. And that, after all, is what GARA is all about."[12]

Since that time, aircraft production has picked up and continues to increase, albeit slowly.

The "Revenue Diversion" Issue

Beginning with the FAA Authorization Act of 1982 and repeated with growing emphasis in the Acts of 1987, 1994, 1996, and beyond, diversion of airport revenue for non-aviation needs has become a major issue. Revenue diversion occurs when a city, presumably with budget problems, uses an airport revenue surplus to fund other departments, such as police, fire or education. The federal position against aviation revenue diversion is based on the realistic premise that federal, state, and local aviation funding is becoming increasingly scarce. Thus, airports seeking federal funds must first make full use of local funds. Airlines also strongly oppose revenue diversion; at air carrier airports, many use what is known as the "residual" method of budgeting, meaning that airport administration and improvement costs are recovered to the extent possible from lease, parking, and concession revenues, with the remainder or "residual" being collected from landing fees set precisely to cover the remaining amount. It is in the airlines' interest for this residual amount to be as low as possible, which in turn limits landing fees. Airlines support the FAA in pushing for all airport revenues to be reinvested back into the aviation system and not elsewhere. Over the years, the focus has not just been on diversion of normal revenues, but on making sure that income is priced at reasonable market levels; long-lasting "sweetheart" lease or concession deals may translate to missed revenue opportunities. For example, if a Chamber of Commerce had a lease on non-aeronautical airport land for $1 per year, the airport authority would seek to bring up the lease rate as soon as the lease expired and needed to be renewed.

The revenue diversion issue does not directly pertain to FBOs as much as it affects airport managers and owners. However, all airport businesses need to understand the stringency with which FAA and the USDOT Inspector General are addressing this issue; in instances that the FBO is also the public airport operator or manager, it must directly comply with these rules.

Threat of Airport Closures

FBOs are in an unusual position of being totally dependent on other agencies, their landlords and providers of essential facilities such as runways and taxiways, for business operations. An FBO facing closure of its airport is akin to a retail merchant under constant threat of having the access road to its store blocked. The uncertainty can be almost as damaging to business as the actual event. Airports are not being built or replaced at a rate near the closure rate, and it remains very difficult to make even modest expansions of existing airports. Thus preserving what already exists is a more effective strategy than creating new fields. Strategies for countering the airport closure trend are discussed in Chapter 12, Physical Facilities.

Airport Improvement Funding

Despite a longstanding system of collecting aviation user fees through passenger ticket taxes, fuel fees and other sources of revenue, Congress has traditionally not funded airport improvements to the extent that is either needed or possible. One reason for this is that, the inclusion of Aviation Trust Fund balances make the national budget look superficially stronger despite these funds already being earmarked for solely aviation purposes. Another problem is that, over the years, progressively more and now all of the cost of running the FAA has come from the Fund rather than from general revenues. The FAA may be able to achieve higher and more reliable funding through various changes in how airport revenues are raised and spent, aggregately strengthening the airport system.

Fractional Aircraft Ownership

The ability to own a portion of a business aircraft as small as 1/32 or 1/16 gives companies flexibility advantages and cost savings in corporate travel. Allowing companies to purchase 25, 50 or 100 flight hours per year is changing the face of business travel. One industry specialist suggests that it is the single biggest development in general aviation since World War II.[13] Flying in a company's fractionally-owned plane may be financially favorable over airlines and full ownership. This trend is likely to continue as traditional airline travel is hindered by time inefficiencies and general inconveniences. This topic is discussed in greater depth in Chapter 3, Marketing.

NASA's Involvement in General Aviation

Through 2005, NASA was involved with the development of a Small Aircraft Transportation system (SATS) program, with planes that can use shorter runways. If this becomes a marketable product, then possibly smaller communities with small landing strips will seek to use their airports as a basis for economic development and growth. As one article says:

> "NASA's vision of a future full of aerial taxis depends on airports that are under siege and rapidly disappearing.
>
> "With the introduction of a fleet of automobile-sized, high-tech, on-demand jet-powered aerial taxis and various airspace enhancements, NASA says business travelers could be taking daylong regional jaunts for about the price of coach air-fare by the end of the decade.
>
> "Advocates say the Small Aircraft Transportation System (SATS) will be a pressure relief valve for the congested hub and spoke transportation system, a tool that will boost business productivity by cutting travel time in half or more. Not only that, but communities connected to such a point-to-point transportation system—those with an airport—would prosper, in theory."[14]

Transportation Security Administration

One result of the September 11, 2001 terrorist attacks was the creation of the Transportation Security Administration and Department of Home-land Security. With a new layer of transportation security regulations and oversight, FBOs' security-related tasks have grown in complexity and cost. Recent TSA requirements or recommendations for general aviation activities, such as the Security Guidelines for General Aviation Airport Operators published in 2017, reflect a shift from one-size-fits-all measures to a risk-based and intelligence-driven approach.

Revolution in Avionics Technology

The trend toward smaller, cheaper, and more integrated electronic devices, many of which aid aviation, continues. Examples include Global Positioning Systems (GPS), Automatic Dependent Surveillance–Broadcast (ADS-B), flight computers, controls, and displays, transceivers, tablets, and Internet-based training material. Given the volume of controlled airspace, pilots may be required to equip their aircraft with advanced electronics to enter certain areas or perform certain maneuvers.

Components of the Modern Aviation Industry

Summary

Today's aviation system has evolved to encompass many distinct areas:

> Pilots;
> Mechanics and other support personnel;
> Airports and supporting infrastructure;
> Air navigation systems;
> Aircraft manufacturers;
> Scheduled air carriers;
> General aviation and service centers;
> Aviation interest groups; and
> Government regulatory authorities.

Pilots

Pilots in the early 1900s entered aviation from an enthusiasm for testing the new technology of powered flight. World War I and II, as well as subsequent Korean, Vietnam, Iraq, and Afghanistan wars have created generations of pilots trained by the military and later employed commercially. General aviation pilots may start with the private license, moving up to higher ratings including the instrument rating and for some, the Airline Transport rating. Today, many military-trained pilots are reaching retirement age and the military system is not supplying aviation with as many pilots as in years past. Thus, general aviation businesses today must increasingly look to their own ability to attract pilots into the system. The simpler and cheaper Sport Pilot certificate, mentioned previously, is expected to attract pilots to more advanced ratings. Nevertheless, general aviation will likely continue to see its most-qualified student pilots move up the career ladder from flight instruction to charter to air taxi to corporate or airline employment. General aviation often trains pilots for the later benefit of other areas and employers in aviation.

Other Aviation Support Personnel

The pilot may be the most visible person that the member of the traveling public observes when flying. However, the aviation business is not possible without the work of maintenance technicians, fueling crews, ground instructors, flight schedulers, aircraft and parts salespeople, and numerous other administrative and secretarial, personnel. Whether a large or small business, these functions are necessary for safe and successful aviation operations. In the typical FBO, many of these support functions are poorly compensated compared with what a qualified person can earn elsewhere in aviation. FBO aviation

Courtesy of Landmark Aviation.

maintenance technicians suffer perhaps the most from low pay levels.

More on personnel issues is provided in Chapter 5, Human Resources.

The Airport System

The U.S. airport system is the most extensive and sophisticated in the world. Airports vary widely in their physical quality and the type of traffic they handle. Approximately 3,310 airports are designated as essential to the national air transportation system, and are listed in the National Plan of Integrated Airport Systems (NPIAS). These airports vary from major international facilities with several runways and many services to single, short-runway fields with limited or no services. Nationally, there are nearly 20,000 landing facilities. However, most of these are small crop-dusting or fire-fighting airstrips with very limited facilities and restricted access. Most do not have any public services and may only have tie-downs. Of an estimated 5,080 public-use airports, roughly 20 percent are privately owned and operated directly by the owner or through a contract with an FBO.

The nature of airport usage varies greatly. The most densely populated states are well served by scheduled commercial service airports such as the Kennedy-LaGuardia-Newark trio in the New York area and the Los Angeles International-Long Beach-Burbank-Ontario-Orange County system in southern California. Rural parts of the country depend more heavily on general aviation airports; the most extreme example is the bush areas of Alaska, where many communities have no road access, and hopping into a small plane is as natural as driving. In this instance the local airport may be a seaplane facility with floats used in summer and skis in winter.

Airports in the NPIAS have had a reliable funding source for several decades through the Airport Improvement Program (AIP). Funds go into this from airline ticket taxes, aviation fuel taxes, and other sources for redistribution to the various classes of airport in the NPIAS.

Airports are classified by a variety of factors, including function, facilities, passenger and cargo volume, ownership, and based aircraft. General aviation airports are functionally classified as such, and may also be designated as "reliever" airports, meaning they relieve nearby commercial-service airports of general aviation traffic. Enhancing capacity and improving safety at larger airfields by separating the smaller, slower aircraft from transport-category jets and their dangerous wake turbulence, the commercial air carrier system subsidizes the general aviation system through AIP fund distribution formulas.

The national airport system, as backdrop to the individual airport location of a particular aviation business, is discussed in more depth in Chapter 12, Physical Facilities.

Traditionally, rural parts of the country depend more heavily on general aviation airports for social and economic exchange.

The Air Navigation System

The U.S. civil airspace and its management come exclusively under the purview of the FAA.[15] Nationwide, there are 21 Air Route Traffic Control Centers (ARTCCs) that keep track of all traffic flying on Instrument Flight Rules. The ARTCCs transfer traffic to individual air traffic control areas, usually tower areas. En-route airways and navigational aids of various types are in place as well as rules for en-route and airport vicinity flying.

Flight Service Stations provide information to pilots and enable them to file flight plans. The number of Flight Service Stations has been decreased. Flight Service Stations have been automated by use of computers and telephone voice systems; three large hub facilities and some smaller satellite stations remain. Pilots may now receive current weather information for their arrival, departure, and en-route phases of flight without speaking to a "live" briefer. The option of a personal briefing is still there for those who want further clarification or who prefer the ability to converse with an FAA representative.

After several years and many hundreds of millions of dollars, a project to upgrade the FAA air traffic computer system was abandoned in the late 1990s, as it was behind schedule and progress was not adequate. Part of the decision to halt the project was also due to discussion, still at a conceptual stage, of privatizing the air traffic control system in the U.S. However, the newest concepts in airspace safety through the FAA's NextGen program provide the individual pilot with much greater control and options through the use of high tech tracking systems.

Figure 1.1 → General Aviation Aircraft Shipments

Airplane	2018	2019	Change
Piston Airplanes	1,137	1,324	+16.4%
Turboprops	592	525	−11.3%
Business Jets	672	809	+15.1%
Total Airplane Shipments	**2,432**	**2,658**	**+9.3%**
Total Airplane Billings	**$20.6B**	**$23.5B**	**+14.3%**
Helicopters	**2018**	**2019**	**Change**
Piston Helicopters	281	179	−36.3%
Turbine Helicopters	754	698	−7.4%
Total Helicopter Shipments	**1,035**	**877**	**−15.3%**
Total Helicopter Billings	**$4.2B**	**$3.8B**	**−9.1%**

Source: General Aviation Manufacturers Association (GAMA), 2020

Aviation Manufacturers

The General Aviation Manufacturers Association (GAMA) represents many of the general aviation Membership includes Textron Aviation (formerly Cessna and Beechcraft), Cirrus, Piper, Gulfstream, and others. As McDonnell Douglas was absorbed by Boeing in 1997 and Lockheed Martin has focused mostly on defense applications, Boeing is the largest U.S. planemaker for commercial carriers. Developers of avionics and other aircraft components are another important group. Much U.S. aircraft component manufacturing is now taking place overseas because of lower labor costs and improved supply chain management.

In terms of new sales, general aviation hit its post-WW II volume peak in 1978. This represented a high number of low-priced aircraft. A high of 17,811 aircraft shipments in 1977 is compared to a low of only 941 in 1992.

Today, the dollar value of sales is higher although the number of aircraft is smaller. Much more expensive aircraft are being sold to a different segment of the market—fractional or direct corporate users—than was the case in the 1970s. With the passage of the General Aviation Revitalization Act (GARA) in 1994 and the growth in fractional aircraft ownership, an upward trend has been developing. In 2019, a total of 3,535 GA aircraft were shipped worldwide, as shown below in Figure 1.1.

As may be seen by the average prices of the smaller GA planes shown in Figure 1.1, the need by the flying public for cost-effective aircraft still exists. Interest in general aviation continues strongly as evidenced by the attendance annually at aviation events around the country. The annual convention sponsored by the Experimen tal Aircraft Association in Oshkosh, Wisconsin attracted 642,000 participants and over 10,000 aircraft from around the world at its annual fly-in in August, 2019.

Scheduled Air Carriers

Since the Airline Deregulation Act of 1978, the hierarchy of air carriers has changed both in nomenclature and in role. Industry giants such as Braniff, Eastern, Pan American, and TWA have gone bankrupt or been merged; smaller new carriers have emerged such as Spirit, Allegiant, and JetBlue. As shown in Figure 1.2, large jet airline aircraft, including all-cargo, in 2020 involved 5,188 aircraft, comprising only 2.36 percent of the civil aircraft fleet. Commuters and regionals for 2,362 aircraft, or 1.07 percent. The distinction between these two levels of carriers, which started in 1969, is fast disappearing. The new nomenclature for the carriers is majors, nationals, large regionals, medium regionals, and small regionals. Nevertheless, airlines comprise, by the numbers (not by seats) only a very small proportion of the civil aircraft fleet, as shown in Figure 1.2.

General Aviation

As Figure 1.2 indicates, general aviation will continue to dominate the civil aircraft fleet, decreasing slightly from 96.89 percent in 2010 to 95.71 percent by 2040, according to FAA FY 2020 forecasts. General aviation has maintained a relatively steady percentage

Figure 1.2 ✈ U.S. Civil Aircraft Fleet, 2010 to 2040

Type	2000 (Actual) #	%	2013 (Estimated) #	%	2034 (Forecast) #	%
Large jet air carriers	4,572	1.98	5,188	2.36	7,101	3.23
> passenger	3,722		4,282		5,310	
> argo	850		906		1,791	
Regionals/commuters	2,613	1.13	2,362	1.07	2,320	1.06
> turboprops	857		486		128	
> jets	1,756		1,876		2,192	
General aviation	223,370	96.89	212,380	96.57	210,380	95.71
> Pistons	155,419		141,245		115,970	
> Turbine	20,853		25,490		36,595	
> Rotorcraft	10,102		10,340		14,295	
> Experimental	24,784		27,970		33,475	
> Light Sport	6,528		2,845		5,430	
> Other	5,684		4,490		4,615	
TOTAL	230,555	100.00	219,930	100.00	219,801	100.00

Source: Table created from data in FAA aerospace forecasts, FY 2020–2040[16]

of the total fleet for decades. While the seating capacity, speed, and range of general aviation aircraft have changed, this proportion illustrate the relative size of general aviation in the national transportation system—a portion often ignored by the public and media. Many corporate aircraft departments and fractional aircraft ownership companies operate fleets larger than those of some countries' air forces or air carriers.

As shown in Figure 1.3, general aviation also constitutes almost two-thirds of the system in terms of hours flown.

Additionally, although also less readily quantified, some industry sources estimate that general aviation carries about one quarter of all air passengers in the United States. This segment of aviation is growing rapidly.

The nation's fixed-base operators provide the airport services to this major segment of the aviation system as well as servicing airlines at a number of the country's major airports. General aviation is a complete array of types of users and varieties of activity. Figure 1.4 suggests a method of organizing general aviation into a meaningful system. It divides general aviation activity into (1) methods of travel and (2) methods of undertaking an operation from the air. This second role of general aviation is a key distinguishing characteristic because airlines provide only transportation, whereas general aviation provides aerial functions of diverse types, many of which are very important to the economy.

Aviation Industry Groups

Many pilot and industry groups represent aviation interests and are dedicated to the growth and support of general aviation. As a part of the business plan and marketing efforts of a fixed-base operator, a working knowledge of those organizations, their initiatives, and their structures may be useful in achieving the goals of the local aviation business. Appendix I (see website) contains a website listing and brief description of many of these groups and more can be found through links from this database as well as through web searches for specialized topics. The many general aviation industry groups are fairly influential given the diverse nature and interests of the industry and its sheer size. Congress, particularly the Senate, tends to be very conscious of the value of general aviation since many politicians need general aviation aircraft to effectively campaign across such a large country. Thus, given the small number of aircraft in private ownership compared with autos and trucks (under 220,000 of the former and 273.6 million of the latter in 2020), general aviation, through its various interest groups, succeeds relatively well in bringing attention to its issues. General aviation could probably be even more effective if it communicated better with the vast proportion of the American public that knows little to nothing about the industry and may only have been disturbed by a noisy night flight or entertained at an air show.

Figure 1.3 → Thousands of Aviation Hours Flown, FY 2010 to 2040

Type	2010 (Actual) #	%	2020 (Estimated) #	%	2040 (Forecast) #	%
Scheduled carriers	14,467	36.84	N/A		N/A	
> Passenger & Cargo (domestic)						
General aviation	24,802	63.16	26,039		30,205	
> Pistons	13,979		13,497		11,177	
> Turbine	5,700		7,827		11,983	
> Rotorcraft	3,405		3,098		4,682	
> Experimental	1,226		1,242		1,707	
> Light Sport	311		223		496	
> Other	181		153		160	
TOTAL	42,541	100.00				

Source: BTS and FAA aerospace forecasts, FY 2020–2040[16]

Here is a summary of a number of these organizations.

Aircraft Owners and Pilots Association. AOPA represents approximately 400,000 general aviation aircraft owners and pilots who use their aircraft for non-commercial, personal, and business transportation. AOPA members constitute about two-thirds of the active pilots in the nation.

Alaska Air Carriers Association. The Alaska Air Carriers Association was founded March 10, 1966 to help operators in Southeast Alaska have an organized "voice" in aviation regarding state legislation and worker's compensation insurance premiums for high-risk businesses.

American Association of Airport Executives. AAAE is a professional organization representing the men and women who manage general aviation airports as well as scheduled service fields that enplane some 90 percent of the passengers in the United States.

Aviation Insurance Association. The Aviation Insurance Association is a not-for-profit association dedicated to expanding the knowledge of and promoting the general welfare of the aviation insurance industry through numerous educational programs and events. AIA welcomes members of all facets of the aviation insurance industry, including such professionals as: agents/ brokers, claims professionals, underwriters, attorneys, associates, and college students interested in the business.

Experimental Aircraft Association. EAA is a sport aviation association with a worldwide membership of over 200,000 aviation enthusiasts, pilots, and aircraft owners. The organization includes an active network of over 1,000 chapters.

Florida Aviation Trades Association. Since 1946 the Florida Aviation Trades Association has been actively promoting and protecting the rights, interests, and development of Florida's aviation industry.

General Aviation Manufacturers Association. GAMA represents U.S. and international manufacturers of general aviation aircraft, engines, avionics, and related equipment or components.

National Air Transportation Association. NATA represents the business interest of the nation's general aviation service companies that provide refueling, flight training, maintenance, repair, sales, and charter functions. NATA has nearly 3,700 member companies, most of which are designated as small businesses.

Figure 1.4 A Taxonomy of General Aviation Flying

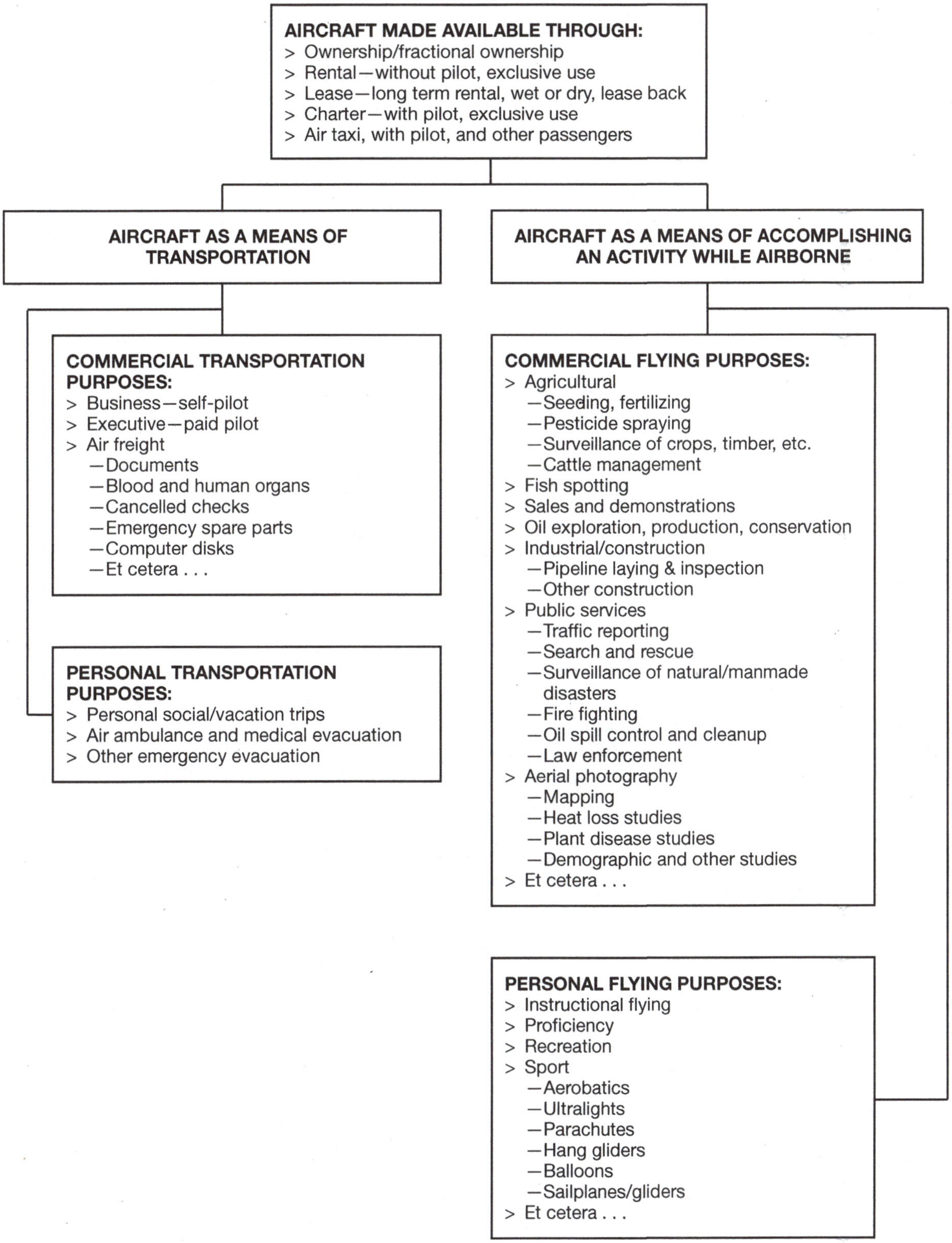

AIRCRAFT MADE AVAILABLE THROUGH:
> Ownership/fractional ownership
> Rental—without pilot, exclusive use
> Lease—long term rental, wet or dry, lease back
> Charter—with pilot, exclusive use
> Air taxi, with pilot, and other passengers

AIRCRAFT AS A MEANS OF TRANSPORTATION

COMMERCIAL TRANSPORTATION PURPOSES:
> Business—self-pilot
> Executive—paid pilot
> Air freight
 —Documents
 —Blood and human organs
 —Cancelled checks
 —Emergency spare parts
 —Computer disks
 —Et cetera . . .

PERSONAL TRANSPORTATION PURPOSES:
> Personal social/vacation trips
> Air ambulance and medical evacuation
> Other emergency evacuation

AIRCRAFT AS A MEANS OF ACCOMPLISHING AN ACTIVITY WHILE AIRBORNE

COMMERCIAL FLYING PURPOSES:
> Agricultural
 —Seeding, fertilizing
 —Pesticide spraying
 —Surveillance of crops, timber, etc.
 —Cattle management
> Fish spotting
> Sales and demonstrations
> Oil exploration, production, conservation
> Industrial/construction
 —Pipeline laying & inspection
 —Other construction
> Public services
 —Traffic reporting
 —Search and rescue
 —Surveillance of natural/manmade disasters
 —Fire fighting
 —Oil spill control and cleanup
 —Law enforcement
> Aerial photography
 —Mapping
 —Heat loss studies
 —Plant disease studies
 —Demographic and other studies
> Et cetera . . .

PERSONAL FLYING PURPOSES:
> Instructional flying
> Proficiency
> Recreation
> Sport
 —Aerobatics
 —Ultralights
 —Parachutes
 —Hang gliders
 —Balloons
 —Sailplanes/gliders
> Et cetera . . .

Source: Julie F. Rodwell, 2009

National Agricultural Aviation Association. NAAA is the voice of the aerial application industry, working to preserve general aviation's place in the protection and production of America's food and fiber supply. Aerial application is a safe, fast, efficient, effective, and economical way to apply pesticides.

National Association of State Aviation Officials. NASAO represents the 50 states' aviation departments, usually within state departments of transportation, as well as the aviation departments of Puerto Rico and Guam.

National Business Aircraft Association. NBAA represents the interests of over 11,000 companies that operate or support general aviation business travel. Since its founding in 1947, NBAA has been the leading advocate for the business aviation community.

Professional Aviation Maintenance Association. PAMA is a national professional association of aviation maintenance technicians, with some 4,000 individual members and 250 affiliated company members.

Helicopter Association International. The not-for-profit trade association for the civil rotocraft industry. Its mission is: "To provide our membership with services that directly benefit their operations and to advance the civil helicopter industry by providing programs to enhance safety, encourage professionalism, and promote the unique societal contributions made by the rotary flight industry."

Women in Aviation, International. WAI is a professional organization representing the interests of women in all facets of aviation—general, commercial, and military. The organization provides networking opportunities, educational outreach programs, and career development initiatives.

Aviation Regulation

Introduction

Virtually all aspects of the aviation industry are heavily regulated—aircraft production, pilots, airspace usage, airport planning, design and construction, and aircraft repair and maintenance. While passenger airlines operating under Federal Aviation Regulations (FAR) Part 121 are under more stringent controls than the rest of the system (mainly due to rules about pilot duty hours and passenger safety), the entire system must budget time and resources for the understanding of and compliance with regulations.

It would be impossible to describe all requirements here; they change from time to time, also. This section summarizes the main considerations and indicates how to get more information.

Federal Aviation Administration (FAA)

The Federal Aviation Administration is the primary regulatory and safety agency for the aviation industry. A list of Federal Aviation Regulations is contained in Appendix II (see website). In addition to the FARs, the FAA issues Advisory Circulars on numerous topics, updated from time to time. These are shown in Appendix III (see website).[17] These updates are generally treated like new rulemakings and are published in the Federal Register for comment, as are new FARs. To the extent possible, each chapter touches on the pertinent FAA or other regulations.

At first glance, it may seem confusing as to why the FAA regulates different types of flights under three sections of the FARs—part 121, part 135 and part 91. In its discussion about appropriate regulation for fractional aircraft companies, the Fractional Aircraft Ownership Aviation Rulemaking Committee (FOARC) succinctly summarizes the differences.[18]

"In general, airline passengers exercise no control over, and bear no responsibility for, the airworthiness or operation of the aircraft aboard which they are flown. Because the traveling public has no control over, or responsibility for, airline safety-of-flight issues, an optimum level of public safety is provided by the FAA's imposition of very stringent regulations and oversight under part 121 and the sections of part 135 applicable to scheduled service.

"In general, on-demand or supplemental air charter passengers exercise limited control over but bear no responsibility for the operation of the aircraft aboard which they are flown. On-demand or supplemental air charter passengers negotiate the point and time of origin and destination of the flight, and may have the ability (subject to the pilot's supervening authority) to direct or redirect the flight. Under these circumstances, the optimum level of public safety is provided by the FAA's imposition of stringent regulations and oversight under part 121 or part 135.

"In general, aircraft owners flying aboard aircraft they own or lease exercise full control over and bear full responsibility for the airworthiness and operation of their aircraft. Under these circumstances, the optimum level of public safety is provided by the FAA's imposition of general operating and flight regulations and oversight under part 91.

"These policies and differing levels of responsibility were reflected in the development of part 91, subpart D, subsequently subpart F, which governs most business aviation today. On July 25, 1972, the FAA promulgated Amendment 91–101 to 14 CFR part 91 (37 FR 14758, July 25, 1972). This Amendment added to part 91 a new subpart D, applicable to large and turbojet-powered multi-engine aircraft. Subpart D was the predecessor to the current subpart F of part 91 (54 FR 34314, Aug. 18, 1989). Section 91.181 of subpart D was the predecessor of current Section 91.501 (54 FR 34314)."

National Transportation Safety Board (NTSB)

The NTSB is an independent government agency that reports directly to Congress on safety and accidents, and provides interpretations of "probable cause" for accidents and serious incidents. The NTSB is responsible for investigating domestic aircraft accidents, but it often delegates these responsibilities to the FAA in cases of nonfatal general aviation accidents. NTSB reports also provide recommendations on corrective actions aimed at preventing similar problems in the future. Due to this, the FAA and NTSB are sometimes at odds. More on the NTSB is in Chapter 11, Safety, Security, and Liability.

Transportation Security Administration (TSA)

The TSA was established in late 2001 as a result of concerns raised by the 9/11/01 terrorist attacks. It has taken over responsibility for ensuring aviation security. The TSA has enforcement authority similar to that of the FAA in terms of requiring compliance with Transportation Security Regulations (TSRs) and Security Directives (SDs).

Fixed-Base Operators

Fixed-Base Operators: Their Role

For many, the term "fixed-base" operator conjures up images of a gas station for aircraft. While providing quality fuel is a key service, it is far from the only contribution FBOs make to general aviation. FBOs provide the avionics, airframe, and powerplant repair and overhaul services needed by the aviation community, not only for routine maintenance but also for unexpected problems. Some specialized locations may also perform major overhauls for large aircraft. FBOs provide other services as well, such as aircraft hangaring, flight training, air taxi, and charter services, which help stimulate interest in general aviation, serve the general public, and facilitate business aviation.

There are approximately 3,000 FBOs in the U.S., including franchises, chains, and independent locations. The first companies labeled fixed-base operators were easily distinguished from their counterparts known as the field hopping, post World War I "barnstormers." Although exciting and colorful, these transient characters had no fixed base of operations and were pilots first, last, and always. Those who started the first FBOs were also generally pilots. Those who survived financially became, out of necessity, businesspeople committed to providing professional flight and ground services to all customers from a permanent, or fixed, base of operations.[19]

There is great diversity in services provided. Many airports define how many of the following services must be provided for an aviation service business to be defined as a full-service FBO at that airport. The array of possible aeronautical services listed by the Federal Aviation Administration is as follows:[20]

> Charter operations;
> Pilot training;
> Aircraft rental and sightseeing;
> Aerial photography;
> Crop dusting/aerial application;
> Aerial advertising and surveying;
> Passenger transportation;
> Aircraft sales and service;
> Sale of aviation petroleum products;
> Repair and maintenance of aircraft; and
> Sale of aircraft parts.

Specifically excluded by the FAA's definition are ground transportation (taxis, car rentals, limousines); restaurants; barbershops; and auto-parking lots—although, of course, an FBO may operate such items.[21]

One Cessna dealer lists the following array:

Transportation:

> Air ambulance;
> Aerial pick-up and delivery service;
> Air cargo in;
> Air cargo out;

- Emergency medical supply;
- Fishing and hunting trips;
- Holiday and vacation travel;
- New business calls;
- Out-of-town conventions and meetings;
- Personal transportation to special events;
- Plant-to-plant company transportation;
- Regular sales calls;
- Transportation of time-sensitive articles; newspapers, mail, reports, proposals, retail merchandise;
- Transportation of perishables;
- Trouble-shooting with tools;
- Transportation to hub air terminals;
- Ski trips; and
- VIP passenger service for clients, customers or prospects.

Aerial Functions:

- Aerial inspection and patrol;
- Aerial mapping;
- Aerial photography; and
- Management field inspection.

Scale and Prospects of the Industry

Despite the advent of large FBO chains, there are still many individual airport service companies. The future will likely see increased FBO specialization, more focus on corporate jet customers, and single FBOs joining chains. It is anticipated that the number of locations of existing, large FBO networks such as MillionAir, Landmark, Signature, Atlantic Aviation, and Jet Aviation will continue to rise.

Airport Management

FBOs that also have the responsibility for the management of the airport are tasked with an especially challenging, multifaceted role. The best interests of the FBO and the (publicly owned) airport may not always coincide; conflict of interest situations are easily developed. Little is documented about aviation businesses that wear dual hats; however, it appears that about half the public use airports (often the smaller, non-NPIAS half) are run by FBOs. These figures do not distinguish between FBOs running airports for either a private or a public owner.

FBO Industry Trends and Issues

Maturity and Professionalism

Early aviation managers or operators undoubtedly faced many technical and financial problems associated with simply keeping aircraft airborne. The challenge of the job and the associated thrills may have kept them fully occupied, with little focus on business administration. Historically, general aviation businesspeople have tended to be primarily pilots or mechanics, whose love of aviation and preoccupation with technical problems took precedence over sound business practices. Conventional wisdom indicates, almost without exception, that inventors and other creative geniuses make poor business people, no matter what the field of endeavor. The early decades of technical genius in aviation were indispensable, but the industry has matured, and needs other skills.

The 1980s first showed signs of being a transition era in the management of many aviation enterprises. The early flyer-turned businessperson, often a retired military pilot, was now becoming a manager with more awareness of the need for a businesslike approach. In many instances, the manager simply acquired this additional insight through training and the necessity for survival. In other situations, the second generation in a family business influenced the business and accomplished changes. There has also been, and continues to be, an influx of new professional managers bringing new ideas to the general aviation field. Last but far from least, the original post-World War II and Korean War "mom and pop" businesses are largely disappearing. While some independent FBOs have succumbed to competition or a lack of demand, many others have been acquired by large FBO networks with more stream-lined central management procedures and training.

When surveying the general aviation business scene, one must remember that most FBOs are small businesses in a very competitive and volatile industry. Despite this scale mismatch, the FBO manager is called upon to provide an uncommonly wide array of complex skills and services usually found in much larger enterprises. It is hard to be a manager in any industry—a post that requires wearing many hats and dealing with many types of problem solving; being a manager in an aviation service business is perhaps one of the hardest managerial roles of all.

An examination of many general aviation businesses reveals four unsound business practices:

> Continued investment of money in a business for a love of flying rather than a pursuit of profitability;
> Low rates of return for costly qualification training, capital investments, and operating expenses;
> Lack of adequate record-keeping and procedures for monitoring and evaluating performance; and
> Weak marketing and promotional activities.

The anticipated growth in general aviation and accompanying needs for high-quality aviation services and highly-qualified aviation professionals likely mean that, in order to financially survive in future years, FBOs will become more profit-driven. A passion for aviation is a necessary but not sufficient condition for success in modern aviation business.

Public Awareness

When the various aviation markets are examined, it seems each acquires a different image. The airlines have been commonly accepted forms of transportation—about 88 percent of the adult U.S. population had flown at least once by 2017, as shown in Figure 1.5.[22] Air cargo traffic is beginning to be perceived by the public as a vital service, particularly as express package delivery grows in importance.

Nevertheless, general aviation still seems to have the silk-scarf-and-goggles image of the barnstorming days. Even major general aviation airports have been referred to by airport commissioners as "just marinas for planes." Some elements of the flying community may heighten this frivolous image by taking unnecessary chances that can result in spectacular accidents and incidents. The negative image may also be heightened because general aviation airports are often located in smaller communities. In such locations, people expect a quiet living environment. There is likely a large percentage of the general public who lack awareness of general aviation's key role in the transportation system. As population has grown continuously for the past 60 years, constructing residential neighborhoods next to airports, aircraft noise and air pollution has long solicited negative community responses from people living or working near airports.

For the past two decades, the U.S. has been losing an average of one airport per week. Nearly all such losses are of GA fields. Furthermore, it is increasingly difficult to site new airports and expand existing fields. For example, a decade-long effort was made in Washington State to find a "satellite" airport to assist Seattle-Tacoma International airport (Sea-Tac) and siphon off 5–10 percent of its flights. Some thirty possible locations were considered, which is in itself a challenge, as each must have many precise qualities including flat land, reasonable access to the population to be served, relatively unobstructed approaches, and so on. The 30 were narrowed down to 12, and then narrowed further. Not a single site was found to be feasible, because of community opposition and an effective political protest effort at the state level. The project was abandoned; another effort, completed in 2009, determined that Sea-Tac has sufficient capacity through 2025 and there was thus no need to further examine the issue. This topic is discussed in Chapter 12, Physical Facilities.

A lack of public awareness or support affects general aviation's future and, by extension, the future of the FBOs with complaints regarding:

> Pollution;
> Noise;
> Land use including encroachment of runway protection areas by development, and addition of noise-sensitive activities in airport surrounds;
> Community safety;
> Transportation availability;

Figure 1.5 → Expanded Market for Flying

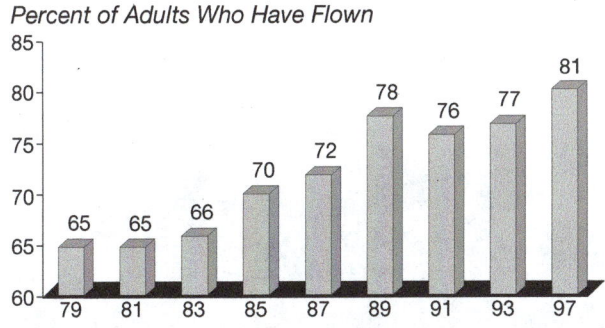

Percent of Adults Who Have Flown

Year	Percent
79	65
81	65
83	66
85	70
87	72
89	78
91	76
93	77
97	81

Source: Airlines for America

> Allocation of the costs of the airway system in equitable fashion;
> Access to the airport system for all users; and
> The image of general aviation.

Many organizations have campaigns attempting to introduce more people to aviation. Of particular note are the Let's Go Flying! program, Young Eagles program, and Project Pilot.

Let's Go Flying! is a program sponsored by the Aircraft Owners and Pilots Association (AOPA) to increase understanding and knowledge about general aviation, and to encourage people to learn to fly. Participating Let's Go Flying! flight schools offer introductory flights for a special low rate and Let's Go Flying! runs a media campaign to let people know about the flights and the reasons to learn to fly.[23]

The Young Eagles program, initiated by the Experimental Aircraft Association in 1992, has thus far introduced over 2.2 million young people (ages 8–17) to the world of aviation. The full program teaches youth how to build their own aircraft. Members of the Association volunteer their time and aircrafts to take the youngsters on an introductory aircraft flight.

AOPA has been promoting its "Project Pilot" program since its creation in 1994. Members of AOPA are asked to bring their friends to the airport, bring them on a flight, and bring them to a local pilot's meeting in hopes of attracting more people to aviation operations.

All of these programs are efforts to help market aviation to the general public. Obviously, the hope is that many of these individuals will go on to obtain pilots certificates and ratings, becoming "consumers" of aviation. But the other benefit is an educated pro-aviation citizenry; local members of the community who will not complain about aircraft noise or block expansion of their local airport, and so on.

Technical Issues

Fuel prices and availability will periodically be major concerns for general aviation, as they will be for all transportation systems. The possible switch to alternative fuels appears less easily achieved than for stationary fuel consumers. More fuel-efficient engines and lighter aircraft will continue to be major technical objectives. Quieter aircraft engines present another major technical challenge, although much has already been done in this area. Improved navigational safety and enhanced crash survivability are substantial domains of development of the coming years. Pioneers tackled the basic aerodynamic problems leading to successful flight, and the same brand of innovative spirit and technical skill will be needed for aviation to maintain its role in the transportation system of the future.

Conclusion

The nation's fixed-base operators are central to serving general aviation as well as many airlines. The skill of their pilots and the FBOs' sensitivity to the concerns of the non-flying public, as well as the way they

Existing airports are increasingly close to full capacity and need relief. Source: Air Transport Association

bring new users and advocates into the system, will have much to do with the prosperity of general aviation in coming decades.

Conversely, failure to develop a more understanding and enthusiastic general public will likely lead to more restrictive zoning, more lawsuits about noise, failure to pass needed bond issues and appropriations, restrictive leases, closure of airports, lack of business support in the community, and limited access for general aviation to the nation's airports and airspace.

Thus, the nation's FBOs have more than just the challenge of running a profitable and safe business in an ever-changing economy. Rather, they have wider and deeper obligations concerning:

> Technical improvement in the performance and safety of flight vehicles and equipment;
> Improved standards of professionalism and business orientation throughout the industry;
> Reduced fragmentation of the industry; and
> Better public understanding and acceptance of the general aviation industry and its benefits to country.

In sum, the typical FBO owner/manager should be budgeting only about 70 percent of his or her time for internal running of the business, and about 30 percent on outside liaison and educational work.

Summary

A brief review of aviation history reveals a background of inspirational individual accomplishment in overcoming tremendous technological challenges. The civil aviation industry began to divide in the 1920s into scheduled passenger service and "general aviation." Particularly since the Airline Deregulation Act of 1978, airlines have functioned similarly to other big businesses, by focusing on market opportunities and the bottom line. They have developed management practices as specialized and sophisticated as most other large organizations. By contrast, general aviation is still closer in spirit and functions to the exciting days of barnstorming and record-breaking. The industry went through rapid growth until the late 1970s when a variety of factors caused a decline. Since that time, the surviving operators have achieved success largely because of their ability to evolve from aviation enthusiasts to professional managers. As the general aviation industry matures, these trends will likely continue, although there may and should still be a strong experimental and innovative aspect in general aviation design that will affect the whole industry. In short, general aviation will be predominantly a business activity, but there will also be a segment that flies for the sheer enjoyment.

DISCUSSION TOPICS

1. What did the Wright brothers achieve that was unlike previous successful flights?
2. Where does the term "Fixed Base Operator" come from? Can you think of a better term?
3. What aspects of aviation are encompassed by the term "general aviation"? Can you think of a better term?
4. What is the future growth potential of the general aviation fleet? Why?
5. Why would a business person remain in a low-profit aviation activity?
6. What are the major elements of the U.S. aviation industry?

Endnotes

1. Kane, R. M. (2002). *Air Transportation*. Dubuque, IA: Kendall-Hunt.
2. Ibid.
3. Ibid.
4. Constant, E. W., II. (1980). *The Origins of the Turbojet Revolution*. Baltimore, MD: John Hopkins University Press.
5. Serling, R. J., & Editors of Time-Life Books. (1982). *The Jet Age*. New York: Time-Life Books.
6. See *http://www.britannica.com/technology/Tupolev-Tu-144*

7. Retrieved from *http://www.space.com/8099-hypersonic-rocket-test-launched-australia.html*
8. See *https://www.leonardocompany.com/en/air/helicopters*
9. For a discussion of the various ways to get licensed for sport flying, see *http://www.sportpilotexaminer.com*.
10. See *https://www.aopa.org/advocacy/advocacy-briefs/issue-brief-aviation-security*
11. AAAE news article about British Airways, April 2002.
12. Retrieved from *http://www.generalaviation.org/gara/index2.html*.
13. Telephone conversation between author Rodwell and Bill de Decker, May 13, 2002.
14. Croft, J. (2002, April 15). Small Airports: To Be or Not To Be? *Aviation Week and Space Technology, 156*(15), 58–61.
15. See *http://www.faa.gov/air_traffic/*
16. Ibid.
17. See *https://www.faa.gov/regulations_policies/advisory_circulars/*
18. Regulation of Fractional Aircraft Ownership Programs. The Recommendation of the Fractional Ownership Rulemaking Committee (FOARC), presented to FAA on February 23, 2000. See also NATA comments on Docket #FAA-2001-10047: Notice No. 01–08, Regulation of Fractional Aircraft Ownership Programs and On-demand Operations, November 16, 2001.
19. Cook, B. (1997). *The Evolution of FBOs—Seeking the Perfect Mix.* Airport Magazine, 9(2), 16–25.
20. U.S. Department of Transportation, Federal Aviation Administration. (2006). *Minimum Standards for Commercial Aeronautical Activities.* (FAA Publication No. 150/5190-7). Washington, DC: U.S. Government Printing Office. This publication makes available to public airport owners, and to other interested persons, basic information and guidance on FAA's policy regarding exclusive rights at public airports on which Federal funds, administered by FAA, have been expended.
21. Ibid.
22. Air Travelers in America Survey: Conducted by Ipsos on behalf of Airlines for America, this detailed 2019 survey describes current trends in air travel demographics, including a breakdown by purpose of trip, number of trips taken annually, and destination data.
23. Learn more about the program including all the flight schools that participate by going to the YOU CAN FLY web page at *https://youcanfly.aopa.org*.

2

Management Functions

OBJECTIVES

> Identify the "four functions" of management as they are traditionally set forth, how they specifically relate to the operation of an FBO, and what else is involved in management.

> Recognize some of the common managerial errors found in business today and how to correct them.

> Describe the elements of a successful business plan.

> Understand the "dos and don'ts" of delegation.

> Show how good time management techniques relate to good business practice, and demonstrate good time management concepts that you can implement now.

"Some are born managers, some achieve management, and some have management thrust upon them."[1]

Introduction

The Four Traditional Functions of Management

Management can be defined as getting things done through others. This means that a manager's (or supervisor's) primary responsibility is to keep his or her staff effectively occupied with the priority activities of the department or company, reserving to him or herself those tasks that only he or she can do. Its successful accomplishment may require different techniques of delegation for different tasks and people. The practice of management is both an art and a science. The FBO owner/manager may be the only manager in the company, depending on its size. In a larger organization the owner/manager will have other departmental managers reporting to him or her.

An early philosophy of the role of management in organizations was developed by a French engineer, Henri Fayol, who in 1916 published a book called *Administration Industrielle et Generale*. This book was translated into English in the late 1940s and its basis was that management activities are divided into four functions:

1. Planning;
2. Organizing;

3. Directing and Coordinating;
4. Controlling.

Yet, it can be very difficult to divide managerial functions in this manner. At the end of any given day the typical manager will not likely be able to say, "I spent three hours planning, one hour organizing, two hours directing, an hour coordinating, and two hours controlling." An article in the *Harvard Business Review* (Mintzberg, et al., 1998) puts the challenge of management as follows:[2]

> "If you ask managers what they do, they will most likely tell you that they plan, organize, (direct), coordinate, and control. Then watch what they do. Don't be surprised if you can't relate what you see to these words.
>
> "When a manager is told that a factory has just burned down and then advises the caller to see whether temporary arrangements can be made to supply customers through a foreign subsidiary, is that manager planning, organizing, coordinating, or controlling? How about when he or she presents a gold watch to a retiring employee? Or attends a conference to meet people in the trade and returns with an interesting new product idea for her employees to consider?
>
> "These four words, which have dominated management vocabulary since the French industrialist Henri Fayol first introduced them in 1916, tell us little about what managers actually do. At best, they indicate some vague objectives that managers have when they work."

The article goes on to discuss what managers *really* do, which cannot be packaged as neatly as M. Fayol perceived. Peter Drucker has said that the purpose of business management is to get and keep customers profitably.[3] That being the case, the functions of planning, organizing, directing, and controlling are simply tasks performed to accomplish these goals of happy customers and a profitable enterprise.

The typical manager is a problem-solver on a tactical level, ensuring that the various aspects of a business function smoothly. However, it is increasingly important that managers also serve as leaders, developing an all-important strategic vision for their organizations. Such an approach is necessary, because new and often unpredictable external events affect the general aviation business almost daily.

Regardless, some framework is needed for discussing managerial work, and so these four traditional functions are used in what follows. The manner in which these functions are interpreted and executed will have major impact on the success or failure of the enterprise. Management theory has been evolving rapidly in the last few decades, and there are numerous books, articles, websites, and professional associations that can assist the aspiring manager.[4] For example, there has been much emphasis in recent times on the manager as leader. Yet, if a manager is a visionary leader but impractical and disorganized, the vision will not come to pass unless subordinates take care of the practicalities—and are empowered to provide the consistency and persistence that are needed.[5,6]

The general approach to management tends to look at the internal operations of a company. In the aviation service industry, the typical owner/manager of an FBO will need to budget his or her time to spend perhaps as much as 30 percent of each week on external affairs. This external involvement includes dialogue with local businesses about issues such as the merits of aviation and their potential use of that airport, time spent with community members concerned about airport noise, and time with local elected officials, especially the local jurisdiction that controls land use around the airport and receives noise complaints, and the airport owners and policy makers, who have it in their control to make or break the FBO's business by actions they take to support or limit the activity that it depends on for its livelihood.

Common Managerial Errors and How to Address Them

It is useful to consider the traits of the ineffective manager so that they can be avoided.

There are a number of problem areas that managers must overcome if they are to be successful.[7]

1. Failure to Anticipate Industry Trends

We are part of a rapidly-changing economy, and new technology and other changes in a related field may affect part of our sales. For example, the facsimile machine became available in the mid 1970s, but it did not offer good quality until the late 1980s. This led to a decrease in the demand for delivery of documents by air, which had a corresponding impact on the FBOs that depended on that market as an income source. As Internet-based technology has vastly increased in ubiquity and available bandwidth, the fax machine itself has been replaced to a large extent by e-mail attachments, web-based project management, and Internet-based teleconferencing.

Managers can address this problem by staying abreast of industry periodicals (e.g., *Business & Commercial Aviation, FBO Management, Aviation*

International News), by thinking "outside the box" about how new societal trends and technologies may affect their business, and by investigating and pursuing emerging markets early in their development. The latest tools include social networking media (Twitter, Facebook, etc.), and electronic conferencing applications (e.g., GoToMeeting, Zoom, Microsoft Teams) to support the business.

2. Lack of Priorities

We live in an increasingly information-based society, making it harder to decide what to read and do. Greater knowledge about the economy, potential markets, and new business tools can make the business owner feel indecisive about where to begin. Technology is changing more rapidly than ever; and obsolescence can arrive before a new device is paid for and sometimes before we even know how to use it properly. As many business owners have said, "I'm so busy doing the urgent things that there isn't time to do the important things."[8] Going in all directions causes exhaustion and lack of results.

Focusing first on what is most important, and letting non-important and non-urgent matters diminish in priority can help address this, assuming the manager IS making some kind of priority list and working through it each day.

3. Indecisiveness and Lack of Systems

As some businesses grow, they resist the establishment of policies and procedures. The owner/ manager decides how to proceed each time, even though similar situations arise again and again. Worse yet, the manager makes a decision, the staff begins to implement it, and a week or two later he or she revises it.

Companies of any size need rules about who can make what level of decision, and routine or repetitive occurrences need standard procedures, manuals, and policy guidelines. It is a managerial responsibility to break work down into steps and memorialize these procedures into written formats so that they do not have to be continually reinvented.

4. Poor Time Management

The average manager gets interrupted once every eight minutes. This means that even large tasks must be managed in smaller-than-eight-minute segments. The manager must expect, and welcome, interruptions because answering questions from staff members about their assignments in an efficient and timely manner enables those employees to be fully productive and to stay on task.

The manager must be extremely disciplined about getting back on task after each interruption. Failure to do so is the most time-consuming consequence of interruptions. It may be desirable to set up specific drop-in times when interruptions are invited, in order to prevent them from happening throughout the day. This topic is discussed in more depth later in this chapter, with many more suggestions for solutions.

5. Poor Communication Skills

Since the manager's results are manifested significantly through the work of his or her subordinates, the ability to communicate clearly what one wants or expects is crucial. This is discussed more fully in Chapter 5, Human Resources.

6. Lack of Personal Accountability and Ethics

Even though the manager delegates a task, he or she is still responsible for the result. If the delegatee fails to produce, it becomes the manager's responsibility, including finding a timely remedial action, so that the project can still be salvaged.

By modeling personal accountability and ethics, the manager encourages similar behavior among staff. The business ethics issue is discussed at greater length below.

7. Failure to Develop, Train, and Acknowledge People

A knowledgeable manager may have a better understanding than any of his or her staff about the desired outcomes of the proposed project. However, unless staff members are given a chance to grow to a comparable level of knowledge, understanding, and initiative, the manager will not obtain satisfactory results from them. This starts with a realistic assessment of the staff's current capabilities, and the assignment of tasks and projects they are reasonably ready to handle. It continues with mentoring, coaching, and on-the-job training, so that each person gradually acquires skills akin to those of the supervisor making the assignments. More on how to "stretch" people is discussed below under "Delegation", and Chapter 5, Human Resources, reviews concepts in training and recognition. In addition, even modest performers need appreciation and acclaim.

8. Failure to Support Company Policy in Public

When talking to one's team, a manager's attitude that says, "We've got to do this even though we don't like it because the higher-ups say so," is more likely to engender reluctance than enthusiasm.

Good managers express their reservations and concerns to their superiors in private, and sometimes they are persuasive enough to get their way. However, if they lose the argument, they must support the supervisor's decision gracefully (or resign).

9. Failure to Acknowledge and Accommodate People's Workstyles

Numerous testing instruments divide people into four or more personality types.[9] Each personality type has styles of thinking, learning, and working that differ from the others. People of a particular personality type may have difficulty respecting the cognitive characteristics of those of other types. This can create conflict and lack of progress on assigned work.

Effective supervisors, regardless of their own style, carefully (and formally or informally) identify the styles of their people and use this knowledge both in what they assign to whom, and how they manage the assignees. A good manager knows this is the most effective way to get the most out of everyone. Management also needs to teach people of differing workstyles how to recognize and respect each other's strengths and work together effectively as a team.

10. Failure to Focus on Profit

Ignorance of measurable business results and lack of awareness about the comparative success of one's different products and services can be a major downfall.

In most companies, sales and profits are the key objectives, and everything else undertaken must be in keeping with these goals. Individual employees may contribute to the goal of profitability in different ways; problems should be resolved primarily because lack of resolution will negatively affect profits.

11. Failure to Recognize the Needs People Fulfill by Working

There are diverse viewpoints about the needs people hope to fulfill by working. Chapter 5, Human Resources, discusses these in more depth. Many studies about motivation indicate that people want to do a good job and to be shown greater challenges by a supportive supervisor. The manager who behaves unpredictably, who manipulates people for his or her own ends, who is dishonest about use of time, who is not accountable, or who attempts to be a buddy with subordinates will generally create a counterproductive level of discomfort.

By contrast, a supervisor who sets clear expectations, recognizes equal work with equal (and honest) praise, and acts equitably in other ways, will generally get more than adequate performance from employees.

12. Failure to Establish and Adhere to Standards

Lack of standards means that staff members do not know how high the bar is set and what good performance is supposed to look like. Not knowing may mean not delivering.

Clearly articulated standards permit all staff to be treated alike. In recurrent task areas, standards must be established and then all "violators" treated the same way (see the discussion of the "red hot stove" rule in Chapter 5, Human Resources). Incompetence must not be tolerated, no matter who is the source.

Planning and Organizing

The Need for a Business Plan

The first two functions of management—planning and organizing—should be a daily process for the manager. Much can also be gained by formally structuring these processes into a written business plan. The aviation manager wears the many hats of the entrepreneur and does so in an organization that has more than the average number of service lines, each with different levels of sophistication. The FBO manager must also cope with the very rapid changes that continually occur in aviation. New equipment, changing regulations, advancing technology, and a volatile market all contribute to a rapidly-changing environment.

A sound planning process is a necessity in this kind of environment. The effective utilization of a planning process becomes a way of life for the successful manager. Planning falls into two areas: objectives planning and operational planning. Objectives planning concerns the things the business owner wants to accomplish: the vision, the long-range ideas, and the business mission. Operational planning concerns the methods used to achieve them. Objectives planning deals with specific long-and short-range targets for the organization. It is the subject of the following paragraphs. Operational planning deals with the development of policy, procedures, rules, and standards used to run the business. This is the topic of Chapter 6, Organization and Administration.

Some business owners will argue that in a rapidly changing world it is impossible to plan effectively, so it is better to simply seize opportunities as they present themselves and not formulate any kind of plan. Here are some considerations in relation to this approach:

1. If the business has not specifically selected a market niche, it will tend to pursue all opportunities without priority. Time and energy may be wasted pursuing activities that are not central to the company's areas of strength. In the meantime, more valuable opportunities may be overlooked. Furthermore, the company will present no clear image in the marketplace.
2. Customers who know the direction in which the company is going will tend to plan their own needs around the future availability of services. For example, a businessman thinking of learning to fly may choose a flight school based on the fact that starting next year it will be offering instrument instruction. The flight school needs to not only plan, but also let its plans be known.
3. Customers who know the service area priorities of the business will refer their acquaintances who need those services.
4. Setting ambitions through expression of achievable goals is often a self-fulfilling prophecy. The very process of setting higher goals tends to shift the employees' "comfort zone" of what is attainable. Well conceived, well expressed goals seem to create a synergy of their own, such that fortuitous occurrences will support the goals. In other words, planning can unlock synergies that might not otherwise be available.
5. A business plan can be written for each project and any new business project, such as a new product line. The analysis required for a written feasibility study or plan will help to make it very clear what resource commitments will be needed to implement the new line profitably.
6. Last, but not least, any business that may ever need outside financing must have a written business plan in order to attract suitable lenders.

A business plan is only valuable if it is examined and evaluated frequently and its results compared to the original objectives. It is a living process, not a beautifully bound report on a shelf.

Most business experts place great emphasis on having a written business plan. Even for a new company or one that is in a very volatile market, such as general aviation, it should be specific, made available to all managers, and periodically reviewed and updated.

Business Plan Outline

A suggested outline for a full business plan is shown in Figure 2.1. The numbers after each section indicate a practical order in which to write the sections. It is vital to have a one-page, or at most two-page, executive summary with 1–2 figures. That section may be the only one that some audiences read.

Mission Statement. The mission statement should answer the questions "Who are we, what is it we do, and why are we different?" It can be quite short, such as the following:

"To be the best quality repair, instruction, and flight service facility within a fifty-mile radius of XYZ"; or

"To provide the lowest-cost fuel and services and maximize transient traffic."

There can be different mission statements for each segment or profit center in the business. Another way of considering the mission statement is to answer the question "What business(es) am I in"? The functions enjoyed most by the owner may be neither the most needed nor the most profitable functions. The manager should consider why customers use aviation to meet their needs and whether the business is primarily service-or product-oriented.

Values. The most successful businesses seem to be those that (1) stay close to the customer and (2) operate with real concern for employees.[10,11,12] A value statement combining these two concerns might be "Our top priority in XYZ Company is customer satisfaction, and our only means of achieving it is through the contributions of each and every employee. We want happy employees so that we will have happy customers."

Other values could apply, such as quality, durability, reliability, speed of service, and friendliness. The important thing is to spell out the paramount value or values so that all members of the firm know what comes first. Companies with a strong "corporate culture" have a better chance of dealing with difficulties.

Market Niche and Goal. This part of the business plan should discuss the total market for the company's products or services, and the precise niche or segment of that overall market the company hopes to serve. It will include whatever ways the company's products or services differ from those available elsewhere, and should relate closely to (and support) the mission statement.

Figure 2.1 ✈ Business Plan Outline

A. EXECUTIVE SUMMARY (12)
   - *Mission Statement*—definition of business purpose, product, and market.
   - Method of sales and distribution.
   - Values.
   - Brief description of management team.
   For Seeking Capital or Loans:
   - Summary of financial projections.
   - Amount, form, and purpose of money being sought.

B. COMPANY HISTORY AND BACKGROUND (10)
   - Date and state of inception, form of company.
   - Principals, and functions of each.
   - General progress to date.
   - Successful strategies to date.
   - Most urgent issues to be addressed.
   - Other general context.

C. INDUSTRY OVERVIEW (2)
   - Current status and outlook for industry.
   - Specific industry-related issues (economic, social, technological, regulatory).
   - Areas of growth and opportunity.
   - Performance of primary participants.

D. PRODUCT(S)/SERVICES (1)
   - Description.
   - Research and development.
   - Future development.
   - Environmental factors.
   - Policies and warranties.

E. MARKETING ANALYSIS (3)
   - Market definition.
   - Market size.
   - Market trends.
   - Competition.
   - Competitive advantages/disadvantages.

F. MARKETING PLAN (4)
   - Potential target markets, estimated sales, market share.
   - Strategies.
   - Pricing.
   - Sales and distribution.
   - Suitable promotional methods and costs.
   - Promotional mix and total budget.
   - Evaluation/effectiveness plans.
   - Stationery, logos, and image.

G. LEGAL REQUIREMENTS (9)
   - Legal structure of the business.
   - Licensing, trademarks, patents.
   - Certification.
   - Insurance.
   - Building codes, zoning, and regulations affecting business.
   - Anticipated changes affecting business.
   - Company name.

H. PERSONNEL (6)
   - Number and type of employees.
   - Labor issues and compliance.
   - Sources of labor.
   - Changes anticipated.

I. OPERATIONS (7)
   - Location.
   - Plant and/or office facilities.
   - Equipment.
   - Methods of production and manufacture.

J. MANAGEMENT (5)
   - Organization.
   - Key people and résumés.
   - Strengths and weaknesses.
   - Professional advisors.

K. FINANCIAL INFORMATION (8)
   - Funding requested (if appropriate).
   - Desired financing.
   - Capitalization.
   - Use of funds.
   - Future financing needs.
   - Current financial statements.
   - Financial projections.

L. APPENDIXES (11)
   - Goals.
   - Objectives.
   - Functional Schedules, Gantt Charts, etc.
   - Information Systems.

Source: Julie F. Rodwell, 2009

Company History and Background. This section, primarily for readers outside the company, can also be useful for orientation of new employees.

Industry Overview. The FBO may be in the mainstream of industry trends or may already have one or more specialty niches. Periodic evaluation of major factors and trends affecting the whole industry is important. It may also be valuable to examine national economic and demographic trends that do not at first appear to have much to do with aviation, (e.g., the aging of, and ethnic changes within, the population, discussed further in Chapter 3). Some open-ended brainstorming on such issues may yield both threats and opportunities. An example: an aerial application firm may want to assess in its market area the trend toward organic foods, which means fewer farms using pesticides and thus less demand for the aerial applicator's services. Another example: with the advent of such tools as Google Earth and digital cameras, the demand for traditional aerial photography is changing, although with the increased use of Geographic Information Systems (GIS) in cartography, new opportunities may arise.

Products. This section provides the opportunity to examine not just the present product and service mix, but also key technological, institutional, environmental, and market changes that may affect that mix.

Marketing Analysis. The marketing analysis will likely be addressed separately for each product line. Some products and services may have a highly local market (e.g., a flight school). Others may have national markets, such as a maintenance shop certified to do major overhauls of specific types of aircraft engines, or major avionics installation and repair.

Marketing Plan. Some new activities may not need much marketing. Perhaps a new product or service is being offered based on market research in terms of customer feedback, which is an excellent way to find out what the existing client group wants. Methods of marketing that product or service might then be to enclose a flyer with billings, to tell each customer as they come in, and so on. But these alone may not be sufficient to spread the word and gain new customers. The choices of marketing and sales techniques are discussed more fully in Chapter 3, Marketing, and the precise approach utilized will probably need to be developed on a case-by-case basis. It is useful to keep track of what worked or didn't in a promotional campaign so that the plan can be more finely honed the next time. Again, the plan requires segmentation into individual tasks with deadlines, quotas, and staff assignments. One fundamental component of marketing that is extremely important is an attractive (and up-to-date) company website with easy-to-find contact information and a summary of the various services offered. The issue of *branding* needs to be addressed—the use of a business name or family/series of names and creation of a unified visual look through logo and color scheme on all materials—stationery, brochures, business cards, invoices, web site, equipment, as well as the repetition of key company themes in advertising.

Legal Requirements. In many cases, such as obtaining business licenses, legal requirements are a one-time process. However, the FBO manager should not assume that this is always so. An example of the past few decades is the underground fuel storage tank compliance issue, for which tough environmental requirements have caused FBOs (and others) to look very carefully at both past and future operations relating to fuel storage. The legal structure of the business should also be re-examined periodically with regard to succession planning. This becomes especially important as the business changes and key personnel move closer to retirement.

Personnel. Written organization charts and job descriptions are an important starting point, although in some cases they are only that. Staff in a small business may need to perform more than one job, and special task forces, quality circles, and the like are increasingly supplementing organizational structures. But even these more nebulous arrangements generally lend themselves to some kind of description and narrative. This topic is discussed more fully in Chapter 5, Human Resources, and Chapter 6, Organization and Administration.

Operations; Production Plan. This part of the business plan reviews physical requirements, supplies, materials, labor, office equipment, and other items needed to actually produce the goods and services. These areas are reviewed more fully in Chapters 8 through 10 and in Chapter 12. Plans describe the overall functions of each division of the firm, and tasks describe individual work assignments to accomplish them.

Management. This may include a review of key people, strengths and weaknesses, and available resources, such as a financial or management consultant used on a periodic basis. These topics are crucial to any business plan seeking outside investment, but should not be ignored even if the business plan is for internal use only.

Management information can have other uses, such as selling company services to a major corporation and developing public relations campaigns. A small brochure may be appropriate, especially one separate from the rest of the business plan document.

Ownership information may or may not be appropriate in a plan to be shared with all employees; in any case, it requires some consideration, especially in a very small and/or family-owned business where the sudden absence of key people might cause the firm to struggle or fail. As previously suggested, the succession plan needs to be stated explicitly, and key-person insurance may be a consideration.[13]

Financial Information. In some cases, it may be desirable to set up a financial plan for a period of 20 or 30 years, if this is the length of a new FBO lease or the anticipated retirement date of the owner. In other cases, a five-year period may be sufficient. In a new business, particularly one that is seeking financing, the first year should be presented on a month-by-month basis. In the case of a new project or firm, the financial plan should include a profit-and-loss statement, a balance sheet showing assets and liabilities, and a cash-flow analysis showing how long it will take to reach the break-even point. This will also show the cumulative amount of cash needed before break-even is reached and a positive cash flow begins. Another useful feature of the financial plan can be a listing of threshold criteria or conditions under which new expenditures can be made; for example, should a new firm with multiple owners have written agreements about the conditions under which each owner will start taking a salary, the conditions under which benefits will be offered to employees, or the conditions under which company cars can be acquired? These topics are discussed more fully in Chapter 4, Profits, Cash Flow, and Financing.

Strategy and Objectives. For a long-established firm, the strategy for becoming established in a market niche may already have been accomplished. However, new market opportunities arise constantly, and if a decision is made to pursue them, a strategy is needed. The strategy statement must discuss the new product in terms of its features, benefits, and pricing, its means of production, the sales and promotional activity, and the cost of getting it launched. Setting objectives means choosing specific, measurable targets with associated dates.

Growth Strategy. Once the target level of activity is reached for a product or service, several growth strategies are possible for successive time periods. For example, the sequence or choice of strategies might be:

> No growth. (Note that in a growth market this necessarily means declining market share);
> Maintenance of market share, i.e. growth at the pace of the total market;
> Growth to a specific, higher dollar volume or market share by a certain year, with annual rates of growth until then;
> Growth until some perceived point of diminishing returns. For example, a repair shop might grow only until utilization of the maintenance hangar reaches the maximum point, stopping short of the need for, and expense of, a new hangar.

Growth will be limited by the competition, national trends in aviation, and by the business cycle. A detailed growth strategy will need to examine carefully the actual and potential profitability of each service or product, a topic discussed more fully in Chapter 4, Profits, Cash Flow, and Financing. Note that a conscious choice not to seek growth beyond a certain level is perfectly valid, and may enable the company to select only the best customers–those who appreciate what it offers, and who pay on time.

Functional Schedules. In the appendix section of the business plan, it may be appropriate to develop more detailed schedules than in the main text of the business plan, by using PERT (flow diagram) and Gantt (timeline) charts, as well as matrices that show the assignments of staff to tasks. Software such as Microsoft Project is available to produce these.

Information Systems. The best plan in the world is of little value if there is no way of telling whether it is being successfully implemented. Information on the competition, industry sales, new clients acquired, costs, productivity, profit by area, promotion results, and so on must be collected regularly and organized into quickly-usable formats. As when flying an aircraft, the quicker one knows the aircraft is off course, the easier it is to make a correction to return to course; the sooner one knows that turbulent conditions are ahead, the easier it is to change heading or altitude to remain in smooth air. This subject is reviewed in depth in Chapter 7, Management Information Systems.

Directing, Coordinating, and Controlling

Managing, Versus Doing

Poor delegation is probably the typical manager's greatest weakness when it comes to Fayol's last two management functions—directing and controlling. The more management responsibilities an employee has, the more that employee must obtain results through the work of others. And yet, some technical knowledge and interest in specific aviation activities are generally what bring a person into the business, and indeed, equip them to delegate with some knowledge about what it takes to perfrom the task being delegated. Moreover, particularly in the FBO business, employees tend to be promoted up the ranks and may well become managers and supervisors themselves without ever having had any major exposure or training in how to accomplish work through others, as opposed to doing it oneself.

Most individuals start off in the working world by acquiring a technical skill or specialty. Most aviation managers start with an interest in aviation and some professional qualifications. Their career track may have included being a corporate pilot, mechanic, flight instructor, aircraft production worker, or simply a private pilot and aviation enthusiast. When the opportunity came along, they moved into a managerial position, either as an owner or an employee. Such a move required that they begin performing different activities, such as planning, organizing, directing and controlling the work of others, rather than doing wholly technical or line activities. In an FBO, this tends to mean more office work and less time in an airplane, classroom, or hangar. This changing role requirement is illustrated in Figure 2.2.

Figure 2.2 → Managing Versus Doing

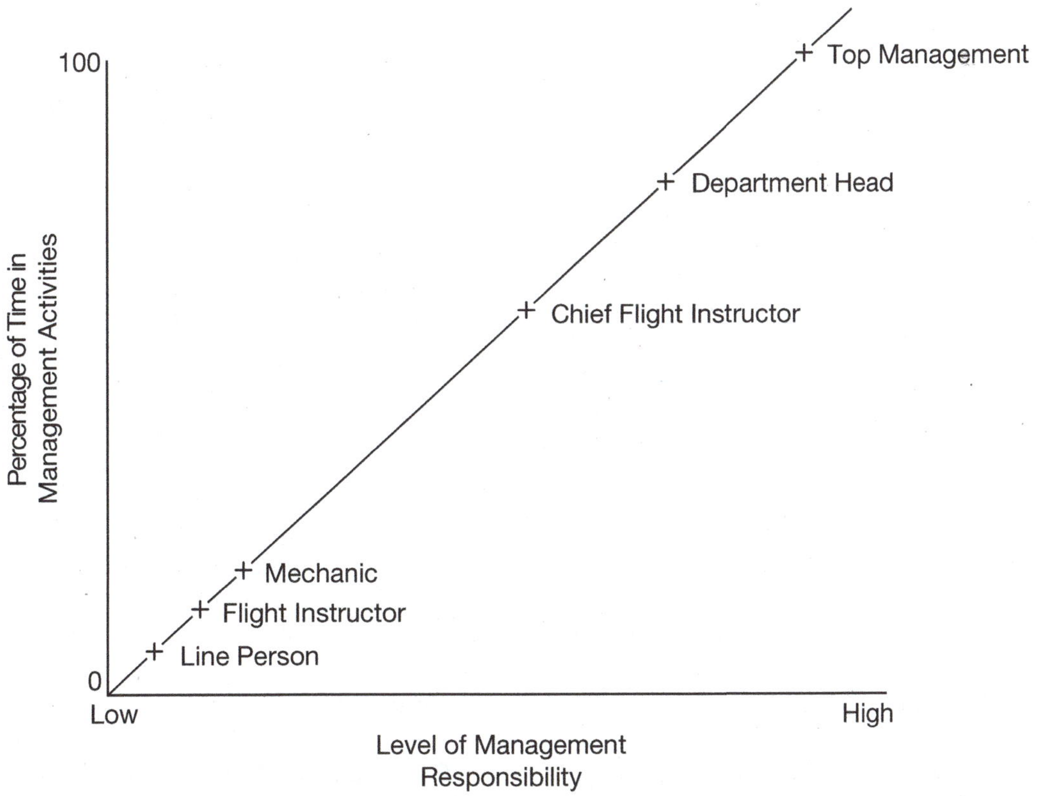
100
0
Percentage of Time in Management Activities
Top Management
Department Head
Chief Flight Instructor
Mechanic
Flight Instructor
Line Person
Low
High
Level of Management Responsibility

Source: Tom Carney

Studies of successful and unsuccessful managers indicate that the successful ones pay close attention to planning, organizing, directing, and controlling others, even though, as discussed, it is not always possible to know what balance of those functions is occurring at any given time. The less successful managers spend most of their time doing other things. In a study by the Small Business Administration, the success of a business was also found to be in direct proportion to the owner's possession of these talents:

> Alertness to change;
> Ability to adjust or to create change oneself;
> Ability to attract and hold competent workers;
> 180-degree vision with respect to operating details; and
> Knowledge of the market—customers and their needs.

A difficulty for the FBO manager in a very small firm, one with ten or fewer employees, is that he or she never will be able to devote 100 percent of the time solely to managing, but must generally spend part of the time in line functions. This is because of the hours in which most FBO services must be available and the variation of those services. Managers in such a position may need to specifically divide each day into "doing" and "managing" times, making sure that their precious and limited managerial time is blocked out to minimize interruptions and assigned to issues of the highest priority.

Style of Problem-Solving and Delegation

In line with the array of overall management styles, discussed in the next section, there are many styles of delegation, and any that accomplishes both short- and long-range results is successful. Much has been written about whether the autocratic or the democratic approach is most suitable. The choices exist on a continuum, as follows:

> The manager makes a decision and announces it;
> The manager obtains buy-in for a premade decision
> The manager presents ideas/decisions and invites questions;
> The manager presents a tentative decision subject to change;
> The manager presents the problem, gets suggestions, and then makes his or her decision; or
> The manager—or another staff member—presents the issue and the entire work team brainstorms solutions and chooses how to proceed, using a consensus approach.

In general, U.S. management style appears to have shifted from the autocratic toward the democratic or participatory during the past four decades. The advent of Internet-based information sharing means it is more difficult for management to operate autocratically than was previously possible.

Leadership Styles

The successful leader is one who wants others to be successful in the business and who is a super-coach, most of the time, rather than a star player.

One myth of leadership is that leaders are born. People are born with a degree of health and intelligence, but leadership can be learned. The following is intended to be an introduction to some basic leadership styles and their definitions and is derived from the foundational work of psychologist Kurt Lewin. It is by no means a comprehensive examination of this topic.

Authoritarian Leadership. These leaders are also known as autocratic leaders, and typically provide clear expectations on the nature of the work required, as well as how and by when it is to be performed. The delineation between an authoritarian leader and followers is usually clear, and this implies that decisions are made by the leader with little input from the rest of the group.

Researchers have found that decision making is often less creative under such leaders, and that it is difficult to transition from an authoritarian style to a participative style. The authoritarian style is best applied in situations when time is a constraint in allowing group input, or when the leader is the most knowledgeable group member. Some managers who aspire to an authoritarian style mistakenly believe that being periodically or regularly harsh, critical, demanding, and coercive are positive elements of the authoritarian style. Such managers will generally find that a positive approach, yet still within their authoritarian style, achieves better results and more job satisfaction and motivation among the staff.

Participative Leadership. Participative leaders, also known as democratic leaders, function as members of the group and solicit input from other group members, which they use to develop their guidance of those members. These leaders typically retain the final decision in the process.

Research has shown that members of groups guided by this style are less productive than those under authoritarian leaders, but their contributions are of higher quality. Members of participative groups generally feel engaged in the decision-making process and are more motivated and creative than members of autocratically led groups.

The participative leadership style is best applied when sufficient time is available for input from the larger group, or when the leader is not the most knowledgeable group member.

Delegative Leadership. Delegative, or laissez-faire, leaders provide little guidance to group members and leave final decisions up to those members. This style of leadership may be appropriate for groups with highly-qualified members, but can easily lead to diminished motivation due to confusion over roles and responsibilities of members. Research indicates that this is the least productive of the three leadership styles, with members who are often unable to work effectively either independently or as a team.

Summary. Each of the three styles of leaders has implications for the decision-making process and for the manager's role as a change agent. No one person fits neatly into any one of these styles. There are fragments of each style in virtually all administrators. Any kind of leader is accepted by some and rejected by others. It depends on the needs of the followers or employees.

The effective leader must strive to meet the needs of all employees, which is why it is so difficult for a leader to be truly effective. A good leader and manager will also to some extent adapt his or her style to the context for the particular issue being tackled.

Objectives of Delegation

An autocratic view of management might be described as a system set up with only the manager's immediate personal needs in mind; thus, its objective is mainly to free the manager's time for other important tasks. The staff is to do the manager's bidding, preferably carrying out tasks as closely as possible to the manner in which the manager would carry them out. This manager is happier, the more closely his or her subordinates are able to function as "clones."

An alternative perspective is that "freeing up" the manager is only a minor purpose of delegation. A more complete list of the purposes of delegation includes:

> Leaving the details to others;
> Getting the job done;
> Allowing key management staff to take initiative;
> Keeping things going in the supervisor's absence;
> Raising the level of employee motivation;
> Increasing the readiness of subordinates to accept change;
> Improving the quality of all managerial decisions;
> Developing teamwork and morale; and
> Furthering the individual development and growth in skills of employees.

The degree of delegation depends on the level of skill available, the central or peripheral nature of the task, its urgency, and the manager's other priorities.

Managerial Control

Delegation of responsibility for a task does not mean giving away control or accountability. In delegating work, a manager must:

> Spread the work realistically among available aides;
> Ensure that the job is done correctly;
> Budget enough time to take corrective action should something go wrong; and
> Develop subordinates' talents and abilities.

To accomplish these ends a process of auditing the progress of the task is needed, but without constantly peering over the person's shoulder or being rigid about how the task is performed. Methods for doing this include:

> Verbal reporting at prearranged points or dates;
> Written reports on progress;
> Scheduled conferences;
> Setting deadlines for results;
> Checking results; and
> Measuring results in related areas (e.g., changes in the number of complaints).

When the task is clear from the outset, product inspection and deadlines, with some intermediate status reporting, may be successful. When the task itself still requires definition, the manager may assign various specific tasks to different people and schedule a conference at some reasonable future date to brainstorm and decide what the problem is, and to design and assign a course of action. In the case of new problems or new employees, where untrodden ground is to be covered, it is advisable to delegate only small, highly specific tasks with short deadlines. One can then evaluate quite soon the person's response to the issue, and whether more supervision of the newcomer and the provision of additional expertise are needed. Throwing people "into the deep end" to see how they will respond is often unproductive and demoralizing. It is also easier to gradually give out more responsibility based on good results than it is to remove it once it has been granted.

Choosing the Areas to Delegate

One of the most crucial questions for a small business is what areas to delegate on a consistent basis. Lower-level managers constantly encounter the need for decisions, ranging from trivial to those of a more serious nature. It is possible for such decisions, because they are so numerous, to consume an entire day without some basic guidelines. The general level of responsibility given to line managers must be tailored to each company, but some possible criteria include:

> The owner approves all decisions related to expenditures (e.g., overtime, new promotions, equipment), or all such decisions above some specified amount;
> The owner is the final decision-maker with regard to employment and termination at or above a particular level of management.
> The owner decides about new services to be offered; and
> The owner makes or approves all public statements to the media, politicians, and so on.

Dos and Don'ts of Delegation

One of the biggest complaints of managers is that they cannot "get the monkey off their back"; that is, the task delegated to a subordinate is brought right back and for various reasons becomes the manager's problem again. Successful delegation requires that the delegatee actually perform most of, if not the entire, delegated task. How can this be accomplished? Delegate tasks to those employees who appear reasonably prepared to handle them. Give clear instructions and establish mechanisms and timelines for checking how things are going. While a reasonably prepared employee will rarely be able to fully accomplish with 100% accuracy a task that has never before been performed, that employee will be able to handle the majority of the assignment, will develop additional knowledge and skills in the process, and will be better prepared the next time the task is to be completed. At lower levels of readiness, a person will be intimidated, overwhelmed and anxious, ask too often for help, and may make serious mistakes—or do nothing until right before the deadline. It is the manager's responsibility to know how ready each employee is for specific new functions. Figure 2.3 offers some dos and don'ts for successful delegation.

The Decision-Making Process

In an established enterprise where basic authority has been clearly delegated to certain people for specific things, few if any questions should arise about who should decide what. Indeed, nothing but routine decisions should be taking place. Such periods of stasis are, however, the exception rather than the rule in small businesses and in aviation businesses in particular. The need to react frequently to rapidly-changing conditions is what one author considers the crucial difference between entrepreneurs and executives. The latter are characterized as having a custodial role and temperament while caring for an existing enterprise.

A sound decision-making process involves eight steps:

1. Setting company goals and objectives as a context for the decision.
2. Diagnosing the problem or issue precisely—not the symptoms, but the root causes.
3. Collecting and analyzing data about the problem or issue.
4. Developing an array of possible alternative solutions, including a do-nothing option.
5. Through brainstorming, mind-mapping, and the like, using the rule that "all ideas are good" at this stage.[14]
6. Evaluating and weighing the pros and cons of each alternative, quantifying wherever possible, and perhaps combining elements of several alternatives into a new one.
7. Selecting and implementing the alternative(s) that best fit(s) with objectives to solve the problem.
8. Obtaining Feedback—did it really work?

Figure 2.3 → Dos and Don'ts of Delegation

- Select the right person for the job. Don't just dump it on the first person that comes to mind. The right person will depend on how urgent the job is and whether you can afford the time to give the assignment to a less-experienced employee. Urgent tasks must normally be assigned to old hands.
- Set the climate for a comfortable briefing. Don't delegate on the run. Encourage the delegate to ask questions.
- Encourage the free flow of information. Don't forget to impart everything you know about the assignment, including any existing materials.
- Focus on the results, the what. Don't stress the how, unless there's absolutely only one right way. Describe the finished product as clearly as possible, and acknowledge any risk that may evolve while the work is being done.
- Delegate through dialogue. Don't do all the talking yourself.
- Set firm deadlines, both interim and final. Don't leave due dates uncertain, but work them out with the delegatee's input, given other projects and plans. Interim dates allow time to rescue the project, if necessary.
- Be certain the person is pointed toward all the necessary resources. Don't leave them wondering where to start.
- Turn over the entire job to one person. Don't give bits of it to several different people.
- Give the person the full authority to do the job, and make this clear to others, through staff meetings, memos, and phone calls to peers in other departments. Don't set them up to fail.
- Offer guidance without interfering. Don't fail to point out the minefields. Don't cross wires by making your own contacts with people you have told the delegatee to consult.
- Establish a system of controls beforehand. Don't do it as an afterthought. A written week-by-week schedule is desirable.
- Support your people if they need help or are involved in disputes. Don't leave them to succeed or fail on their own.
- Follow up along the way. Set the task up to minimize surprises. Don't wait until deadline day to see if the job is done.
- Give credit to the delegatee for a job well done. Put his or her name on the cover of the report; take him or her to the meeting with higher-ups to present the work, and praise the person to your bosses. Don't hog the glory. You will shine much more by publicly acknowledging your team. Give away each and every task that someone on your staff is ready to handle. Reserve for yourself only the tasks that you alone can do.

The manager may choose to involve the staff in all, or none of, these stages, depending on his or her theories of effectiveness. However, the numerous studies of group dynamics in recent years, coupled with motivational analysis, suggest that much better results will be obtained by involving key personnel in the process. They may have other interpretations of the performance data so that the problem may ultimately be defined quite differently. Their staffs may need to be enlisted to help gather better data on the problem, requiring both an understanding of what is going on and a time commitment. The process of developing alternatives almost always goes through several iterations. The manager might develop what he or she considers to be a complete array of possibilities, and in a management conference find new options proposed. Similarly, when everyone has carefully considered the issues and perhaps chatted with outsiders (e.g., with spouses and customers, if appropriate) and presented the options to their own departments, it is almost a certainty that constructive and creative new alternatives and sub-alternatives will spring up on any big issue. In the task of quantifying the pros and cons of the possibilities, the staff's time and understanding may be needed. The manager may reserve the final selection of the best alternative, but if this is after a participatory process, everyone will know why it is being selected, why certain alternatives were rejected, and what is going to be involved. After this groundwork, it will be much easier to delegate the plan's implementation successfully.

The decision-making process just described can be accomplished in five minutes or five months, depending on the complexity and importance of the issue. One of its most difficult elements is the second step—being sure one has properly identified the problem, and is not just treating symptoms. It is essential to define the root cause. Treating only the symptoms will not resolve the problem. It is suggested that the problem-solving team seek to isolate the critical factor that has to be changed, moved, or removed before anything else can be accomplished.

The process of developing alternatives does not imply that these are necessarily exclusive possibilities. More likely they are "program packages" of compatible solutions to related issues. The selected alternative may include program elements from several other alternatives, as long as the program as a whole is internally consistent.

Decision-Making Tools

Some decision-making tools are straightforward and commonsense applications of knowledge in an aviation business. Figure 2.4 depicts some typical problem areas and available techniques for decision-making.

Other tools are more complex. The field of operations research is devoted to the study of management problem-solving tools. A few of the basic approaches are summarized here.

Sampling theory involves the use of random samples to represent the population being considered. Sampling techniques can be of great value in an aviation business to conduct surveys of customer opinion regarding service, products, or potential business. Other applications include the sampling of inventory, work activity, or production quality.

Time distribution can be a useful piece of data when examining staffing shifts and hours of operation.

Figure 2.4 → Practical Decision Tools for the Aviation Manager

Areas	Typical Problems	Available Techniques for Decision-Making
1. Human Resources	Selecting personnel	Application blanks, interviewing guides, tests, reference checks, job specifications, job descriptions.
	Evaluating employee performance	Job requirements, performance appraisal system and forms.
	Compensating employees	Job evaluation program, job surveys, performance appraisal system.
	Disciplining employees	Clear rules and regulations, positive discipline environment, progressive discipline system, grievance procedure.
	Organizing work activity	Organization chart, manual; planning goals, organization.
2. Financial	Determining profitability	Financial information system, financial analysis, goals and objectives.
	Capital investment problem	Analysis by pay-back, rate of return, present value, company investment requirements.
	Departmental development and control	Goals and objectives, budget, analysis of activity.
	Cash requirements	Cash budget, short-term loan source.
	Accounts receivable	Credit policies, credit application, account aging process, collection procedure.
3. Material	Supplies on hand	Inventory system, information system, goals and objectives.
	Pricing material	Pricing policy, pricing guidelines and procedures.
	Maintaining physical assets	Maintenance policy, preventive maintenance procedure.
4. Aviation Operations	Marketing activity level	Company goals and objectives, economic situation, product requirements, outside assistance with promotion.
	Flight activity guidelines	Operating procedures manual, pilot training and flight standards.
	Maintenance quality control	Statement of policy, guidelines and procedures, inspection check lists, inspector personnel.
	Fuel contamination	Company policy, rules, procedures, fuel sampling, clear markings, training.
	Safety	Policy statement/SMS, rules, regulations and procedures, reinforcement and recognition, involvement.

Arrival time, service time, average wait time, maximum parking accumulation, and similar time-related data sampled over different days of the week and seasons can be a great help in making decisions about the best allocation of resources. A sign at the front desk (or on the company's website) about avoiding peak times of the week (as is sometimes done by the postal service and by transit systems) can even help customers with choices about their schedules.

Simulation of scenarios can be useful, and while simulations can be conducted manually, computer software allows the rapid performance of even complex simulations. Simulation of scenarios alters one factor at a time in an analysis to examine "what-ifs." For example, an analysis of acquiring a new aircraft might examine the consequences of different interest rates on monthly repayments. A drop in prices could be tested using different assumptions about elasticity to see what would happen to revenue in worst-and best-case situations. Computer models or spreadsheets can be built to test the results of changing variables.

Decide Something!

"Not to decide is to decide," as the saying goes. That is, the world will not stand still even if you do—there will be changes to deal with as a result of not making a decision. This generally seems to hold true, except for the most compulsive type of manager, who may benefit once in a while from letting things run their course for a few days, as opposed to constantly "pulling up the plant to check on the roots". Serendipity sometimes steps in and resolves things without any action on the manager's part. However, relying occasionally on serendipity still means making a decision to address the issue at a specific future date, if it hasn't resolved itself by the simple passage of time.

Time Management

"Work expands so as to fill the time available for its completion," says Parkinson's First Law. "Granted that work (and especially paperwork) is thus elastic in its demands on time, it is manifest that there need be little or no relationship between the work to be done and the size of the staff to which it may be assigned."[15]

While written in a whimsical vein, there is a very large measure of truth to these remarks. The converse is also true: work compresses to get done in limited time, if that's all there is; an urgent or important project always seems to be done by the deadline no matter how few people are assigned to it. Witness how even the most lethargic organization regularly gets its paychecks out on time.

Sources of Time-Management Problems

Various studies and books available on time management seem to agree on the key causes of problems:[16,17,18]

> Procrastination—because you dislike a task or because it is overwhelming;
> Telephone calls and other interruptions;
> Failure to get back to the task being worked on before the interruption;
> Excessive e-mail requiring responses;
> Meetings—too many, too long, too unspecific, no agenda, no decisions made in the meeting as to who will do what next, and by when;
> Reports—too many, too many contributors, too long, too many iterations, unclear purpose or audience;
> Unplanned visitors;
> Lack of delegation; poor delegation;
> Failure to make decisions on incoming items;
> Preoccupation with operational crises rather than preventive activities;
> Special requests—interruptions without adequate warning;
> Delays;
> Too much unfocused reading; and
> Lack of priorities.

Learning to plan time is one of the tasks that all workers need to do, but managers need it most because they are not on a production line or customer service position, with tasks coming at them constantly, but rather must exercise a choice about how to spend every minute of the day.

Time Management Strategies

In anything but a one-person business, clear delegation is the best means of gaining time. Moreover, all tasks that someone else can handle should be delegated, whether or not the person to whom they are assigned can do them quite as fast, quite as well, or in exactly the same way as the manager would. As the mother whose three-year-old washed the supper dishes every night said, "Sure I could do it faster, but she enjoys it and has a sense of achievement; besides, it frees me up to do the things a three-year-old can't do, such as paying bills or vacuuming." It pains many managers immensely to delegate to people they sometimes feel have the competence of a three-year-old, but as long as the job gets done, the manager can get on with something else that only he or she can do. Moreover, the person given a task slightly ahead of their skill level will learn, feel pride, and be ready for even more difficult assignments in the future.

Other recommended methods of better time management include:

> "Plan your work and work your plan." Plan each day's activities; don't just stumble through;
> Use a "tickler" file to organize papers relating to future dates and projects;
> Sort tasks into urgent and important, or not urgent and not important.[19] Aim to do two to three important tasks a day. Many of the other tasks will simply disappear, and fewer important tasks will be left till they become urgent;
> Work out six to seven goals with your manager;
> List the tasks that relate to the goals, along with start and finish dates;
> Use an electronic task organizer to track tasks and due dates.
> Use a notebook or electronic note software to save ideas or things to remember;
> Eliminate avoidable distractions—set up a routine where someone working for you handles them;
> Identify long-winded individuals in your work group, and set up short times to speak to them (e.g., just before a meeting or shift end, or shortly before you have another appointment);
> Communicate with "chatterers" by e-mail instead of orally and have them do the same to you;
> Use a desk needle or other simple organizer for messages and calls to be returned/responded to;
> Manage by wandering around and asking questions—forestall problems before they develop;
> Don't interrupt others, but plan your use of their time;
> Stick to one task until it's done; try not to interrupt yourself;
> Keep your supervisors posted on your results before they have to chase you down and ask about them;
> Make sure your employees know their authority and discretion;
> Develop a section/division procedures manual so everyone knows what he or she should be doing, how, and to whom to go for help; institutionalize procedures so that they don't have to be redeveloped for routine situations
> Use the phone instead of going to see people;
> Use a written checklist for shift transfer issues;
> Set up "red flags" for certain personnel or work groups on things they must get help on;
> Graduate your people to fewer "red flags" as they learn new skills;
> Take advantage of the POSITIVE side of deadline pressure by getting adrenaline going and allowing less time for reviewers of the product to change their minds;
> For tasks that are done well under deadline pressure, set up more (artificial) deadlines to increase productivity;
> If a task must be stopped in the middle, try to leave it where you know exactly what you were going to do next; write yourself a note;
> Coordinate and aggregate errands and tasks whenever possible;
> Delegate all possible tasks and set up reporting-back arrangements;
> Set goals for your people with check-up dates;
> Personal prime time—identify your best working time and block it out for no interruptions at least twice a week. Block out time several times weekly when you don't take calls or visitors, except in dire emergencies. (If you were out of the office at a business meeting, people would understand and accept not being able to reach you, so be adamant);
> Don't put in too much time at one stretch on high-priority tasks;
> Do low-priority tasks at low-output periods of the day;
> If unable to reach a decision, examine the need for more information and delegate getting it;
> Divide overwhelming projects into small pieces and delegate parts. Recognize that nearly any new project at first looks like a tabletop piled with random jigsaw puzzle pieces and even missing pieces. Once this jumble has been sorted into separate puzzles and the "corners" put in place, most of the team knows what to do next. It's your job as manager to foster an atmosphere where that first daunting situation can be broken down into workable parts;
> Do a first, appealing, or random small task on a big project to develop momentum and ideas and to get over the intimidation that a new project can create. Break big projects down into small, manageable steps; delegate some parts;
> Set intermediate milestones for big projects and acknowledge getting there;
> Identify why you are procrastinating on a project, and tackle the cause;

- Make all projects as simple as possible to get the job done—remember that *opportunity cost* is the cost of what doesn't get done while you are doing what you do;
- To get yourself started, do the task first that you like best; it will be easier then to do the distasteful part;
- Use small parcels of time (e.g., while waiting to see someone) to check lists, add to lists, write someone a note, brainstorm a new assignment, or use your phone or tablet
- Be assertive—assert the right to refuse more assignments;
- Exercise telephone discipline; be task-oriented;
- Reduce paper flow to essentials. Handle correspondence only once, by assigning it to someone else, handling it, filing it, or tossing it out;
- Be selective in reading;
- Utilize travel time for catching up on projects, when it is safe to do so;
- Finish and put away one thing before starting another;
- Relax and recharge occasionally by attending conferences and meetings away from the office;
- Limit your re-involvement in tasks already delegated. Use your worst times of the day for trivial matters and low-priority items;
- Work somewhere else on key projects, away from the phone and visitors;
- Write less and use the phone more. Not only do you get instant feedback, but also some things are better floated as trial balloons, discussed informally without being committed to paper. This is especially true today as e-mail is so easy to route beyond its originally intended audience, often causing confusion and misunderstandings;
- If you get stuck in a useless meeting, work on something else to stay alert and productive (while keeping in mind that many senior executives do NOT look with favor on multitasking during meetings);
- Avoid meetings with ineffective chairpersons, when possible;
- Since meetings can be considerable time-wasters, meeting management skills should be a goal of everyone who ever has to call a meeting. Figure 2.5 shows meeting management pointers.

Business Ethics

Good business ethics are essential.[20] In the first decade of the 21st century there were some very public examples of unethical behavior, such as the Bernard Madoff case, in which an unscrupulous owner ran

Figure 2.5 → Meeting Management

Don't hold a general meeting if two individuals can resolve the issue together.

Prepare a written agenda, get it out ahead of time, and identify topics and OBJECTIVES of the meeting on these topics. Consider specifying the time to be spent on each agenda item. Ask for missing items when the meeting begins, but reserve the right to save them for another time.

Call invitees in person or send an email reminder 48 hours before the meeting. Important issues include:
- Remind attendees of meeting time and place;
- Refresh their memory on their role at the meeting;
- Ask for any other agenda items from them;
- Substantial handouts should be distributed (via hard copy or email attachment) by the person responsible for that agenda item, BEFORE the meeting;
- Be sure any needed audiovisual aids are available and working properly;
- Keep most meetings to no more than one hour;
- Start on time. Have someone other than the meeting chairperson take minutes. Record those present and absent in the minutes. Do not allow side conversations. End on time!
- Listen to input, but firmly steer discussion back to the subject if it wanders. Give the talkers tasks to complete; invite involvement of the quiet ones;
- Take action on every item, even if the action is not final. For example, assign responsibility for more research, with deliverables and deadlines;
- Use a location with no chairs to speed up meetings, if necessary;
- Get minutes out promptly; highlight each recipient's new tasks and deadlines;
- If votes are required or other formal protocol, establish rules (Robert's Rules of Order, for example).

a pyramid scheme that caused his investors to lose billions when it was finally uncovered. In contrast are examples of highly ethical businesses such as:

> Malden Mills, inventor of Polartec fleece, whose owner paid over $25M in salaries to his employees while the mill was being reconstructed after a devastating fire;[21]
> Ben & Jerry's Ice Cream, which gives a percentage of profits to charity;[22]
> And many more.

Whether the philosophical underpinning of these positive ethical choices is deontological (because it is the right thing to do) or consequential (because it achieves desired results) in nature, the outcomes are similar.

Ethics applies to at least the following areas: treating employees fairly and honestly; ensuring that the company's accounting and fiscal practices are legal and accurate; paying vendors on time; not using potentially substandard generic parts; ensuring that all company environmental practices are sound and legal; and being honest and straightforward with customers, especially if there is a disagreement about a product or service.

A strong ethical foundation for the business will have positive results, including:

> Increased customer loyalty;
> Motivation of employees to also conduct themselves with integrity;
> Increased employee loyalty, because it will be evident that supervisors are committed to fair dealings within and outside the company;
> Keeping the business out of legal trouble, which is potentially costly and time-consuming; and
> Creating a business with higher resale value.

Summary

Management is defined as getting things done through others. It therefore involves planning, organizing, directing, coordinating, and controlling the work of others. The problem for most managers, especially those promoted up the ranks and not formally trained in management, is that despite their titles, they are also "do-ers." This has become even more true as the prevalence of computers has meant fewer secretarial positions. A business plan can help the manager lay out the long-range activities and rationale of the business. The plan will contain many elements of the manager's job. In so doing, he or she is constrained by time in how much "doing" takes place. In a small business, a manager wears many hats and spends significant portions of time delivering the goods or services of the business; however, this activity should be delegated as much as possible. Various styles of delegation and levels of decision-making are available. The manager, by these choices, achieves good or poor use of not only his or her own time, but also the time of all the staff.

DISCUSSION TOPICS

1. Describe the manner in which a manager's functions will change as he or she advances up the promotion ladder.
2. Tasks should be delegated to the most junior staffer remotely capable of handling them. Discuss.
3. What subjects should a business plan cover? Why?
4. What are the pros and cons of preparing/not preparing a written business plan?
5. What are some of the choices about styles of delegation? How should a manager set about deciding which style to use?
6. What are the pros and cons of participative decision-making?
7. What are the greatest causes of time mismanagement? How can they be dealt with?
8. Why should a business have an ethics policy and what should it include?

Endnotes

1. With apologies to William Shakespeare's Julius Caesar: "Some are born great, some achieve greatness, and some have greatness thrust upon them."
2. Mintzberg, et al. (1998). *Harvard Business Review on Leadership*. Cambridge, MA: Harvard Business Publishing.
3. Peter Drucker has been writing on the subject of business for 60 years. More about him can be found at various web sites such as *http://en.wikipedia.org/wiki/Peter_Drucker*.
4. See, for example http://www.deming.org/.
5. Goleman, D., Boyatzis, R., & McKee, A. (2016). *Primal leadership: Unleashing the Power of Emotional Intelligence*. Cambridge, MA: Harvard Business Review Press. ISBN13: 9781633692909.
6. Collins, J. C. (2001). *Good to great: Why some companies make the leap—and others don't*. New York, NY: Harper Business. ISBN 0-06-662099-6 (hc).
7. Brown, W. S. (1987). 13 *fatal errors managers make and how you can avoid them*. New York: Berkley Books.
8. Covey, S. R. (2004). *The 7 habits of highly effective people: Restoring the character ethic*. New York: Free Press.
9. One of the most frequently used is the Myers-Briggs personality classifier, which classifies people based on self-testing, into 16 major types. For an example, see *http://www.humanmetrics.com/cgi-win/JTypes2.asp*.
10. Bygrave, W. D., & Zacharakis, A. (ed.) (2010). *The Portable MBA in entrepreneurship*. Hoboken, NJ: John Wiley & Sons. ISBN: 978-0-470-48131-8
11. Collins, J. C., & Lazier, W. C. (1992). *Beyond entrepreneurship: Turning your business into an enduring great company*. Englewood Cliffs, NJ: Prentice Hall. ISBN: 0-13-381526-9.
12. Drucker, P. F. (1999). *Innovation and entrepreneurship*. Oxford: Butterworth-Heinemann.
13. Brown, D. (1980). *The entrepreneur's guide*. New York: Macmillan.
14. A right-brain process of recording ideas, developed originally by Tony Buzan. See http://www.buzan.com.au/
15. Parkinson, C. N. (1980). *Parkinson, the law*. Boston: Houghton Mifflin.
16. Morgenstern, J. (1998). *Organizing from the inside out: The foolproof system for organizing your home, your office, and your life*. New York: Henry Holt.
17. Lagatree, K. M. (1999). *Checklists for life: 104 lists to help you get organized, save time, and unclutter your life*. New York: Random House.
18. Blanchard, K. H., & Johnson, S. (1982). *The one minute manager*. New York: Morrow. ISBN: 0-688-01429-1.
19. This is the basis of the time management skills taught by Covey. Non-urgent and non-important things either fall off the bottom of the list or get done in the "interstices" of time. Important things must come before urgent things.
20. For resources, see the following: Business New Haven Consulting Group, Inc. offers customized workshops in business ethics and other related training at *http://www.nhcg.com*; SAI Global: the leader in online business ethics training for employees; offers courses in over 130 ethics and compliance topics at *https://www.saiglobal.com/risk/solutions/ethics-and-compliance-learning*
21. See *http://en.wikipedia.org/wiki/Malden_Mills*
22. *http://www.benandjerrysfoundation.org/* "The mission of the Ben & Jerry's Foundation is to engage Ben & Jerry's employees in philanthropy and social change work; to give back to our Vermont communities; and to support grassroots activism and community organizing for social and environmental justice around the country".

3

Marketing

OBJECTIVES

> Understand what forecasting sources are available to the FBO in planning for each market area.

> Show how businesses must be able to identify new prospects and relate that process to customer needs.

> Discuss the differences between cost-based pricing and price-based costing.

> Realize the importance of location and promotion to a marketing strategy.

> Describe three elements of distribution and how they function.

> Describe some reasons why a marketing plan aimed at the mass market would not be beneficial to a general aviation business manager.

Introduction

The Need for Marketing in Aviation

The aviation industry has evolved from being largely an experimental, exploratory, and record-setting sport to serving as an important component of modern business operations and productivity. In its early years, the excitement created by air travel and the barnstorming stunts of pioneer aviators was often enough to attract huge crowds of patrons. Today, however, the general aviation enterprise cannot simply wait for people to come to the business; rather, the business must actively seek its customers. If those customers are business travelers, they will constantly be evaluating competing modes of travel, such as the scheduled airlines, rail, and the automobile. If they are recreational fliers, many non-aviation activities compete for their support and dollars.

The general aviation business does not necessarily have a captive audience. If the staff is unfriendly, service is inefficient, or the desired fuel is unavailable, most pilots will plan their enroute fuel stop or destination at a different airport or operator on subsequent flights. Within metropolitan areas, studies have shown that pilots are willing to travel some distance to hangar or tie down their aircraft at a favored airport.

Aviation businesses must continually seek to improve their marketing skills if the business is to prosper. The most effective companies in the United States expend a good deal of effort "staying close to the customer." An online survey for customers may gather only complaints, whereas training all employees to invite feedback will generate more discussion and ideas of a positive nature. Aviation customers, especially those who travel extensively, are also excellent sources of information on innovations introduced elsewhere.

The entire business strategy must be focused on anticipated sales. The money to run the business, make improvements, and expand comes from gross sales. A realistic sales forecast must include a planned schedule of daily and weekly marketing activities.

Modern aviation business leaders have learned that a specific marketing plan and its timely implementation are essential to a healthy business. This is an area of expertise for which the manager and employees should be continually exploring and experimenting. Marketing is still as much art as science, and what works for one business, geographical location, or product may not work for another.

Marketing Orientation

Having a marketing orientation in a business means running the entire business with a focus on the customer.[1] It is a viewpoint that recognizes the dependence of the business upon customers and the sale of products and services to them. The manager who has no marketing orientation makes no real effort to attract customers, and by many subtle and even obvious indications, may actually suggest that customers go elsewhere. Such a manager's employees are likely to be unconsciously mirroring and reinforcing the same message. By contrast, the manager with a marketing orientation is totally conscious of customers, their needs, and how to attract and keep them, spending considerable time emphasizing this through formal and informal staff training and by constantly modeling it in his or her own behavior.

Such a manager recognizes that the business depends upon sales and that the first step is getting the customer inside the door (or on the ramp). Once on the premises and face-to-face with a company representative, the customer should continue to experience the positive marketing orientation of the business from initial welcome through the conclusion of the sale and in any subsequent encounters, particularly with the handling of any problems. The marketing-oriented manager sees to it that each employee behaves likewise, whether or not that employee's job calls for dealing directly with customers and whether or not that employee knows all the answers those customers seek.

Definition of Marketing

Marketing, as treated in this chapter, covers the entire process of identifying customer needs, purchasing or producing goods and services to meet those needs, determining the price and place to dispense the items, and prospecting, promoting, and selling the items. In short, it deals with the four Ps: Product, Price, Place, and Promotion.

The marketing function in a company may be handled by the owner/manager, or by each department. For example, the maintenance manager might be responsible for planning sales in his area, the line service manager might be responsible for her area, and so on. In a large company, a separate department, as shown in Figure 3.1, may handle the marketing function. Whether these functions are handled by one individual as a part-time responsibility, or whether

---

Figure 3.1 → **Typical Marketing Department**

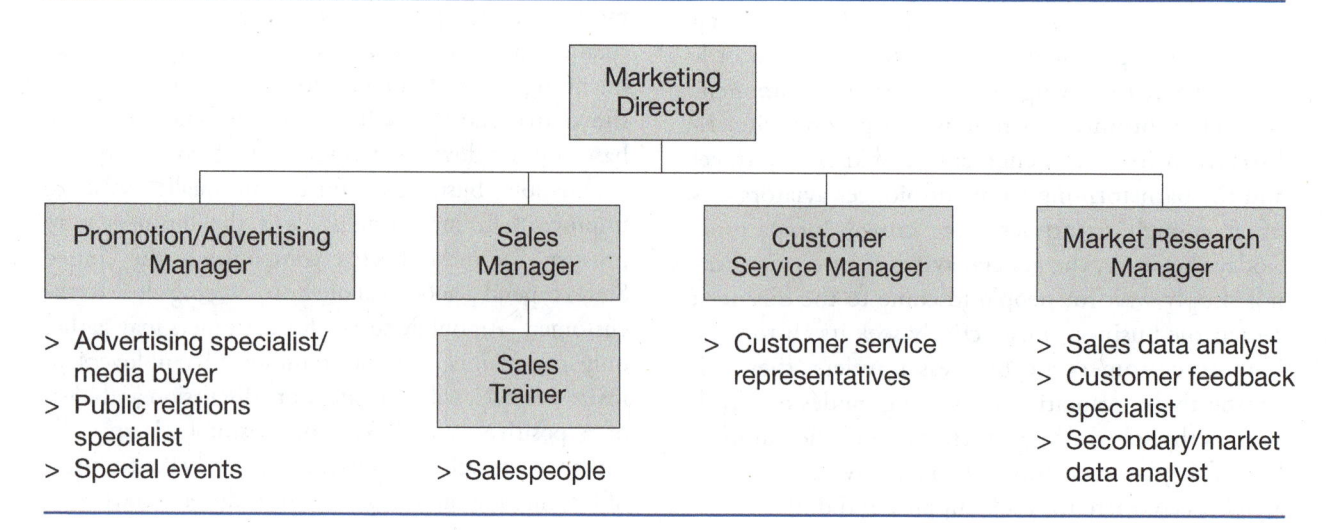

```
                        Marketing
                        Director
      ┌────────────────┬─────────┴────────┬──────────────────┐
Promotion/Advertising  Sales         Customer           Market Research
     Manager          Manager      Service Manager         Manager
```

- Promotion/Advertising Manager
  > Advertising specialist/media buyer
  > Public relations specialist
  > Special events

- Sales Manager
  - Sales Trainer
  > Salespeople

- Customer Service Manager
  > Customer service representatives

- Market Research Manager
  > Sales data analyst
  > Customer feedback specialist
  > Secondary/market data analyst

they are each handled by a separate person, they are all necessary functions in the marketing area.

FBOs and other general aviation businesses are unusual in at least two distinct ways. First, their entire function depends heavily on infrastructure over which they often have little or no control—the airport and its facilities. Without the airport, there could be no aviation business. Its quality affects the general aviation business customers, even though there may be no direct quality control exercised by the business.

Second, full-service FBOs are sellers of both products and services, and both types of goods encompass a wide array of choices. Even the smallest FBO business generally caters to a variety of perhaps incompatible market segments (e.g., the corporate chief executive chartering a flight, juxtaposed with the euphoric sixteen-year-old in half a T-shirt who has just soloed). The ambiance and facilities of the general aviation business must accommodate both ends of this spectrum, and an array of markets and functions in between.

Since services are one product of a general aviation business, the word "product" is generally used here to refer to both goods and services.

Natural Segmentation of Aviation Markets

Not all goods and services are of interest to all people. The U.S. aviation customer base may be generally divided as follows:

1. The lowest-earning 20 per cent of the population. This group has little or no discretionary income.
2. The mass market—the next 21 percent to roughly 88 percent of the population.
3. The next 10 percent of the market, or the top 88 to 98 percent, in terms of wealth.
4. The wealthiest or most highly-specialized market (the top 2 percent).[2]

Each of these markets has different interests and needs, and each lends itself to different types of promotion. For clarity, they break down as follows:

Percent of population	Aviation buying power/ discretionary income
Poorest 0–20%	Almost none.
Middle 21–88%	Mass market; modest aviation buying power.
Upper 88–98%	Greater discretionary income—good potential customers.
Wealthiest 98–100%	Excellent aviation prospects.

What, therefore, is the target market for the general aviation business based on this population segmentation? An evaluation of national statistics suggests that general aviation businesses should target the top two market segments, namely the upper 88–98 percent and the wealthiest 98–100 percent groups.

As was indicated in Chapter 1, the top assets in the U.S. are shared as follows:

Households with net worth of:	Number
$1–$5 million	10,380,000
$5–$25 million	1,397,000
Over $25 million	173,000

Statistics on the use of general aviation support this focus on the top two market segments. About one person in a thousand (or .01 percent of the population) owns an aircraft. The FAA U.S. Civil Airman Statistics listed 664,565 active certificated pilots as of December 31, 2019, up from 633,317 at the end of 2018 (retrieved from https://www.faa.gov/data_research/aviation_data_statistics/civil_airmen_statistics/ on July 11, 2020).

Non-pilots traveling by air are thought to use general aviation transportation about one quarter of the time. Although the U.S. population totals nearly 330 million people, less than one million Americans hold a pilot's license or permit. General aviation business owners must understand that their products do not appeal to the mass market. Mass marketing techniques are normally not appropriate for general aviation businesses. The general aviation business requires custom-tailoring of its services, so that each client is treated differently and has different needs met. However, despite providing far fewer seat-miles than airlines, general aviation comprises over 96 percent of the U.S. civil aircraft fleet and is still a large sector of the aviation industry.

Market Research

Marketing and market research are often used synonymously, but they are by no means the same. Market research helps the business owner understand the trends—and the underlying factors causing them—that then can help him or her focus on the products and services to emphasize. Without market research, the organization's attempts to increase sales will tend to be haphazard and unfocused. With market research, they can be closely targeted because one knows what is going on in each segment of the market. Today, through the use of tools such as Survey Monkey (http://www.surveymonkey.com), it is relatively easy to design and conduct regular online surveys of both existing and prospective customers.

The Nature of Aviation Market Research

In the U.S. as a whole, the last two decades of the 20th century saw a relatively continuous upward trend in prosperity. There were geographic areas, socio-economic groups, and limited time periods for which this was less true, but in general, discretionary income rose, consumer spending rose, and many new products and services arose to tap these funds. The aviation market operates in this wider context, but is also heavily influenced by matters unique to the industry. It has been a matter of some frustration over the years that no simple tool exists for projecting general aviation growth in relation to growth in the overall economy.[3] However, since this correlation is difficult to determine, it is important that the FBO manager maintain a focus on the factors that affect the growth of general aviation.

National Trends That May Affect General Aviation

Introduction

General aviation has been a relatively volatile industry for the past 50 years or more. Thus, accurate interpretation of past causes of volatility is important because it can help to project future development. Chapter 1, Introduction, touched on some of these issues; some of the key issues are covered here in greater depth.

Terrorism and Its Aftermath

Significant milestones in general aviation include not only the General Aviation Revitalization Act of 1994 and the recreational and sport pilot certificates, but also the events of September 11, 2001. For a period, these attacks severely curtailed aviation activity and harmed many businesses. The resulting delays and periodic airport terminal evacuations that have occurred since then are encouraging many more companies to look at air taxi, charter, and fractional or full aircraft ownership as a means of enabling their employees to make rapid business trips. This can be encouraging news for general aviation businesses seeking to serve these new customers.

The events of September 11, 2001 led to a massive increase in attention to airport security, at both large and small airports.[4] This, coupled with lost airline revenues, is placing a financial burden on the system that may mean other investments, which help aviation businesses—such as runway, and taxiway improvements—may be curtailed or delayed. In addition, current security measures for the typical GA airport can be a mixed blessing, as they tend to decrease ease of airport access for regular customers while making access more difficult for undesirable individuals. These problems may be resolved as new security technology is brought into use; however, it may take time for the necessary comprehensive security plans and related funding programs to take effect.

The events of September 11, 2001, as well as vast improvements in teleconferencing technology, have also triggered consideration by many companies of whether travel to allow face-to-face meetings needs to take place at all. An *Airport Magazine* article addresses this, among other issues.[5] Researchers for several decades have been asking whether telecommunications methods of various kinds would start to replace face-to-face gatherings, but only by the 1990s did technology really begin to be available that makes this widely possible. Today, almost every company has access to high-speed data lines with significant bandwidth for transmission of large amounts of data. "Virtual meetings" can take place on desktop equipment rather than requiring special conference rooms. Videoconferencing tools such as Skype and Zoom now put this option on every desk at little incremental cost to the user. The availability of the technology, coupled with the post-September 11, 2001 time loss and security concerns, have led many companies to not only promote this technology but often, to require a justification for why electronic communication would NOT work, before approving air travel for a face-to-face meeting.

Fractional Aircraft Ownership

Fractional aircraft ownership companies and equally importantly, individuals, apparently now account for more than 40 percent of all new corporate jet orders, according to a 2001 article in *Airport Magazine*.[6] The number of companies and individuals

More often, companies are using corporate jets when their people need to travel.

with fractionally-owned aircraft jumped 62% from 3834 in 2000 to 6217 in 2003, according to those tracking this fast-growing industry. The article states:

> "Now, private jet service can be cost-effective for companies with travel budgets of less than $200,000 per year, particularly if they are headquartered near a large city and travel to other large cities. In fact, the minimum aircraft hours in a fractional ownership contract has fallen to only 50 hours per year. Just as importantly, companies using NetJets and other large providers can now fly on four hours' notice, with multiple staff to multiple destinations. Airport check-in time can be a mere 15 minutes at each end, with the return scheduled comfortably to eliminate connections and overnight stays."

The origins of fractional aircraft ownership lie with corporations owning aircraft and seeking to utilize them more fully to share costs. However, passengers for hire cannot be carried in a corporate aircraft unless it is certified to operate under FAR Part 121, which regulates air carriers, or (more frequently) Part 135, which regulates air taxi operators. Most corporate flights operate under FAR Part 91. In October 1999, a working group, the Fractional Aircraft Ownership Aviation Rulemaking Committee, or FOARC, was convened by FAA to explore the issue of appropriate safety rules for fractionals.

The FAA accepted the FOARC's recommendations, which had been drafted as a preliminary Notice of Proposed Rulemaking (NPRM), almost verbatim. The main thrust was to add a new section to FAR 91 to address fractionals, and to also make similar changes to FAR 135. The rules allow for as little as a 1/16 share of some aircraft and 1/32 share of others. Over 200 comments were received, and the FAA published the final rule as Subpart K of 14 CFR 91, with a compliance date of December 17, 2004.

The regulatory/safety debate, however, has not in any way slowed the development of the fractional market. The primary question for FBOs (and the reason that this topic is addressed most fully in this Chapter) is how will exponential growth in fractional aircraft ownership affect our business?

At first glance, one might think that FBOs would be hurt by the growth in fractionals. The fractional management companies such as Net-Jets, Flexjet, Planesense, Flight Options, and others, operate their own flight departments and maintenance facilities. Consequently, this may mean less work for FBO pilots and mechanics, as well as enticement away to higher-paying jobs (ever FBOs' lament that they are near the bottom of the aviation "food chain"). Yet it appears so far that fractional aircraft operation is developing its own market niche. To a small extent, its customers are coming from charter operators, but according to NATA, it is still generally cheaper to charter than to become a fractional owner.[7] This is true based on aircraft usage that does not exceed a certain maximum number of hours per year, and will be discussed more fully in Chapter 9, Flight Operations.

According to the aviation consulting firm IMG, fractional operations can be cost-effective for companies with travel budgets under $200,000 per year.[8] Another study indicates that an individual with an annual income of $10m or a company with gross sales of $30m would find fractional ownership cost-effective, and that the range of annual flight hours for which this mode is most useful is about 145 to 390 hours per year. With a 1/2 share equating to 200 flight hours per year, the individual's or company's travel needs can be met without owning an entire aircraft

Fractional aircraft clients come primarily from one of three sources:

> Smaller corporations that have closed their flight departments in favor of fractional ownership;
> Business people who have not previously used general aviation and are realizing its efficiencies. This group has no doubt increased due to the events of September 11, 2001, and the greater time burden imposed by security processes during airline travel; and
> Frequent charter fliers switching modes within GA.

Benefits of fractional ownership include:

> Lower cost than operating a corporate flight department and no managerial hassles of overseeing it;
> No aircraft maintenance responsibilities;
> Much lower travel times than using airlines;
> Wide choice of aircraft, through the fleet pools owned by fractionals; and
> NO direct costs for "deadheading" or aircraft positioning flights.

The NBAA web site (http://www.nbaa.org) offers several other partial aircraft ownership options, and every FBO should pay close attention to this still-emerging approach to aircraft ownership and operations. Major changes with critical implications for a specific FBO can occur in a relatively short period of time.

Encouraging New Pilots

A number of national programs continue to have a degree of success at encouraging individuals to explore aviation as a hobby or for business purposes. Examples of such programs include the AOPA Let's Go Flying program and the EAA Young Eagles program. It is encouraging that FAA Civil Airmen Statistics indicate a strong increase in active student pilots over a recent four-year period, from 128,501 in 2016 to 197,665 in 2019. This trend, if it continues, is positive for the industry as a 2001 Cessna survey found that 40 percent of individuals buying their piston aircraft are new pilots, according to GAMA.[10]

A certain segment of general aviation customers are first introduced to the experience through military service. However, in later life they may or may not continue. FBOs should be aware of this and other promising sources of new customers.

Sport Pilot Certificate

Another example is the sport pilot certificate as shown in Figure 3.2, which enables beginning pilots to become qualified with fewer barriers in terms of required training time and, consequently, cost.[11] As of 2019 the FAA reported a total of 6,467 active sport pilots, which is up from 5,889 in 2016, and the FAA projects as many as 10,615 sport pilots by 2035.

Another positive trend in terms of flight training has been the increasing percentage of pilots who are adding advanced certificates and ratings, this trend has been especially true of women pilots. The percentage of domestic pilots with instrument ratings totals 67.2 percent, as of 2019.

Airport Closures

An area for which state and local governments may be able to influence the aviation market is in protecting or not protecting "threatened" airports. Such airports may be encroached on by urban development, or may have severe noise complaint levels, causing local government to favor the needs of the community at large over the needs of the aviation community. An aviation business located at such an airport needs to be aware of its situation and ideally, be proactive in creating a more favorable local climate for the airport as a whole and thus for that business in particular. This is discussed more fully in Chapter 12, Physical Facilities.

Technology

Technology is of key importance in aviation. As electronics become increasingly sophisticated, this leads to changes in avionics that will affect the sale of parts as well as the avionics maintenance requirements that a business may seek to meet. It is also very important to keep the latest technology in mind when considering marketing strategies and continued long-term business. In the computer industry, a company on the

Figure 3.2 ✈ Sport Pilot Certificate

The 2004 sport pilot regulations established new FAA certification categories for:

Aircraft
Light-sport aircraft are simple, low-performance aircraft that are limited to 1,320 lbs. (600 kg) maximum weight, two occupants, a single non-turbine powered engine, stall speed of 45 knots, maximum airspeed of 120 knots, and fixed landing gear. Aircraft categories include airplanes, weight-shift-control aircraft, powered parachutes, gyroplanes, gliders, balloons, and airships. Due to their complexity, helicopters and powered-lifts are not covered by the LSA regulations. Light-sport aircraft standards meet the "Voluntary Consensus Standards" of OMB Circular A-119.

These are the two new airworthiness certificates that have been established:

> - A new experimental light-sport aircraft airworthiness certificate for existing light-sport aircraft that do not meet the requirements of Part 103 (ultralight vehicles) of the Federal Aviation Regulations;
> - New special, light-sport aircraft airworthiness certificates for light-sport aircraft that meet airworthiness standards developed by industry.

Pilots and Flight Instructors
New pilot and flight instructor certificates have been established:

> - New airmen certificates include a student pilot certificate for operating light-sport aircraft, a sport pilot certificate, and a flight instructor certificate with a sport pilot rating;
> - Two new aircraft category and class ratings—weight-shift-control (with land and sea class ratings) and powered parachute;
> - New training and certification requirements for these new ratings;
> - A current and valid U.S. driver's license or an FAA airman medical certificate are required to operate a light sport aircraft.

Repairman
A new repairman certificate and ratings have been established. A new repairman certificate, and ratings, with a maintenance or inspection rating to maintain and inspect light-sport aircraft, have been established.

Source: http://www.faa.gov/licenses_certificates/airmen_certification/sport_pilot

---

cutting edge today can find their product basically obsolete in one or two years. For example, both VOR-based RNAV systems and LORAN technology, which were the most advanced flight navigation systems in the 1980s, are now considered "dinosaurs" by many, in view of the latest GPS and RNP (Required Navigation Performance) systems available.

Tort Reform

Since aviation is such a heavily-regulated industry and relies for its facilities so greatly on government agencies, part of the market context is created by government actions. Being aware of both pending and new regulations, and of the federal and local budgeting and funding processes, can help shape marketing decisions. A prime example is the General Aviation Revitalization Act of 1994 (GARA), which basically encouraged a rebound in small aircraft production in an industry that otherwise might have all but ceased to exist.[12] It was paralleled by the growth in fractional aircraft ownership, which also had a major impact on the new aircraft market.

While many feel that the turnaround in aircraft sales since 1994 is due primarily to GARA, others feel that the growth of fractional aircraft ownership is also accountable since almost half of all new aircraft on order are destined for fractional aircraft companies. Figure 3.3 provides an alternative view of past industry trends in summary form that can be reviewed in more detail if desired.[13]

Sources of Aviation Forecasts and Market Projections

For many years, FAA has been a leading source of aviation forecasts. These are produced primarily for their own use in forecasting personnel and facility

Figure 3.3 → An Alternative View of GARA

Prior to GARA

Before 1979, the industry's growth was fueled by a number of factors that artificially boosted demand and led to a flooding of the market with new aircraft, including growth in the nation's air transportation system, large numbers of ex-military pilots, and the GI Bill program that fueled the demand of former members of the military for flight training.

The general aviation industry's decline in the early 1980s was based on a number of economic and institutional factors, such as rising inflation and the reduction of the GI Bill program.

After the early 1980s, while other industries bounced back, the small aircraft market—one segment of the general aviation industry—remained in the doldrums due to several factors other than product liability lawsuits, including the industry's own behavior:

> Limited demand—the industry's success at building long-lasting products, coupled with stringent FAA maintenance requirements;
> Decline in pilots; fewer military pilots returned to civilian life;
> Other factors—some aircraft enthusiasts turned their attention and money to the experimental and kit aircraft market.

While the small aircraft market remained depressed, the general aviation industry began to grow on the strength of high-priced turboprop and jet aircraft.

After GARA

Small-aircraft prices have not dropped as the general aviation industry promised, because the industry has realized no product liability "savings" due to GARA. The small aircraft market has experienced a very modest revival over the last two years, but nothing that even approaches the robust demand of 30 years ago.

Because small-aircraft prices have not dropped, this modest increase in demand can be attributed to other factors.

The effects of GARA cannot be isolated from the effects of other efforts by government, industry, and organizations of aircraft owners and operators to revitalize the industry.

Cessna's decision to resume single engine manufacturing in 1994 was not the result of any financial savings resulting from GARA, but because Cessna's then-chairman, Russell W. Meyer, Jr., promised Congress that production would resume if GARA were enacted.

Source: Embry-Riddle Aeronautical University[13]

---

needs for the airspace system, but are increasingly being tailored to meet the needs of state and local users. FAA produces a guide called "Forecasting Aviation Activity" which provides a how-to for any party wishing to prepare a forecast.[14]

FAA forecasts appear in the following documents:

> The National Plan of Integrated Airport Systems (NPIAS)—a biennial document that sets out airport needs and improvement plans;
> Terminal Area Forecasts—the official FAA forecast of aviation activity for U.S. airports; and
> FAA Long-Range Aerospace Forecasts—an annual document that sets projections for the national airport system as a whole.

The FAA website (www.faa.gov) should be consulted for the most current data, which can generally be downloaded.[15] Figures 1.2 and 1.3 in Chapter 1 contained summaries of the most recent forecasts of fleet mix and flight hours. These forecasts are by no means cast in stone and are revised every year; however, they are a reliable backdrop to a particular general aviation business's situation. Some key figures, produced by the FAA forecasting branch, are shown here in Figures 3.4 through 3.9.

More fine-grained analysis of the market can be obtained from other public studies, such as the FAA Terminal Area Forecasts that contain airport-by-airport forecasts for every facility in the National Plan of Integrated Airport Systems. An example of the data provided is shown in Figure 3.10. Individual airports also periodically conduct airport master plans, usually with the help of a consultant. These plans provide a guide not only to demand for growth at a particular airport, but also to capacity limitations that could affect those plans and the projects identified to improve the limitations.

Historic trends in one sector of aviation can be a useful backdrop to the individual general aviation business situation. For example, general aviation aircraft production and sales began to fall off in the late 1970s. It was predictable from that time—though not widely observed—that general aviation business in the early 1980s would decrease as a result.

Figure 3.4 ✈ **Active GA Fleet, 2020**

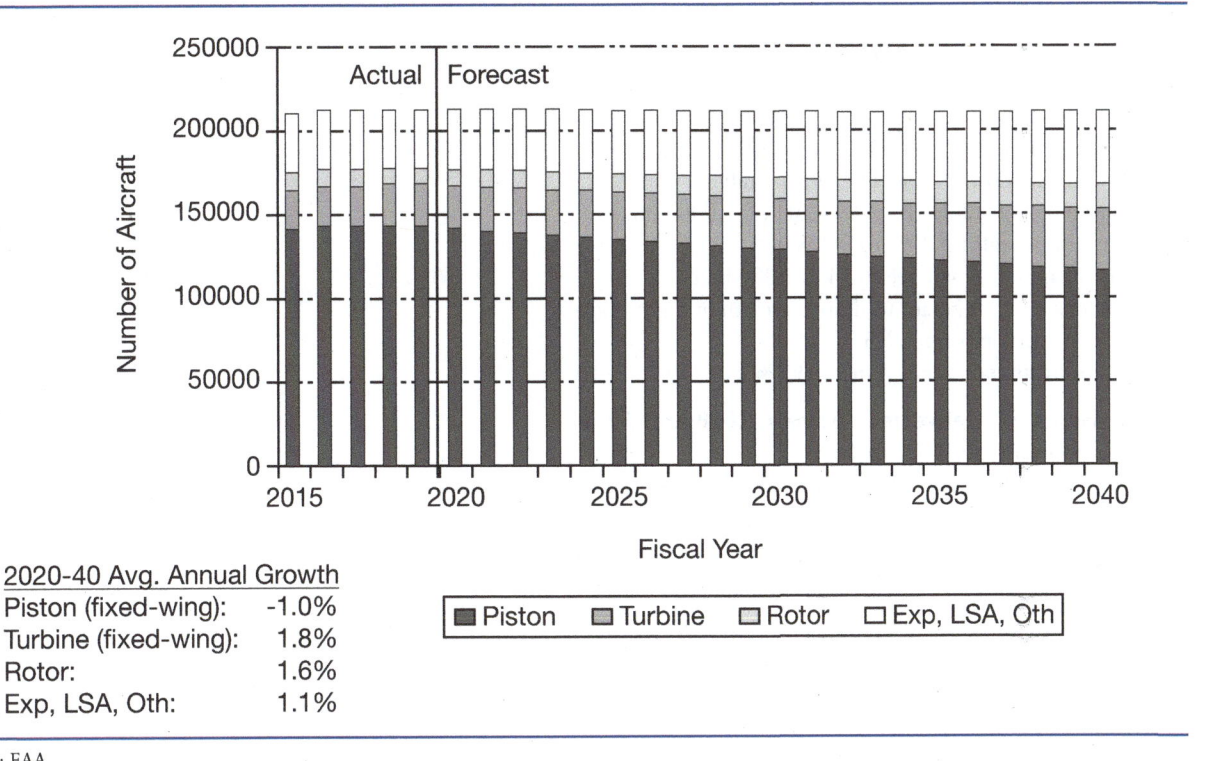

2020-40 Avg. Annual Growth
Piston (fixed-wing): -1.0%
Turbine (fixed-wing): 1.8%
Rotor: 1.6%
Exp, LSA, Oth: 1.1%

Source: FAA

Figure 3.5 ✈ **GA Flight Hours, 2020**

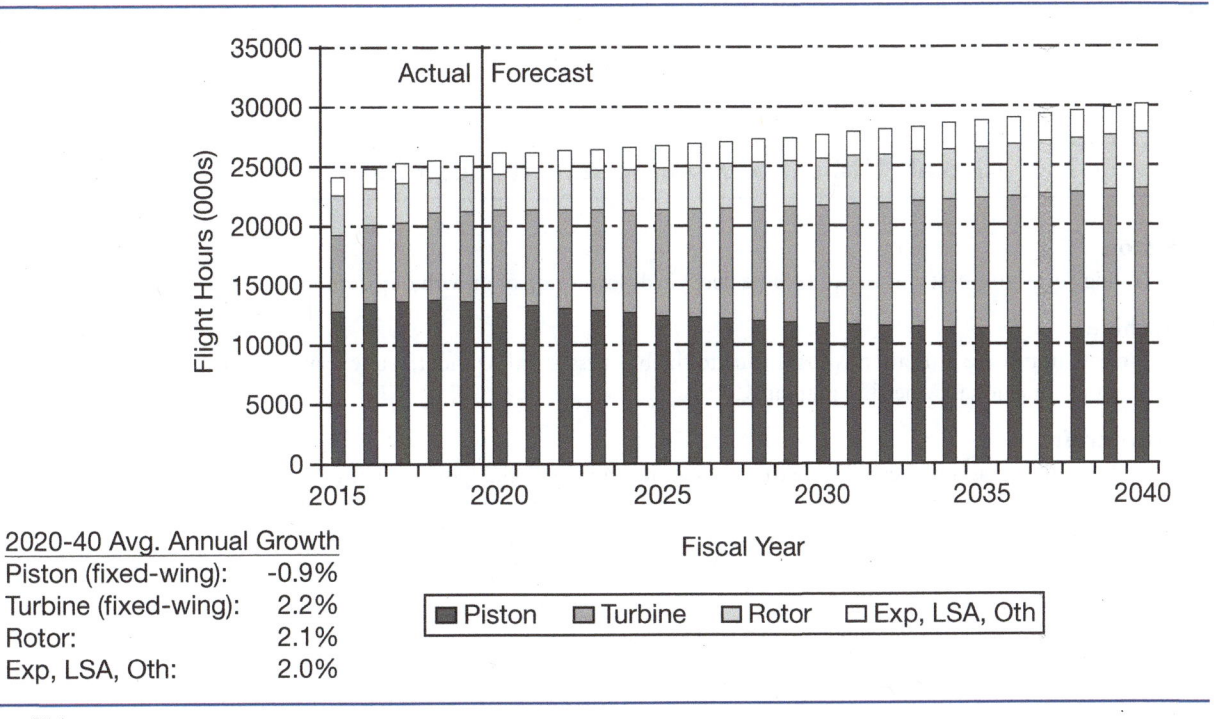

2020-40 Avg. Annual Growth
Piston (fixed-wing): -0.9%
Turbine (fixed-wing): 2.2%
Rotor: 2.1%
Exp, LSA, Oth: 2.0%

Source: FAA

Figure 3.6 → FAA's Assessment of General Aviation Trends Through 2040

Mixed news for GA from 2010–2019
- Shrinkage in fleet but growth in hours
- Total fleet 4.9% lower (decrease of 11,035 aircraft)
  - Turbine fleet up 17% to 32,035 aircraft
  - Piston fleet down 9% to 145,465 aircraft
- Flight hours 4.2% higher (increase of 1,051,000 hours)
  - Turbine aircraft up 20% to 9,978,000 hours
  - Piston aircraft down 3% to 14,321,000 hours
- Mixed changes in operations despite growth in flight hours
  - Tower operations up 2.9%
  - En-route aircraft handled down 3.7%

Slight rise in GA deliveries between 2018 and 2019
- Overall deliveries up 1.4% to 1,771
  - Single-engine piston deliveries up 7.0%; multi-engine deliveries flat
  - Turbojet deliveries up 6.3%; turboprop deliveries down 13.3%

Forecast Overview Through 2040
- Growth in high-end of the market exceeds low-end retirements
- Total fleet 0.9% lower by 2040 (decrease of 1,955 aircraft)
  - Piston fleet decrease by 25,280 aircraft (down 1.0% per year)
  - Turbine fleet increase by 14,640 aircraft (up 1.8% per year)
- Flight hours 16% higher by 2040 (increase of 4,352,000 hours)
  - Piston aircraft decrease by 2,132,000 hours (down 0.7% per year)
  - Turbine aircraft increase by 5,674,000 hours (up 2.1% per year)
- Operations growth reflects growth in turbine aircraft fleet
  - Tower operations increase by 6.1% (0.3% per year)
  - En-route aircraft handled increase by 10.4% (0.5% per year)

Source: FAA

Figure 3.7 → Aircraft Utilization Assumptions through 2040

- **Piston**
  - "Aging" of piston fleet and rise of turbine fleet yields lower rate of growth of piston utilization
- **Turbine**
  - Increase in turbine utilization is largely due to the business travel segment, namely an increase in the number of these aircraft in fractional ownership programs
- **Rotorcraft**
  - Utilization increases moderately due to change in fleet mix and changing use patterns

Source: FAA

Figure 3.8 → Active Pilots by Type of Certificate

Active Pilots by Type of Certificate

As of Dec. 31	Students[1]	Recreational	Sport Pilot	Private	Commercial	Airline Transport	Rotorcraft Only	Glider Only	Total Pilots	Total Less at Pilots	Instrument Rated Pilots[2]
Historical*											
2010	119,119	212	3,682	202,020	123,705	142,198	15,377	21,275	627,588	508,469	318,010
2013	120,285	238	4,824	180,214	108,206	149,824	15,114	20,381	599,086	490,291	307,120
2014	120,546	220	5,157	174,883	104,322	152,933	15,511	19,927	593,499	472,953	306,066
2015	122,729	190	5,482	170,718	101,164	154,730	15,566	19,460	590,039	467,310	304,329
2016	128,501	175	5,889	162,313	96,081	157,894	15,518	17,991	584,362	455,861	302,572
2017	149,121	153	6,097	162,455	98,161	159,825	15,355	18,139	609,306	460,185	306,652
2018	167,804	144	6,246	163,695	99,880	162,145	15,033	18,370	633,317	465,513	311,017
2019	197,665	127	6,467	161,105	100,863	164,947	14,248	19,143	664,565	466,900	314,168
Forecast											
2020			125	6,740	161,700	100,950	166,900	14,100	19,350		469,865	316,300
2021			122	6,994	160,933	100,920	167,732	14,179	19,488		470,405	317,570
2022			119	7,258	160,169	100,890	168,567	14,259	19,627		470,945	318,845
2023			116	7,531	159,409	100,860	169,407	14,339	19,767		471,486	320,125
2024			113	7,815	158,653	100,830	170,252	14,419	19,908		472,028	321,410
2025			110	8,110	157,900	100,800	171,100	14,500	20,050		472,570	322,700
2026			106	8,354	156,433	100,660	172,245	14,732	20,100		472,695	323,990
2027			102	8,605	154,980	100,519	173,397	14,969	20,150		472,820	325,284
2028			98	8,864	153,540	100,379	174,557	15,209	20,200		472,945	326,584
2029			94	9,130	152,113	100,240	175,724	15,452	20,250		473,070	327,890
2030			90	9,405	150,700	100,100	176,900	15,700	20,300		473,195	329,200
2031			86	9,635	149,439	99,939	178,181	15,952	20,330		473,622	330,411
2032			81	9,872	148,189	99,779	179,472	16,208	20,360		474,050	331,627
2033			77	10,113	146,949	99,619	180,772	16,468	20,390		474,478	332,847
2034			74	10,361	145,719	99,459	182,081	16,732	20,420		474,906	334,071
2035			70	10,615	144,500	99,300	183,400	17,000	20,450		475,335	335,300
2036			65	10,820	143,711	99,160	184,721	17,252	20,480		476,250	336,472
2037			61	11,029	142,927	99,019	186,051	17,509	20,510		477,168	337,648
2038			57	11,242	142,147	98,879	187,391	17,768	20,540		478,087	338,828
2039			53	11,459	141,371	98,740	188,741	18,032	20,570		479,007	340,012
2040			50	11,680	140,600	98,600	190,100	18,300	20,600		479,930	341,200
Avg. Annual Growth:											
2010–2019	5.2%	-5.5%	6.5%	-2.5%	-2.2%	1.7%	-0.8%	-1.2%	0.6%	-0.9%	-0.1%
2019–2020		-1.6%	4.2%	0.4%	0.1%	1.2%	-1.0%	1.1%		0.6%	0.7%
2020–2030		-3.2%	3.4%	-0.7%	-0.1%	0.6%	1.1%	0.5%		0.1%	0.4%
2020–2040		-4.5%	2.8%	-0.7%	-0.1%	0.7%	1.3%	0.3%		0.1%	0.4%

*Source: FAA U.S. Civil Statistics.
[1] In 2010, the duration of student certificates for pilots under age 40 was raised to 60 months; in 2016, the expiration date was removed.
[2] Instrument rated pilots should not be added to other categories in deriving the total number of pilots.

Figure 3.9 → Factors Affecting Future GA Trends

> Increases in air traffic congestion, due partly to Urban Air Mobility and Unmanned Aerial System integration
> Greater emphasis on environmental concerns, namely air and noise pollution
> Strength of demand, given the effects of instability in oil prices and a recessionary economy
> Success of on-demand air taxi operators and the use of Very Light Jets (VLJs)

Source: FAA

Figure 3.10 → Example of Data Provided in FAA Terminal Area Forecasts

Region State: ACE-IA
City: DUBUQUE
LOCID: DBQ
Airport: DUBUQUE REGIONAL
Contract Tower

AIRCRAFT OPERATIONS

Year	Scheduled Enplanements Air Carrier	Commuter	Total	Itinerant Operations Air Carrier*	AT & Comm#	GA	Military	Total	Local Operations GA	Military	Total	Total OPS	Based Aircraft
2010	460	33,618	34,078	14	2,907	23,188	157	26,266	20,942	21	20,963	47,229	74
2011	1,290	34,103	35,393	26	2,921	21,592	174	24,713	23,399	41	23,440	48,153	74
2012	1,041	32,830	33,871	21	2,206	20,096	142	22,465	19,602	151	19,753	42,218	75
2013	1,158	31,518	32,676	37	1,815	21,370	265	23,487	23,563	130	23,693	47,180	64
2014	1,117	33,816	34,933	26	1,795	20,395	52	22,268	24,926	52	24,978	47,246	64
2015	1,229	36,154	37,383	19	1,899	22,154	90	24,162	26,882	128	27,010	51,172	64
2016	1,068	37,316	38,384	27	1,980	20,748	109	22,864	25,799	58	25,857	48,721	64
2017	1,440	38,423	39,863	32	2,000	22,750	82	24,864	30,426	78	30,504	55,368	64
2018	1,309	37,425	38,734	23	1,899	23,273	135	25,330	30,637	23	30,660	55,990	64
2019	1,510	37,207	38,717	28	2,003	25,307	99	27,437	34,483	16	34,499	61,936	67
Forecast													
2020	1,510	37,766	39,276	28	2,019	25,442	99	27,588	37,662	16	37,678	65,266	68
2021	1,510	38,332	39,842	28	2,035	25,433	99	27,595	37,774	16	37,790	65,385	70
2022	1,510	38,907	40,417	28	2,051	25,424	99	27,602	37,886	16	37,902	65,504	72
2023	1,510	39,490	41,000	28	2,067	25,415	99	27,609	37,998	16	38,014	65,623	74
2024	1,510	40,081	41,591	28	2,085	25,406	99	27,618	38,112	16	38,128	65,746	77
2025	1,510	40,682	42,192	28	2,103	25,397	99	27,627	38,226	16	38,242	65,869	80
2026	1,510	41,289	42,799	28	2,121	25,388	99	27,636	38,340	16	38,356	65,992	83
2027	1,510	41,906	43,416	28	2,139	25,379	99	27,645	38,454	16	38,470	66,115	86
2028	1,510	42,534	44,044	28	2,157	25,370	99	27,654	38,569	16	38,585	66,239	89
2029	1,510	43,173	44,683	28	2,175	25,361	99	27,663	38,684	16	38,700	66,363	92
2030	1,510	43,820	45,330	28	2,194	25,352	99	27,673	38,799	16	38,815	66,488	95
2031	1,510	44,476	45,986	28	2,213	25,343	99	27,683	38,914	16	38,930	66,613	98
2032	1,510	45,143	46,653	28	2,232	25,334	99	27,693	39,031	16	39,047	66,740	101
2033	1,510	45,820	47,330	28	2,251	25,325	99	27,703	39,148	16	39,164	66,867	104
2034	1,510	46,506	48,016	28	2,270	25,316	99	27,713	39,265	16	39,281	66,994	107
2035	1,510	47,205	48,715	28	2,289	25,307	99	27,723	39,381	16	39,397	67,120	110
2036	1,510	47,915	49,425	28	2,308	25,298	99	27,733	39,499	16	39,515	67,248	113
2037	1,510	48,634	50,144	28	2,328	25,289	99	27,744	39,617	16	39,633	67,377	116
2038	1,510	49,362	50,872	28	2,348	25,280	99	27,755	39,735	16	39,751	67,506	119
2039	1,510	50,101	51,611	28	2,368	25,271	99	27,766	39,855	16	39,871	67,637	122
2040	1,510	50,852	52,362	28	2,388	25,262	99	27,777	39,975	16	39,991	67,768	125
Forecast Growth Rate:	0.00%	1.50%	1.45%	0.00%	0.84%	−0.04%	0.00%	0.03%	0.29%	0.00%	0.29%	0.18%	3.09%

*This category includes both scheduled and unscheduled air carrier (60 seats or more) operations. Operations may include passenger or cargo aircraft.
#This category includes both scheduled and unscheduled commuters and air taxi (less than 60 seats) operations. Operations may include passenger or cargo aircraft.

Almost all states produce aviation forecasts as part of their state airport system plans, and most local airports update their airport Master Plans every 5–10 years and include a forecasting section.

Beyond all of these government sources, several industry organizations such as GAMA and AOPA produce industry outlook reports on an annual basis. Again, web sites should be consulted for the most current documents.

The need for activities performed from the air, such as aerial photography or aerial application, is a function of the economic conditions of the corresponding industry, such as real estate or farming, respectively. This means general aviation businesses also need to be aware of overall business and recreational trends and competing products. Aviation tends to be substantially affected by the health of the overall economy. It has also been radically affected by the industry's insurance climate, most specifically product liability insurance. By the early 1990s, the insurance cost of a new aircraft accounted for such a high proportion of the total cost that prices became almost prohibitive, nearly ceasing sales and production. This issue is discussed more fully below. Suffice it to say that an aviation businessperson who wants to foresee the future for his or her organization must understand macroscopic trends in the global economy as well as microscopic phenomena unique to general aviation.

It is impossible to review here the entire economy from the general aviation business standpoint; moreover, regional variations make such a task of doubtful value at the national level. However, the national general aviation landscape is well documented, and is a useful guide to regional or local trends.

Understanding the Local Aviation Market

Trade magazines and newspapers provide a more detailed review of technology and market shifts. Some major periodicals are listed in Figure 3.11.

Another shift in markets that general aviation businesses should anticipate is the "ripple effect" in metropolitan areas. As major airline airports become more congested, there is a tendency to encourage general aviation traffic to use reliever general aviation airports that in some cases are located further out in the suburbs of metropolitan areas. As these airports, in turn, become busier, the proportion of business traffic tends to increase, and recreational and instructional pilots may prefer to go out to more rural airports where smaller, lighter aircraft dominate the pattern and where they are not constrained by the requirements of an air traffic control tower.

Location will tend to determine the market of an airport, and a general aviation business must be prepared for these shifts and adapt to them. The process of working downwards from national trends to local and individual business prospects is described in Figure 3.12.

Forecasting Techniques and the Individual General Aviation Business

The individual general aviation business must examine the local and regional aviation markets and determine what share of local sales will be captured for each area of service. As reviewed above, national events and trends play a role, as well as the policy of the general aviation business. For example, suppose a new market opportunity is observed. Suddenly this market area looks promising. How promising? How long until government rules supporting the new program are approved and underway? What are likely to be the plans of other general aviation businesses in the area? Is this a market that our general aviation business should pursue? Can our business be sufficiently profitable undertaking the new program?

Some research and analysis will be needed before forecasting new sales for the next year. A contingency plan may even be appropriate: "If the new federal program reaches a certain state of readiness by this date, we can initiate our program at this later date and, by the end of the year, we will expect to have these results." Sales will be forecasted under various startup scenarios.

Such forecasting is not very sophisticated because many factors are outside the control of the general aviation business. However, sensitivity analyses of the forecast to internal and external factors is feasible and recommended.

Other areas of the business may be more predictable. For example, suppose the general aviation business has a well-established maintenance shop that has increased its business by 5 percent per year over the past 5 years. Is it reasonable to predict that sales will be up 5 percent this year, too? Yes and no.

First, you need to know whether that history of 5 percent growth includes inflation. If the cost of labor and parts also increased five percent per year,

54 Essentials of Aviation Management

Figure 3.11 → General Aviation Publications

Aero Magazine, Macro-Comm Corp., P.O. Box 38010, Los Angeles, CA 90038. *http://www.boeing.com/commercial/aeromagazine*

Agricultural Aviation, National Agricultural Aviation Association, Suite 103, 115 D Street, Washington, D.C. 20003. *http://www.agaviation.org/agmag*

Aircraft Maintenance Technology Magazine, 1233 Janesville Ave., Fort Atkinson, WI 53538. *http://www.amtonline.com*

AOPA Pilot, 421 Aviation Way, Frederick, MD 21701. *http://www.aopa.org/pilot*

The Aviation Consumer, P.O. Box 972, Farmingdale, NY 11737. *http://www.aviationconsumer.com*

Aviation Safety, 1111 East Putnam Avenue, Riverside, CT 06878. *http://www.aviationsafetymagazine.com*

Aviation Week and Space Technology, Mc-Graw Hill, 1221 Avenue of the Americas, NY, NY 10020. *http://aviationweek.com*

AvWeb (online only), Publisher Tim Cole. *http://www.avweb.com*

Flight Global, Reed Business Information, Quadrant House, The Quadrant, Sutton, Surrey, England, SM2 5AS. *http://www.flightglobal.com*

Flight Training Magazine, 405 Main St., Parkville, MO 64152. *http://flighttraining.aopa.org*

General Aviation News & Flyer, 5611 76th St., W., Tacoma, WA 98467. *http://www.generalaviationnews.com*

In Flight USA, PO Box 620447, Woodside, CA 94062. *http://inflightusa.com*

Pilot's Web, Pilot's Web editorial (631) 736–6643. *http://www.pilotsweb.com*

Plane and Pilot, Werner & Werner Corp., Ventura Boulevard, Suite 201, Encino, CA 91436. *http://planeandpilotmag.com*

Professional Pilot, West Building, Washington National Airport, Washington D.C. 20001. *http://www.propilotmag.com*

Sport Aviation, Experimental Aircraft Association, Wittman Airfield, Oshkosh, WI 54903–2591. *http://www.eaa.org/sportaviation*

Customer Service for the following publications can be reached via email at the addresses below:

Aviation Week and Space Technology: *avwcustserv@cdsfulfillment.com*

Business and Commercial Aviation: *buccustserv@cdsfulfillment.com*

Overhaul and Maintenance: *omtcustserv@cdsfulfillment.com*

World Aerospace Database: w*adcustserv@cdsfulfillment.com*

Figure 3.12 → Industry Market Research and Sales Planning Process

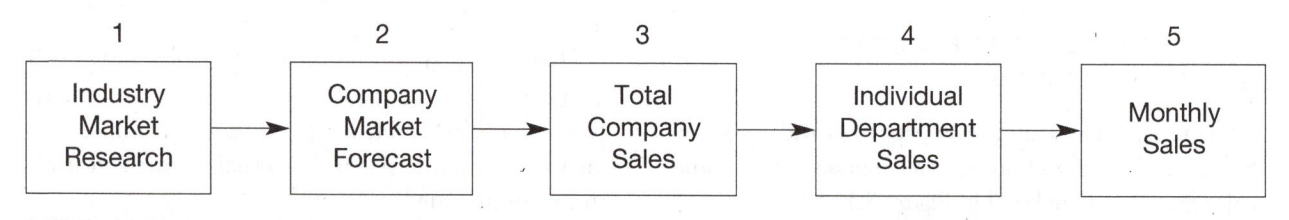

1. Industry Market Research → 2. Company Market Forecast → 3. Total Company Sales → 4. Individual Department Sales → 5. Monthly Sales

then the actual volume of business remained constant, even though the dollar volume increased. If the price increased more than 5 percent in a single year, then the volume of business actually declined.

Are there any reasons why the coming year might be other than "business as usual"? Is a competitor about to get into or out of the aircraft maintenance business? Why? Or does the general aviation business have a good chance of winning a new fleet maintenance contract for a regional airline, fractional ownership company, or a corporate flight department? You must examine whether the total market "pie" is growing or shrinking in your market area and what share of that pie you can realistically expect to garner.

It should be apparent from this discussion that forecasting based on historic trends is somewhat superficial and risky unless some evaluation is made of the underlying causes of those trends and whether any of them will change. Nevertheless, predictions are often generated from historical trends. This is most accurate at the large-scale level and least accurate at the individual company level. As the saying goes: "Trying to forecast the future by looking just at the past is like trying to drive a car by using only the rear-view mirror."

Market share is another method of forecasting. In looking at individual airport-based aircraft operations, one approach is to use national forecasts and aircraft per capita data, to disaggregate to the local or regional level. This method requires a reliable national forecast and does not allow for local variations in aviation activity.

The most sophisticated approach to forecasting is to create a multiple regression model. The factors contributing to historical activity are identified, as well as the scale of each factor's contribution. The relationships that held in the past are postulated to hold true in the future. Future values for the causal factors, such as disposable income, fuel prices, and population by age group, are identified from published sources, and predictions developed.

Such a method still has weaknesses, especially since relationships do change over time; therefore, the time period over which a factor is examined may not be representative, and good forecasts may not be available for the variables considered relevant. Moreover, such a method is difficult to use for such a small unit as an individual general aviation business; it will be much more accurate at the regional or national level.

In summary, there is no single best way to create aviation sales forecasts. They should be developed with as much knowledge as possible about what the competition is doing, with as much understanding as possible about national aviation trends and their causes, and with company goals and policies in mind. Forecasts can often become self-fulfilling prophecies. They should be tracked against actual results and adjusted as new events occur.

Customer Needs and Identification of New Prospects

There are a number of ways to identify and sell to more customers, and for most businesses these methods must all be applied continuously. Every business continually loses clients through relocation, death, competition from other businesses, and changing customer needs. Therefore, every business requires new clients just to maintain current levels of sales. A growth plan requires more than just replacements. Growth of the client base is possible to achieve in the following ways:

1. Maintain market share; grow or decline in step with the total market.
2. Obtain a bigger market share, that is, take business away from the competition.
3. Offer new products to existing clients.
4. Offer existing or new products to clients not previously in this marketplace (e.g., new general aviation enthusiasts attracted away from the airlines or from other hobbies).

Effective ways of selling more, or different, items to the same clients include maintaining close rapport and seeking feedback in person, by surveys, by complaint forms, and by direct-mail advertising. It is recommended by most marketing experts that both a list and a wall map of existing clients be maintained.

Lists of prospective clients served by competitors are obtainable from many sources. In today's sophisticated direct-mail market, lists of pilots and aircraft owners; and lists by area, income, and industry type (and thousands of other classifications) may be rented from list brokers in all major cities.

To attract customers who are not currently involved with general aviation, new techniques may be needed. For example, there is substantial evidence of a growing reliance on executive travel among the top companies (Fortune 500). For many years, *Business Week* magazine tracked the financial performance of the aircraft owning and non-aircraft owning companies in the Fortune 1000, and found that aircraft owners had higher performance on virtually every measure. In the 1990s, GAMA and NBAA commissioned an analysis by Arthur Andersen to investigate this issue in more depth.[16] The study found that of the 214 operators and 121 non-operators in their sample, operators earned 141 percent more between 1992 and 1999. The study stated that, "according to the CFOs interviewed, aircraft helped improve performance in the areas of greatest importance in today's fast-paced economy (e.g., identifying and executing strategic opportunities for new relationships and/or alliances, reaching critical meetings and closing transactions; expanding into new markets; and increasing contact with customers)." Armed with some of the statistics correlating company performance and use of general aviation aircraft, a general aviation business might present the benefits to business people at Chambers of Commerce, Jaycees, Kiwanis, and Rotary clubs. Industry groups can assist in this respect.

To encourage recreational flying and instruction, a general aviation business may need to publicize the initiatives organized by the GAMA, AOPA, and other groups, as well as introduce tailored approaches.

While the overall economic climate for general aviation in the past two or more decades has been discouraging, many enterprising general aviation businesses nationwide have developed healthy markets in such areas as:

> Corporate aircraft maintenance;
> Restoring antique aircraft;
> Airline flight training;
> Propeller balancing;
> Airport hotels and motels;
> Air charter;
> Servicing airline aircraft, including deicing;
> Airport snow removal;
> Overnight package express;
> Air ambulance services;
> Airport barbershops; and
> Retail merchandise.

For the general public, awareness of the benefits of the general aviation business's services and related fly-ins, open houses, and air shows needs reinforcement. However, care should be taken not to present a barnstorming image if that is not the airport's target market.

Product and Service Definition

The starting point of any business is the identification of the products and services it wishes to sell. In the aviation business, this means the identification of several products and services, such as transporting passengers, providing hangar facilities, servicing aircraft, renting space, and so on. It is extremely important that the manager clearly identify each of the products that he or she expects to include as part of the business. Each product or service offered by a business must be identified because decisions related to other marketing variables, such as promotional techniques, will depend upon identification and knowledge of the product. In addition, the manager may want to identify services that will be offered by request only. What makes up such items, for a particular general aviation business, is hard to predict. They will be things that require more training and special equipment than will normally prove feasible or profitable. For many businesses, such an item might be banner towing or aerial photography.

By the same token, there may be activities that are in demand, but the general aviation business might want to refer customers for these to another specialist in the field. Some examples include specialized avionics, radio work, and the whole maintenance function. Some general aviation businesses do not pump fuel, except perhaps for their own use, either because the entity owning the airport runs this operation or because another general aviation business is better placed to do it profitably. Each general aviation business must decide its own service and product lines based on perceived customer desires and the potential profit in each area.

Finally, there are some activities, that although non-aviation, are profitable and interconnected areas. Examples include restaurants, car rentals, and limousines. A quality airport restaurant with good views of active aircraft operations may attract many non-fliers and help introduce them to the services and facilities at the airport, providing a positive public image. Even boat and RV storage can be useful services to offer if ramp and hangar space are available.

Total Product

Even for consumer goods, the typical customer is not just looking for the product but for the experience of fulfilling a want. He/she is concerned with the total product, including the experience of obtaining it, as well as the assurance of related functional and aesthetic features. The typical customer is also concerned with necessary accessories, installation guidance, instructions, packaging, dependability, and assurance that good support is available in the future. In making his or her purchase, he/she is looking for indications that all needs in regard to the item will be met.

Product Classification

Traditional marketing theory divides consumer goods into four categories:

1. **Convenience goods** are those the purchaser wants to buy frequently and with minimum effort; therefore, comparison-shopping is not employed. An example might be aviation fuel, unless there is a discount fuel operator in the area.

2. **Shopping goods** are those items that shoppers do compare. Examples might be a used aircraft or a flight instruction course. The general aviation business will have to offer some special features of quality or price to attract customers.
3. **Specialty goods** are those items for which the customer wants one specific brand name or custom item and will go to great lengths to obtain it.
4. **Unsought goods** are those items that the customer is not seeking and may not have thought he or she needed. New products, service contracts, preventive maintenance, and novelty items fall into this category.

The Competition

Convenience goods will be sold to local customers and to transients who are at that point a captive market. If you have what they need, they'll buy it. Shopping goods are the subject of strong competition and there may not be any customer loyalty for these. Specialty goods can help a general aviation business to establish a reputation either as carrying a good inventory, or as being known to go to any lengths to find what a customer wants. Unsought goods require skills to make a potential customer aware of the benefits of something new.

Market Niche

Based upon the total market in the area and the competition's share of it, each general aviation business must seek to determine its own market niche or share. This will also require review from time to time. Questions to address are "What do we do?" and "Why are we different?"

Price

There are three basic methods of pricing: cost-based, demand-based, and price-based. These are discussed in the following sections.

Cost-Based. In order to understand cost-based pricing, we must first understand the difference between the terms "margin" and "markup," and their relationship to the cost and sale price of an item. This relation is expressed as follows:

Margin = 1 − (1 / (Markup + 1))
Sale Price = Cost * (Markup + 1),

where margin and markup are expressed as decimals. Retail stores usually mark up wholesale prices about 100 percent; that is, an item costing ten dollars wholesale will be sold at twenty dollars retail. This markup is a cost-based approach to pricing in which the overheads such as storage, retail showroom space, labor, bookkeeping, and financing costs are covered by a predetermined overhead rate. Assuming you purchase wholesale at competitive rates, and your overhead costs are within industry norms, you can price competitively using this method.

Cost-based pricing is a straightforward approach with a basic appeal to most aviation managers. There are, however, two typical difficulties that exist in many situations: (1) many managers are not aware of, nor do they have available, adequate unit cost data, and (2) even where some data are available, the manager may not adequately recognize the various ways that costs change. In order to use cost-based pricing, the manager must understand the following types of costs:

1. *Total fixed cost* is the sum of those costs that are fixed in total, regardless of output level. Typical fixed costs include rent, property taxes, insurance, depreciation, and administrative salaries. Such expenses must be paid even if business activity ceases temporarily.
2. *Total variable cost* is the sum of those costs that are closely related to volume of products or services. Variable expenses include materials used, wages paid, and sales commissions. At zero business, total variable cost is zero. As output increases, total variable cost increases linearly.
3. *Total cost* is the sum of total fixed cost and total variable cost. With total fixed cost set, growth of total cost is dependent upon the increase in total variable cost.
4. *Average cost* is figured by dividing the total cost by the volume of business; the average cost per unit of business is then obtained.
5. *Average fixed cost* is when the total fixed cost is divided by the quantity of business.
6. *Average variable* cost is the total variable cost divided by the related quantity of business.

The following example illustrates the above cost structure for a line department engaged in aircraft fueling operations in an aviation business: Figure 3.13 illustrates the cost data for such an activity. Notice that the average fixed cost decreases as the quantity of fuel increases and that the average variable cost remains constant with changing quantity. Note also that the average cost is decreases to a plateau. The behavior of the three average-cost curves (average variable, average fixed, and average cost) has been graphed in Figure 3.13. The aviation manager could use this type of graph in setting prices. First, he/she must include an element for profit either as a fixed amount in the total fixed costs, or as a fixed amount per unit in the average variable cost. The next step is to simply decide, for example, how many gallons of fuel the department will sell. If the goal is to sell 30,000 gallons, then by referring to the cost curve (Figure 3.13), the price is determined per gallon. Therefore, for whatever quantity is desired, the price can be identified. The quantity selected can be related to previous levels of activity or to target goals for future periods; thus, it becomes a useful managerial tool.

Demand-Based. Another method of pricing is simply to charge whatever the traffic will bear. As long as this is higher than the price arrived at by cost-plus, there will be a higher profit. For highly-desirable or scarce items, there will be a very considerable profit that will bear little relationship to the actual cost of making that item.

Price-Based. Price-based costing is based on the concept that business traffic is a result of the potential customer checking out your competitors and making a choice based on relative prices.[17] Therefore, in order to be competitive, cost-based pricing will not work if your costs are higher than the competition's. Rather, you need to also check out the competition's prices, and then determine how to get your own prices down to that level. A small difference in price won't

Figure 3.13 → **Shape of Typical Fuel Cost Curves when Average Variable Cost Is Assumed Constant**

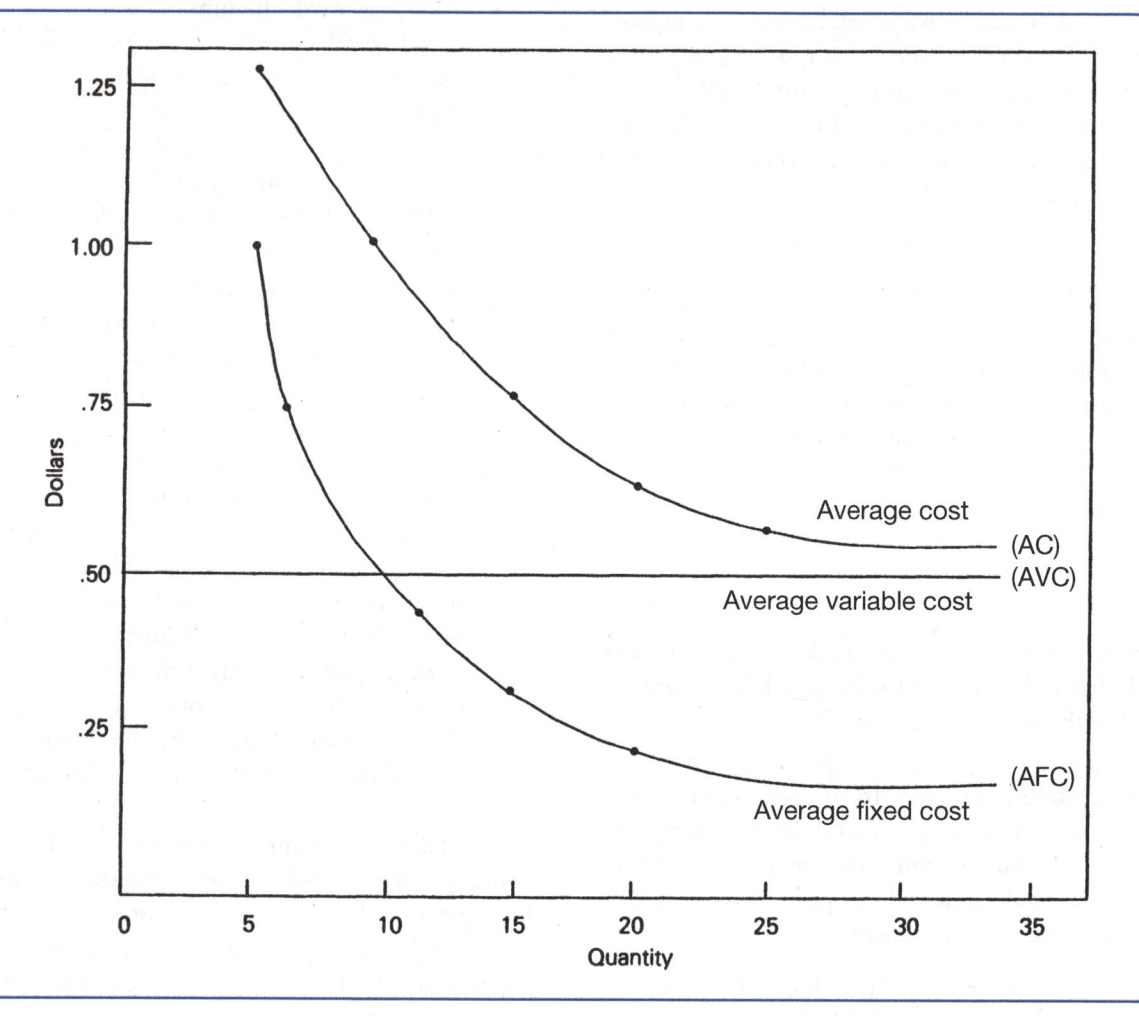

make a difference to most customers, but you would have to offer a tangible difference in quality to entice a customer to pay much more than the competition wants. Such points of differentiation are hard to both develop and then convey to a customer who is shopping for the best price.

How to reduce one's costs is then the challenge and the manager will likely examine every component of cost that goes into that particular commodity or service.

Elasticity

Pricing that reflects what the traffic will bear conforms to the economist's true pricing theory. That theory states that price is determined as the point at which the seller's supply curve and the buyer's demand curve intersect. As price goes down, the consumer is assumed to want more; and as the price goes up, the consumer wants less. For a price increase of 1 percent, a drop in demand of 1 percent would follow under conditions described as unit-price elasticity. Price elasticity is less than one or less than unit if demand goes down less than 1 percent for a 1 percent price increase. Due to its essential nature, aviation fuel has a fairly low price elasticity. Where price elasticity is small, a seller would have to lower prices considerably to sell more to the same customer.

In theory, each buyer has his or her own demand schedule for each product at different prices. This is expressed most commonly as a curve, as shown in Figure 3.14. As a customer comparison-shops for a product or service, it seems likely that his or her demand curve will shift depending on the asking prices of suppliers for various levels of quality.

The seller behaves in the opposite way. The higher the price of the product he/she is selling, the more he/she wants to sell. Such a supply curve is shown in Figure 3.15. The buyer's demand curve and the seller's supply curve intersect at the market price and volume, as shown in Figure 3.16. In practice, most prices are not negotiated with each buyer but are predetermined and based on assumptions about regional aggregate supply-and-demand conditions. Moreover, in times of scarcity, when traditional economic theory would indicate raising the price as the market's best form of allocating supply, there are usually regulations in effect about price gouging. Offering more of something for sale when costs increase is questionable, because the profit margin probably will not go up at all. In order to preserve profit margins, a general aviation business may pass on inflated costs to the customer. However, trying to make an exceptional profit when demand is high will probably alienate customers and be self-defeating.

Pricing is, therefore, an inexact science in which basic economic theory must be combined with common sense and knowledge of the competition. Year-end or other sales to get rid of old inventory, as well as discount pricing for promotions, must be added to this complex picture before any revenue calculations can be made.

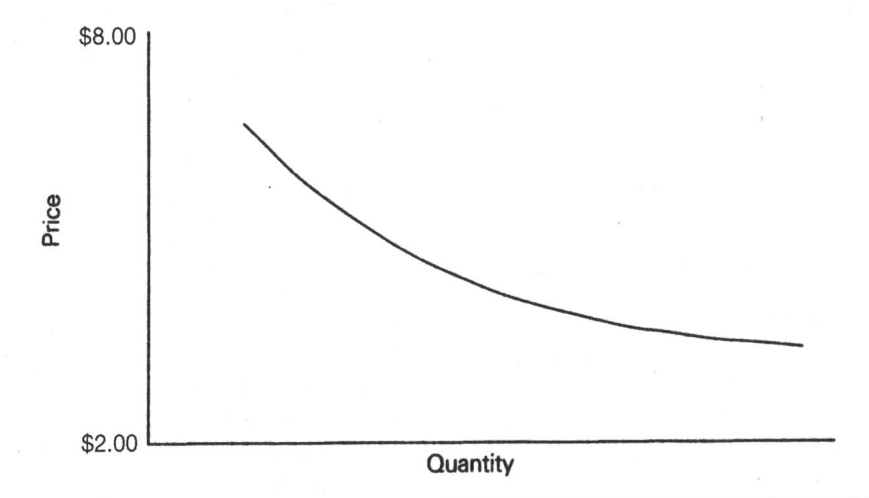

Figure 3.14 → **Demand Curve for Aviation Fuel**

Figure 3.15 → Supply Curve for Aviation Fuel

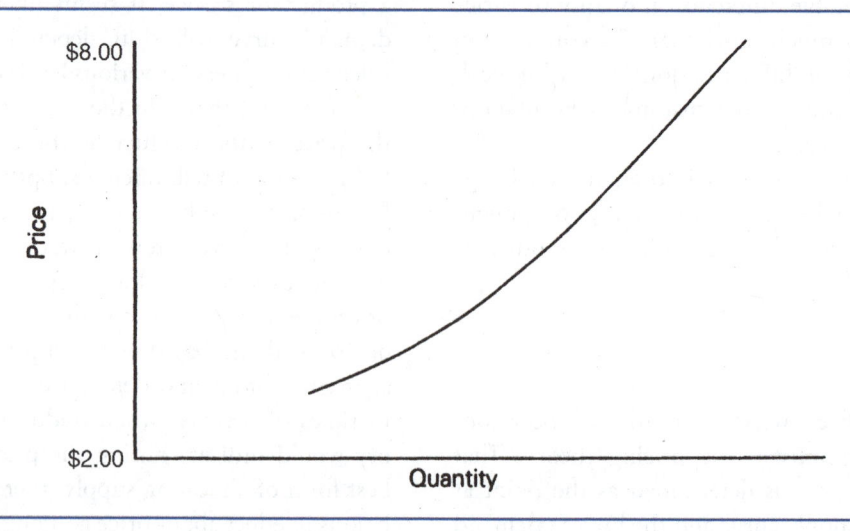

Figure 3.16 → Equilibrium of Supply and Demand for Aviation Fuel

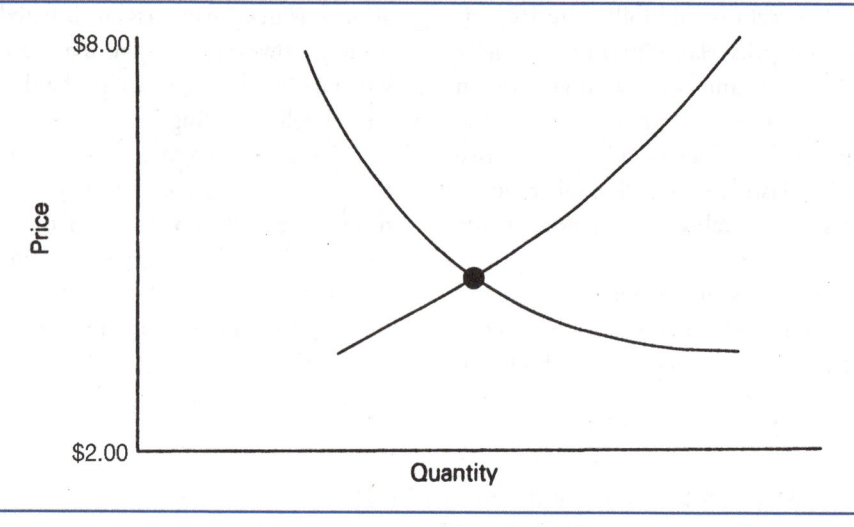

Pricing Policy

Most businesses set their prices according to specific guidelines rather than adjust their prices to constantly changing market pressures. For the pricing practices of an organization to follow a logical pattern, for the employees to have a pricing guide to follow, and for the customer to experience understanding and goodwill toward the business, it becomes essential to have pricing policies established in at least seven areas. Knowledge of each of these areas will allow the operator to capitalize successfully on various opportunities that arise.

Area 1: Fixed vs. Flexible Pricing. One policy option is to have one price to all customers, versus a flexible-price policy where different customers may pay different prices. This area—that of fixed or flexible prices—can be applied to many aviation business activities. For example, consider the charter business. Will you charge a flat $250.00 per hour for charter of an aircraft, or will you promote a lesser rate of $175.00 per hour for overnight hours to enhance the aircraft utilization? Although some managers are reluctant to use flexible pricing due to complexity concerns, others are quick to deploy such policies for financial advantage.

Area 2: Pricing in Relation to Market Level. A second policy area deals with the decision to price above or below the market level. Many factors must be considered in making this decision and in reviewing it for necessary changes. Higher prices may well be associated with better quality products, lower prices with cheaper products. At the same time, the decision to price below market level may very well be influenced by the desire to increase the volume of business. Finally, your price level may well determine the level of support/services you may provide in your business for your products. For example, higher rates for maintenance labor may enable you to have better equipment and better-qualified mechanics.

Area 3: Pricing in Relation to Product Life Cycle. A third policy area relates to product life cycle. Will prices be high initially in order to "skim" the market "cream" or will prices be lower in order to "penetrate" the market and get a larger share? Pricing related to the life cycle of a given product means normally that when the product is new and in great demand, the price may be higher. Later, when the product is not as new or when competition has become plentiful, the price may, out of necessity, be reduced.

Area 4: Pricing in Relation to Product Line. The fourth policy area deals with the creation of pricing policies in relation to the various product lines. Is the relationship among the various product lines a logical one? Are price classes to be set, rather than individual prices? Pricing by product line may be hard to introduce in the typical aviation business. For example, pricing fuel lower than competitors may unwittingly help establish a low-price (and low-quality) image for the entire business.

Area 5: Promotional Pricing Policies. A fifth policy area deals with promotional pricing, such as the use of coupons, multiple unit pricing, loss leader pricing, and setting of introductory prices. Promotional pricing activities are a familiar feature of the aviation business world. Coupons and bonuses have been used at many airports. Airline special or promotional rates are commonplace and accepted. Introductory price cuts by new businesses are sometimes used to ease their entries into a competitive market.

Area 6: Geographic Pricing Policies. Geographic pricing policies may not have to be considered by many smaller aviation businesses, but these businesses are frequently the targets of such policies set by large national distributors of aviation products. The primary issue is how freight costs on products for resale are split between buyer and seller. FOB (Free On Board) pricing means that the closer you are to the source, the cheaper the product becomes. Zone pricing is used to smooth out the delivered prices in sections of the country. Uniform delivered pricing means that there is one price to all customers, wherever they may be. Freight absorption pricing is another example of geographic pricing policy. This technique is used by businesses that are shipping their goods over considerable distances. In those distant markets they may absorb some of the freight costs in order to allow their salespeople to become more competitive.

Area 7: Channel Pricing. The last pricing policy area deals with distribution channel activities. Most aviation businesses do not set channel policies, but they should be aware of the approach. By viewing a distribution channel as a unit, they can consider applying many of the above policies to the channel as a unit. The secret to the success of this approach lies in every member of the channel feeling that commensurate profits are being realized. See Figure 3.17.

Place

No product is commercially valuable if it cannot be brought to the right place at the right time. The general aviation business is one in which the customer comes to the business, rather than the other way around. While the location of the business is critical, the location of a general aviation business is fixed and relocation is difficult, if not impossible. "Place" is a given for the general aviation business. By contrast, off-airport aviation supply stores and mail-order avionics firms are much less tied to geography and have considerable advantage over a general aviation business in this respect.

Distribution Systems

Place also refers to the channels and means of distribution that moves products into the general aviation business inventory for use or sale. The channels between original producer and end consumer may involve no intermediaries (e.g., Beech Aircraft and Cessna sell factory direct and no longer use dealers). Or the process may require one or more wholesalers and the general aviation business as a retailer. Figure 3.17 shows these four basic channels of distribution.

The general aviation business as retailers must understand the distribution process for all items sold since it affects the goods' cost, availability, and supply schedules. The business must also be involved in merchandising, which means the display of retail goods to best advantage, and inventory control, which means keeping track of fast- and slow-moving items and setting up appropriate reorder schedules. These functions, to some extent, will also be necessary in the parts department, where display and merchandising can perhaps produce many more sales than many parts and service managers realize. Chapter 10, Aviation Maintenance, discusses how to calculate the "Economic Ordering Quantity" of a given item.

Computerized inventory control can be useful in keeping track of stock from the point of sale and preparing summaries of product turnover.

Promotion

The object of promotional activities is to inform, persuade, and remind existing and new customers about what the business offers them. *Broadcast* methods of promotion include television, radio, newspapers, trade publications, and websites. *Focused* or *targeted* methods include e-mail sent to customers, list serves, LinkedIn, and other electronic tools, as well as direct mail, novelty items, air shows, tours, signs, and referrals from happy customers (with or without incentives). Figure 3.18 shows a year-long promotional planning chart for a typical general aviation business.

Advertising. Advertising is but one method of promotion, and its cost-effectiveness is generally greater when it can be targeted toward the desired market. Thus, advertising in the mass media, such as radio and television, may not be very appropriate for general aviation, whereas advertising in aviation publications may be. One common, targeted mode of advertising is cable television, where detailed information is available on the types and locations of audiences watching certain shows.

Direct Mail. For general direct mail, such as to all households in certain zip codes, a "rule of thumb" in the industry is that a two percent response constitutes a good result. The use of aviation mailing lists and preselected business lists could be much more successful than general lists for aviation purposes. Carefully-designed and targeted direct-mail advertising with a

Figure 3.17 → **Four Basic Channels of Distribution**

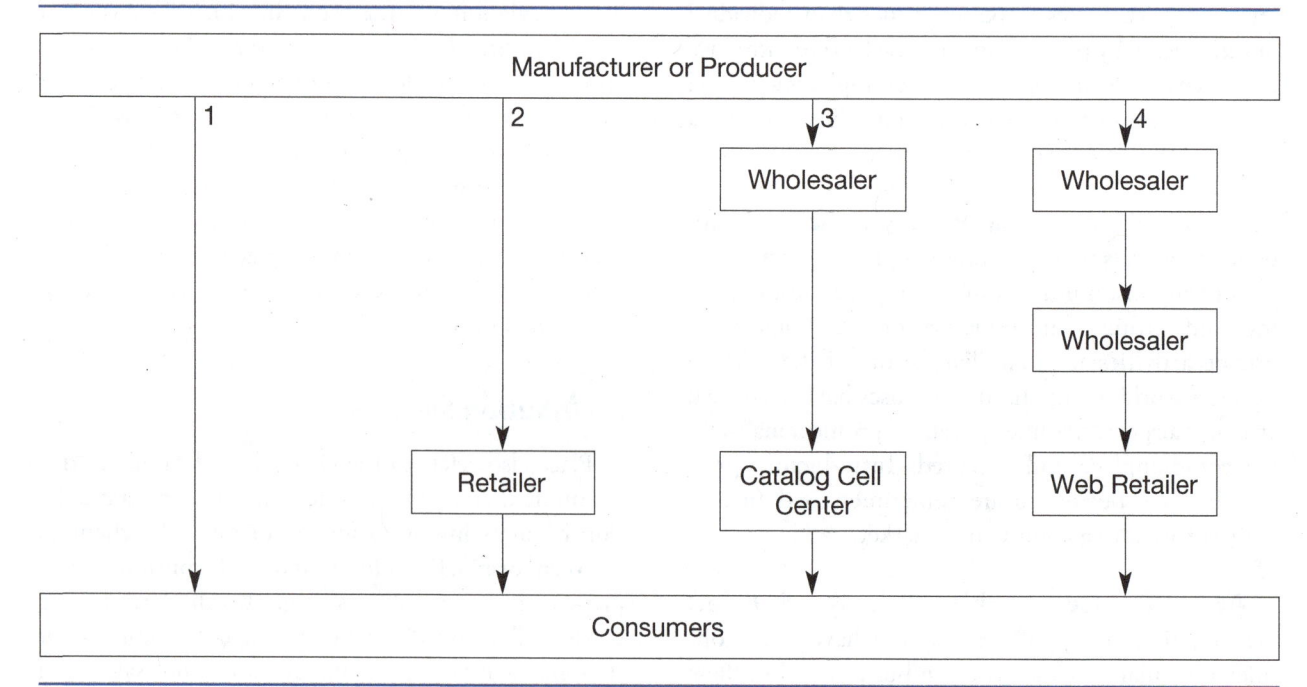

Chapter 3 Marketing 63

Figure 3.18 ✈ **Promotion Planning Chart for a General Aviation Organization**

Planning Chart

ACTIVITY		January	February	March	April	May
LINE OPS	PLAN		Check line needs			
	ACTION		Refurbish line, exterior paint, spruce			
INSTRUCTION	PLAN	Plan Spring Learn to Fly Program Develop Prospect List Films - etc.			Flight School Air cond. June & July	Local Learn to Fly Promotion
	ACTION	Advanced Ratings	Go Twin Learn to Fly Two Time	High School Promotion	←— National Adv. Learn to Fly $25 into flights —→	
FLIGHT SVCS	PLAN					
	ACTION	←————— Twin Invitational —————→				
ACFT SALES	PLAN			Plan for Arrow II		
	ACTION	National Adv. Business Twins Local Promotions Travel Analysis			←— Arrow II Nat'l Adv Local Promotion PFC	
PARTS	PLAN					
	ACTION			Promote Pilot Supplies		
SERVICE	PLAN	Review Annual List				Plan programmed maint. effort
	ACTION		Promote Engine Work	Promote Radio Work		Promote Aircraft Clean-up
OTHER General	PLAN	Review Plan for Year!!		Prepare Open House		
	ACTION		Visit: News T.V.	Visit: Real Estate	Annual Open House	
OTHER	PLAN					
	ACTION					

Continued

64 Essentials of Aviation Management

Figure 3.18 → **Promotion Planning Chart for a General Aviation Organization—Cont'd**

Planning Chart

	June	July	August	September	October	November	December
	←——— Special Service Program ———→			Winterize & Safety Program			
		Plan Fall Instrument Rating Course	Plan PIP Trng. Fac.				
	←——— Flight School—Air Cond. ———→		Follow Up Learn to Fly		Local Instr. Course Go Twin Natl. Adv.	PIP New Model Aircraft	Ground School
				Order Calendars Navajo Now!		Plan local Twin Promotion	PFC Refresher
	←——— Piper Nat'l Adv Fortune 500 ———→				Navajo now!	Calendar mailing	2011 material & literature
				Prepare Xmas List			
	←——— Promote Cleaning Materials ———→				December Gift Mailing		
	←——— Piper Nat'l Adv. For Programmed Maint. Local Program ———→			Maint. Seminar			
		Airport Promotion			Fall Fly-In		
	Movie Showings					Dealer Meeting!!	

no-risk "offer" to encourage a quick response can be an excellent source of leads. E-mail advertising has become a new form of direct solicitation that can reach more yet cost less than traditional mail.

Referrals. Word-of-mouth advertising through existing customers is one of the most effective promotional tools in an aviation business. The individual general aviation business will need to test different concepts in limited ways before deciding which particular techniques work in each of the targeted market segments. What appeals to the business market is more likely to relate to efficiency and quality, while price may be the overriding consideration for the flight instruction market. The business should follow up with every customer to collect negative feedback as a prompt follow-up allows a negative experience to be addressed so that the customer makes positive referrals. If not contacted, the unsatisfied customer will likely be persuaded by the competition the next time. Conversely, firms that get a reputation for dealing fairly and promptly with problems get good word-of-mouth press that can continually strengthen their client base.

Institutional Promotion. Institutional promotion or advertising means the promotion of a positive image about the business without stressing any particular product or service. It ties closely to branding, discussed earlier. In many cases it is very successful, because people want to buy an experience and some assurance of ongoing service, rather than a product, per se. Institutional promotion can use mass media or targeted marketing, depending on the situation. As general aviation businesses begin to compete with airlines for business travel, they need to consider emulating airline institutional advertising techniques. A major step has been taken in this direction through the NBAA "No Plane, No Gain" plan, designed to promote the importance of business aviation through a national advertising campaign. Similarly, AOPA's "GA Serves America" campaign has been launched to educate both policymakers and the public about the important role general aviation plays in local communities and the national economy.[18]

Consulting as a Promotional Tool. In the aviation world a great deal of selling relates to the selling of aircraft, particularly to new business users that have been identified as a growth market. The general aviation manufacturers, such as Gulfstream, Beechcraft, Cessna, and Piper, offer to prospective buyers a pro forma analysis comparing current methods of travel for time and costs with travel by means of a corporate aircraft. The astute general aviation business that seeks to sell aircraft as part of its product line can conduct this analysis on behalf of prospective buyers. The study helps the general aviation business achieve a better rapport with the prospective client and helps to specify exactly what type of aircraft will best meet the company's needs. Such a study will include:

> Current and future company travel patterns;
> Current travel costs, including overnight expenses and the value of time spent by personnel, particularly executives;
> Prior years' travel costs and present value of savings if general aviation was used;
> Costs of current travel by general aviation, including deductions for savings of time and overnight expenses;
> Costs of aircraft ownership, minus tax benefits; and
> Annualized net positive or negative value resulting from the use of corporate aircraft.

The study may be manual or computerized; the latter permits the testing of more assumptions and "what if" scenarios. A general aviation business might offer to conduct such an analysis and refund its cost if the client buys an aircraft.

Sales

Needs Assessment. "Staying close to the customer" should be interpreted in part as listening to what problem the customer is trying to solve or what benefit he or she is seeking. Particularly for major purchases, a first discussion should likely focus only on needs and not on products.

Alternatives. A second, or second-stage discussion will present the prospect with some alternatives that the general aviation business has available to address the needs already expressed. The dialogue will give the buyer the opportunity to comment about each alternative and the seller the opportunity to pivot or improve products.

Closing. Closing may involve satisfactorily addressing objections raised by the customer and presenting the best alternative in terms that suggest a purchase. Much has been written elsewhere about sales techniques and much of the necessary skill appears teachable if the salesperson has the right attitude. Selling should be a win-win proposition for the buyer and the seller, especially assuming that follow-up sales and service and positive referrals are desired. The same approach should be used whether the sale is large or small because the next purchases may grow in dollar value.

Collecting

Payment is, of course, essential. Part of the sales process is to determine whether the buyer is financially qualified for credit or is going to pay cash. Terms must reflect your costs of financing. Chapter 6, Organization, discusses credit policies and procedures.

Marketing Controls

Marketing Plan

The marketing effort must be part of the overall business plan, for it is that portion of business planning that helps estimate sales and hence revenue to the business. Measures of marketing effectiveness may include sales volume, area coverage, new clients, product establishment, or other metrics unique to marketed goods or services.

Contribution Analysis

Contribution analysis allows the manager to examine what contribution each sale makes to profit or overhead. For example, each sale of a small item might result in a cost of sixty cents to purchase and ship the item, and yield forty cents to pay commission, overhead, and profit. If a certain advertising promotion results in another 100 sales, the advertising cost per sale can be calculated and the contribution to profit compared with previous results. In this way, the merits of altering advertising tactics for a certain product can be calculated in terms of the return over a baseline yield. See Chapter 4 for more information.

Performance Evaluation

An important area of sales performance evaluation is the collection and analysis of sales data. Routine analysis of sales data for selected operating periods provides benchmarks that can be used to measure progress toward the selected goals of the

This "follow-me" truck is ready and waiting for the arrival of an aircraft.

firm as well as to analyze some of the fundamental assumptions used in setting these goals. The focus of measurement may be on sales per territory, quantities of individual products, salesperson productivity, or cost per sale.

Quality Control

Pilots who have flown to many parts of the United States may note that in some regions there seems to be more of a "pro-aviation" attitude. When an airplane arrives, there may be a "follow-me" van to direct it to parking, or a team of line personnel waiting at the designated parking spot to provide a red carpet treatment that provides whatever is required in terms of fuel, ground transportation, and other needed services, and information. In other locations, pilots may find it difficult to obtain even basic services. Continuous upholding of high-quality standards is a goal for the best businesses.

Budgeting

The marketing budget is part of the cost of doing business and must be set up by product or by division of the firm on an annual or quarterly basis. Chapter 4, Profits, Cash Flow and Financing, discusses budgeting in more depth.

Information Systems

In order to keep track of the sales of numerous goods and services, computerized record-keeping is heavily recommended. Moreover, computers also allow this data to be organized and analyzed in useful periodic summaries or specialized reports. Data collection and management systems are discussed in Chapter 7, Management Information Systems.

Integrated Marketing

Realizing the maximum profit potential in a given business requires a marketing effort that is developed with an integrated, consistent, and systematic approach. The first step is to clearly identify the business goals, the products, and the market. All available tools should be utilized. Specific short-range objectives should be set. Budgeting may use costs that are a fixed proportion of past or forecast sales, or may be based on other considerations such as matching competitors' performances or reaching a sales goal.

Coordination of the marketing tasks to achieve the desired schedule is a key step. Work schedules, flowcharts, and similar tools may be helpful. Careful analysis at each step not only checks on progress, but also may suggest corrections and alternative marketing actions.

Marketing the Airport

Since the success of the general aviation business also depends on its home airport and not just the part the company leases, the marketing function must address the issue of collaboratively crafting a positive image for the airport among non-user neighbors who are likely more numerous and hostile than aviation consumers. Even a privately-owned airport must pay increasing attention to this because of:

> Land-use zoning and noise ordinance decisions around the airport;
> Property tax questions;
> Availability of public federal funds for selected private airports;
> Public actions needed to protect airspace;
> Needs for highway signs and roadway access to the airport; and
> Needs for support from local social, political, and economic groups.

The general aviation business owner relies on a good public image for the whole airport, whether he or she owns the field or is a tenant. Regardless, a general aviation business can create a positive attitude about the airport. For example, the owner can:

> Ensure that flight instructors and students know the noise-sensitive areas to avoid and the proper noise-abatement techniques for flight operations;
> Set up a documented complaint system for airport neighbors, and meet personally with any who are the victims of "buzzing" or other inappropriate flight behavior;
> Invite airport neighbors to open houses and demonstration flights, and include giveaways;
> Present the operation of the airport to business groups;
> Invite local public officials to visit the general aviation business and make presentations to flight school classes;
> Conduct tours for school children and others;
> Organize a fly-in or air show and invite the local community to attend;

- Organize group meetings and seminars for local pilots or other groups, including local business owners, letting them utilize airport facilities; and
- Maintain liaison with the airport administration and with the airport commission, city, or county council, and ensure any negatives about the airport or the general aviation business are quickly addressed.

This topic is discussed further in Chapter 12, Physical Facilities. Such activities are a form of institutional marketing, except that the general aviation business will be trying to convey the value of the airport as a whole, not just the part of it being leased. Even if other general aviation businesses on the field are not willing to spend the time to do this, you should try to make it a priority. As a result, goodwill will flow to the business.

Summary

The successful aviation manager develops a marketing orientation; he or she operates the business with a major focus on the customer. Specific target customers and markets are identified. The product and service, price, place and promotion mix is developed to meet customer needs. To accomplish this effectively, the first step involves the collection of information on the market and the many variables that affect business activity. From these data, a forecast must be developed that will enable the manager to formulate a marketing mix to satisfy buyer needs. The product element is concerned with firm identification of the "total" product and the market-related concepts of product categories, demand curves, supply curves, and product planning. Place deals with physical distribution: getting the right product to the target market in order to provide the customer with the right time-place-possession value. Promotional activities are designed to inform target customers about the available products and services and persuade them to purchase those products and services. This may be done through personal selling, mass advertising, direct mail, and other techniques. Price is based upon cost considerations and may either inflated or tempered by demand changes and competitor actions. Predicting market opportunities requires thorough study and constant vigilance for necessary adjustments resulting from economic and technological changes as well as changes in the competition. Marketing is more than sales; a sound marketing plan should guide the planning and execution of a business model through market research and adaptation.

DISCUSSION TOPICS

1. Why is a "marketing orientation" important to the aviation manager, and how can it be acquired?
2. Identify three key activities associated with a comprehensive marketing strategy.
3. Identify the "four Ps" of marketing and discuss the ways in which they are different for general aviation businesses than for most other types of businesses.
4. Why are demand and supply curves useful in relation to market planning?
5. Describe three elements of distribution and how they function.
6. What methods would one select for (a) mass marketing, and (b) targeted marketing? How would each work?
7. How can one determine the effectiveness of promotional activities? Discuss two methods.
8. What are the pros and cons of setting prices, based simply on your costs plus a profit margin?
9. Identify five types of market-related input that can be obtained from either sales records or customer feedback. How can each be used?
10. What is meant by "integrated marketing" and why does it matter?

Endnotes

1. Peters, T. J. & Waterman, R. H. (1982). *In Search of Excellence—Lessons from America's Best-Run Companies.* New York: Warner Books.
2. Stoltenberg, J. (1984, May). Determining Your Natural Market: Practical Market Research for Entrepreneurs. *Fifteenth Annual Entrepreneurship Symposium.* Symposium conducted at the School of Business Administration, University of Washington, Seattle, WA.
3. In the 1970s, the FAA undertook major research through Battelle Laboratories in an effort to build a General Aviation Dynamics Model that would correlate GA growth to the growth of key economic factors. After about a decade this attempt was abandoned.
4. Discussed in more depth in Chapter 11.
5. Steckler, S. A. (2001). The Transformation of the U.S. Air Travel Market. *Airport Magazine, 13*(5).
6. Ibid.
7. Telephone conversation on May 9, 2002 between author J. Rodwell and NATA's Jackie Rosser, who provided staff support to FOARC.
8. Ibid.
9. Worrells, D. S., NewMeyer, D. A. & Ruiz, J. R. (2001). The Evolution of Fractional Ownership: A Literature Review. *Journal of Aviation/Aerospace Education & Research, 10,* 41–59.
10. Bolen, E. M. (2002). *2001 Annual Industry Review.* Washington, DC: General Aviation Manufacturers' Association.
11. See *http://www.faa.gov/licenses_certificates/airmen_certification/sport_pilot*
12. General Aviation Revitalization Act (GARA), August 17, 1994.
13. Stolzer, A. M. (1998). *GARA: An Overview of Tort Reform. http://www.citizen.org/congress/civjus/prod_liability/general/articles.cfm?ID=562*
14. See *https://www.faa.gov/data_research/aviation_data_statistics/forecasting/media/af1.doc*
15. See *http://www.faa.gov/data_research/aviation/* for a complete list of FAA forecasts.
16 Arthur Andersen, Inc. (2001, Spring). *Business Aviation in Today's Economy: A Shareholder Value Perspective* (The White Paper Series, Number 4). See *https://nbaa.org/wp-content/uploads/2018/02/AndersenPart01.pdf*
17. de Decker, B. (1999). Price Based Costing-Gobbledygook or Good Business? *Aviation Maintenance Technology, 1999*(8).
18. See *https://www.aopa.org/advocacy/take-action/aopa-aviation-advocacy-fund*

4

Profits, Cash Flow, and Financing

OBJECTIVES

› Discuss the meaning of the word "profit."

› Enable the manager to distinguish between lack of profit and lack of cash flow and take appropriate corrective actions for each.

› Understand the relationship between social responsibility and profit in business decisions.

› Give examples of the differences between "fixed" and "variable" costs.

› Recognize methods of improving the FBO's cash position.

› Realize the conditions needed for effective budgeting.

› Describe the benefits and disadvantages of extending credit to customers.

"Lack of cash can drive a firm into bankruptcy even though its products are first rate and its operations are profitable."[1]

Introduction

Historic Context

The economy of this country has been built by entrepreneurs who started out small, provided a vitally needed product or service, made often very large profits, and plowed them back into the business to finance further growth. Profits were the key to a successful business. In former times, profits provided cash flow for two reasons. First, cash was far more prevalent than credit, and customers expected to pay cash on the spot. Second, the economy was growing rapidly, so demand was always outstripping supply and "supernormal" profits were often possible.

While this is a greatly simplified view of the United States economy in pioneer days, it is conceptually significant. The main concept is that cash flow was not much of an issue. A second concept is that businesses were less likely than today to seek outside sources of financing to support growth, since internally generated funds arrived copiously and quickly.

Profit and Cash Flow Today

Due to slower growth in most markets and to virtually instant communications regarding the competition, "supernormal" profits tend now to be only evident in new industries and then only in the beginning.

Due to the role of credit in the economy, cash flow lags are an issue in almost every type of business. In many industries, customers expect at least thirty days to pay. Money has to be spent producing and delivering the products or services, well before the producer gets paid. Profits alone, when they are eventually collected, may not be sufficient to finance growth. Even when they are, financing may be needed to get a business out of seasonal or startup cash flow "holes."

One facet of cash flow is often not well understood: that the faster a company is growing today, the greater its cash flow problems are likely to be. It may be growing rapidly simply because it has found a highly profitable market niche. Nevertheless, more raw materials, people, and other inputs are generally needed to respond to the growth market with quality and promptness. Costs increase quickly; and even though receipts will eventually come in, any lag can result in a cash flow "hole."

A successful small business will do well to consider, therefore, that both positive profit and positive cash flow are essential for continued operation. As indicated in the opening quote, the two do *not* necessarily go hand-in-hand. Each is a necessary, but not a sufficient, condition for the survival of the business.

If a business is unprofitable, outside financing or funds obtained from selling off assets will only delay, and not prevent, its demise. Borrowing to ensure a positive bank balance does not constitute an adequate financial survival plan.

Yet, having more money coming in than going out does not mean a business is healthy, if the revenues show only on financial statements (in terms of receivables) and not in the bank account (in terms of payments actually received).

This chapter examines how a financial plan for an FBO business can positively impact both profit and cash flow in such a way as to provide the basis for long-term viability.

Definitions of Profit

Reward for Effort

Some business owners regard profit simply as what is left after all the bills are paid. It is a bonus on top of their salary. Some do not even take a salary but look on profit, if there is any, to be their compensation. Yet the theoretical economist says that the very reason for the existence of business is the "profit motive." Is the possibility of cash remaining when the bills are paid a sufficient motivation?

Reward for Risk

A more precise definition of profit is reward for the risk of entrepreneurship. It therefore should be ranked with similar rewards from higher or lower risks that can be made with one's time and money. These range from speculating in highly risky but highly profitable investments, to putting one's money into a low-interest savings account.

Return on Investment

Profit measured in this way is examined as a percentage earned on the equity the owner(s) hold in the business. In other words, it is simply net profit divided by cost of the investment or equity. For example, if net profit was $5 and equity was $100, the return on investment would be 5 percent ($5/$100). This ratio indicates the organization's net profit in relation to dollars invested. A business owner should be able to get at least as high a return on total investment as he or she could get by investing elsewhere, or have some acceptable reasons why that rate of return is unavailable.

Profit to Sales Ratio

This ratio is calculated by dividing the net profits of the business by its total sales. For example, if net profits are $5, and net sales are $20, the Profit to Sales Ratio would be 25 percent. It indicates the size of profits in relation to total dollars brought into the company through sales.

Profit ratios should be examined over time, as well as used to compare the company with industry norms. As will be seen below, this is difficult to do, as the aviation service industry has tended to be an industry with few norms.

Profit Objectives

A business person has many possible choices regarding the plan for profits he or she wishes to pursue. The only inappropriate course, which happens more often than it should, would be to not plan for profit at all and just hope for the best.

Profit Maximization

In theory, this is the only profit motive. The point of maximum profit can be calculated by examining the point of diminishing returns for the business's production; that is, that production volume at which the marginal cost of producing one more item exceeds the marginal revenue from selling it.

In reality, however, production increases tend to be step functions. That is, new economies of scale and higher rates of profit might be realizable by expansion. This expansion—such as a new maintenance hangar—tends to come in large units. Also, for any given set of production factors there is a point of diminishing returns. In a labor-intensive business, this might be the point reached when everyone has worked 15 hours a week overtime and is getting too tired to continue working productively. More output will require hiring additional employees, and unit costs will go up again. Maximizing profit could mean doing less, not more.

Satisfactory Profit

More for the sake of more is increasingly less consistent with the lifestyle of many of today's managers, who may seek more leisure rather than more stress. Setting a profit goal in terms of a certain satisfactory dollar level, whether or nor it represents maximum profits, is a choice that many managers may want to make.

Non-Monetary Profit

A variation on the theme of modest profit is that many people in aviation are in it as much for the fun and excitement as for the money.

> An old joke is told about the "mom and pop" FBO owners who won the lottery. "What are you going to do with the winnings?" ask their friends. "We're just going to keep running our FBO till all the money's gone." This is a variation on another "old saying" that "if you want to make a little money by operating an FBO, start with a lot!"

This is fine, except that it does not provide the return on investment one could get elsewhere. But a conscious *reduction* of one's profit goal because of enjoyment of the business is a possible approach.

Hobby/Business

A love of one's occupation to the point where profit is not a concern puts that enterprise in the category of a hobby rather than a business. If one's business is bordering on being a hobby, a consultation with a tax specialist would be in order. Passive loss limitations and other tax consequences may be a factor in a decision to operate in this manner.

Social Responsibility

There is a body of opinion, which appears to be growing as a result of government program cutbacks, that says business has a duty to support the needs of the community. Most managers do show considerable concern for the social environment that affects their employees and the community at large. Some of the things a good manager does perhaps cannot be justified on the grounds of profitability alone or even at all. However, good managers build loyalty and goodwill and help to ensure a long-range future for the airport. Such activities come under the topic of enlightened self-interest, and are a cost of doing business. They do not need to erode profit, but do need to be planned in the budgeting process.

Profit Levels

In planning for profit, what guidelines should an FBO manager use? In addition to the return he or she could obtain through a bank, a manager may want to know how others in industry set their profit goals, or what profits they actually realize.

FBO Reports

When a number of FBOs were asked many years ago what their profits actually were, these were some of the replies:[2]

> "I'm lucky not to be losing money";
> "Hope to break even";
> "One to three percent";
> "Five percent";
> "Ten percent";
> "Fifteen percent"; and
> "Twenty-five percent".

The range of the stated profit levels and an analysis of the respondents' statements suggest several things. Possibly some of these owners did not have good tracking systems and were guessing. One group felt that it was lucky to be breaking even. These people were in the one to five percent rate of return category. These may be new businesses, marginal operations, those experiencing temporary setbacks, or perhaps those that do not know their true profit.

Another group, in the 10 to 15 percent category, attempted to be realistic when comparing the aviation business with other possible investments for their money.

The last group (25 percent) was apparently concerned with the risky nature of the aviation business and felt that higher compensation for risk was appropriate.

The industry has become more business-like since this informal survey was taken; however, today it is difficult to obtain system-wide profit information on FBOs because of confidentiality, the competitive nature of the industry, and the fact that some FBOs still do not keep adequate records to determine their profit.

A person could invest in mutual funds and generate a six to eight percent return over the long term. The 10 to 15 percent rates of return discussed here, are the rates that Fortune companies seek, and are only reasonable, given the hard work and risk involved. So if a person does not think they can make at least 10 percent return on the business investment, then they should seriously consider whether to proceed. This is the profit level that the more effective aviation businesses have traditionally reached. However, recent global economic trends have negatively impacted this estimate and should be taken into account.

What one does with profit is a matter of choice: it can be taken as part of the owner's pay, passed on to employees, distributed to shareholders, reinvested in the business, used for corporate philanthropy, or a blend of all these.

What Is Your Profit?

It seems apparent from this discussion that there may have been successful FBOs making 15 percent or more profit. It also appears that lack of effort to ensure profit may contribute to the failure in reaching such a level. In addition to being the reward for risk, a solid profit margin has the added advantage of providing a cushion, should extra expenses be suddenly required, as well as being available for planned long-term development of the business. Each business must make its own profit target and then, even more important, monitor to see if the desired results are being produced. Profit should be planned in the same way as fixed costs.

Realizing Profit

Profit Orientation

Profit orientation refers to a positive attitude toward profit generation, the effective utilization of managerial tools and techniques in achieving desired profit levels, and the permeation of this orientation throughout the organization.

Key techniques used for staying focused on the dollar aspects of profits include cost control, break-even analysis, financial statement analysis, budgeting, and pricing. These are discussed in subsequent sections.

Cost Control

A major portion of profit orientation is cost containment orientation. The difference between profit and loss in a given time period may not be a higher volume of sales or different services, but simply reducing the cost of each unit of output through recycling, comparison shopping, thrifty approaches to getting work done, and minimization of overhead costs. These steps can be taken in major degree without harming quality.

Planning

Planning for profit is part of the overall business plan; and within the business plan, it is part of the financial plan; another part is cash flow planning. An adequate accounting system to track and report all financial activity in the firm is necessary for this purpose. Software such as "TotalFBO," "QuickBooks" or "FBO Director" should be used; these are discussed more fully in Chapter 7, Management Information Systems.[3]

Landmark FBO employee edits customer data in Microsoft Excel

Marketing Orientation

As discussed in Chapter 3, Marketing, a marketing orientation is also essential to a healthy business and to profits. It includes:

> Understanding the customer's needs, and
> Meeting those needs.

Information System Design

A poor information system design can prevent profits because the manager is not getting information needed to answer certain questions. The manager needs the right information at the right time, although she/he does not need data collection on "everything". With computerized financial management, analysis of even small time periods or small segments of the business is possible with very little effort, and thus the manager should be able to keep needed profit data and trends at his or her fingertips.

Records

Record keeping is the first step in implementation of the information system. Poor-quality or absent records can be a hidden weakness in the system. Examples include:

> Fuel readings not taken and recorded;
> Sales slips not filed;
> Maintenance charges not entered for company aircraft;
> Accounts receivable not organized by age; and
> Balance sheets being available only many months after the period they cover.

With a completely computerized financial system these problems should not arise.

Depreciation Practices

It is possible to treat depreciation of capital items in a number of ways, with each having different long- and short-term tax implications that can affect income after taxes for a given year. Seeking the advice of a Certified Public Accountant who understands aviation is strongly recommended.

Inventory

In calculating the assets and liabilities of the business, inventory is often a major asset. However, it is important to know the true value of inventory, which may not be the same as what was paid for it. Replacement costs may change, and existing stock may deteriorate during storage, due to obsolescence or decay. Storage and insurance costs must also be attributed to existing inventory.

Bad Debts

Not all accounts receivable—another major asset of most businesses—will be collected in full. Total losses, and those where only a percentage is realized because a collection agency is involved, must be subtracted from income. This can have a significant effect on profits. The manager will need to allow some estimate for the level of bad debts, based on past history and on policy.

Managerial Decisions

There are numerous other areas in which managers can affect profit without necessarily even realizing it. These may include:

1. Excess allocation of overhead could turn a profitable department into an unprofitable department.
2. A department receiving labor or supplies at cost is receiving a cash benefit. The department supplying it is losing because the same goods and services could have been sold outside the business, at full markup.
3. "Pot mixing" or combining resources from several departments for a special project can make it hard to allocate overhead costs by department.

Cash Flow

The U.S. has shifted to a credit economy. Accepting credit cards for payments will cost the business owner a few percentage points, which are usually passed on to the customer, and will create a slight lag in payments compared with cash.

Allowing customers to have accounts that are invoiced monthly is a greater source of cash flow lags. When payments lag 30 to 90 days after delivery of goods and services, the resulting cash flow lag may require outside financing. If so, the cost of money can erode profits. However, some customers, longstanding and/or large, insist on company credit, and you may still need to be prepared for this.

Setting Your Cash Flow and Profit Goals
Part 1—Planning for Positive Cash Flow

Forecasting Sales and Revenues

For each product and service, an annual forecast of units sold must be made. This will be based partly on the past year's performance, partly on market research about needs and trends in the area, and partly on expectations about the economy. The aviation business manager should consider both national and local projections before deciding on the next year's volumes of business for each of his/her lines of activity, as was discussed in Chapter 3, Marketing.

After predicting volume for each item, the price must be multiplied by volume to get expected revenue for each line, and total revenue must be calculated. (Pricing was reviewed in depth in Chapter 3 and will not be discussed here.) Sales by month should be estimated, using seasonal factors common to the area. December, January, and February tend to be much slower aviation months nationwide than during the rest of the year, except in certain resort areas, and most activity shows peaking over the summer months.

Forecasting Expenses

Expenses are generally defined as *fixed*, which have to be met merely for the business to open its doors, or *variable*, which increase as sales volume increases. Fixed expenses or overhead include such items as:

> Rent;
> Utilities;
> Property taxes;
> Office supplies;
> Labor;
> Professional dues and publications;
> Janitorial and routine maintenance;
> Advertising (part);
> Inventory (part); and
> Insurance.

Of course, when inflation is more than zero, so-called fixed expenses also will tend to increase, and while some are predictable, such as rent, governed by escalator clauses in a lease, some are not. A worst-and best-case figure may be appropriate.

Some so-called fixed costs actually do increase because of growth. For example, more space or inventory may be needed. Heavy use of office machinery may cause it to wear out faster, or more inventory may be needed. More personnel often means new office furniture, another computer and other overheads, and if overhead costs are figured on top of the payroll cost of a new person, it may be two to three times his or her salary. The costs of expansion must always be calculated before deciding whether it is feasible. This is discussed in a subsequent section, "Break-Even Analysis."

Expenses and sales of all types should be calculated at least on a monthly basis, and can readily be calculated more frequently if desired. This should be done allowing for seasonality of production needs and variations in overheads, such as utility costs.

Month-by-Month Cash Flow Analysis

Particularly in a new business, but also in any business where cash flow lags or is erratic, a monthly cash flow for at least a year is a very valuable tool.

Figure 4.1 depicts a sample cash flow analysis for a very small company taking in $8,000 to $12,000 per month. The cash position shown on the bottom line indicates whether and when the firm will encounter a cash flow problem and how great the need for cash will be. It can also predict for a new business how many months will pass before the firm has a permanent positive cash flow. Some businesses swing from negative to positive cash flow and back each year. Such may be the case with a small airline serving an island resort, for example. Almost all the revenue is made between May and September. During the rest of the year the revenue is less, because of much less activity producing much less income.[4] Borrowing and repayment are part of the year's plan.

Figure 4.1 provides a dramatic illustration of how a business can show a profit ($3,400 over seven months) yet still encounter serious cash flow problems (a cash flow "hole" of $4,750 by April).

In order to test a variety of cash flow scenarios, a spreadsheet program such as Excel may be used to change one figure at a time and automatically recalculate the bottom line. It is important to use realistic assumptions in order to create realistic expectations. It is also a good idea to make pessimistic assumptions

Figure 4.1 ✈ Monthly Cash Flow Chart

	Jan.	Feb.	March	April	May	June	July	7 months
Cash In								
Product 1	2,000	6,000	3,000	4,000	5,000	5,000	5,000	30,000
Product 2	1,000	2,000	1,000	2,000	2,000	3,000	3,000	14,000
Product 3	5,000	5,000	4,000	4,000	5,000	5,000	5,000	33,000
Total cash in	$8,000	13,000	8,000	10,000	12,000	13,000	13,000	77,000
Cash Out								
Materials	2,000	2,500	1,500	1,000	1,000	1,000	1,000	10,000
Labor	3,000	3,000	3,000	3,000	3,000	3,000	3,000	21,000
Other inventory	500	600	700	400	500	500	500	3,700
Subtotal variables	5,500	6,100	5,200	4,400	4,500	4,500	4,500	34,700
Management	3,000	3,200	3,200	3,200	3,200	3,200	3,200	22,200
Rent	1,500	1,500	1,500	1,500	1,500	1,500	1,500	10,500
Utilities	250	250	250	250	250	250	250	1,750
Marketing/ads	200	1,000	1,000	750	1,000	250	250	4,450
Subtotal fixed	4,950	5,950	5,950	5,700	5,950	5,200	5,200	38,900
Total cash out	10,450	12,050	11,150	10,100	10,450	9,700	9,700	73,600
Net Cash Out	(2,450)	950	(3,150)	(100)	1,550	3,300	3,300	3,400
Starting Cash	–0–	(2,450)	(1,500)	(4,650)	(4,750)	(3,200)	100	–0–
Cumulative Cash or Deficit "hole"	(2,450)	(1,500)	(4,650)	(4,750)	(3,200)	100	3,400	3,400

about payment lags, bad debts, and costs, in order to see the difference in cash position between that and the most likely case.

Improving the Cash Position

In order to improve cash flow lags, a firm can:

> Seek as many low rate credit cards as possible for the company and use these to be the company's "banker";
> Obtain cash on the spot for more services;
> Collect deposits and advance payments for a percentage of the cost (e.g., collect a deposit on a charter flight rather than waiting till the customer returns to get paid);
> Require larger deposits on aircraft sales;
> Seek regular advance deposits from regular customers;
> Offer memberships paid in advance of service delivery, such as annual flying dues;

> Apply interest charges on accounts over 30 days past due;
> Require flight instruction courses to be paid in advance;
> Consider factoring receivables—selling the accounts receivable at a discount;
> Borrow against receivables.

A firm can also delay its own cash outflow by such techniques as:

> Buying supplies on consignment—which are only paid for, if and when sold; and
> Negotiating more favorable terms for purchases, including discounts and extended payments.

However, putting off payment will not cure the problems in a firm that never has a positive cash flow and is basically operating at a loss. Payments should only be deferred until the cash position is expected to improve. "Robbing Peter to pay Paul" is a slippery slope.

Part 2—Planning for Profits

Profit Objectives

The desired profit in percentage terms should be converted into dollar terms for the year, so that a precise profit figure is part of the budget.

Break-Even Analysis

Break-even analysis is the tool that identifies whether doing more business will result in more profit, or a loss. The first few units sold by a business probably do not bring in enough to cover even fixed costs. As sales go up, first variable costs, and subsequently fixed costs, are covered. The point at which all costs (variable plus fixed) are exactly covered (but no more) is the break-even point. Beyond that point, all revenue is profit, as long as fixed costs do not increase. This is shown in Figure 4.2.

A break-even level of sales can also be calculated after insertion of a figure for profit. In this case, profit is treated just like any other expense:

$$\text{Break-Even Sales} = \text{Fixed Costs} + \text{Variable Costs} + \text{Profit}$$

When a business owner is trying to decide whether to expand, it is apparent that any expansion requiring only higher variable costs will be profitable because each extra sale must cover only the extra or marginal cost of producing that sale, not the full cost. For example, an aviation manager plans to sell model airplanes at a convention. He can purchase the planes at $2.00 each on consignment (with the privilege of returning all unsold models). The booth rental is $800.00 and payable in advance. He feels that he can sell the aircraft models at $4.00 each. Assuming he is not going to count any labor costs, how many models must be sold to break even (zero profit)?

$$\text{Break-Even Sales} = \text{Fixed Costs} + \text{Variable Costs} + \text{Profit}$$

Let x = number of units to be sold to break even

$$\$4.00x = \$2.00x + \$800.00 + 0$$
$$2.00x = 800 + 0$$
$$X = 800 / 2$$
$$X = 400 \text{ units}$$

Or in dollars:

$$X = \$1600.00$$

Besides the Break-Even Point, three other important terms are the Contribution Margin, Unit Contribution Margin, and Contribution Margin Ratio. The Contribution Margin is what's left of revenue to offset fixed costs after variable costs have been subtracted:

$$\text{Contribution margin} = \text{revenue} - \text{variable costs}$$

Figure 4.2 → **Break-Even Chart**

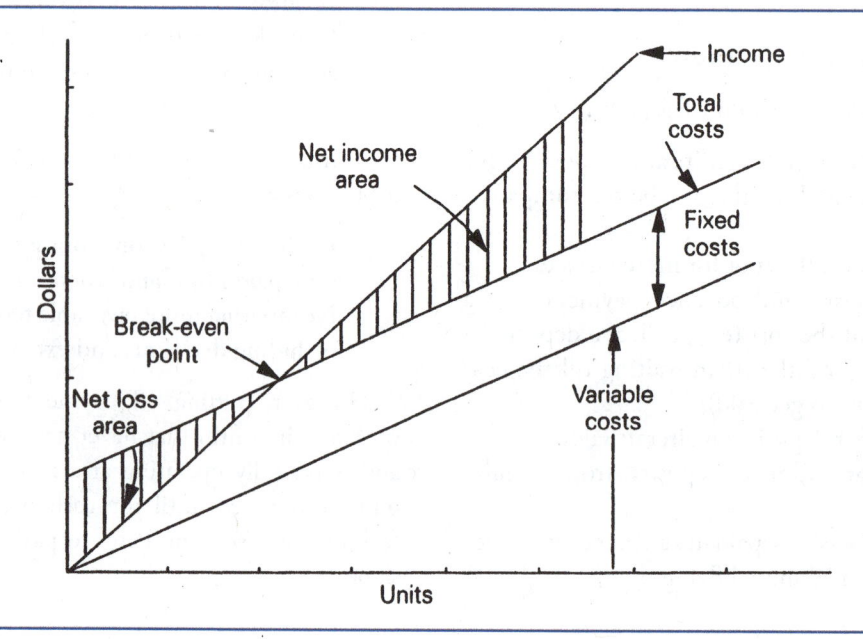

The Unit Contribution Margin is the unit selling price less the unit variable cost. The Contribution Margin Ratio is the unit contribution margin divided by the selling price. In the case just described it was .5 ($2.00/$4.00).

The Break-Even Point (BEP) in units is fixed costs / unit contribution margin.

Break-even analysis can be used for a single product line or department, or for the business as a whole. It can apply to:

> Fuel servicing activity;
> Billed service hours of a maintenance department;
> Instructional hours of a flight department;
> Activity volume of a parts department;
> Revenue hours of air charter or air taxi work;
> Route planning; and
> Acquisition of new operating equipment.

A more complex example is the decision of whether or not to start a new flight school operation. The FBO has the following estimates:

Fixed expenses:

Building lease, utilities and taxes	$21,600
Aircraft insurance and depreciation	$18,000
Flight instructor base pay	$15,000
Subtotal, fixed expenses	$54,600

Variable expenses:

Aircraft operating expenses—$60/hr, for 2600 hours	$156,000
Instructor commission, $30/flight hour	$78,000
Classroom time, $30 per classroom hour, 200 hours	$6,000
Supplies/materials $300/student, 75 students	$22,500
Subtotal, variable expenses	$262,500
Total costs,	$317,100

Estimated income

Ground School, 75 students @ $1,050	$78,750
Flying Lessons, 2600 hours @ $135.00	$351,000
Instruction Kit, 75 @ $600	$45,000
Total income,	$474,750

Using these figures we can examine the break-even point for each product:

Ground school selling price, $78,750/200 hours; per hour	$393.75
Average variable cost per hour	$30.00
(Unit) Contribution Margin	$363.75
(Unit selling price of $393.75 less unit selling cost of $30)	

Break-Even Point = Fixed Expenses/ Contribution Margin
$54,600/$363.75 = 150.09 hours
150.09 hours × $393.75 = $59,099.44

Thus, the ground school alone could pay for all fixed costs and for its own variable costs if $59,099 worth of enrollments or 56 students were mustered, compared with an estimated sales potential of $78,750 or 75 students.

Suppose the FBO decides to examine only added flying lessons, without the ground school.

Selling Price per Hour	$135.00
Average variable cost per hour ($60 + $30)	$90.00
Contribution Margin	$45.00

Break-Even Point = Fixed Costs/Contribution Margin
$54,600/$45.00 = 1,213 hours
1,213 × $135.00 = $163,755.00 income

Thus, the flying lessons must make $164,000 (rounded) to break even, compared with an estimated $351,000.00 in prospective sales.

Suppose now that all fixed costs are double the previous estimate, without changing the price:

Break-Even Point = Fixed Costs/Contribution Margin
$109,200/$45.00 = 2,427 hours
2,427 × $135.00 = $327,646

This is a much more marginal proposal because 93 percent of forecast sales must actually be achieved just to break even, which leaves very little room for sickness, bad weather, lack of demand, and other factors beyond the FBO's control.

Profitability Variations among Product Lines

In the above example, the ground school was more profitable than the flying lessons because one instructor was handling one class of 75 students, whereas the flying lesson is a one-on-one situation, with substantial aircraft and instructor costs. It is possible to take all fixed expenses and allocate them among the divisions of the business and estimate which areas are most profitable. For example, one FBO manager may decide that since there is ample extra space in the front desk area, he or she is going to emphasize the sale of aviation books and accessories, for which the markup can be 100 per cent over wholesale for relatively little variable cost in handling. Another aviation business may see the strongest profit line in aircraft sales, for which even a few percentage points in commissions can mean many dollars because of the high value of the item. An FBO manager may decide not to even compete in such areas as specialty avionics because there is another operator on the field and the cost of startup and getting established in the market would be excessive, for limited profit. Whatever the manager's inclinations, break-even analysis will provide the tools to find out whether something is really worth doing, from a financial standpoint. Both break-even and cash flow analysis should be applied to all prospective new ventures by using conservative assumptions. Keep in mind the "garbage in, garbage out" rule; that is, one's results are only as useful and accurate as the quality of one's assumptions and input data.

Profit Centers

Owing to different profitability levels, it may be desirable to divide the various functions of the business into separate profit centers, each having its own profit goal. This avoids hidden cross-subsidies among areas, and also permits conscious subsidy of certain items that may persuade people to make greater purchases (e.g., free aircraft detailing, coffee, air, and other items).

In practical terms, setting up profit centers means that each revenue-generating area of the business will be treated as its own mini-business. Each must carry a share of the company's overhead or fixed costs (e.g., rent, heat, and telephones). Each may also perform many staff functions such as reception, Information Technology (IT) specialist, bookkeeper, and CEO.

How this overhead is allocated should be examined as annual budgets are set. Under the profit center model, one department performing work for another must also then be billed internally, just as an external customer would be billed.

Profit and Loss Statement/Income Statement

This is the operating income statement that represents the picture annually, monthly, or quarterly. A skeleton income statement is shown in Figure 4.3.

The gross margin is calculated as sales less cost of goods sold before any of the internal or overhead costs

Figure 4.3 ✈ Income Statement

Run: 10/28/2009 10:40AM

Horizon Business Concepts, Inc.
General Ledger Monthly Budget Review
For The Year: 1/01/2009 Through: 12/31/2009

Page: 1

Income

Account	Description	01/2009	02/2009	03/2009	04/2009	05/2009	06/2009	07/2009	08/2009	09/2009	10/2009	11/2009	12/2009	Total
4000.00	Parts Sales	1000	1000	1000	1000	1000	1000	1000	1000	1000	1000	1000	1000	12000
4010.00	Fuel Sales	5000	5000	5000	5000	5000	5000	5000	5000	5000	5000	5000	5000	60000
4020.00	Oil Sales	150	150	150	150	150	150	150	150	150	150	150	150	1800
4060.00	Solo Aircraft Rental Revenu	2000	2000	2000	2000	2000	2000	2000	2000	2000	2000	2000	2000	24000
4080.00	Charter Revenue	4000	4000	4000	4000	4000	4000	4000	4000	4000	4000	4000	4000	48000
4100.00	Flight Store Revenue	500	500	500	500	500	500	500	500	500	500	500	500	6000
4180.00	Shop Labor Revenue	5000	5000	5000	5000	5000	5000	5000	5000	5000	5000	5000	5000	60000
4311.00	GSE Service	500	500	500	500	500	500	500	500	500	500	500	500	6000
	Total Income:	18150	18150	18150	18150	18150	18150	18150	18150	18150	18150	18150	18150	217800

Cost of Sales

Account	Description	01/2009	02/2009	03/2009	04/2009	05/2009	06/2009	07/2009	08/2009	09/2009	10/2009	11/2009	12/2009	Total
5000.00	Parts Cost Of Goods Sold	750	750	750	750	750	750	750	750	750	750	750	750	9000
5000.60	Cost Of Pilot Supplies Sold	250	250	250	250	250	250	250	250	250	250	250	250	3000
5010.00	Fuel Cost Of Goods Sold	3500	3500	3500	3500	3500	3500	3500	3500	3500	3500	3500	3500	42000
5020.00	Oil Cost Of Goods Sold	100	100	100	100	100	100	100	100	100	100	100	100	1200
5050.00	Charter Costs	3000	3000	3000	3000	3000	3000	3000	3000	3000	3000	3000	3000	36000
	Total Cost of Sales:	7600	7600	7600	7600	7600	7600	7600	7600	7600	7600	7600	7600	91200

Expense

Account	Description	01/2009	02/2009	03/2009	04/2009	05/2009	06/2009	07/2009	08/2009	09/2009	10/2009	11/2009	12/2009	Total
7110.10	Salaries & Wages - Shop	3000	3000	3000	3000	3000	3000	3000	3000	3000	3000	3000	3000	36000
7110.60	Salaries & Wages - Flight S	1000	1000	1000	1000	1000	1000	1000	1000	1000	1000	1000	1000	12000
7140.00	Taxes-FICA	500	500	500	500	500	500	500	500	500	500	500	500	6000
7150.00	Taxes-Medicare	100	100	100	100	100	100	100	100	100	100	100	100	1200
7210.00	Utilities	200	200	200	200	200	200	200	200	200	200	200	200	2400
7220.00	Telephone Expense	200	200	200	200	200	200	200	200	200	200	200	200	2400
7420.60	Aircraft Maintenance Expen	1500	1500	1500	1500	1500	1500	1500	1500	1500	1500	1500	1500	18000
	Total Expense:	6500	6500	6500	6500	6500	6500	6500	6500	6500	6500	6500	6500	78000
	Net Budgeted Profit:	4050	4050	4050	4050	4050	4050	4050	4050	4050	4050	4050	4050	48600

Courtesy of TotalFBO Accounting and Business Management Software by Horizon Business Concepts, Inc.

(usually fixed costs) have been subtracted. The net profit (or loss) is what is left after all expenses have been subtracted.

Balance Sheet

A balance sheet tells not only about the operating account, but also about the capital account, as shown in Figure 4.4. It tells "What you own and what you owe." Both the operating statement and the balance sheet need to be projected two to five years ahead. A company can have a healthy-looking balance sheet and still have cash flow problems if its assets are tied up in plant and equipment needed for the business operation and not readily liquidated.

Part 3—Budgeting

Introduction

Popular misconceptions about budgeting can detract from its use and effectiveness. Some individuals feel that a budget is like a straitjacket, while others feel that it wastes time; and still others feel that it is a mechanical, futile activity. Those who have experienced the practice in some organizations of heavy end-of-year spending in order to make the budget appear accurate or to ensure they get at least as much in the next cycle ("use it or lose it") may lack confidence in the process. Most of these misconceptions are the result of personal experiences in situations where budgeting was not handled properly or where the individual may have overreacted to the process.

Budgeting involves setting expenditure limits for each unit of the business or cost center. These limits are based on the projected sales and profit desired for the line departments, and include overhead as a percentage of business costs or sales for the administrative or staff departments. Budgets can be changed in the course of the year if circumstances warrant, and break-even analysis can help decide if a budget should be increased to capture a new business opportunity. The budget need not be a straitjacket, but should achieve the following:

> Set specific goals for each department and the firm as a whole;
> Establish limits within which managers know they are to operate;
> Establish company needs and priorities during the budget period;
> Foresee potential operational problems and allow preventive action;
> Provide management with increased flexibility;
> Provide management with an early warning system if costs are getting too high or revenues too low;
> Inform line departments about improvements planned for later years;
> Provide standards to measure performance and results; and
> Enable intended profits to be actually realized.

Aviation Budget Development

Developing a budget for an aviation business, like any other business, is really profit planning. From a practical point of view, it is a technique for managing the business. It involves the formation of definite plans (budgets) for a limited future period. These plans are normally expressed in financial terms. A complete budget develops and includes standards to measure and evaluate the actual performance of management and the business.

In order for a budget to serve as an effective managerial tool, several conditions should exist. It is perhaps the absence of these conditions that creates the misconceptions and apprehension previously mentioned. Advocates of the budgeting process suggest that:

1. A suitable organizational structure for the business activity should exist.
2. An adequate accounting system should be in operation.
3. The budget should have the interest and support of top management.

Preliminary preparations for the budget, already discussed in planning for profit, include:

> Forecasting business activity;
> Preparing budget proposals;
> Assembling and approving the budget;
> Using supporting budgets; and
> Designing budget operation and control.

Steps in Budget Development

Preparing for the budget. The first and most important preparations include setting goals, defining objectives, and creating long-range plans for

Figure 4.4 ✈ Balance Sheet

Run: 10/27/2009
5:03PM

Horizon Business Concepts, Inc.
General Ledger Balance Sheet Report
As Of: 5/31/2009
All Departments Consolidated

Page: 1

Assets

Current Assets

State Bank Checking #780017315	18,774.94	
Petty Cash	425.00	
Undeposited Receipts/Cash On Hand	6,923.22	
Accounts Receivable	27,992.45	
A/R Credit Cards	244.72	
Contract Fueling Receivable	3,819.42	
Employee Advances	450.00	
Parts Inventory	13,785.35	
Fuel Inventory	196,377.63	
Oil Inventory	307.43	
Aircraft Inventory	31,000.00	
Total Current Assets	300,100.16	300,100.16

Fixed Assets

Furniture And Fixtures	17,500.00	
Automobiles	68,000.00	
Computer System	17,000.00	
Computer Software Costs	3,989.00	
Leasehold Improvements	18,900.00	
Total Fixed Assets	125,389.00	125,389.00

Total Assets:		425,489.16

Liabilities

Current Liabilities

Accounts Payable	175,944.91	
Unearned Revenue (Customer Deposits)	780.72	
N/P - Short-term	(468.65)	
Employee Benefits Payable	(186.00)	
Payroll Taxes/Premiums Payable	1,527.09	
Sales Tax Liability	1,448.95	
Fuel Taxes/Fees Payable	317.41	
Charter Taxes/Fees Payable	1,266.38	
Landing Fees Payable	49.00	
Total Current Liabilities	180,679.81	180,679.81

Long-Term Liabilities

N/P - Long-term	14,708.93	
Suspense	182.00	
Total Long-Term Liab	14,890.93	14,890.93

Total Liabilities:		195,570.74

Figure 4.4 ✈ **Balance Sheet —Cont'd**

```
Run: 10/27/2009                Horizon Business Concepts, Inc.                   Page:    2
     5:03PM                   General Ledger Balance Sheet Report
                                     As Of: 5/31/2009
                                 All Departments Consolidated
```

Equity

Owner's Equity
Owner Capital	500,000.00	
Retained Earnings	(250,000.00)	
Current Earnings	(20,081.58)	
Total Owner's Equity	229,918.42	229,918.42

Total Equity:	229,918.42

Total Liabilities and Equity:	425,489.16

Notes:
　All Departments Consolidated.

Courtesy of TotalFBO Accounting and Business Management Software by Horizon Business Concepts, Inc.

---

the organization. As discussed in Chapter 2, Management Functions, under planning, it is essential that the manager have a good grasp of the present organizational situation. He or she then develops a plan that is complete with goals and objectives. These goals and objectives should then be related to financial and other specific goals for the coming year. The objectives that are developed should express the desired results in specific measurable terms. Developing the long-range plan is considered an important prelude to the development of the budget, for there is obviously a very close correlation between the two. A budget may be considered the first year of a long-range plan—the implementation of that plan.

Forecasting business. Chapter 3, Marketing, discussed other business forecasting issues and sources of published aviation forecasts. What follows here is in brief.

Next, the manager develops some careful estimates or forecasts of business conditions that are expected to exist during the budget period. Although all external conditions are important, the critical task is the selection of conditions facing the individual business. Many managers feel that this is the chief difficulty in budgeting. They indicate major problems in: (1) obtaining meaningful figures in advance; and (2) making good use of the information available. Forecasting future business is a fundamental and necessary process that must be accomplished for all businesses. It is not, however, an activity approached enthusiastically by all managers, partly because of unfamiliarity with the procedure, and partly due to apparent failures of past attempts. It is as much art as science; that is, the assumptions made cannot all be derived and defended technically. However, they should be based on solid research and spelled out. Then if conditions change, it will be clear what assumptions underlying the calculations need to be changed.

It is believed that more meaningful figures can be developed by using a procedure that considers:

> Identification of the general economic situation;
> Key regulatory factors that influence the business environment;
> Recent industry developments;
> Specific trend and forecast data that might be acquired;
> Evaluation of business opportunities that may exist;
> Local economic trends and factors; and
> Competitive elements facing the business.

Several iterations will likely be required.

Budget preparation. With the forecast data in hand, the manager's next (and very important) step is the preparation of the budget. The development of the detailed blueprints showing how the company intends to reach the desired goals and objectives in the business conditions identified by the forecast is best accomplished in the following sequence:

1. Set a timetable for activity.
2. Develop a sales budget.
3. Formulate a purchases budget.
4. Develop an expense budget.
5. Formulate the income statement budget.
6. Calculate a balance sheet budget.

Timetable. It is very important to prepare a timetable for the development of a budget. Such a schedule is necessary because budgeting is a planning activity and is very likely to be postponed if not established on a fairly regular schedule. The following illustrates a timetable for a small company budget based on a calendar-year cycle. Larger companies, with more review layers, will need to start earlier. Many public agencies using a calendar year budget cycle start their annual budget process in about April of each year, aiming at adoption by Thanksgiving. Of course, it should be recognized that a budget is a living, flexible document that must meet the variable conditions facing the business enterprise. The possibility and reality of change must become part of the timetable and the budgeting process.

Date	Activity
15 October	Develop plans and objectives
1 November	Complete business forecast
10 November	Develop preliminary budget
15 November	Final budget review
1 December	Budget acceptance and approval
1 January	Initial implementation
As required	Budget modification

Sales budget. The manager begins the development of this budget by working on the sales estimate. Since most of the activities of a business are geared to the level of expected sales, budget preparation begins with sales forecasting. The business forecasting procedures mentioned earlier should provide the basic material needed to start this first step. In some instances, it may even provide the exact information for a sales budget, if the historical information is available and the forecast has been prepared in adequate detail. The initial step is a thorough market analysis. Normally, this is composed as an analysis of the overall marketplace (as is outlined in the earlier section on forecasting), estimates by salespersons of the market situation, general sales trends in the industry and the area, and a study of previous years' sales.

The second step is the development of the company market forecast. Using a great deal of the data acquired in step 1, a forecast is prepared for the company's market potential. The structure of the market in terms of types, units, time, and prices is set forth.

The third step involves pricing out the company forecast in order to acquire the expected total sales (revenue) of the business. Adjustments are made to reflect the goals and objectives of the business, as well as the market situation.

The next step is the breakdown of total sales into the departments or work centers of the business. The expected incomes for parts, service, flight line, aircraft sales, instruction, and miscellaneous are estimated for the accounting year. Naturally, this reflects the forecast data used in the development of total sales.

The final step is the development of a monthly sales budget for each of the identified departments in the business. This enables the manager to identify and anticipate the monthly variations in sales due to seasonal and other fluctuations.

Figures 4.5 and 4.6 illustrate worksheets that may be used to calculate the sales budget for an aviation business. Figure 4.5 deals with the total and departmental sales budget, while Figure 4.6 shows the monthly budget report with budgeted activity, actual results, and variances (deviations) from that budget. In using Figure 4.5, the previous year's budget and actual sales are filled in for the total business and the departmental components. Using this as a guide, along with the other elements for developing a forecast, the current year budget is projected.

Purchases budget. Next is the development of the purchases budget. In order to achieve the desired sales level, it will obviously be necessary to purchase the products that will be sold to meet the sales budget goals. Figure 4.7 illustrates a worksheet for developing a purchases budget. Note that in calculating anticipated purchases, it is necessary to first consider dollar volume of sales. Then, the number of units or quantity of material needed to arrive at the right sales

Figure 4.5 → Annual Sales Budget Worksheet

	Previous Year Budget	Previous Year Actual	Current Year Budget
Total Sales of Company			
Aircraft sales (300)			
Parts Sales (400)			
Service Sales (500)			
Flight Sales (600)			
Line Sales (700)			
Misc. Activity (800)			

volume is calculated by using the markups for the various product areas. These annual purchase figures (which are expressed in dollars and units of quantity) are then divided into monthly allocations.

Expense budget. The manager next constructs an expense budget. Figure 4.8 illustrates a worksheet that may be used in this process. In this instance, the expense budget categories are identified as employment, aircraft, occupancy, and other. These four major groups are divided into subcategories, which are the expense accounts contained in the ledgers of the accounting system. In general, these expenses are incurred by operating the business. The worksheet is completed, and filling in previous-year expense figures and then projecting the current-year expenses forms the budget. Of course, this projection is based upon anticipated sales and the effect upon expenses required to support the sales. The annual expense for each account is divided into a monthly expense item according to an appropriate system. It may be a straight percentage or it may be based on workload, number of people, borrowing schedule, marketing plan, or some other factor according to the elements that dictate or control the expenditures.

Income statement budget. The next key step in developing a complete budget is bringing the sales and expenses together into an income statement budget. Figure 4.9 suggests an income statement budget worksheet. This form follows the income statement format and uses it for budgeting purposes with both previous year and current year figures. Sales, purchases, and expenses all come from the completed worksheets in those areas. The basic objective of this worksheet is to bring them all together with the primary focus on profit. At this stage, the goal is profit planning-budgeting for the desired net income for the operating period. This is the ultimate objective of budgeting: to identify planned operating profit.

Balance sheet budget. The final step in the budgeting process is to transfer the overall outcome of the operating budget to the balance sheet and to develop a projection for the total financial condition of the company. By using a worksheet comparable to those for sales, purchases, and expenses, a budget can be constructed to show anticipated changes in assets, liabilities, and net worth. Desired business activities will influence accounts in each of these areas, and a budget will assist in testing the feasibility of such action.

Figure 4.6 → **Monthly Budget Report**

Run: 10/28/2009
10:42AM

Horizon Business Concepts, Inc.
One-Year Budget Performance Report
All Departments Consolidated

Page: 1

Account	For the Month Ending: 5/31/2009 Actual	Budgeted	Variance	Pct	Year-To-Date: 1/01/2009- 5/31/2009 Actual	Budgeted	Variance	Pct
Income								
4000 Parts Sales	1,578.26	1,000.00	578.26	57.8 %	15,059.44	5,000.00	10,059.44	201.2 %
4010 Fuel Sales	2,982.75	5,000.00	-2,017.25	-40.3 %	26,995.81	25,000.00	1,995.81	8.0 %
4011 Minimum Fueling Surcharge	17.22	0.00	17.22	0.0 %	72.92	0.00	72.92	0.0 %
4015 Into-Plane Fees	816.66	0.00	816.66	0.0 %	2,571.17	0.00	2,571.17	0.0 %
4020 Oil Sales	6.63	150.00	-143.37	-95.6 %	26.43	750.00	-723.57	-96.5 %
4050 Miscellaneous Shop Supplies Sales	126.98	0.00	126.98	0.0 %	355.43	0.00	355.43	0.0 %
4060 Solo Aircraft Rental Revenue	260.64	2,000.00	-1,739.36	-87.0 %	26,036.80	10,000.00	16,036.80	160.4 %
4070 Dual Aircraft Rental Revenue	658.60	0.00	658.60	0.0 %	24,225.50	0.00	24,225.50	0.0 %
4080 Charter Revenue	12,085.80	4,000.00	8,085.80	202.1 %	21,128.01	20,000.00	1,128.01	5.6 %
4081 Charter Income - Aircraft Standby	1,190.00	0.00	1,190.00	0.0 %	2,545.00	0.00	2,545.00	0.0 %
4082 Charter Revenue - Pilot Standby	150.00	0.00	150.00	0.0 %	740.00	0.00	740.00	0.0 %
4083 Charter Revenue - Extra Charges	0.00	0.00	0.00	0.0 %	35.00	0.00	35.00	0.0 %
4085 Charter Revenue - Catering	260.00	0.00	260.00	0.0 %	665.00	0.00	665.00	0.0 %
4086 Charter Revenue - Other Expenses C	300.00	0.00	300.00	0.0 %	620.00	0.00	620.00	0.0 %
4090 Headset Rental Revenue	39.50	0.00	39.50	0.0 %	130.00	0.00	130.00	0.0 %
4100 Flight Store Revenue	7.50	500.00	-492.50	-98.5 %	48.35	2,500.00	-2,451.65	-98.1 %
4130 Hangar And/or Tie Down Revenue	385.00	0.00	385.00	0.0 %	1,155.00	0.00	1,155.00	0.0 %
4140 Instruction - Primary Revenue	226.10	0.00	226.10	0.0 %	16,107.80	0.00	16,107.80	0.0 %
4150 Instruction - Advanced Revenue	78.40	0.00	78.40	0.0 %	4,123.60	0.00	4,123.60	0.0 %
4160 Ground School Revenue	111.15	0.00	111.15	0.0 %	288.75	0.00	288.75	0.0 %
4170 Sublet Work Revenue	2,000.00	0.00	2,000.00	0.0 %	3,250.00	0.00	3,250.00	0.0 %
4175 Outside Repairs Revenue	-2,000.00	0.00	-2,000.00	0.0 %	-1,925.00	0.00	-1,925.00	0.0 %
4180 Shop Labor Revenue	2,232.50	5,000.00	-2,767.50	-55.4 %	27,628.50	25,000.00	2,628.50	10.5 %
4200 Off-Site Fuel Credit Revenue	-138.35	0.00	-138.35	0.0 %	-173.44	0.00	-173.44	0.0 %
4230 Flight Package Sales	0.00	0.00	0.00	0.0 %	15,825.00	0.00	15,825.00	0.0 %
4311 GSE Service	0.00	500.00	-500.00	-100.0 %	0.00	2,500.00	-2,500.00	-100.0 %
4900 Finance Charge Revenue	177.44	0.00	177.44	0.0 %	384.11	0.00	384.11	0.0 %
4930 Miscellaneous Revenue	50.00	0.00	50.00	0.0 %	100.00	0.00	100.00	0.0 %
4950 Freight Revenue	232.75	0.00	232.75	0.0 %	1,061.75	0.00	1,061.75	0.0 %
Total Income:	23,835.53	18,150.00	5,685.53	31.3 %	189,080.93	90,750.00	98,330.93	108.4 %
Cost Of Sales								
5000 Parts Cost Of Goods Sold	1,006.73	1,000.00	6.73	0.7 %	11,178.47	5,000.00	6,178.47	123.6 %
5010 Fuel Cost Of Goods Sold	1,939.17	3,500.00	-1,560.83	-44.6 %	18,777.42	17,500.00	1,277.42	7.3 %
5020 Oil Cost Of Goods Sold	1.98	100.00	-98.02	-98.0 %	3.11	500.00	-496.89	-99.4 %
5040 Purchases Discounts	-9.14	0.00	-9.14	0.0 %	-9.14	0.00	-9.14	0.0 %
5050 Charter Costs	2,057.93	3,000.00	-942.07	-31.4 %	5,352.06	15,000.00	-9,647.94	-64.3 %
5070 Fuel & Oil Expense - Flight School Ac	1,005.45	0.00	1,005.45	0.0 %	2,630.99	0.00	2,630.99	0.0 %
5080 Ramp Vehicle Fuel/Oil Cost	0.00	0.00	0.00	0.0 %	28.89	0.00	28.89	0.0 %

Figure 4.6 → Monthly Budget Report—Cont'd

Run: 10/28/2009
10:42AM

Horizon Business Concepts, Inc.
One-Year Budget Performance Report
All Departments Consolidated

Page: 2

Account	For the Month Ending: 5/31/2009 Actual	Budgeted	Variance	Pct	Year-To-Date: 1/01/2009- 5/31/2009 Actual	Budgeted	Variance	Pct
Total Cost Of Sales:	6,002.12	7,600.00	-1,597.88	-21.0 %	37,961.80	38,000.00	-38.20	-0.1 %
Gross Profit:	17,833.41	10,550.00	7,283.41	69.0 %	151,119.13	52,750.00	98,369.13	186.5 %

Expense

Account	Actual	Budgeted	Variance	Pct	Actual	Budgeted	Variance	Pct
6160 Advertising Expense	0.00	0.00	0.00	0.0 %	87.50	0.00	87.50	0.0 %
6170 Freight Expense	460.90	0.00	460.90	0.0 %	2,110.55	0.00	2,110.55	0.0 %
7100 Administrative Salaries Expense	7,025.24	0.00	7,025.24	0.0 %	44,478.72	0.00	44,478.72	0.0 %
7110 Salaries/Wages Expense	17,737.02	4,000.00	13,737.02	343.4 %	91,329.53	20,000.00	71,329.53	356.6 %
7130 Payroll/Employee Benefit Expenses	1,998.28	0.00	1,998.28	0.0 %	5,992.91	0.00	5,992.91	0.0 %
7140 Taxes-FICA	1,400.52	500.00	900.52	180.1 %	3,982.65	2,500.00	1,482.65	59.3 %
7150 Taxes-Medicare	327.56	100.00	227.56	227.6 %	931.44	500.00	431.44	86.3 %
7160 Taxes-FUTA	144.84	0.00	144.84	0.0 %	482.46	0.00	482.46	0.0 %
7170 Taxes-SUTA	523.19	0.00	523.19	0.0 %	1,859.70	0.00	1,859.70	0.0 %
7200 Building/Office Rent Expense	1,785.00	0.00	1,785.00	0.0 %	3,570.00	0.00	3,570.00	0.0 %
7210 Utilities	0.00	200.00	-200.00	-100.0 %	0.00	1,000.00	-1,000.00	-100.0 %
7220 Telephone Expense	0.00	200.00	-200.00	-100.0 %	0.00	1,000.00	-1,000.00	-100.0 %
7240 Office Supplies Expense	0.00	0.00	0.00	0.0 %	1,284.00	0.00	1,284.00	0.0 %
7250 Taxes-Fuel	0.00	0.00	0.00	0.0 %	1.75	0.00	1.75	0.0 %
7260 Taxes-Sales	0.00	0.00	0.00	0.0 %	35.27	0.00	35.27	0.0 %
7360 Flight School Supplies	0.00	0.00	0.00	0.0 %	6.25	0.00	6.25	0.0 %
7370 Insurance-General	0.00	0.00	0.00	0.0 %	2,743.00	0.00	2,743.00	0.0 %
7380 Insurance-Auto	0.00	0.00	0.00	0.0 %	500.00	0.00	500.00	0.0 %
7390 Insurance-Aircraft	2,175.00	0.00	2,175.00	0.0 %	10,025.00	0.00	10,025.00	0.0 %
7420 Maint-Aircraft	178.12	1,500.00	-1,321.88	-88.1 %	957.37	7,500.00	-6,542.63	-87.2 %
8520 Interest Expense	445.85	0.00	445.85	0.0 %	556.68	0.00	556.68	0.0 %
8530 Credit Card Fees Expense	144.63	0.00	144.63	0.0 %	265.93	0.00	265.93	0.0 %
Total Expense:	34,346.15	6,500.00	27,846.15	428.4 %	171,200.71	32,500.00	138,700.71	426.8 %
Net Income:	-16,512.74	4,050.00	-20,562.74	-507.7 %	-20,081.58	20,250.00	-40,331.58	-199.2 %

Courtesy of TotalFBO Accounting and Business Management Software by Horizon Business Concepts, Inc.

Figure 4.7 → Combined Annual and Monthly Purchases Budget Worksheet

Guidelines:
Includes those goods that are the basis for the sales budgets. Materials on hand, plus units required for sales, resulting in the desired ending inventory needed for each product, by month.

Procedure:
1. Determine the total amount for each activity area, that is, sales, parts, service, and so on as a portion of the total sales volume.
2. Allocate the amount for the twelve months, making adjustments for seasonal variances.
3. Determine the units required for each month where appropriate.
4. Adjust the annual budget, as required.

		Annual Budget	Jan.	Feb.	Mar.	Apr.	May	June	July	Aug.	Sept.	Oct.	Nov.	Dec.
Aircraft	$													
Sales	#													
Parts	$													
Sales	#													
Service	$													
Sales	#													
Flight	$													
Sales	#													
Line	$													
Sales	#													
Misc.	$													
Activity	#													

Budget assembly and approval. All the business activities should be represented in the budgeting process and in the various divisions of the budget. The final budget should be presented to identify clearly the organization's subdivisions and assigned budget goals. The budget's figures should represent reasonable, but not easily attainable goals. The personnel involved in, and responsible for, achieving business results should also be involved in the development of the budget. Backup schedules should be developed as necessary features to augment the budget. The budget should not be developed with the expectation of cutting; rather, it should be considered a realistic contribution to achieving the overall company objectives.

In addition to these guidelines, the human aspects of budget building should be recognized and every effort made to create a structure as free from personal distortion as possible. Some of the likely distortions that might detract from the desired objectivity include:

> Preparing a low budget in order to play the role of the "successful, driving leader";
> Preparing a budget with excessive figures to get "my fair share" of the funds;
> Allowing personal biases for a favorite aspect of the business to have too great an influence;
> Allowing a "halo" effect associated with one department or business activity to distort the overall company budget beyond realistic expectations; and

Figure 4.8 ✈ Expense Budget Worksheet

EXPENSE BUDGET WORKSHEET

Procedure:
1. Fill in previous year annual expense figures
2. Based upon anticipated level of business activity, project the current year annual expense figures
3. Allocate the annual expense figures to months.

		EXPENSES:		Annual Previous Year	Annual Current Year	Jan.	Feb.	Mar.	Apr.	May	June	July	Aug.	Sept.	Oct.	Nov.	Dec.
6	E	Salaries – Officers	20														
7	M	Salaries/Wages – Employees	21														
8	P	Commissions	22														
9	L	Taxes – Payroll	23														
10	O	Employee Benefits	24														
11	Y																
12	M																
13	E N T																
14		TOTAL EMPLOYMENT															
15		Aircraft Delivery	30														
16		Aircraft Miscellaneous	31														
17	A	Depreciation – Aircraft	32														
18	I	Fuel & Oil – Aircraft	33														
19	R	Insurance – Aircraft	34														
20	C	Interest – Aircraft	35														
21	R	Inventory Adjustment – Aircraft	36														
22	A	Maintenance – Aircraft	37														
23	F T	Warranty	38														
24																	
25																	
26		TOTAL AIRCRAFT															
27	O	Rent	44														
28	C	Depreciation/Amortization – Bldg.	45														
29	C	Taxes – Property	46														
30	U	Utilities	47														
31	P	Maintenance – Bldg./Land	48														
32	A																
33	N C Y																
34		TOTAL OCCUPANCY															
35																	
36		Advertising	50														
37		Bad Debts	53														
38		Depreciation – Other	56														
39		Donations & Dues	59														
40	O	Freight & Postage	62														
41	T	Insurance – Other	65														
42	H	Interest – Other	68														
43	E R	Inventory Adjustment – Other	71														
44		Maintenance – Other	74														
45	E	Professional Services	77														
46	X	Supplies	80														
47	P	Taxes – Other than P.R./Prop./Inc.	83														
48	E	Telephone & Telegraph	86														
49	N S	Travel	89														
50	E S	Vehicle	92														
51																	
52																	
53																	
54		Miscellaneous	98														
55																	
56		TOTAL EXPENSES															

Figure 4.9 → **Budget Worksheet Used to Formulate Complete Budget**

INCOME STATEMENT BUDGET WORK SHEET

LINE NO.		ACCT. NO.	PREVIOUS YEAR BUDGET	PREVIOUS YEAR ACTUAL	CURRENT YEAR BUDGET
1	TOTAL SALES				
2	COST OF SALES				
3	GROSS PROFIT				
4	Gross Profit – % of Sales				
5	EXPENSES:				
6	Salaries – Officers	20			
7	Salaries/Wages – Employees	21			
8	Commissions	22			
9	Taxes – Payroll	23			
10	Employee Benefits	24			
11					
12					
13					
14	TOTAL EMPLOYMENT				
15	Aircraft Delivery	30			
16	Aircraft Miscellaneous	31			
17	Depreciation – Aircraft	32			
18	Fuel & Oil – Aircraft	33			
19	Insurance – Aircraft	34			
20	Interest – Aircraft	35			
21	Inventory Adjustment – Aircraft	36			
22	Maintenance – Aircraft	37			
23	Warranty	38			
24					
25					
26	TOTAL AIRCRAFT				
27	Rent	44			
28	Depreciation/Amortization - Bldg.	45			
29	Taxes – Property	46			
30	Utilities	47			
31	Maintenance – Bldg./Land	48			
32					
33					
34	TOTAL OCCUPANCY				
35					
36	Advertising	50			
37	Bad Debts	53			
38	Depreciation – Other	56			
39	Donations & Dues	59			
40	Freight & Postage	62			
41	Insurance – Other	65			
42	Interest – Other	68			
43	Inventory Adjustment – Other	71			
44	Maintenance – Other	74			
45	Professional Services	77			
46	Supplies	80			
47	Taxes – Other than P.R./Prop./Inc.	83			
48	Telephone & Telegraph	86			
49	Travel	89			
50	Vehicle	92			
51					
52					
53					
54	Miscellaneous	98			
55					
56	TOTAL EXPENSES				
57	Expenses – % of Sales				
58	OPERATING PROFIT				
59					
60	Other Income				
61					
62	Other Expense		( )	( )	( )
63					
64	INCOME – Before Special Charges				
65					
66					
67					
68					
69	INCOME – Before Income Taxes				
70	Provision For Income Taxes	290			
71					
72					
73	FINAL NET INCOME				

Procedure:
1. Using the sales, purchase and expense worksheets, complete the previous year budget and actual columns
2. Complete the current year column
3. Compare projected net income with selected business goals

> Distorting the sound budget concept into a game of trying to outguess the approving level of management.

Using supporting budgets. In developing and utilizing a budget system, a manager will undoubtedly need to establish and use several additional supporting budgets to meet specific needs. He or she may need one for cash, one for people, one for fixed assets, or for a variety of items already in the budget as a single line, but in need of support. One such budget with universal application and of special interest is a cash budget or a cash flow projection. Since cash flow is so very important to a business, we will review the cash budget in some detail. It is a schedule over time of cash inflows and outflows. This is used in an attempt to pinpoint cash surpluses and shortages so that the manager can always be certain to have sufficient money in the bank to cover obligations or have a plan to invest surpluses. There are a variety of uses of the cash forecast, including:

> Determining operating cash requirements;
> Anticipating short-term financing needs;
> Investing excess cash;
> Planning reductions in debt;
> Scheduling capital payments;
> Taking advantage of cash discounts; and
> Supporting credit policies.

The primary method of developing a cash flow forecast is through the receipts and disbursements comparison. This procedure generally adheres to the following pattern:

1. Develop forecasts for sales, purchases, expenses, and related schedules.
2. Analyze the forecasts carefully to identify the cash receipts and payments of each particular period.
3. Using the cash budget worksheet (Fig. 4.10), list the receipts and disbursements for the periods that cash actually changes hands.
4. Review the projected results of the worksheet. Note the periods requiring additional outside funds and periods providing excess funds for use in retiring obligations or for investment.

With the cash position identified, the manager can take action to control the cash flow through day-by-day transactions. Three such actions are (1) speeding up collections, (2) controlling payables, and (3) controlling bank balances.

Budget Operation and Control

Once a budget period begins, the function of budgeting becomes one of exercising control over business operations as they progress. The word "control," as used here, means making certain that actual performance is going according to plan. This usually requires:

> Periodic reporting of performance;
> Comparing performance with the budget (plan);
> Pinpointing reasons for variations;
> Taking action to correct unfavorable variances; and
> Reevaluating the original budget goals.

Frequently, an examination of a budget in operation will suggest problems in the operation.

When the budget is examined closely, one or more of the following deficiencies may be noted:

1. There has been a poor determination of budget standards.
2. Not all key business activities have been included in the budget.
3. The budget has not been revised when the need is indicated.
4. There is a lack of understanding of the budget process throughout the organization.
5. There is a lack of acceptance of the budget by management and employees.

The budget area should not be closed without at least a mention of the flexible, or variable budget. The budgets discussed thus far are static, in that the plan is geared to a single target level of sales. A flexible budget is geared to a *range* of activity, rather than one level, and will supply a more dynamic basis for comparison than will a fixed budget. As aviation managers and organizations become better versed in budgeting, they might choose to utilize flexible budgeting for some or all of their needs.

Part 4—Other Considerations

Tax Planning

Tax planning can affect cash outlays in a given year, especially in relation to depreciation and tax credits. Taxes are an expense like any other, but their prediction requires the skills of an accountant with specific knowledge regarding the organization's financial situation.

Figure 4.10 ✈ Cash Budget Worksheet

Procedure:
1. Identify the cash receipts and payments expected throughout the year.
2. List receipts and disbursements for the periods that cash actually changes hands.
3. Note difference between cash available and cash required.
4. Plan for the source of needed funds or the utilization of excess.

Item	Jan.	Feb.	Mar.	Apr.	May	June	July	Aug.	Sept.	Oct.	Nov.	Dec.
1. Cash Balance, Beginning												
2. Add Receipts from Customers												
3. Total Available												
4. Less Disbursements												
5. Payroll												
6. Aircraft												
7. Occupancy												
8. Misc.												
9. Income Tax												
10. Total Disbursements												
11. Difference between Cash Available and Required (Line 3 minus Line 10)												
12. Source or Utilization												

Competition

At all times, an FBO must watch the competition and determine whether to match a competitor's new service, reduce prices, or be innovative by trying something new. These contingencies should be allowed for in the business plan.

Retained Earnings

These are profits plowed back into the business or kept liquid for a rainy day. The cyclical nature of aviation suggests the desirability of retained earnings, if only to even out the good and bad years.

New Revenue Sources

Many public airport owners are seeking new sources of revenue at airports and often finding that the most profitable ones are not directly aviation-related. The feasibility of each depends on the size and clientele of the airport, but an FBO might be able to tap these just as readily as an airport manager. These are some possibilities:

> Vending machines;
> Luggage lockers;
> Copying machines;
> Cash machines;
> Stores and newsstands;
> Lattè stands, coffee shops, and restaurants;
> Car rentals;
> Parking lots;
> Rental of unneeded space for commercial use;
> Agricultural (for example, sale of hay, turf farming, worm farming);
> Video games; and
> Travel agent commissions.

While leases may preclude some of these, any potential to increase profits should be given consideration, especially if little or no expense is involved.

Financing

Types of Money

There are basically four types of money for running a business: short-term cash, long-term cash, equity, and employee ownership. Each has different conditions. The last two involve giving away some of the ownership, and probably control, to others in exchange for their cash or their willingness to forgo short-term cash.

Sources of Money

Family and friends. Many businesses start this way. It is important to be extremely specific about ownership, control, profit sharing, and inheritance when this approach is used. All loans should be written in as businesslike fashion as if they are for strangers.

Accounts receivable financing. Accounts receivable financing, usually involving factoring, can help a business that is already operational but suffering from lags in collection. The factoring company normally takes title to invoices and collects directly from your customers. Rates charged by factors may be fixed or may vary with the collection period. However, your administrative costs can be cut or even eliminated; and once a relationship with a factor is set up, you can get cash at the time of invoicing.

Banks may also occasionally finance accounts receivable. A bank will not buy the invoices; rather, it will loan a percentage of the face value of invoices to the business on the day that the invoices are sent out. The business owner pays receipts from them into a special account from which the bank first pays itself and then any remainder goes into a business checking account. This will normally cost less than factoring, but the business owner is still responsible for collection of any bad debt. A bank may want some collateral to set up this arrangement.

Leasing. Leasing and lease purchasing are good ways to keep monthly costs down when compared with buying outright. Lease purchase terms, especially where the lessor is responsible for maintenance, may be extremely cost effective.

Venture capital. Venture capital sources grew in the 1990s, particularly for high-tech areas, but for an FBO they remain an unlikely source. This is because investors are looking for five to ten times their investment within five years; also, they usually will only consider large potential sales. The dramatic growth path of any product in its early years is ideal for venture capital. One drawback is that ownership of 40 percent or more of the business will pass to the venture firm.

SBA loans. Small Business Administration (SBA) loans are usually loan guarantees. Being turned down twice for regular bank financing makes one eligible. Points above prime are charged, making this a very costly source of aid. Owing to the paperwork and conservatism involved in the banking industry, it is uncommon for new businesses to get SBA loans.

Customers. A common practice among FBOs is to invite steady customers to place several hundred dollars on account with a small bonus of credit for doing so. This method can encourage a close rapport with a set of regular customers and keep the bank balance healthy. However, the true cost of this financing depends on how fast the customer uses up his credit account. If he uses it in a month, you may have provided a rather high annual rate of interest.

Banks. A bank will give a secured line of credit relatively easily for a few points above market rate. For an unsecured line of credit or a loan against receivables, the bank will want to know everything about the business, its managers, and its performance. The more polished and detailed the presentation, the better chance of success. Moreover, presentation style is important when making the request.

Public markets. Going public is complex and expensive, but some states have limited variations on full public ownership that could be feasible for an FBO.

Small Business Investment Companies (SBIC). These are companies licensed by the Small Business Administration (SBA) to provide equity capital or long-term loans to small businesses.[5] They are restricted to businesses with a limited net worth and net income. Rates and equity are determined by negotiation, but the SBIC does not generally seek control.

Credit Management

Nature and Reason for Credit

Credit is derived from the Latin word *credo*. This means "I believe" or "I trust." Today, credit provides a measure of the trustworthiness of an individual to receive goods and services, while deferring payment. The seller (creditor) recognizes this confidence or trust by allowing the customer (debtor) to receive the goods immediately, but to postpone payment until later.

The individual consumer uses credit to:

> Raise his standard of living or increase his enjoyment;
> Realize the convenience offered by use of credit; and
> Meet the pressures of economic necessity.

The manager or the business uses credit to:

> Enable the business to sell more goods and services;
> Permit the business to purchase goods on terms that will normally allow the business to resell and thus gain funds for repayment and profit; and
> Provide the opportunity for cash loans from some financial institutions.

Creating a Credit Policy

The easiest way to grant credit is through the acceptance of credit cards. Numerous companies are now available to assist the small business not only to accept credit cards face-to-face, but also on-line through secured channels. Every aviation business should also have a credit policy for charge accounts. Figure 4.11 displays a sample credit policy from The Conference Board.

Although it may have many possible variations, such a credit policy should consider the following areas:

> Authority for granting credit;
> Terms of the sale;
> Dealing with delinquent customers;
> Classification of risks; and
> Helping customers obtain financing.

In some cases, such as new and used aircraft sales, the FBO may simply provide loan search assistance to its customers and let various other lenders deal with the actual financing. However, should you decide to permit credit accounts, the following procedures should be a guide.

Functions of Credit Management

The purposes of credit management are:

> To maximize sales and profits;
> To minimize bad debt losses;
> To achieve efficient utilization of invested funds;
> To develop full coordination with the operating departments;
> To establish an effective credit policy;
> To set standards for credit operation;
> To collect information on credit activity;
> To analyze and evaluate credit activity information;
> To make decisions on credit issues and matters; and
> To carry out the credit process.

Credit Process

The total credit process in a business is made up of five elements:

1. Determining the acceptable credit risk.
2. Setting the credit limits.
3. Handling credit transactions.
4. Maintaining controls and efficiency.
5. Collections.

Acceptable credit risk. In determining acceptable credit risk, the manager needs to answer two questions: "Can the customer pay?" and "Will the customer pay?"

In arriving at an answer to these questions, the manager frequently considers four key factors, known as the four Cs of credit. The credit risk is felt to be indicated by character, capacity, capital, and conditions (economic).

$$\text{Credit Risk} = \text{Character} + \text{Capacity} + \text{Capital} + \text{Conditions}$$

These four factors sound logical. Unfortunately, they are hard to determine in actual cases, because they are difficult to identify through direct inquiry. There are, however, some credit qualities that can be

Figure 4.11 → Sample Credit Policy

Credit Policy Statement

1. *General policy.* It is the policy of the company to extend credit and grant terms to customers in relation to the credit risks involved. The responsibility for implementation of this policy rests with the Vice President, Finance. Administration and control of this policy is centralized in the Credit Department.

2. *Extension of credit and limits of authority.* Credit may be extended to new customers as follows:

Extended value of each order	Authorized by
a. Up to $100	Any order-processing unit or salesperson
b. Up to $750	Division or Subsidiary Sales Managers
c. Up to $25,000	Credit Manager or his or her delegate in certain locations as specified
d. Over $25,000	Treasurer

3. *Procedure*
   a. Credit approvals made by authorized personnel in paragraphs 2a and b must be reported to the Credit Department by memorandum, which will include customer name, address, class of trade, and amount of credit granted.
   b. Credit limits on specific customers will be established by the Credit Department and will be adjusted from time to time as conditions warrant.
   c. Orders received from customers with whom the company has had unfavorable credit experience must be cleared for credit approval through the Credit Department, regardless of the amount of the order, until such time as credit limits have been reestablished.
   d. Collection of accounts is the responsibility of the Credit Manager or his or her delegate. Other personnel may be asked to assist by the Credit Manager.
   e. Standard payment and discount terms are established by the Vice President, Finance and are published in sales literature and other documents. Deviations from standard payment and discount terms must have approval of Vice President, Finance or his or her delegate prior to granting such terms.
   f. All orders received from customers outside of the United States and Canada must be forwarded to the Credit Department for credit approval and determination of method of payment prior to release of the order for processing.

Courtesy of The Conference Board

investigated and used to infer a standing on the four Cs. Among the typical items reviewed are:

> Payment record;
> Income level;
> Employment history;
> Residence: ownership and value;
> Marital status;
> Age and health;
> References and reputation;
> Reserve assets;
> Equity in purchases; and
> Collateral available.

In arriving at a decision on the acceptability of a credit applicant, the manager will normally consider information supplied by the applicant, information supplied by direct inquiry, and information supplied by in-house records.

The applicant must provide much of the credit evaluation data. The use of a credit application form similar to Figure 4.12 provides a great deal of the needed information, and should be included routinely as part of the process. An interview can be used to expand and verify the data received in the application and from other sources.

The organization should routinely check each applicant by making direct inquiry of all appropriate sources. This inquiry may be made by mail or by telephone, and should include references, bankers, business associates, and previous credit sources. Credit bureaus and groups should also be considered and used, if appropriate. They are a ready source of credit information and are designed for that specific purpose by having established procedures and sources.

In-house records may be used in many instances to either verify or obtain additional insight regarding

Figure 4.12 ✈ Credit Application Form Available to Piper Aircraft Dealers

APPLICATION FOR CREDIT

PIPER Management Services

Social Security Number _____

Full Name _____ Age _____

Spouse's Name _____ Single _____ Number of Dependents _____

Home Address _____ Own _____
                                          Rent _____
_____   Tel. No. _____
_____ Zip How Long _____

Previous Address _____ How Long _____
_____
_____

Employed By _____
  Address _____
Position _____ How Long _____
  Monthly Salary $ _____
Former Position _____ How Long _____
Landlord or
Mortgage holder _____
  Address _____

Nearest Relative (other than Husband or Wife)
_____
  Address _____
_____

Personal Reference: _____
  Address: _____
_____

Banks: _____ Regular Checking ☐
       _____ Savings ☐
                                                  Loans ☐
Make of Car _____ Year _____

CREDIT REFERENCES

Name	Address	Type Credit

The above information is for the purpose of obtaining credit and is warranted to be true. I agree to pay all bills upon receipt of statement or as otherwise expressly agreed.

Date _____ Signature _____

Courtesy of Piper Aircraft Corporation

the credit risk of an applicant. Records of previous business transactions (cash) are useful indicators to consider, along with pertinent correspondence or business association.

One technique that has been found useful in evaluating an application for credit is credit grading. By using a form such as the one illustrated in Figure 4.13, each credit element or quality is consciously and separately reviewed and given its proper appraisal. Thus, a higher degree of objectivity is achieved and a better overall decision is obtained. Measurement of a number of specific factors is likely to be more accurate than a single, overall judgment. Credit quality is a relative evaluation, not an absolute measurement.

Setting the credit limits. Based upon evaluation of the risk and other factors such as economic environment, company position, and supplier policies, the manager must set a credit limit for the customer. This is normally expressed as a specific dollar figure and is used to guide administrative or accounting personnel responsible for monitoring the credit program. It is difficult to recommend any one method of fixing the credit limit. Some managers use the technique of limiting the account to a certain time (e.g., a month's purchases). Another manager might start with a low credit limit and raise its level with satisfactory experience. Some organizations encourage the customer to fix his/her own limits (within bounds, of course).

Handling credit transactions. The actual handling of credit transactions involves identification, authorization, recording, and billing. The first step is identifying the customer as an approved credit customer. This may be done through the use of a credit card, by signature, or by personal recognition. The approval or authorization of each transaction should be based upon the positive identification of the customer and a verification of the records indicating that he or she is an approved credit customer. This must be done by the salesperson at the point of sale.

Recording and billing are covered in Chapter 7, Information Systems.

Maintaining controls and efficiency. A successful manager should be concerned with maintaining positive control over the credit program to ensure its effectiveness. A number of tests can be applied by the manager in controlling credit activity. The generally accepted tests include:

> Review of bad debt loss record;
> Analysis of credit sales;
> Collection percentage records;
> Days required to collect charges;
> Turnover of receivables;
> Number of new accounts;
> Aging of accounts; and
> Cost analysis of the credit program.

The use of these techniques will greatly assist the manager in working toward previously determined goals and objectives.

Collections. Normally, collection problems are not major, if sufficient attention is paid to the investigation and analysis of a credit risk. If, however, the credit analysis is weak, collection problems will multiply. There is a need for close supervision and administration of the collection policy. In accomplishing this it is desirable to understand the different types of debtors, to have a systematic collection system, to use the appropriate collection tools, and to understand the reasons for slow payment by customers.

Debtors vary considerably, but the different types can be classified as those who are prompt to pay, those who are sure but slow, and those who are undesirable risks because they are either dishonest or unfortunate. The process of collecting requires a systematic approach (steps and procedures) and a schedule to ensure timely action. Many organizations have found that a tickler file is most useful to ensure correct follow-up action.

The most frequently used collection tools in their sequence of normal use are:

> Invoices on transaction;
> Statements on account;
> Collection letters;
> Follow-up notes;
> Telephone reminders;
> Possibly a personal visit; and if necessary
> Sending the account to a collection agency.

Throughout the entire credit process there should be an awareness of credit problems and every effort should be made to spot trouble as soon as possible and to adjust the credit exposure.

Figure 4.13 → **Credit Check Form**

CREDIT CHECK FORM

Date

Name

Address

Credit qualities	GRADE			Verified:
	GOOD 1	FAIR 2	POOR 3	
Income				
Employment				
Residence				
Marital Status				
Age				
References				
Reserve Assets				
Payment Record				
Reputation				
Summary - Overall				
Appraisal				

Decision: Accept _____   Refuse

Limit: $ _____   Single Item

   $ _____   Monthly

   $ _____   Total

Special Conditions: _____   _____   _____

_____   _____   _____   _____

_____   _____   _____   _____

Signed

Date

Source: Dr. J. D. Richardson

Terms and Definitions

There are a number of terms and procedures that should be understood by those involved in developing and administering a successful credit program.

Credit period is the length of time allowed the buyer before payment becomes due; it is usually computed from the date of the invoice.

Example: Net 30 (payment of the full amount within 30 days).

Cash discount is the deduction from the invoice amount allowed the customer for payment within a specified time prior to the expiration of the net credit period.

Example: 1 percent 10, Net 30 (payment within 30 days with a 1 percent discount for payment within 10 days).

Prepayment terms include COD-cash on delivery, CBD-cash before delivery, CWO-cash with order.

Individual order terms are the stated terms applying to each order individually and start with the invoice date.

Lumped-order terms are used when sellers allow buyers to accumulate obligations over a period of time, usually one month, and submit one payment after the end of each period.

Example: Net 10 EOM (invoice on last day of the month, payment due on the 10th day of the next month); Net 10th Proxy (payment of all previous month's invoices on the 10th of the following month).

Cash discount value is the value of a cash discount and can be equated to the annual interest on the money involved.

Example:

1/2 percent 10 days, Net 30	9 percent per annum
1 percent 10 days, Net 30	18 percent per annum
2 percent 10 days, Net 60	14 percent per annum
2 percent 30 days, Net 60	24 percent per annum
2 percent 10 days, Net 30	36 percent per annum

Collection period is calculated by

365 × receivables = average annual net sales collection period

In general, the collection period should not exceed the net maturity indicated in the selling terms by more than 10 to 15 days.

Cash or Credit Card?

With increasing frequency, many aviation businesses are giving their customers only two alternatives in paying for products or services—cash or credit card.

This trend stems from the realization that a small business is not as efficient as a "lending" agency, as will be evident from the description just given above. Without training, experience, procedures, and volume, the real cost of extending credit and billing clients monthly is relatively high for the benefit received. As a result, many businesses have turned to credit cards from a few selected sources, such as fuel companies, MasterCard, and Visa. Some organizations still provide open credit for local and carefully screened customers of long standing, but these are increasingly at a minimum. The tendency has been to give the credit business to the "experts."

It should be recognized, however, that it costs the aviation business to use a credit card service. Typically, the charge is expressed as a percentage of the sales, with the actual percentage based upon the volume of business. This cost will vary, but for the small business it is likely to be around 3 to 5 percent. The actual figure may be negotiable, so the prudent manager will explore options.

When using credit cards, it is necessary to train issuing personnel in the correct procedures in order to minimize delays in receiving payment or the chances of error voiding a voucher.

It appears likely that the number of fuel companies that issue fuel credit cards or those accepting cards of other companies will continue to decrease. This trend will reduce the options available to aviation businesses for extending credit and place a heavier burden on the remaining sources.

Summary

Business profits should be planned for and not left to chance. Planning techniques include ratio analysis, financial statement evaluation, budget projection, break-even analysis, and cash flow projection. The sales planning, pricing, and marketing orientation discussed in Chapter 3 underlie sound profit planning. Net profit levels vary in aviation businesses; a goal of 10 to 15 percent is not unrealistic given the high-risk nature of the industry. Non-monetary profit measures such as "satisfactory" profits, social goals, and personal satisfaction may, however, play an important part in setting the overall profit goal. The manager's job is to guide the organization to an explicit and acceptable level of profitability. In doing this, he or she must also pay close attention to credit policies, bad debts, and the cost of finance.

DISCUSSION TOPICS

1. What are some meanings associated with the word "profit"?
2. How should a business trade social responsibility against profit? Is it always an either/or situation?
3. Describe a manager with a "profit orientation."
4. What techniques should be used in profit planning?
5. What are the conditions needed for effective budgeting?
6. How should one distinguish between fixed and variable costs? Why does the distinction matter?
7. Why is cash flow important? How can it be improved?
8. What is a break-even point? How is it calculated, and what does it tell you about profit?

Endnotes

1. Murphy, John F. (1984). *Sound Cash Management and Borrowing*. Washington D. C.: SBA Management Aids, 1(016).
2. From an informal survey conducted by Dr. J. D. Richardson, the original author of the 1st and 2nd editions of this book.
3. See *https://totalfbo.com/* and *https://www.fbodirector.com/*.
4. This information was derived from New England Airlines, Block Island, RI and is typical of any resort or tourist area.
5. See *https://www.sba.gov/category/lender-navigation/sba-loan-programs/sbic-program-0*

5

Human Resources

OBJECTIVES

› Discuss 21st-century labor market trends in aviation.

› Understand some of the legal implications involved in hiring practices due to federal statutes.

› Review a variety of methods utilized in recruiting qualified candidates for aviation positions.

› Identify issues that should be covered in an exit interview.

› Relate Maslow's hierarchy of needs theory to leadership styles.

› Describe the importance of having a personnel policy manual.

Pipeline Concept

A common tendency of small business owners is to assume (or hope) that staffing can be done just once and stability will follow. This is an illusion brought about by the many pressures of the job and, in some cases, it is accentuated by the manager's discomfort in dealing with personnel issues.

A more realistic view, especially during periods of labor shortages, and considering that general aviation is thought by many to be the lowest rung on the aviation career ladder, is that most staff members pass in a steady flow into, through, up, and out of the organization. Some degree of turnover is not only inevitable but can also be beneficial to both the company and the individual. Some individuals may not be adequate performers or may have insufficient interest in a career with a general aviation service business. Others may have ambitions beyond what an FBO can offer. In such cases, the employer is better off, for the good of the company, when such employees choose to move on. Managers may even have to terminate a person with a poor attitude and lack of enthusiasm. An influx of new people provides the opportunity for fresh ideas to be brought into the organization.

The flow of personnel can be envisioned as a pipeline having various "valves" or control points along its length for the manager to utilize. Dealing efficiently with personnel matters is one of the many control functions that are an essential part of the management process. The pipeline is illustrated in Figure 5.1.

Control Points

In dealing with the personnel pipeline, the manager has key control points for successfully operating

Figure 5.1 → Human Resource Pipeline

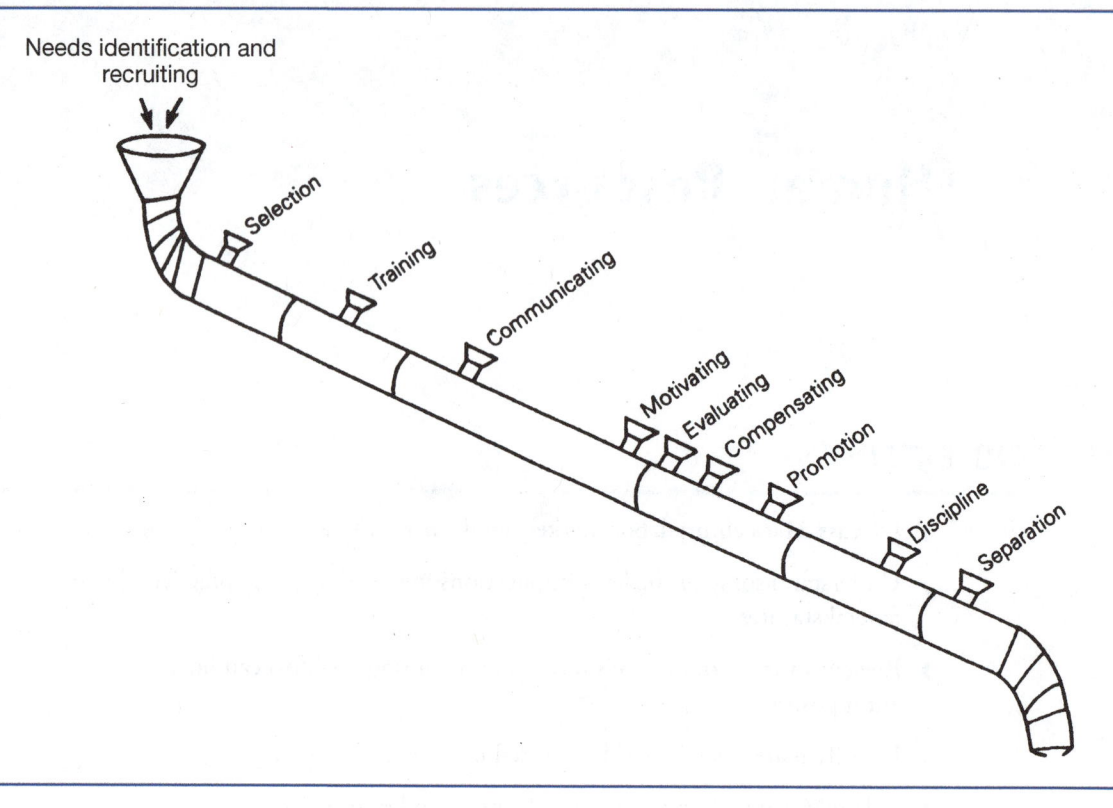

the personnel management program. These points include:

> Identifying human resources needs;
> Recruiting qualified candidates;
> Screening and selecting applicants;
> Orientation and training;
> Communications;
> Motivation;
> Evaluating employees;
> Promotion;
> Compensation adjustments;
> Discipline; and, ultimately
> Separation (when required).

Scope of Chapter 5

This chapter discusses the key elements of human resource management by considering all of these control points and the ways in which they can be used to manage the flow of people through the company in the most beneficial manner.

The pipeline is a good description of the career progression for some percentage of a company's employees, but it is neither desirable nor inevitable for all personnel. People work for reasons other than advancement and money. For example, they also work for a sense of belonging to a team that produces quality products and services. The reasons a person might spend his or her entire working life in the same company include shared values about the goals of the company and a perception of its overall direction and purpose, along with recognition and rewards, job security, and continued challenge.[1] Some of the ways in which these qualities have been developed in a company's staff include participatory decision-making, interchangeability of most staff among the jobs in the business, and a willingness of senior management to encourage the best contribution from each employee, even from the least capable workers in the firm. Particularly in a small business, there may indeed be little choice but to run the firm this way. Its positive aspects need to be considered.

This chapter also discusses personnel matters in the context of related laws and regulations. It reviews how the key control points are combined into a personnel policy manual and considers appropriate ways to handle employee organizations. For each of the topics, the manager needs to have information on currently accepted tools and procedures, which are referenced in the text and the sources provided.

Aviation Industry Trends

Labor Market Trends in Aviation

Smaller FBOs have a variety of pay structures that range from minimum wage for line personnel through hourly commissions for flight instructors and hourly wages for maintenance technicians. Some build in a bonus system for instructors so that they earn more per hour with more bookings. Many aviation service center employees "wear several hats". People on the flight line may also work on front desk and maintenance assignments; the secretary may also fuel aircraft, order parts, and do bookkeeping; and, of course, the owner/manager is often a jack-(or jill-)of-all-trades who fills in for anyone who is unavailable and solves problems, large and small, as they occur.

Perhaps the single most important issue in the area of FBO personnel matters is that many of the people in the business are in it for their love of aviation. They may accept low and sometimes erratic pay patterns for a few years until the need to be more financially established takes them into other areas of aviation. Positions in these other areas, such as corporate pilot or mechanic, airport manager, airline pilot or mechanic, and aircraft manufacturing technician, tend to offer better pay, substantially better fringe benefits, and often greater job stability and better working conditions. As the FBO industry becomes more mature and more professional, it is addressing the issues of pay levels in competing areas, and the questions of high turnover and high levels of part-time involvement that have traditionally characterized (and challenged) smaller FBOs.

The Bureau of Labor Statistics (BLS) publishes reports on the job outlook in all industries in the U.S. and is a good source of pay and other data.[2]

Recent studies and forecasts (e.g., Boeing Pilot and Technician Outlook 2020–2039, ICAO Economic Impacts of COVID-19 on Civil Aviation, Airbus Global Workforce Forecast) have projected that qualified labor for the worldwide aviation/aerospace industry as a whole, and general aviation operators in particular, is in short supply and in future years will become even shorter in meeting projected demands. The unprecedented economic effects and industry downturn as a result of the COVID-19 pandemic caused a sudden oversupply of pilots and other professionals as the current edition of the book was being written; however, the Boeing 2020–2039 Outlook noted (in regards to pilots) "Given the current oversupply of qualified pilots, labor shortages may seem a distant memory. However, as the industry positions itself for recovery, adequate qualified pilot supply remains an important consideration as a large contingent of the workforce approaches mandatory retirement age. Positions left vacant because of retirements will need to be filled, which is likely to coincide with industry recovery, fleet growth and efforts by other operators to recruit new pilots for similar purposes."[3] Moreover, as countries such as India, China, and others in Asia and the Middle East continue to undergo rapid development of their aviation systems, there is likely to be a return to a growing worldwide shortage of qualified aviation professionals. Historically, the job market for pilots in the U.S. has been highly cyclical, tending to lag the state of the economy and its effect on air travel. The following discussion addresses pilots first, then mechanics.

Employment Outlook for Pilots. In its Occupational Outlook Handbook, updated on September 1, 2020, the BLS provides the following information:

> Airline and commercial pilots fly and navigate airplanes, helicopters, and other aircraft.
>
> Airline pilots work primarily for airlines that transport passengers and cargo on a fixed schedule.
>
> For all but small aircraft, two pilots usually make up the cockpit crew. The captain or pilot in command, usually the most experienced pilot, supervises all other crew members and has primary responsibility for the flight. The copilot, often called the first officer or second in command, shares flight duties with the captain. Some older planes require a third pilot known as a flight engineer, who monitors instruments and operates controls. Technology has automated many of these tasks, and new aircraft do not require flight engineers. With proper training, airline pilots also may be deputized as federal law enforcement officers and be issued firearms to protect the cockpit.
>
> Commercial pilots are involved in unscheduled flight activities, such as aerial application, charter flights, and aerial tours. Commercial pilots may have additional nonflight duties. Some commercial pilots schedule flights, arrange for maintenance of the aircraft, and load luggage themselves. Pilots who transport company executives, also known as corporate

pilots, greet their passengers before embarking on the flight. Agricultural pilots typically handle agricultural chemicals, such as pesticides, and may be involved in other agricultural practices in addition to flying. Pilots, such as helicopter pilots, who fly at low levels must constantly look for trees, bridges, power lines, transmission towers, and other obstacles.

Pilots plan their flights by checking that the aircraft is operable and safe, that the cargo has been loaded correctly, and that weather conditions are acceptable. They file flight plans with air traffic control and may modify the plans in flight because of changing weather conditions or other factors.

Takeoff and landing can be the most demanding parts of a flight. They require close coordination among the pilot; copilot; flight engineer, if present; air traffic controllers; and ground personnel. Once in the air, the captain may have the first officer, if present, fly the aircraft, but the captain remains responsible for the aircraft. After landing, pilots fill out records that document their flight and the status of the aircraft.

Some pilots are also instructors using simulators and dual-controlled aircraft to teach students how to fly.

Pilots held about 127,100 civilian jobs in 2019. About 85,500 worked as airline pilots and 41,600 worked as commercial pilots.

In 2019, most airline pilots—about 86 percent—worked for airline companies; of the remainder 4 percent worked for the federal government and 2 percent in nonscheduled air transportation.

In 2019, 28 percent of commercial pilots flew in nonscheduled air transportation, 12 percent were employed with technical and trade schools—private, 10 percent flew in support activities for air transportation, 10 percent in ambulance services, and 3 percent in manufacturing.

Employment Change. The BLS (Occupational Outlook Handbook, online data current as of September 1, 2020) provides the following:

Overall employment of airline and commercial pilots is projected to grow 5 percent from 2019 to 2029, faster than the average for all occupations.

Employment of airline pilots, copilots, and flight engineers is projected to grow 3 percent from 2019 to 2029, about as fast as the average for all occupations. Employment of commercial pilots is projected to grow 9 percent from 2019 to 2029, much faster than the average for all occupations. The number of commercial pilot jobs is projected to increase in various industries, especially in ambulance services, where pilots will be needed to transfer patients to healthcare facilities. Most job opportunities will arise from the need to replace pilots who leave the occupation permanently over the projection period.

Job prospects may be best with regional airlines and nonscheduled aviation services because entry-level requirements are lower for regional and commercial jobs. There is typically less competition among applicants in these sectors than there is for major airlines.

Pilots seeking jobs at the major airlines will face strong competition because those firms tend to attract many more applicants than the number of job openings.

Pilots' Wages and Benefits. The BLS provides the following:

The median annual wage for airline pilots, copilots, and flight engineers was $147,220 in May 2019. The median wage is the wage at which half the workers in an occupation earned more than that amount and half earned less. The lowest 10 percent earned less than $74,100, and the highest 10 percent earned more than $208,000.

The median annual wage for commercial pilots was $86,080 in May 2019. The lowest 10 percent earned less than $45,480, and the highest 10 percent earned more than $179,440.

Airline pilots usually begin their careers as first officers and receive wage increases as they accumulate experience and seniority.

In addition, airline pilots receive an expense allowance, or "per diem," for every hour they are away from home, and they may earn extra pay for international flights. Airline pilots and their immediate families usually are entitled to free or reduced-fare flights.

Federal regulations set the maximum work hours and minimum requirements for rest between flights for most pilots. Airline pilots fly an average of 75 hours per month and work an additional 150 hours per month performing other duties, such as checking weather conditions and preparing flight plans. Pilots have variable work schedules that may include several days of work followed by some days off.

Airline pilots may spend several nights a week away from home because flight assignments often involve overnight layovers. When pilots are away from home, the airlines typically provide hotel accommodations, transportation to the airport, and an allowance for meals and other expenses.

Commercial pilots also may have irregular schedules. Although most commercial pilots remain near their home overnight, they may still work nonstandard hours.[4]

Using data for May 2019 the BLS website showed median annual wages of airline pilots, copilots, and flight engineers to be $147,220. The lowest 10 percent earned less than $74,100, and the highest 10 percent earned more than $208,000. The median annual wage for commercial pilots was $86,080 in May 2019. The lowest 10 percent earned less than $45,480, and the highest 10 percent earned more than $179,440.[5]

Employment Outlook for Mechanics. The BLS provides this overview of positions for aircraft mechanics and avionics technicians:

Airplanes require reliable parts and maintenance in order to fly safely. To keep an airplane in operating condition, aircraft and avionics equipment mechanics and technicians perform scheduled maintenance, make repairs, and complete inspections. They must follow detailed regulations set by the Federal Aviation Administration (FAA) that dictate maintenance schedules for different operations.

Many mechanics are generalists and work on many different types of aircraft, such as jets, piston-driven airplanes, and helicopters. Others specialize in one section, such as the engine, hydraulic system, or electrical system, of a particular type of aircraft. In independent repair shops, mechanics usually inspect and repair many types of aircraft.

The following are examples of types of aircraft and avionics equipment mechanics and technicians:

Airframe and Powerplant (A&P) mechanics are certified generalist mechanics who can independently perform many maintenance and alteration tasks on aircraft. A&P mechanics repair and maintain most parts of an aircraft, including the engines, landing gear, brakes, and air-conditioning system. Some specialized activities require additional experience and certification.

Maintenance schedules for aircraft may be based on hours flown, days since the last inspection, trips flown, or a combination of these factors. Maintenance also may need to be done at other times to address specific issues recognized by mechanics or manufacturers.

Mechanics use precision instruments to measure wear and identify defects. They may use x rays or magnetic or ultrasonic inspection equipment to discover cracks that cannot be seen on a plane's exterior. They check for corrosion, distortion, and cracks in the aircraft's main body, wings, and tail. They then repair the metal, fabric, wood, or composite materials that make up the airframe and skin.

After completing all repairs, mechanics test the equipment to ensure that it works properly and record all maintenance completed on an aircraft.

Avionics technicians are specialists who repair and maintain a plane's electronic instruments, such as radio communication devices and equipment, radar systems, and navigation aids. As the use of digital technology increases, more time is spent maintaining computer systems. The ability to repair and maintain many avionics and flight instrument systems is granted through the Airframe rating, but other licenses or certifications may be needed as well.

Designated airworthiness representatives (DARs) examine, inspect, and test aircraft for airworthiness. They issue airworthiness certificates, which aircraft must have to fly. There are two types of DARs: manufacturing DARs and maintenance DARs.

Inspection authorized (IA) mechanics are mechanics who have both Airframe and Powerplant certification and may perform inspections on aircraft and return them to service. IA mechanics are able to do a wider variety of maintenance activities and alterations than any other type of maintenance personnel. They can do comprehensive annual inspections or return aircraft to service after a major repair.

Repairmen certificate holders may or may not have the A&P certificate or other certificates. Repairmen certificates are issued by certified repair stations to aviation maintenance personnel, and the certificates allow them to do specific duties. Repairmen certificates are valid only while the mechanic works at the issuing repair center and are not transferable to other employers.

Overall employment of aircraft and avionics equipment mechanics and technicians is projected to grow 5 percent from 2019 to 2029, faster than the average for all occupations. Employment growth will vary by occupation, with avionics technicians' numbers increasing by 4 percent and aircraft mechanics and service technicians growing by 5 percent, during the period 2019–29.

Air traffic is expected to increase gradually over the coming decade, and will require additional aircraft maintenance, including that performed on new aircraft. Job opportunities are expected to be good because there will be a need to replace those workers leaving the occupation.[6]

Mechanics' Wages and Benefits. The BLS also provides data on earnings:

> The median annual wage for aircraft mechanics and service technicians was $64,090 in May 2019. The median wage is the wage at which half the workers in an occupation earned more than that amount and half earned less. The lowest 10 percent earned less than $37,890, and the highest 10 percent earned more than $101,070.
>
> In May 2019, the median annual wages for aircraft mechanics and service technicians in the top industries in which they worked were as follows:

Scheduled air transportation	$89,820
Aerospace products and parts manufacturing	67,180
Nonscheduled air transportation	60,350
Federal government, excluding postal service	60,070
Support activities for air transportation	54,920

> The median annual wage for avionics technicians was $65,700 in May 2019. The lowest 10 percent earned less than $40,350, and the highest 10 percent earned more than $97,150.
>
> In May 2019, the median annual wages for avionics technicians in the top industries in which they worked were as follows:

Aerospace products and parts manufacturing	$74,860
Professional, scientific, and technical services	72,810
Federal government	58,530
Support activities for air transportation	56,020

> Aircraft and avionics equipment mechanics and technicians usually work full time on rotating 8-hour shifts. Overtime and weekend work are common.

Aircraft and avionics equipment mechanics and technicians work in hangars, in repair stations, or on airfields. They must meet strict deadlines while following safety standards.

Most of these mechanics and technicians work near major airports. They may work outside on the airfield, or in climate-controlled shops and hangars. Civilian aircraft and avionics equipment mechanics and technicians employed by the U.S. Armed Forces work on military installations.

As airline operations in developing countries, business aviation, and fractional aviation return to anticipated strong growth following the worldwide recovery from the effects of the COVID-19 pandemic, competition for new employees already faced by fixed base operators seems likely to become more acute.[7]

Industry Maturity and Professionalism

In general, "tightening" of aviation labor markets is welcome news for aviation students and other early-career professionals and usually suggests that working conditions and compensation may become more competitive. A career with an FBO may be a viable choice for even the best students, rather than just for the mediocre students or for those who find flying a paid avocation. Increased professionalism, in terms of the quality of position and career ladder for employees, is part of an increasing level of professionalism in the entire FBO industry.

During the 1980s, high insurance costs, coupled with high costs of fuel and high interest rates, caused stagnation in aviation activity. Many FBOs went out of business during this period, leaving an area of great opportunity over the next few years for the remainder. Not only do FBOs have to operate with the highest standards of professionalism in order to compete for scarce skills, but also certain markets for FBO services are strong enough for them to do so. The business and corporate aviation markets should continue to be particularly strong. Excellent and well-prepared employees will continue to be key to meeting the demands of the market.

Identifying Human Resource Needs

The Human Resources Component of the Business Plan

Like the rest of the business plan, the personnel issue should be considered when the manager first gets into the business, and then should be updated periodically. In a fast-changing enterprise it is difficult to predict needs more than a few months ahead; in a static operation, more advanced planning may be possible. In most small aviation businesses, the personnel function is complicated by the fact that many or all of the staff must serve in more than one capacity. The industry also tends to be characterized by extensive use of part-time labor and, in many parts of the country, by large seasonal variations. Nevertheless, at any given time the manager needs to know the anticipated personnel needs, the skills, certifications, and training required to meet those needs, the probable cost for each position, and the best sources for recruiting new talent.

The overall company planning behind the human resources strategy includes:

1. Determine company objectives.
2. Formulate policies, plans, programs, and procedures designed to attain these objectives.
3. Develop a budget that will serve as a short-range plan of operations to meet the objectives.

With the plan, specific objectives, and the means for achieving goals established, the next step is the process of relating the selected plan to human resource needs. The following questions should be asked:

1. What types of personnel are needed?
2. What will each person be required to do?
3. What number of personnel is required? full-time? part-time?
4. What skills will be required of each employee?
5. Can two positions be combined?
6. When will the new staff be required?
7. How long will each position be required?
8. What potential for development is desirable or possible?
9. Can the type of person needed be successfully recruited?
10. Can the type of person needed be afforded?
11. Are the types and numbers of personnel desired available?

Permanent or Temporary Needs

When the need for extra help is short-lived, the business may benefit from using temporary employees. This can be from a temporary agency, from a nearby college or university, or by direct hire with the understanding that the position is of limited duration. Depending on the workload, the tasks, and whether or not overtime is paid, an estimation of whether to hire from outside or have existing staff work longer hours may be worthwhile. Many companies are increasingly using contracted labor obtained through an agency. Such personnel are employees of the agency, not the place where they work. However, there have been several successful lawsuits in recent years in which long-term "temporary" or seasonal workers claimed that their jobs were in fact really permanent and thus that they had experienced discrimination in terms of not being provided with the fringe benefits of regular employees. An employer considering the use of temporary workers needs to be familiar with these legal trends, and adjust staffing arrangements to accommodate them.

Skills Required

The traditional approach to job skills is to look at the tasks performed by the organization, group them together into job descriptions, and seek people that appear to match them. A classic management text suggests otherwise: in his book, *Entrepreneuring: The Ten Commandments for Building a Growth Company*, Steven Brandt presents the fifth commandment as "Employ key people with proven records of success at doing what needs to be done in a manner consistent with the desired value system of the enterprise."[8] In an aviation business, a combination approach may be needed since some of the positions require specific licensure and certifications and the owner/manager cannot just hire people with good energy and values; they must have the technical credentials, as well.

In addition to a concern with values, an employer may also look at a prospective employee's transferable skills from other jobs. In a world that continues to change increasingly rapidly, the ability to transfer one's experience into a new context may be more valuable than years of doing the same thing over and over. Such an approach does have implications for training, as discussed later in this chapter.

Job Descriptions and Specifications

Each position requires specific tasks and definite functions that are all related to the overall plan. The following suggests an approach to be taken when determining the requirements of specific positions in an aviation activity.

Figure 5.2 describes the types of tasks that different aviation positions might require; Figure 5.3 describes skills required for the tasks in Figure 5.2, while Figure 5.4 describes briefly the training needed for aviation maintenance work.

Special Activities

In addition to traditional flight line, flight operations, and aircraft maintenance functions, there are many and varied additional activities for aviation businesses, such as auto rental, property rental, insurance,

Figure 5.2 → Tasks Required in FBO Positions

Mid-Manager This position is required to accomplish a wide variety of activities depending upon the size of the business, its specific functions, and its organizational structure. Typical situations could include:
1. Serving as the manager in his or her absence
2. Performing specific, delegated duties for the manager
3. Representing the manager at certain functions
4. Being responsible for assigned technical or functional areas of the business; for example, department head or area manager
5. Serving on coordinating, developmental, or special project committees

Accountant This position is typically required to establish a bookkeeping and information system, to ensure that the system is performing to the organization's needs, to develop the desired reports, and to provide information and guidance of a financial nature to management.

Office Manager/Administrator This position is responsible for directing the administrative activities of the organization. This includes filing, record keeping, operation of computer equipment, lease administration, insurance programs, reports, and the various activities of a clerical nature that might also include personnel programs.

Front-Desk Manager This person is responsible for reception of visitors, answering questions and directing personnel, scheduling aircraft, and telephone answering. In meeting the responsibilities of dealing with the public, this person must be knowledgeable about the internal organization, services, and activities of all departments.

Line Service Personnel These employees are responsible for meeting, directing, servicing, and seeing off aircraft that land and taxi into the ramp area of the business. They are responsible for operating various aircraft support vehicles and equipment and dealing with aircrew personnel of visiting aircraft. They are usually responsible for fueling.

Flight Instructor This person is responsible for: (1) operating characteristics and performance data for the type(s) of aircraft used for instruction, (2) the required maneuvers and flight information, (3) instructional techniques and testing, and (4) administrative and promotional activities relating to the functions of flight instruction. These responsibilities vary with the size of operations and the type of instruction, such as private pilot, commercial pilot, multi-engine rating, instrument rating, jet type-training, helicopter, and so on.

Aircraft Pilot This person may be engaged in charter work, ferry, aerial application, cargo flights, forest or pipeline patrol, or other similar activities. Pilots are responsible for conducting flights safely and efficiently.

Aircraft Mechanic This position is responsible for a broad range of duties dealing with the inspection and repair of aircraft, maintaining records, completing administrative functions, and assisting customers.

Avionics Technician This person is responsible for servicing the various communication and navigation equipment used in aircraft. Also, he or she is responsible for test equipment, necessary records, and administrative activities.

Powerplant Mechanic This position is responsible for the maintenance of reciprocating and turbine aircraft engines and related accessory equipment such as propellers, starters, generators, fuel pumps, carburetors, and fuel-injection systems.

Custodial Worker This person has a variety of duties that will be determined by the needs of an individual location. A typical worker would be responsible for a number of duties related to security, cleaning, upkeep, and repair, and, in some instances, servicing and maintenance of vehicles, and support equipment.

Sales Personnel Sales personnel can be new or used aircraft salespeople, ticket salespeople, parts salespeople, flight training salespeople, or flight store sales personnel. They are responsible for functions associated with selling. These include product knowledge, determining customer needs, explaining benefits, closing sales, and the administrative or support activities associated with selling.

Figure 5.3 → Skills Required of Various FBO Positions

Mid-Manager Knowledge of the overall business and the specific technical portion assigned. Managerial skills of planning, organizing, directing, and controlling to achieve specific objectives.

Accountant Technically qualified through training and experience to provide the level of accounting services required by the business. Includes normal accounting training, skills in aviation accounting, information system development, and financial reporting.

Office Manager Qualified to direct the administrative office. Normally requires skills in personnel programs, insurance, record keeping, filing systems, computer equipment operation, and perhaps lease administration.

Front Desk Manager Knowledge of the internal organization and the activities of all departments. Skillful in meeting and dealing with the public. Normally, must be knowledgeable and skilled in communications radio operation and dispatching.

Line Service Personnel Technically qualified to operate the assigned fueling and support vehicles. Knowledgeable about the systems and requirements of the aircraft to be serviced. Skilled in dealing with aircrew personnel of visiting aircraft.

Flight Instructor Must possess a flight instructor certificate, with ratings appropriate to the type of instruction assigned. Must be skillful in communicating with and training students.

Pilot Qualified in the assigned aircraft. Must possess the required operational experience and the technical knowledge of the type of flying, such as crop dusting, power line patrol, and so on.

Maintenance Skill requirements depend upon the level of maintenance accomplished by the organization. Generally will include knowledge related to airframe, powerplant, avionics, propeller, and other specific qualifications. FAR Part 65 includes specific requirements for these and other positions.

Custodial Skill or knowledge requirements vary, depending upon the job as developed. Generally includes skills in minor facility repair, security, cleaning and painting, and after-hours duties.

Sales Personnel Must have knowledge of the specific product being sold, be skillful in selling and in the marketing and administrative activities related to selling.

Combination Positions Skills necessary to meet the minimum requirements of the various jobs most likely to be assigned.

---

airline baggage handling, airline fueling, limousine service, auto parking, coffee shop or restaurant administration, and bar operation. Each of these activities may call for a separate position with individual responsibilities to fit the specific functional areas, or a group of these activities may be combined into one position.

Most aviation managers in a small organization are well aware of the necessity for individuals to have a high degree of flexibility and efficiency in order to perform several positions. There are simply not enough people employed to compartmentalize each activity neatly and positively. This problem should not, however, overshadow the need to think through each functional area and consciously plan for its satisfactory completion. It is not desirable to create artificial, impassable barriers around separate positions. Such barriers can reduce motivation as well as the overall flexibility of management and the organization. Employees, for the most part, need to see a path or several paths open to them within the company as they advance in their careers.

Laws and Regulations

Introduction

Employment law has become increasingly stringent over the years, and workers' rights generally get a more favorable treatment in the courts than employers' rights. Any employer thus needs to know not only the law, but also how to document a given situation so that the odds are improved of having the laws interpreted favorably in the courts. Employment law is divided into three primary areas:

> Discrimination in the workplace;
> Workplace Safety; and
> Workmen's Compensation.

Since case law evolves rapidly, as well as new legislation being passed, readers are advised to consult the Cornell web site mentioned below, and their own attorneys, for the most up-to-date information. What follows is a brief summary.

Figure 5.4 → Training Required for Aviation Maintenance Positions

- The FAA requires at least 18 months of work experience for an airframe, power plant, or avionics repairer's certificate. For a combined Airframe and Powerplant (A & P) Mechanic certificate, at least 30 months of experience working with both engines and airframes is required.
- Completion of a program at an FAA-certificated mechanic school can substitute for the work experience requirement. Applicants for all certificates also must pass written, oral, and practical tests and demonstrate that they can do the work authorized by the certificate.
- To obtain an inspector's authorization, a mechanic must have held an A & P certificate for at least 3 years. Most airlines require that mechanics have a high school diploma and an A & P certificate.
- Although a few people become mechanics through on-the-job training, most learn their job at one of about 200 trade schools certified by the FAA. About one-third of these schools award 2-and 4-year degrees in avionics, aviation technology, or aviation maintenance management.
- FAA standards established by law require that certificated mechanic schools offer students a minimum of 1,900 actual class hours. Courses in these trade schools normally last from 24 to 30 months and provide training with the tools and equipment used on the job.
- Aircraft trade schools are placing more emphasis on technologies such as turbine engines, composite materials—including graphite, fiberglass, and boron—and aviation electronics, which are increasingly being used in the construction of new aircraft. Less emphasis is being placed on "old" technologies, such as woodworking, fabric repair, and welding.
- Additionally, employers prefer mechanics that can perform a variety of tasks.
- Some aircraft mechanics in the Armed Forces acquire enough general experience to satisfy the work experience requirements for the FAA certificate. With additional study, they may pass the certifying exam.
- In general, however, jobs in the military services are too specialized to provide the broad experience required by the FAA. Most Armed Forces mechanics have to complete the entire training program at a trade school, although a few receive some credit for the material they learned in the service.
- In any case, military experience is a great advantage when seeking employment; many employers consider trade school graduates who have this experience to be the most desirable applicants.
- Courses in mathematics, physics, chemistry, electronics, computer science, and mechanical drawing are helpful, because they demonstrate many of the principles involved in the operation of aircraft, and knowledge of these principles is often necessary to make repairs.
- Courses that develop writing skills also are important because mechanics often are required to submit reports.
- FAA regulations require current experience to keep the A & P certificate valid. Applicants must have at least 1,000 hours of work experience in the previous 24 months or take a refresher course. As new and more complex aircraft are designed, more employers are requiring mechanics to take ongoing training to update their skills.
- Continuing technological advances in aircraft equipment necessitate a strong background in electronics—both for acquiring and retaining jobs in this field.
- FAA certification standards also make ongoing training mandatory. Every 24 months, mechanics are required to take at least 16 hours of training to keep their certificate. Many mechanics take courses offered by manufacturers or employers, usually through outside contractors.
- Aircraft mechanics must do careful and thorough work that requires a high degree of mechanical aptitude. Employers seek applicants who are self-motivated, hard working, enthusiastic, and able to diagnose and solve complex mechanical problems. Agility is important for the reaching and climbing necessary to do the job. Because they may work on the tops of wings and fuselages on large jet airplanes, aircraft mechanics must not be afraid of heights.
- As aircraft mechanics gain experience, they may advance to lead mechanic (or crew chief), inspector, lead inspector, or shop supervisor positions. Opportunities are best for those who have an aircraft inspector's authorization.

Source: BLS. See www.bls.gov/ooh/Installation-Maintenance-and-Repair/Aircraft-and-avionics-equipment-mechanics-and-technicians.htm

Employment Discrimination

The following overview of laws relating to employment discrimination is provided courtesy of the Legal Information Institute at Cornell University, a free legal website.[9]

"Employment discrimination laws seek to prevent discrimination based on race, sex, religion, national origin, physical disability, and age by employers. There is also a growing body of law preventing or occasionally justifying employment discrimination based on sexual orientation. Discriminatory practices include bias in hiring, promotion, job assignment, termination, compensation, and various types of harassment. The main body of employment discrimination laws is composed of federal and state statutes. The United States Constitution, and some state constitutions, provide additional protection where the employer is a governmental body or the government has taken significant steps to foster the discriminatory practice of the employer.

"The Fifth and Fourteenth Amendments of the United States Constitution limit the power of the federal and state governments to discriminate. The Fifth Amendment has an explicit requirement that the federal government not deprive individuals of "life, liberty, or property," without due process of the law. See U.S. Const. amend. V. It also contains an implicit guarantee that each person receive equal protection of the laws. The Fourteenth Amendment explicitly prohibits states from violating an individual's rights of due process and equal protection. See U.S. Const. amend. XIV. In the employment context the right of equal protection limits the power of the state and federal governments to discriminate in their employment practices by treating employees, former employees, or job applicants unequally because of membership in a group (such as a race or sex). Due process protection requires that employees have a fair procedural process before they are terminated if the termination is related to a "liberty" (such as the right to free speech) or property interest. State constitutions may also afford protection from employment discrimination.

"Discrimination in the private sector is not directly constrained by the Constitution, but has become subject to a growing body of federal and state statutes.

"The Equal Pay Act amended the Fair Labor Standards Act in 1963. The Equal Pay Act prohibits paying wages based on sex by employers and unions. It does not prohibit other discriminatory practices bias in hiring. It provides that where workers perform equal work in jobs requiring 'equal skill, effort, and responsibility and performed under similar working conditions,' they should be provided equal pay. The Fair Labor Standards Act applies to employees engaged in some aspect of interstate commerce or all of an employer's workers if the enterprise is engaged as a whole in a significant amount of interstate commerce.

"Title VII of the Civil Rights Act of 1964 prohibits discrimination in many more aspects of the employment relationship. It applies to most employers engaged in interstate commerce with 15 or more employees, labor organizations, and employment agencies. The Act prohibits discrimination based on race, color, religion, sex, or national origin. Sex includes pregnancy, childbirth, or related medical conditions. It makes it illegal for employers to discriminate in hiring, discharging, compensation, or terms, conditions, and privileges of employment. Employment agencies may not discriminate when hiring or referring applicants. Labor organizations are also prohibited from basing membership or union classifications on race, color, religion, sex, or national origin.

"The Nineteenth Century Civil Rights Acts, amended in 1993, ensure all persons equal rights under the law and outlines the damages available to complainants in actions brought under the Civil Rights Act of 1964, Title VII, the Americans with Disabilities Act of 1990, and the Rehabilitation Act of 1973.

"The Age Discrimination in Employment Act (ADEA) prohibits employers from discriminating on the basis of age. The prohibited practices are nearly identical to those outlined in Title 7. An employee is protected from discrimination based on age if he or she is over 40. The ADEA contains explicit guidelines for benefit, pension, and retirement plans.

"The Rehabilitation Act's purpose is to 'promote and expand employment opportunities in the public and private sectors for handicapped individuals,' through the elimination of discrimination, and through affirmative action programs. Employers covered by the act include agencies of the federal government and employers receiving federal contracts over $2500 or federal financial assistance. The Department of Labor enforces section 793 of the act, which refers to employment under federal contracts. The Department of Justice enforces section 794 of the act, which refers to organizations receiving federal assistance. The EEOC enforces the act against federal employees and individual federal agencies promulgate regulation pertaining to the employment of the disabled.

"The Americans with Disabilities Act (ADA) was enacted to eliminate discrimination against those with handicaps. It prohibits discrimination based on a physical or mental handicap by employers engaged in interstate commerce and state governments. The type of discrimination prohibited is broader than that explicitly outlined by Title VII.

"The Equal Employment Opportunity Commission (EEOC) interprets and enforces the Equal Payment Act, age Discrimination in Employment

Act, Title VII, Americans with Disabilities Act, and sections of the Rehabilitation Act. The Commission was established by Title VII. Its enforcement provisions are contained in section 2000e-5 of Title 42, and its regulations and guidelines are contained in Title 29 of the Code of Federal Regulations, part 1614."

State laws also provide extensive protection from employment discrimination. Some laws extend similar protection as provided by the federal acts to employers who are not covered by those statutes. Other statutes provide protection to groups not covered by the federal acts. A number of state statutes provide protection for individuals who are performing civil or family duties outside of their normal employment.

Workplace Safety

Again, the Legal Information Institute of Cornell University provides a summary:[10]

"Workplace safety and health laws establish regulations designed to eliminate personal injuries and illnesses from occurring in the workplace. The laws consist primarily of federal and state statutes. Federal laws and regulations preempt state ones where they overlap or contradict one another.

"The main statute protecting the health and safety of workers in the workplace is the Occupational Safety and Health Act (OSHA). Congress enacted this legislation under its Constitutional grant of authority to regulate interstate commerce. OSHA requires the Secretary of Labor to promulgate regulations and safety and health standards to protect employees and their families. Every private employer who engages in interstate commerce is subject to the regulations promulgated under OSHA.

"In order to aid the Secretary of Labor in promulgating regulations and enforcing them the act establishes the National Advisory Committee on Occupational Safety and Health. The Secretary of Labor may authorize inspections of workplaces to ensure that regulations are being followed, examine conditions about which complaints have been filed, and determine what regulations are needed. If an employer is violating a safety or health regulation, a citation is issued. The act establishes the Occupational Safety and Health Review Commission to review citation orders of the Secretary of Labor. The Commission's decision is also subject to judicial review. The Secretary of Labor may impose fines with the amounts varying according to the type of violation and length of noncompliance with the citation. The Secretary of Labor may also seek an injunction to restrain conditions or practices which pose an immediate threat to employees. The act also establishes the National Institute for Occupational Safety and Health, which, under the Secretary of Health and Welfare, conducts research on workplace health and safety and recommends regulations to the Secretary of Labor. Federal agencies must establish their own safety and health regulations. The regulations that have been promulgated under OSHA are extensive, currently filling five volumes of the Code of Federal Regulations.

"Under OSHA states are not allowed, without permission of the Secretary of Labor, to promulgate any laws that regulate an area directly covered by OSHA regulations. They may, however, regulate in areas not governed by federal OSHA regulations. If they wish to regulate areas covered by OSHA regulations they must submit a plan for federal approval. The amount of state regulation varies greatly. California is an example of a state that has chosen to adopt many of its own regulations in place of those promulgated under OSHA."

Businesses with eight or more employees are required to record work-related illnesses or injuries. Other OSHA requirements are detailed, complex, and variable depending on the nature of the work-site and work. Consultation with OSHA or state department of labor representatives is advised. As stated on the U.S. Department of Labor website (see www.osha.gov/stateplans) "State Plans are OSHA-approved workplace safety and health programs operated by individual states or U.S. territories There are currently 22 State Plans covering both private sector and state and local government workers, and there are six State Plans covering only state and local government workers. State Plans are monitored by OSHA and must be at least as effective as OSHA in protecting workers and in preventing work-related injuries, illnesses, and deaths". In these (State Plans) jurisdictions, therefore, contact with the state labor department, as implementer of OSHA requirements, is recommended.

Workers Compensation: An Overview

This material also is supplied courtesy of the Legal Information Institute of Cornell University:[11]

"Workers' Compensation laws are designed to ensure that employees who are injured or disabled on the job are provided with fixed monetary awards, eliminating the need for litigation. These laws also provide benefits for de pen dents of those workers who are killed because of work-related accidents or illnesses. Some laws also protect employers and fellow workers by limiting the amount an injured employee can recover from an employer and by eliminating the liability of coworkers in most accidents. State Workers Compensation statutes establish this framework for most employment. Federal statutes are limited to federal employees or those workers employed in some significant aspect of interstate commerce.

"The Federal Employment Compensation Act provides workers compensation for non-military, federal employees. Many of its provisions are typical of most worker compensation laws. Awards are limited to 'disability or death' sustained while in the performance of the employee's duties but not caused willfully by the employee or by intoxication. The act covers medical expenses due to the disability and may require the employee to undergo job retraining. A disabled employee receives two thirds of his or her normal monthly salary during the disability and may receive more for permanent physical injuries, or if he or she has de pen dents. The act provides compensation for survivors of employees who are killed. The act is administered by the Office of Workers' Compensation Programs.

"The Federal Employment Liability Act (FELA), while not a workers' compensation statute, provides that railroads engaged in interstate commerce are liable for injuries to their employees if they have been negligent.

"The Merchant Marine Act (the Jones Act) provides seamen with the same protection from employer negligence as FELA provides railroad workers.

"Congress enacted the Longshore and Harbor Workers' Compensation Act (LHWCA) to provide workers' compensation to specified employees of private maritime employers. The Office of Workers' Compensation Programs administers the act.

". . . . California's Workers' Compensation Act provides an example of a comprehensive state compensation program. It is applicable to most employers. The statute limits the liability of the employer and fellow employees. California also requires employers to obtain insurance to cover potential workers' compensation claims, and sets up a fund for claims that employers have illegally failed to insure against."

Comparable Worth

The issue of comparable worth means equal pay for jobs of comparable difficulty. The concept had its start in the mid-1970s when a landmark study in the state of Washington reviewed state salaries and found that, on average, women earned 20 percent less than men for comparable jobs. In 1981, a Supreme Court ruling supported this finding. The criteria used by comparable worth studies may be of value in reviewing jobs for any purpose. They include:

> Skill, knowledge, and education required;
> Effort (e.g., mental demands, latitude, judgment and difficulty of problems);
> Accountability-scale of executive decisions; and
> Working conditions.

Each of these four factors is given a weight and the sum total of these weights ranks the job with all other jobs having the same weight, regardless of whether there is any similarity in jobs. Positions with extremely desirable ancillary conditions, such as free or almost free air travel if one works for the airlines, would get a negative weighting for that factor.

The first step in developing an equitable pay system for a business involves job definition.

One way to clarify what each employee's job actually consists of, is to have them fill out a simple form stating:

> Job title;
> Reporting relationship;
> Specifications;
> Primary function;
> Main duties;
> Other duties;
> Job requirements;
> Technical/administrative complexity;
> Responsibility for dollar results;
> Responsibility for supervision and training of others; and
> Unusual working conditions.

Some employees may turn out to be doing more or different tasks than thought, and more than one may claim primary responsibility for the same thing. These and other such issues may arise. The review process should also allow you to determine what functions may need to be covered by new staff.

Payroll Taxes and Deductions

In summary, the likely items will include:

> Withholding of employee income tax;
> Social Security and FICA;
> Unemployment taxes; and
> Medicare taxes.

An FBO manager is advised to obtain the professional assistance and direction of an accountant in establishing the financial records and maintaining the proper deductions. It is vitally important that the business and financial plan for the company include adequate funds for the required items.

Employee Access to Records and Fair Information Practices

Although there are no federal laws standardizing employee information practices, a number of states (e.g., California, Connecticut, Illinois, Maine, Michigan, New Hampshire, Oregon, Pennsylvania, and Wisconsin) in recent years have established laws in regard to employee access to, and ability to change, data in personnel files. There has also been increasing attention to the issue of employee privacy and the possible misuse of personnel data (e.g., in relation to credit, medical, or other agencies). The personnel policy manual should contain a section on employee data, its use, and the company's policy on security and access to files by employees, their supervisors, and other persons or entities, both within and outside the company.

Recruiting Qualified Candidates

The recruiting function for an FBO business will vary in its scope and complexity with the degree of specialization involved. If the business needs a bookkeeper or file clerk, local general business sources of candidates will be appropriate. If a very senior aviation technical position requiring current certification or licensure is necessary, sources for persons such as bookkeepers will normally be inadequate and a more far-reaching search will be necessary, using specialized industry publications and contacts. While some positions require specific certifications and licenses and are thereby constrained, the manager should keep in mind that career paths can change radically and motivated employees with strong values and excellent work habits should have the opportunity to move in unusual directions and learn new expertise. Considering existing employees is always an excellent place to start when there is a vacancy to fill, for it also assures staff that the company potentially offers an upward career track for all.

Industry Contacts

For a position requiring aviation skills and knowledge, likely sources of qualified candidates include:

1. Friends or acquaintances of existing personnel (some businesses value this approach so much that they will compensate staff members who provide contacts leading to a new hire).
2. Individuals in competing businesses.
3. Professional and business associations, either general-business or aviation-related.

Whether your pursuit of these channels is local, regional, or national depends on the urgency, uniqueness, and scarcity of the labor required. Such a search, if extended into a large network, may become almost a full-time job. However, it is something that a business owner or even a full-time personnel director must accommodate, despite other normal demands on his or her time.

Recruiters

Recruiters and executive search or "headhunter" agencies are usually national businesses that, for a fee of 15 to 25 percent of the first year's salary, will seek out qualified specialists, screen them by phone or electronic means, in writing, or in person. Out of dozens of possibilities, they present the employer with three to five individuals to interview over the phone or via electronic meeting and then meet in person. While this is a costly process, a reputable recruiter will present only interested candidates who have met all of your criteria. In most cases, all that is left for the employer to do is pick the person with whom the rapport is best. The employer pays the fee, any travel costs, and the cost of relocation, if necessary. Normally, the employer has 30 days to change their mind and pay nothing if the person does not work out. In the interest of self-preservation and good business ethics, a good recruiter will not make the first move to recruit that person away from a placement he or she has already made.

Employment Agencies

Employment agencies operate in somewhat the same manner as recruiters, except that the candidate, not the employer, may pay all or part of the fee. Their labor pool tends to be more regional or metropolitan. Agencies tend to work in regional or metropolitan labor markets and deal with more general types of skills, such as office workers, accountants, purchasing agents, and IT specialists who have reasonable transferability of skills from one industry to another. There is often intensive competition among agencies. Employer-paid fees are becoming more common, and it may be advantageous to work with just one person in one reputable agency in cases where the national, specialized labor pool of the recruiter is not necessary.

Colleges and Trade Schools

These institutions are often among the best sources of personnel for specialized aviation positions. The school's curriculum and national standing are available as indicators of consistent quality before the potential employer even sees a specific individual. The faculty will, in time, begin to know the type of person desired and the employer's requirements. School schedules often facilitate student employment, as well as part-time and seasonal hiring, which can accommodate an FBO's varying personnel needs. An advantage of hiring student labor is that a person can be evaluated for a considerable period of time before a permanent commitment is made.

Particularly excellent sources of well-prepared and highly motivated employee candidates are the college and university aviation programs accredited by the Aviation Accreditation Board International (AABI). AABI (http://www.aabi.aero/) accredits aviation degree programs at 2-year and 4-year colleges and universities that have completed a rigorous and extensive application and review process, which demonstrates their adherence to the requirements of the AABI Criteria and policies. The AABI Criteria are developed, and kept current and relevant, through a vetting process that involves strong input and guidance from industry representatives and current practitioners.

Cooperative education (usually referred to as co-ops) and internships are another excellent way to utilize college and university students. Co-op programs can be set up with an option for payment or non-payment. In many of these programs, the student receives college credit from his or her institution; the benefits of working in an industry environment while earning college credit are significant. Again, this arrangement gives the employer an opportunity to preview a prospective employee (and the intern/coop student to evaluate the employer) with little investment required.

Advertising

Many employers consider advertising a last resort. Although it is not expensive when compared with some of the alternatives, its results can be very time-consuming. Typically, especially when e-mail responses are encouraged, many people respond who are not qualified. The advertisement may run a day or more after you have selected someone, yielding responses that cannot be used. Methods for obtaining the best results from advertising follow:

1. Use only major newspapers, web sites or trade periodicals.
2. Use only a box number and do not totally describe the firm to the extent that a person can guess its name.
3. Ask all respondents to fill out an application form; only those genuinely interested will do so.
4. Use a very specific job description and essential (vs. "nice to have") skill requirements to screen candidates prior to interviewing them in person.

Owing to the large number of immigrants seeking permanent residency in the U.S., widespread generic advertising searches, especially those that are web-based, may result in many candidates who would require employer sponsorship in order to get the appropriate U.S. visa. Web-based advertising also tends to generate candidates from a very large geographic area. The employer may need to write the ad specifying whether or not sponsorship is under consideration, and specifying relocation benefits (if any), should a non-local candidate be offered the job.

Selecting Employees

To a certain extent, the business manager is always looking for suitable talent, whether there is a specific opening or not. Efficient handling of casual or unsolicited inquiries requires a selection process that existing employees must know about and can help to implement. However, *all* applicants, whether they come to the firm through a formal or informal process, need to be screened as to suitability.

Preliminary Screening

The preliminary screening or initial interview provides a ten-minute opportunity to determine whether the individual is a qualified prospect. Figure 5.5 is the job description format.[12] If this preliminary process suggests that the person is suitable, he or she can be told about any available positions and asked to fill out an application form. If, on the other hand, the preliminary screening suggests the candidate is definitely unsuitable because of qualifications or attitude, the person can be told that the firm will contact him or her if a suitable situation arises. Figure 5.6 below offers a screening guide.

The Application Form

The primary value of an application form is its systematic and impersonal collection of data. It establishes a consistent database for all applicants and represents one step in ensuring equal opportunity. It also provides the means, through names of former employers and references, of conducting a more detailed investigation on a person before hiring them. It should be tailored to your own needs, carefully addressing current employment law, and will probably contain at least:

> Name, address, and phone number;
> Previous employer's address, immediate supervisor, supervisor phone, starting and ending salary, reason for leaving, dates of employment;
> Education, formal and informal;
> Height, weight, physical limitations, or disabilities, (this information should be requested only if applicable to job requirements, such as the ability to lift heavy objects).

The application form provides a basis for interview questions and, on occasion, may be another screening mechanism leading to a decision not to further interview the person. If the person is hired, the form becomes a useful background item in their personnel file. Figure 5.7 is an application blank based on one developed by Piper Aircraft in selecting flight instructors.

The Interview

A full interview may last an hour or more and should rarely last less than 20 minutes if an offer is going to be made. Some experts strongly recommend that more than one interviewer be brought into the process, especially when the manager has a poor record of picking good staff, and/or where he or she has a tendency to pick people like him-or herself, regardless of the job. A number of studies in recent years suggest that philosophy, style, and values may be as critical as technical skills in causing an employee to fit well and be productive. It is important for the employer not to ask leading questions in such areas, nor in areas that could be interpreted as discriminatory, as the candidate may try to present only what he or she thinks is wanted. If certain skills, licenses, or education are prerequisites for the job, it may be appropriate to develop a checklist for use with every candidate. The interviewer should guard against forming such a liking for the candidate that he or she forgets to ask about all the key issues. It will not pay to hire a likable, but incompetent person! The following principles should guide most interviews:

> Make a checklist of questions to ask, preferably including questions about how the person would handle problems and crises that typically appear in your business;
> If several candidates are being interviewed, use the same set of questions for each, and keep interview notes on file; some employers ask applicants to arrive 30 minutes early and provide the questions to them at this time so that they can be prepared;
> Set up a quiet atmosphere with no interruptions;
> Put the applicant at ease and give him or her your full attention;
> Keep the discussion at a level suitable to the applicant;
> Listen attentively;
> Never argue;
> Observe style and manner; for example, ability to answer directly and on the subject;
> Provide the applicant with pertinent information about the job;
> Note key replies and concerns; and
> Encourage candidate questions.

Figure 5.5 → Job Description Format

Date _____

Organization _____

1. *Job Identification:*
   A. Job Title _____
   B. Department _____
   C. Employee Type: Exempt   Nonexempt
   D. Reports to _____
   E. Job Number _____
   F. Grade _____

2. *Work Performed:*

   A. *Summary Statement:* (A statement of the job duties sufficient to identify the job and differentiate the duties that are performed from those of other jobs.)

   B. *Major Duties:* (Should indicate (a) what employee does, (b) how he or she does it and (c) why he or she does it. The description should indicate the tools and equipment employed, the material used, procedures followed, and the degree of supervision received. May also include relationship to other jobs.)

3. *Working Conditions:*
   (General physical environment under which the job must be performed. May include such things as lighting, degree of isolation and conditions such as hot, cold, dusty, or cramped. Job hazards identified.)

4. *Equipment Used:*
   (Identifies the equipment to be used in the performance of normal duties.)

5. *Job Requirements:*

   A. *Education:* (May include the minimum formal education, including special courses or technical training considered necessary to perform the job.)

   B. *Training and Experience:* (Minimum amount and type required in order that employee holds a job, expressed in objective and quantitative terms such as years and months.)

   C. *Initiative and Ingenuity:* (Identifies the degree, level and importance to the job and the relationship to others.)

   D. *Physical Demands:* (Identifies the amount of physical effort required and the length of time such effort must be expounded. Normally includes: walking, stooping, lifting, handling, or talking.)

This statement reflects the general details necessary to describe the principal functions of the job identified and shall not be construed as a detailed description of all the work requirements that may be necessary in the job.

Approved by: _____

Source: J. D. Richardson

Figure 5.6 → Screening Guide to Assist in Hiring Review

Date _____

Phone _____

Name _____

Address _____
  No.   Street   City   State   Zip Code

Ask the following questions; but stop if the applicant gives a wrong answer.

1. How long have you lived at this address? _____
   (Must be at least a month.)

2. What experience have you had? _____
   (Not important unless you want an experienced person.)

3. What kinds of work have you done? _____
   (Is this what you need?)

4. How much pay do you expect? _____
   (Is this too much?)

5. Are you willing to work evenings? _____ Yes _____ No; Saturdays? _____ Yes _____ No;
   Sundays and holidays? _____ Yes _____ No; Split hours? _____ Yes _____ No.
   (Can you use this person?)

6. Do you want full-time work? _____ Full-time _____ Part-time _____ Hours per week
   (Can you use this person?)
   Are you willing to attend training programs conducted by our suppliers? _____

7. Are you over eighteen? _____ If you are hired, you'll need to provide proof
   of your age and visa /nationality status. Will you be able to do that? _____ Yes _____ No

8. Do you have a job now? _____ Yes _____ No; (If no:) How long have you been out of work?
   (Must be less than 2 months unless reasonable reason.)
   How did that happen? _____
   (Reason must be very good—such as sickness, school, etc.)

9. What other experience have you had? _____
   (Will this be helpful?)

10. What experience have you had with aircraft? _____
    (Will this be helpful?)

11. Can you drive a car [if needed for the job]? _____ Yes _____ No; (if yes:) What kind of driver's license
    do you have? _____(OK?). May I see it? _____ Out-of-date, _____ Up-to-date; What notations are
    there on the license? _____
    (Is this person a safe driver?)

12. Can you fly [if needed/desirable for the job? _____ Yes _____ No; (If yes:) What certifcates and ratings
    do you have? _____(OK?)

If the applicant **is not qualified** because the answers to any of these questions means you can't use him or her or because you feel the person won't fit in with your organization, say, "I'm sorry but you don't seem to be the person we need right now. However, I have your name and phone number and I'll be glad to give you a call if we can use you some other time. Thanks a lot for talking with me." **If the applicant seems OK, say,** "Fine, I'd like you to complete an application. You can do it right here. Would that be OK?"

Figure 5.7 ✈ Application for Flight Instructor Position

Application for Flight Instructor Position
(All information treated confidentially)

Date _____

Name (print) _____   Telephone number _____   Is this in your name? _____

Present address _____   How long have you lived there? _____
    No.   Street   City   State   Zip

Previous address _____   How long did you live there? _____
    No.   Street   City   State   Zip

Business address _____   Business Telephone number _____
    No.   Street   City   State   Zip

Are you a citizen or legal worker of the U.S.?   ☐ Yes ☐ No   Soc. Sec. No. _____

Why are you applying to this company? _____

1. What influenced you to enter general aviation? _____

2. Do you plan a general aviation career?   ☐ Yes ☐ No

Education

Type of School	Name of School	Courses Majored In	Check Last Year Completed	Graduate? Degrees Received	Last Year Attended
College			1 2 3 4		19
Other			1 2 3 4		19

Jobs While in School and During Summer

Scholastic standing in H.S. _____   In College _____
(Designate top 25 percent, middle 50 percent, lowest 25 percent)

Awards and honors received _____

Favorite subjects _____   Least liked _____

Extracurricular Activities (exclude racial, religious, or nationality groups)

In high school _____   In college _____

Offices held _____   Offices held _____

What scholarships or fellowships have you received? _____

What hobbies do you have? _____

Service in U.S. Armed Forces

What is your current military service status? _____

If exempt or rejected, what was the reason? _____

Have you served in the U.S. Armed Forces   ☐ No   ☐ Yes; (if yes) Date active duty started _____

Which force?   ☐ Army   ☐ Air Force   ☐ Navy   ☐ Marines   ☐ C.G.:

What branch of that force? _____   Starting rank _____

Overseas: Date(s) _____   Location(s) _____

Date of discharge _____   Rank at discharge _____

What citations and awards have you received? _____

What special training did you receive? _____

Continued

Figure 5.7 → Application for Flight Instructor Position —Cont'd

Work History
Include (a) self-employment and (b) secondary or moonlighting jobs (mark the latter with an asterisk)

Beginning with the most recent, list below the names and addresses of all your employers: a. Company name b. Address and telephone number	Kind of Business	Time Employed From Mo.	Yr.	To Mo.	Yr.	
1. a. b.						
2. a. b.						
3. a. b.						
4. a. b.						
5. a. b.						

Indicate by number_____ any of the above employees whom you <u>do not</u> wish us to contact.

Have you had sales experience? ☐ Yes ☐ No

If "yes" please specify type sales: (Big ticket, over the counter, delivery sales, route sales to homes, technical sales, specialty sales-tangibles; books, etc.; intangibles: insurance or services)

List employer(s) for whom you sold:

Name	Address	Lines Sold	Year(s)

If necessary, are you willing to work nights? _____ Weekends? _____

Personal References (Not former employers or relatives)	Address	Phone Number
1.		
2.		
3.		

Date of last FAA physical exam, if applicable. _____

Class _____

Restrictions _____

Figure 5.7 ✈ Application for Flight Instructor Position —Cont'd

Work History
Include (a) self-employment and (b) secondary or moonlighting jobs (mark the latter with an asterisk)

Nature of Work at Start	Earnings Per Month at Start	Nature of Work at Leaving	Earnings Per Month at Leaving	Reason for Leaving	Name of Immediate Supervisor
					Name Title
					Name Title
					Name Title
					Name Title
					Name Title

Flying Experience

Certificate and Ratings: (Give Date Each Acquired)

A. _____ D. _____ G. _____
B. _____ E. _____ H. _____
C. _____ F. _____ I. _____

Flight Schools Attended   From   To   Course   Graduation

_____
_____
_____

Pilot Experience (Other Than Instructor Time)   **Instructor Experience** (Hours)

Pilot In Command   _____   Total Instructor Time   _____
Copilot Time   _____   Private   _____
Instrument Time   _____   Commercial   _____
Multi Engine Time   _____   Night   _____
Night   _____   Instrument   _____
                    _____   Multi Engine   _____

Aircraft Flown: (Give Approximate Hours In Each)

A. _____ E. _____ I. _____
B. _____ F. _____ J. _____
C. _____ G. _____ K. _____
D. _____ H. _____ L. _____

Continued

Figure 5.7 → Application for Flight Instructor Position —Cont'd

Personal Instruction Activity:

% Passed FAA Exam

A. Number of Pilots Instructed. _____ _____

B. Number of Commercial Pilots Instructed. _____ _____

C. Number of Instrument Pilots Instructed. _____ _____

D. Number of Multi Engine Pilots Instructed. _____ _____

E. Number of Instructors Instructed. _____ _____

F. Number of ATR Pilots Instructed. _____ _____

G. Number of Glider Pilots Instructed. _____ _____

H. Number of Float Plane Pilots Instructed. _____ _____

Accidents:

Date: _____

Circumstances: _____

_____

Damage or Injury: _____

_____

Forced Landings:

Date: _____

Reason: _____

Damage or Injury: _____

_____

Has your pilot certificate ever been suspended or revoked? _____

_____

Other Statements

Are there any experiences, skills, or qualifications that you feel would especially fit you for work with this company?

_____

_____

What are your plans or aims for the future? _____

_____

If your application is considered favorably,
on what date will you be available for work? _____

How much notice will you require? _____ Days

Signature _____

---

*Not to be asked unless job related.

Note that the above factors are somewhat subjective and again, notes should be taken and kept regarding the employer's assessment of each. Using a 3–5 person interview panel can help reduce bias.

It is vital to avoid asking discriminatory questions. The only issues that are pertinent (and acceptable to ask about) relate to the person's ability to do the job. For most pilot positions a current FAA medical certificate would be adequate, although temporary medical conditions, such as a broken limb or pregnancy could be pertinent. It is not illegal to let the candidate volunteer such information that would be illegal to request.

Some questions that may seem harmless enough can be interpreted as discriminatory. For example, asking a potential employee if he or she rents or owns his/her own home, whether he or she is married, whether the candidate has any children, which organizations or societies he/she belongs to (other than professional organizations), or the person's age are all potentially risky discriminatory questions, and should be avoided. Figure 5.8 provides an example of questions that might be considered appropriate or inappropriate to ask during an interview.

Figure 5.9 provides twelve key factors to look for in interviewing or screening. Note that none of these factors relate to technical skill or training. Interviewers should complete the assessment at the end of each interview, separately from any record of how the candidate answered the formal questions.

Testing and Investigation

In many cases, testing for FBO jobs will not be necessary as long as checks are made to confirm that all claimed qualifications are current. For typing, a speed and accuracy test is appropriate. An actual document, such as a letter or memo, should be required, as the ability to lay out work attractively is part of the document production person's job. In other cases, the quality of technical skills should be assessed by discussion with former employers, clients, or teachers, if these parties are willing to assist. An increasing number of employers no longer provide competency-related references, offering only confirmation of the person's dates of employment.

Background checks under pilot record and maintenance requirements are not mandated for Part 91 flight operations. Although not strictly required, many employers do find it prudent, however, to require some checks as a matter of course, including drug and alcohol testing. Insurance companies sometimes require such checks. An inquiry of the National Driver Register in Washington, DC, will indicate any driving-related problems that the candidate may have with alcohol or drugs. As the Transportation Security Administration focuses more on general aviation, owners and managers can expect more stringent background check requirements that are likely to vary with the type of position being sought.

Personality testing is another option. People do not always know themselves very well; they may have picked a technical area of training that does not fit ideally with their basic temperament, or they may have personality and managerial skills well beyond the level they envisage for themselves. Various personality style-testing instruments are available. They generally group people loosely into four categories: leaders, influencers, detail workers, and supporting types.[13] Each of us has characteristics of each category but tend to be predominantly one type. We may also change as we mature. A shy individual who enjoys working with numbers should ideally not be put on the front desk to greet people. Someone friendly should be on the line or the desk, so long as they are also results-oriented and can successfully juggle several clients. Technical skills are generally more teachable at any stage in life than personality traits; changing core personality traits is difficult. A person is happiest and most productive when both personality and technical skills are well-used. A good manager needs strength in all the personality styles and the ability to "switch gears" from one style to another as needed, to be able to interact most effectively with employees who operate in a different mode.

This type of testing requires investment of money and time. If it is carefully employed, it can lead to more successful screening and deployment of personnel. If existing personnel take the test, much will be learned that will be a longer-term guide to personality assessment and appropriate assignments. Excellent materials also exist for training employees having varied personality types to work best with those having different styles.

Background and References

Reference checking is critical prior to making a job offer. The most valuable contacts are generally those who have known the person in a business, rather than

Figure 5.8 → **Appropriate and Inappropriate Questions During Interactions with Candidates**

SUBJECT	APPROPRIATE	INAPPROPRIATE
ADDRESS	How long have you lived "in this area?"	List of previous addresses, how long have you lived at specific address?
AGE	None.	Questions about age or questions that would reveal age. Requests for birth certificate.
ARREST RECORD	State law may permit questions on pending charges if related to job, i.e., security or sensitive jobs.	Questions about pending charges for jobs other than those mentioned.
BIRTHPLACE	None.	Birthplace of applicant or applicant's parent's, spouse or other close relatives.
CITIZENSHIP	May ask questions about legal authorization to work in the specific position if all applicants are asked.	May not ask if person is a U.S. citizen.
CONVICTIONS	May ask if any record of criminal convictions and/or offenses exists, if all applicants are asked.	Questions about convictions unless the information bears on job performance. Questions that would reveal arrests without convictions.
DISABILITY	May ask about applicant's ability to do job-related functions.	Question (or series of questions) that is likely to solicit information about a disability.
EDUCATION	Inquiries about degree or equivalent experience.	Questions about education that are not related to job performance.
FAMILY	None.	Number and ages of children, child bearing/rearing queries.
MARITAL OR FAMILY STATUS	Whether applicant can meet work schedule or job requirements. Should be asked of both sexes.	Any inquiry about marital status, children, pregnancy, or child-care plans.
MILITARY	You may ask if a candidate has served in the Armed Forces of the United States or in a State Militia.	You may not ask about military service in the armed forces of any country except the U.S., nor may you inquire into one's type of discharge.
NATIONAL ORIGIN	May ask all applicants if legally authorized to work in this specific position.	May not ask if person is a U.S. citizen.
ORGANIZATIONS	Inquiries about professional organizations related to the position.	Inquiries about professional organizations suggesting race, sex, religion, national origin, disability, or sexual orientation.
PERSONAL FINANCES	None.	Inquiries regarding credit record, owning a home, or garnishment record.
POLITICAL AFFILIATION	None.	Inquiries about membership with a political party.

Source: Purdue University Faculty Search and Screen Procedures Manual

Figure 5.9 → Twelve Steps to Assessing a Job Candidate's Suitability

1. Has the applicant done some homework about the company?
2. Is the applicant dressed appropriately for the interview (example: one degree more formal than the position's current employees)?
3. Did the person arrive alone for the interview?
4. Did he/she arrive on time or call ahead to explain any delay and ask if the interview can still occur?
5. How did he/she greet you?
6. Is she/he assertive and confident?
7. Is he/she argumentative?
8. Does the person present a positive attitude about the most recent employer?
9. Does the candidate have a positive attitude about this company?
10. Does the applicant seem to be observant and learning during the interview process?
11. Does the person project enthusiasm, confidence, energy and dependability?
12. Does the person project loyalty, honesty, pride in work, and a desire to offer work and service for pay received?

---

a personal, context. Any personal references listed have probably been specifically selected by the candidate because of their likelihood of making pleasant remarks about him or her. Previous employers, business associates, and clients may be willing to give a candid appraisal of a person's strong and weak points (everyone has both) once they know the type of job and company for which a person is being considered. If a negative comment comes up that is unexpected or that contradicts the candidate, it may be desirable to give him or her a chance to clarify.

Telephone reference checking is generally more reliable than written requests because it provides an opportunity to ensure that the person being asked really remembers the candidate and is not just looking up file information. It also allows probing questions and prompting on the part of the reference checker. Figure 5.10 provides a form for use in telephone reference checking. As previously noted, many companies no longer allow their employees to provide references because of legal liability; however, they will verify employment. Despite this difficulty, references should still always be checked prior to hiring.

Credit checks may also be a means of obtaining information on a prospective employee and are particularly appropriate for jobs involving money. In this, as in all hiring matters, it is crucial to use consistent procedures. The request for a credit report must be communicated to the applicant; it is desirable to seek the candidate's permission.

Rules continue to evolve about providing a higher level of immigration, police, and other checks for candidates seeking sensitive jobs in aviation. The business manager must stay abreast of the new regulations as they are promulgated by the Transportation Security Administration, and act accordingly.

The Physical Examination

With the exception of pilot positions, for which a current and appropriate class FAA medical certificate may be sufficient to ensure that the applicant can handle (and be medically qualified for) the job, a physical examination can be important in hiring. Physicals should especially be used when the job requires attributes such as:

> Strength;
> Good vision;
> Ability to stand continuously;
> Freedom from allergies;
> Unusual stamina; or
> Other physical abilities.

However, asking health questions and conducting the company's own physical exam can put the firm at risk in relation to discrimination issues. Obtain legal advice, in particular, for this area.

The Job Offer

The job offer may be made in person, by letter, or by telephone. It should include not only the title of the position, but also the compensation, hours, fringe benefits, supervision, and general working rules. Some firms use a job contract that both parties sign. Most people, when offered a job, ask for a few days to decide,

Figure 5.10 ✈ Telephone Reference Check Guide

Telephone Reference Check on Applicant _____
                                                    Name of Applicant

_____   _____
Person Contacted   Position

_____ _____ _____
Company   City and State   Telephone Number

QUESTIONS

1. I wish to verify some of the information given to us by Mr. or Ms. (name), who has applied for a position with our firm. Do you remember him or her? What were the dates of his or her employment with your Company?

   From _____ 19 ____ To _____ 19 ____
                 Do dates check?

2. What was he or she doing when she started?
   _____
             Did she exaggerate?

   When she left?
   _____
             Did she progress?

3. She says she was earning $ ____ per ____ when she left. Is that right?
   ☐ Yes ☐ No; _____
             Did she falsify?

4. How much of this was salary?
   $ _____ Commission? _____

5. How was her attendance?
   _____
       Conscientious? Health problems?

6. What type of instructing did she do?
   _____
             To whom?

7. How did her efforts compare with others?
   _____
       Industrious? Competitive?

8. Did she supervise anyone else?
   ☐ No, ☐ Yes; How many? _____
             Does this check?

   (If yes) How well did she handle it?
   _____
        Is she a good leader?

9. How closely was she supervised?
   _____
       Was she hard to manage?

10. How hard did she work?
    _____
        Is she habitually industrious?

11. How well did she get along with other people? her students?
    _____
        Is she a troublemaker?

12. Did she have arguments with customers?
    _____
    Does she like selling? Can she control her temper?

13. What did you think of her?
    _____
    Did she get along with her superiors?

14. Why did she leave?
    _____
    Good reasons? Do they check?

15. Would you rehire her?
    ☐ Yes, ☐ No; Why not? _____
        Does this affect her suitability with us?

16. What are her outstanding strong points?
    _____

17. What type of work do you feel she would do best?
    _____

18. What, if any, are her weak points?
    _____

HIRING MANAGERS NOTES

From _____ xx ____ To _____ xx ____

Do dates check? _____

Source: Dr. J. D. Richardson

and a deadline for a decision should be established. Do not send out regret letters to anyone else until your chosen candidate has accepted or even started work. However, the eventual "decline" letter is still an important courtesy, because later you may want to offer positions to one or more of the other candidates.

Orientation and Training

New Employees

New employees need orientation in order to develop a sense of belonging, and to know to whom to turn for help in the early days of their employment. Some firms assign a colleague or "buddy" for the first few days until the newcomer knows their way around. Existing employees should be apprised that hiring is in process and informed of the newcomer's status and functions. The manager needs to be sensitive to fears that a new person will diminish someone else's importance.

Orientation will usually involve both written materials and person-to-person review of tasks. In larger companies, the personnel department usually plays a key role, for example, in telling about the history of the firm, its values and goals, the fringe benefits, and the company-wide activities and information systems. If this is how things are to be done, ideally new employees should report on the first day to the person that hired them, be taken to meet the senior managers, and then go for orientation with personnel department staff. In this way, they will be ready to start work with their new colleagues and not be pulled off new assignments in order to attend to the administrative aspects of orientation. A checklist of possible administrative and cultural items to cover in orientation includes:

> Company history, policies, and practice;
> Company values and goals;
> Facilities;
> Organizational structure;
> Employee conduct and responsibilities to the company;
> Company responsibilities to employees;
> The compensation program, including any awards, contests, and other benefits;
> The benefits program: life insurance, medical, pension, profits, deferred compensation;
> The evaluation program and schedule;
> Promotion policy;

> A tour of premises and the employee's own department;
> Introduction to fellow employees;
> Safety and security program, including any required training, such as CPR;
> Overview of the company's emergency preparedness plan;
> Training opportunities and aid;
> Hours and pay schedule; and
> Work assignments.

The process should also involve a check-back after about three days, to answer any new questions that may have arisen. On the first day, it makes settling in easier if someone is assigned to ensure that the new employee does not eat lunch alone.

Training

Training involves three areas: training in the job for which the person was hired, training for more advanced functions in the company, and training to address technical, social, and economic changes that affect the way the company must operate. All employees need training for one or more of these areas. There are four basic modes of providing training:

> On the job, by supervisors;
> In a classroom at the workplace (e.g., during less busy periods of the week, with inside or outside instructors or vendors);
> Through an apprenticeship; and
> With special courses, seminars and classes online or off the premises, provided by specialists outside the firm.

Training may include both technical and human resource development areas.

On-the-Job Training. On-the-job training is the basic training system used by all aviation businesses to some degree. Its success is strongly related to its being a hands-on process. It generally requires that the immediate supervisor is an effective teacher. One effective program for this is the Job Instructor Training (JIT), which was developed during World War II. Its major elements are:

How to Get Ready to Instruct
Step 1: Have a Timetable

> How much skill do you expect the person to have?
> By what date?

Step 2: Break Down the Job

> List important steps.
> Pick out the key points.

Step 3: Have Everything Ready

> The right equipment, material, and supplies.
> Have the workplace properly arranged just as the worker will be expected to keep it.

How to Instruct

Step 1: Prepare the Worker

> Put him or her at ease.
> State the job and find out what the trainee already knows about it.
> Get him or her interested in learning the job.
> Place in correct position.

Step 2: Present the Operation

> Tell, show, and illustrate one important step at a time.
> Stress each key point.
> Instruct clearly, completely, and patiently, but no more than the student can master.

Step 3: Try Out Performance

> Have the trainee do the job; correct errors.
> Have the trainee explain each key point to you as he or she does the job again.
> Make sure he or she understands.
> Continue until you know the trainee knows.

Step 4: Follow up

> Put the trainee to work alone.
> Designate to whom he or she goes for help.
> Check frequently.
> Encourage questions.
> Taper off extra coaching and close follow-up.

It is very important to understand that if the worker has not learned, the instructor has not effectively taught.

Developing Supervisors and Managers. Some people are natural managers, some can acquire the necessary skills, and some would prefer to spend their careers without getting into management. Given the difficulty of attracting experienced talent and the value of seasoned personnel who know your business, it may often be advantageous to bring in entry-level people and train them for management after a period of evaluating their suitability for advancement. The selected management training process should foster self-development, be tailored to individual needs and educational gaps, and should be a long-range concept.

Virtually by definition, management training will involve more than on-the-job training. Although working with someone on selected tasks is one approach, it will probably need to be supplemented by seminars, courses, workshops, and conferences provided by such groups and organizations as:

> Aircraft manufacturers;
> Aviation associations, such as the National

Air Transportation Association (NATA), the National Business Aviation Association (NBAA), state associations of airport managers, and other aviation groups;

> Chambers of commerce;
> Colleges and universities;
> Banks and other service organizations;
> Professional education groups, such as the American Management Association;
> Small Business Administration; and
> Training consultants and vendors providing customized programs.

Who pays, and whether or not the programs are undertaken on company time, must be resolved in advance. The management trainee can perhaps be encouraged to choose among several courses, with any of them covering the needed material. Some firms retroactively pay a percentage of tuition costs for any relevant program completed by an employee on his or her own time. Other firms pay for tuition and time in exchange for having the trainee teach the new material to the other staff.

The concept of lifelong learning applies to virtually every profession in our society today. In an industry such as aviation, it is ever important to stay abreast of new technologies, revised regulations, and market trends. The term "lifelong learning" encompasses everything from attending college courses to professional workshops and conferences, to reading industry-related periodicals and Web sites.

Training to Keep Abreast of Social and Economic Change. The world is changing at an ever-increasing pace and this can affect the marketplace for an aviation business. If it becomes increasingly difficult to make a profit from the sale of fuel or aircraft, perhaps other products and service lines need to be emphasized. If so, different personnel may be involved and need guidance on merchandising and promoting the new items. If the airport operator is receiving a growing volume of noise complaints from certain areas in the airport traffic pattern, flight instructors may need to be trained in more depth on noise issues and abatement procedures. Then they, in turn, can teach this material in the air and in the classroom. If the business is investing in a new computer system or software for accounting purposes, perhaps all the staff would benefit from a short presentation by the vendor about other functions that could be performed using the system. The manager must stay current with trends, both within and beyond aviation, and must also determine how much of the staff's time to allot to briefings, reading of trade and general business publications, and formal classroom activities.

Evaluating the Training Program. Any training takes time away from revenue-producing work. Although theoretically it will pay off by reducing mistakes and increasing productivity and motivation, this consequence is not automatic. A training program should therefore employ "before and after" measurements. Depending on the specific training that was completed, particular aviation business metrics might include:

> Complaint rates;
> Appreciation rates;
> Invoice errors;
> Accident rates;
> Absenteeism rates;
> Personnel turnover rates;
> Rework rates on repairs;
> Time required to perform a task;
> Cost per task;
> Scores on formal written tests;
> Volume of sales; and
> Number of new clients.

The measures selected for evaluation of the training program must be items readily collectable or already routinely being collected. It may be desirable to also survey employee job satisfaction and to review the individual evaluations turned in by employees at the end of a specific training program.

Communicating

Basic Elements

Considerable successful communication has already occurred when a new member of a firm has been recruited, screened, selected, and hired. But even the strongest candidate has to learn to communicate in various ways about new tasks with new colleagues in order to be successful on the job. Communication skills are starting to be specifically taught in schools and business seminars. Their acquisition and continual improvement are necessities for every employee.

Communication occurs whenever information is transferred from one person to another and understood so that it may be correctly acted upon. Information transfer between or among individuals goes through five phases:

> Encoding—expressing the thought in written, oral, or gesture form;
> Transmission—by sound waves, paper, light waves, electronically, and so on;
> Reception—by the correct party, in a timely manner;
> Decoding—translation of the material into ideas; and
> Understanding—evaluation and analysis of meaning.

Figure 5.11 shows this process.

Figure 5.11 → Model of the Communication Process

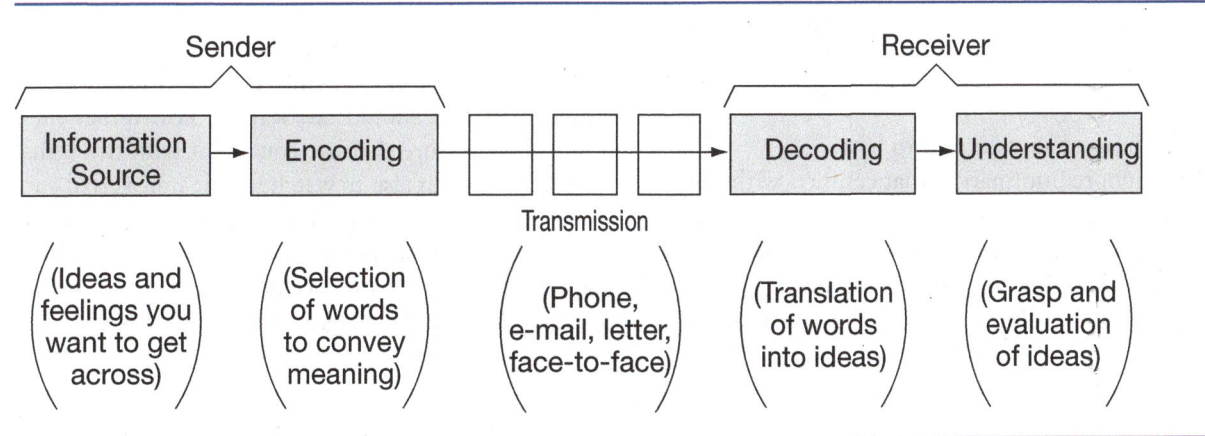

Sender: Information Source (Ideas and feelings you want to get across) → Encoding (Selection of words to convey meaning) → Transmission (Phone, e-mail, letter, face-to-face)

Receiver: Decoding (Translation of words into ideas) → Understanding (Grasp and evaluation of ideas)

Communication can fail because of a problem with any one of these steps. If a supervisor fails to ascertain whether an employee, especially a new employee, has understood a message, that supervisor has not communicated adequately.

People new to an organization may require training in its vocabulary and style of communication. The amount of time it takes for them to pick this up depends not only on their intelligence, but also on their cultural background and the amount of their previous adaptation.

Another possible barrier to effective communication is an array of issues loosely termed "learning disabilities". Some estimates are that over 20 percent of the adult U.S. population suffers from dyslexia. This is also roughly the level of functional illiteracy and the two are closely related. Dyslexia means difficulty with the first step in communication as described in the above list—encoding. Dyslexics have difficulty coding their thoughts into commonly used words. Such difficulties are nevertheless often accompanied by a very high IQ. Dyslexia may show up as slowness to speak, slowness to write, difficulties in reading, especially reading aloud, and errors and transpositions in spelling. It may be accompanied by extreme frustration, as well as by shame and embarrassment because the educational system has traditionally treated such people as having low intelligence. Dyslexics may also use highly ingenious "work-arounds" that may be effective in the new job or may have to be gently replaced with the company's approach, depending on whether the situation can safely allow for divergent approaches. The astute manager will look for signs of learning disabilities and treat them as he or she would any other physical traits (height, weight, strength, agility level, etc.)—that is, as relevant, but not controlling, in the workplace.

Barriers to Effective Communication

These additional problems in the communications process are often referred to as barriers to effective communication. The most common ones are:

1. We hear what we expect to hear.
2. We ignore information that conflicts with what we already "know."
3. We judge the value of information by its source.
4. We are influenced by our group bias.
5. Words mean different things to different people.
6. Groups may have their own jargon or insiders' vocabulary.
7. Our emotional state conditions what we hear or fail to hear.
8. We don't know how the other person perceives the situation.
9. We fail to listen.
10. We ignore body language.
11. We try to convey too much at once.
12. The receiver is not "tuned in."
13. Understanding is assumed, rather than ascertained.
14. Intermediates involved in the transmission filter amend information.

Verbal and Non-Verbal Communications

Verbal communication skills are the most commonly discussed in communications. However, non-verbal communications can be just as important in sending a message. In fact, if the non-verbal cues being given conflict with the verbal message, more often than not the non-verbal message will be the most powerful.

Good verbal communication skills include not only the words used to convey the message but also the tone of voice being used, which can imply a variety of messages. Using a matter-of-fact versus a sarcastic tone will certainly leave different impressions of what the sender is trying to say, even if the words are exactly the same. The rate of speech delivery, the volume of voice, and pitch all affect how the message is received. And, silence sends a powerful message all its own.

Non-verbal communication includes everything from eye contact to gestures, from facial expressions to jingling of keys in the pocket, to pacing. A person who stares at the ceiling while talking with someone is generally perceived to be untruthful. Good eye contact helps instill trust in the listener and implies honesty. Raising your eyebrows or rolling your eyes while shaking someone's hand and saying how pleased you are to meet them sends the opposite message. Good managers should be aware of non-verbal cues, not only to ensure they are consistent with the verbal message sent, but also to watch for the non-verbal cues of others. A common managerial fault is to tell a visitor they have your attention, yet continue shuffling papers on the desk or checking e-mail on a smartphone. The body language contradicts the words.

Improving Communications

A classic text for improving communications in the workplace is *The One-Minute Manager*.[14] Its first recommendation is the joint development by manager and employee of a goal for that employee's work performance that can be written down and that takes a minute or less to read. The manager gives "one-minute praisings" for performances that shine and "one-minute reprimands" for any needed correction, ending with a reminder of the value of the person being reprimanded. Thus, people know what they are supposed to be doing and know they will hear immediately when they do something well or poorly. The manager doesn't just save up grievances ("gunny sacking") or praise for annual review day.

A basic set of tools for improving communications includes:

> Utilizing feedback to make sure the message was not only received but also understood;
> Developing sensitivity to the receiver;
> Being aware of the symbolic meaning of words;
> Being aware of the meaning of words in various cultures;
> Timing messages carefully;
> Reinforcing words with action;
> Using simple, direct language; and
> Introducing the proper amount of redundancy.[15]

Utilizing feedback means being constantly on the alert for clues to whether or not one is being understood. Feedback from the recipient of the message may take the form of a "yes." It could be an understanding look, an attentive nod, or the desired action. Lacking any of these, the communicator should try again. Good supervisors actively encourage questions to make sure the instructions are clear, and instruct employees in the need to give clear feedback. The feedback and evaluation training of Toastmasters International (http://www.Toastmasters.org/) may be a valuable resource for many managers.

Sensitivity to the receiver means attempting to adjust to the receiver's needs, especially possible interpretations of the message. Cultural and emotional factors must be considered.

Awareness of symbolic meanings is important because words can have emotionally loaded meanings in the mind of the receiver that distort and interrupt the communications effort. The communicator should avoid such words or phrases, once they are identified.

Careful timing of messages is important because the receiver must be ready and able to handle the new input. Poor timing can include times when the receiver is already busy on a task, is overloaded with new input, is tired, or has become preoccupied with a distracting problem. It can also include times when prior events mean the person's mind is already made up on an issue. The message sender should anticipate these situations and adjust accordingly.

Reinforcement of words by action will greatly increase the chance of a communication being accepted.

The use of simple, direct language will enhance good understanding. Avoid using technical terms and long, tangled statements.

The proper amount of redundancy is important. Good teachers know that their ability to convey a new idea often depends on their ability to rephrase the same thing in as many different ways as necessary until all the class members have discovered its meaning in their own terms. Sales people are often taught that it takes six different exposure messages about the product before people are generally ready to respond. This repetition of ideas, often in different forms, is needed until it is clear the message has been understood. However, pay attention to ensure that your audience is not tuning out due to the repetition. The message may be more effective if given on several occasions using different media or delivery methods.

The use of communications training can be important. Many people do not know the ways in which they communicate until they see themselves in a recorded role-playing exercise or hear themselves on recorded audio. Guided practice and feedback can almost always be part of a manager's personnel training program.

Last, but not least, it is the job of the message sender, NOT the recipient, to make sure the message has arrived and been correctly decoded and understood. The extensive use of e-mail—a one-way communication that fails to cover all five steps (Encoding / Transmission / Reception / Decoding / Understanding) of the basic communication model just described—places a new burden on the message-sender because it may be necessary to check in person with the recipient that one's requests and instructions were not only received, but also understood.

Motivating

Array of Needs

Many decades of thinking about motivation have been based on a hierarchy of needs first described by Maslow.[16] His theory (1) gave recognition to the fact that motivation comes from within, and (2) depicted the driving forces as a hierarchy with simple physical needs for food, shelter, and safety at the bottom. The typical individual's attention goes to social needs once the basic needs are met. Then their needs become more sophisticated, and esteem and self-respect become the primary motivators. Finally, at the top of the hierarchy—and attended to only if, or when, all preceding needs are met—is the motivator of self-actualization. This hierarchy or staircase of needs is depicted in Figure 5.12.

Other long-standing theories of what motivates workers are described as theories X, Y, and Z. A summary follows:[17]

"Douglas McGregor in his book, *The Human Side of Enterprise*, published in 1960, has examined theories on behavior of individuals at work, and he has formulated two models, which he calls Theory X and Theory Y.

"Theory X assumes that the average human being has an inherent dislike of work and will avoid it if he can. Because of their dislike for work, most people must be controlled and threatened before they will work hard enough. The average human prefers to be directed, dislikes responsibility, is unambiguous, and desires security above everything. These assumptions lie behind most organizational principles today, and give rise both to "tough" management with punishments and tight controls, and "soft" management, which aims at harmony at work. Both these are "wrong" because man needs more than financial rewards at work; he also needs some deeper higher-order motivation—the opportunity to fulfill himself. Theory X managers do not give their staff this opportunity, so that the employees behave in the expected fashion.

"Theory Y assumes that the expenditure of physical and mental effort in work is as natural as play or rest. Control and punishment are not the only ways to make people work, man will direct himself if he is committed to the aims of the organization. If a job is satisfying, then the result will be commitment to the organization. The average man learns, under proper conditions, not only to accept but also to seek responsibility. Imagination, creativity, and ingenuity can be used to solve work problems by a large number of employees. Under the conditions of modern industrial life, the intellectual potentialities of the average man are only partially utilized.

"Comments on Theory X and Theory Y: Assumptions are based on social science research that has been carried out, and demonstrate the potential present in man that organizations should recognize in order to become more effective. McGregor sees these two theories as two quite separate attitudes. Theory Y is difficult to put into practice on the shop floor in large mass-production operations, but it can be used initially in the managing of managers and professionals."

"In *The Human Side of Enterprise* McGregor shows how Theory Y affects the management of promotions and salaries and the development of effective managers. McGregor also sees Theory Y as conducive to participative problem-solving. It is part

Figure 5.12 ✈ **Maslow's Hierarchy of Needs: Two Views**

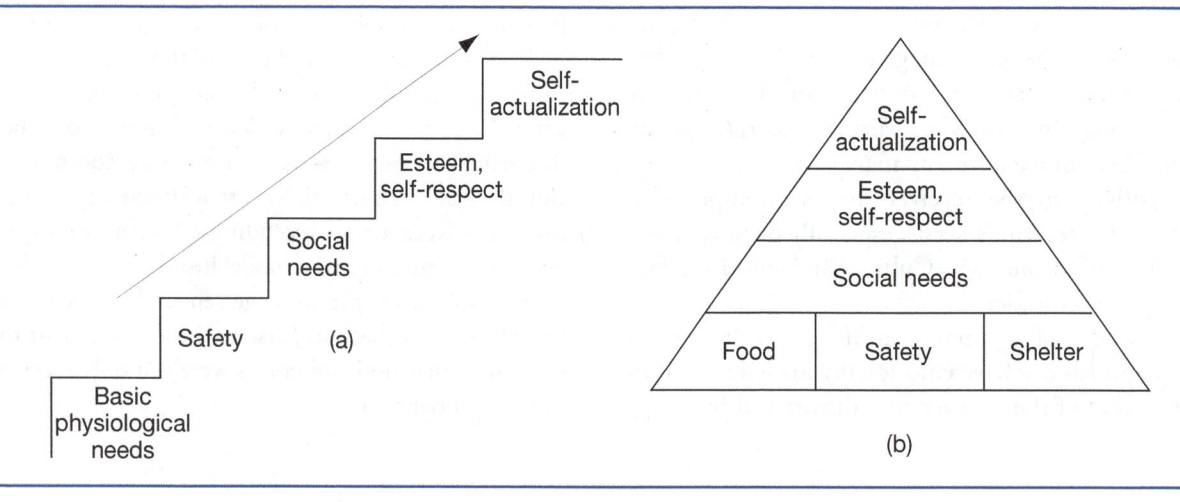

of the manager's job to exercise authority, and there are cases in which this is the only method of achieving the desired results, because subordinates do not agree that the ends are desirable. However, in situations where it is possible to obtain commitment to objectives, it is better to explain the matter fully so that employees grasp the purpose of an action. They will then exert self-direction and control to do better work, quite possibly by better methods than if they had simply been carrying out an order they did not fully understand."

"The situation in which employees can be consulted is one where the individuals are emotionally mature and positively-motivated towards their work; where the work is sufficiently responsible to allow for flexibility; and where the employee can see his own position in the management hierarchy. If these conditions are present, managers will find that the participative approach to problem solving leads to much improved results, compared with the alternative approach of handing out authoritarian orders. Once management becomes persuaded that it is under-estimating the potential of its human resources, and accepts the knowledge given by social science researchers and displayed in Theory Y assumptions, it can invest time, money, and effort in developing improved applications of the theory.

"McGregor realizes that some of the theories he has put forward are unrealizable in practice, but wants managers to put into operation the basic assumption that staff will contribute more to the organization if they are treated as responsible and valued employees."

More recent analysis by several management experts suggests that what motivates people can be any and all of these factors at the same time. As society becomes more mobile and close-knit, family networks are harder to maintain and the workplace often becomes the primary social outlet. Everyone who works is motivated to a greater or lesser extent by money, or the need for more money; however, this is not usually the sole consideration for most people. Job satisfaction takes other forms besides financial.

Needs Satisfied by Working

Some companies appear to be organized only to meet people's lower-level needs. Executives and factory operators alike put most of their creative energy into diverse hobbies outside the workplace, requiring elaborate equipment and planning. However, research such as that described in *In Search Of Excellence* has found that the excellent companies were "organized to obtain extraordinary effort from ordinary human beings."[18] The same feature of motivating even mediocre employees to be the best they can be is a feature of the larger Japanese companies, which in years gone by have offered lifetime employment so that they must find ways of motivating all personnel.[19] Some of the needs that can be met in the workplace include: autonomy and a sense of control over one's situation, being part of a team or a sense of belonging, public recognition and rewards, and a sense of responsibility for vital functions.

Creating a Motivating Environment

If these are the non-monetary rewards that people seek from their work, how can managers arrange working conditions so that these rewards occur?

Autonomy and control mean training a person to know he or she is empowered in specific areas. An atmosphere of trust can be generated by such techniques as abolishing time clocks in favor of an honor system, using flextime, allowing as much freedom as possible over workspace arrangements, and, in general, treating people as adults.

Recognition in the best companies seems to involve several steps. First, clear and specific performance goals are set up. They are set deliberately quite low so that almost everyone can be a winner. Reaching the goal is praised publicly and with varying levels of celebration. Praise and rewards are instant, specific, consist of things important to the performer rather than the rewarder, and may be sporadic and unpredictable, the result of the manager wandering around and particularly looking for things people are doing right.

Money is only one of many factors contributing to job satisfaction. Other important factors include participation in decision-making, challenging assignments, positive feedback from one's superior, and the chance for advancement.

Many employees are motivated by recognition and acknowledgement. The precise mode of giving recognition will vary with the style and value system of the company. Some considerations to include in the design of the recognition system are:

1. The award must have meaning for the recipient (not just for the company).
2. The recipient must understand why the award is being given.

3. The presentation must be dignified, sincere, and conducted before peers and supervisors.
4. The award must have intrinsic value and uniqueness.

Being part of a team of winners in a corporate family is another technique deliberately fostered by some companies. The ways of achieving this can include making the effort to acquire knowledge about each employee's personal situation; company trips and outings; and acceptance of the individual's uniqueness. Some companies have a store of myths and stories about the business, especially perhaps in its early days, that help to perpetuate the sense of belonging to something special. A strong and often repeated corporate philosophy stated in simple ways, such as "product quality and customer service," can be a major contributor. Some firms shift their personnel among functions, and this can strengthen corporate pride because people learn to understand the whole, rather than just a part of the business, and to respect daily challenges handled by their peers in other departments. It gives the rotated employees a hands-on appreciation of the issues and problems facing other divisions of the company.

Finally, delegation means responsibility. If someone is put in charge of a task, then they must be allowed to carry it out in their own way without the boss constantly interfering. However, good delegation does mean that the supervisor is available to answer questions that will inevitably arise; good employees should be encouraged to ask these questions rather than just proceeding along a path that may not be what is wanted. Specific results by specific dates mean specific success or failure. These milestones allow the supervisor to re-delegate the task differently if all is not going well, by using the opportunity to communicate in more detail some of the "how" as well as the "what." Judgment must be used in giving responsibility commensurate with ability; this is one of the key tasks of a manager, since one seldom has at hand exactly all the right talents embodied in one person. This is discussed in Chapter 2, Management Functions.

Leadership

Leadership is an absolute necessity in a successful organization. In a small or new business, it may be almost without effort that the owner/manager conveys to the staff the excitement and drive that motivate him or her. In a bigger or older company, it may take more conscious effort. As previously noted, one author recommends that people with appropriate values be the only ones hired. A possible set of values for all employees follows:[20]

1. Any and every company problem is your problem (Don't pass the buck or point the finger).
2. Glitter on your own time (Don't worry about status symbols or showing off on the job).
3. We price and sell service, not technology or products.
4. Let the customer take advantage of you.
5. Finish, and put your name on, every task you start.
6. Bake the incentives into the work itself.
7. Pace company growth to staff growth.
8. Simplicity rules.
9. Know your customers' shoe size.

Such values, or minor variations thereof, could probably fit almost any company. The same authors recommend that new employees be given at least eight hours of values orientation during their first three months on the job.

Evaluating Employees

The process of employee discipline and evaluation may be best accomplished primarily in the style of *The One-Minute Manager,* that is, by the one-minute reprimand about a specific need for course correction, immediately, but including a reassurance to the employee about his or her worth. This is as an alternative to the "gunnysack" approach in which all the bad things, along with the good, are dumped out on the conference table once every six months or every year at performance evaluation time. The principle of tackling little errors promptly when they occur is aimed at ensuring that big errors do not have a high likelihood of happening. Nevertheless, the manager will need to sit down periodically with each employee and review how things are going. This process will probably be tied to periodic merit, cost-of-living, or other pay reviews.

An effective employee appraisal plan will:

> Achieve better communication between the manager and employee;
> Relate pay to work performance and results;
> Provide a standardized approach to evaluating performance over time and among employees; and

> Help employees better understand what is expected of them and how they can improve.

The appraisal process should consider the following and should be reviewed with all supervisors performing reviews so that there is a reasonably consistent approach to applying these factors:

> Results achieved;
> Quality of performance;
> Volume of work;
> Effectiveness in working with others in the firm;
> Effectiveness in dealing with customers, suppliers, and others;
> Initiative;
> Job knowledge;
> Dependability; and
> Enthusiasm.

A sample performance appraisal form utilized by Midcoast Aviation is shown in Figure 5.13.

Figure 5.13 → **Sample Performance Appraisal**

MIDCOAST AVIATION, INC.

NONEXEMPT PERFORMANCE APPRAISAL / IMPROVEMENT PROFILE

Name: _____

Job Title: _____

Department: _____

Date Assigned to Present Job: _____

Date of This Review Period: From: _____ To: _____

Reviewed By: _____

INSTRUCTIONS:

This form is to be completed for each salaried nonexempt employee:

- **Annually.**

- **On an interim basis, as necessary, to review the Employee's performance.**

This report covers all important aspects of the past performance of the employee, a summary of his/her current status, and a discussion of potential for growth and development. In addition, the report provides a vehicle for your personal discussion of the appraisal results with the employee. Carefully consider the employee's performance during the review period based on assigned duties and level of responsibility. Check the appropriate block in each category which best identifies the employee's current performance. Next, check the block indicating the change in performance since the last evaluation.

Continued

Figure 5.13 ✈ Sample Performance Appraisal —Cont'd

Performance Factor	Current Evaluation					Since Last Evaluation		
	Unsatisfactory	Acceptable	Competent	Competent Plus	Outstanding	Reverse	Same	Improved
1. **ATTITUDE:** Consider the employee's work attitude including overall appropriateness of behavior towards the environment, fellow employees and supervision. Is the individual willing to accept new assignments, policies and procedures? Is he/she flexible in meeting changes, additional requirements, etc.?								
2. **ORGANIZATION:** Consider the individual's ability to effectively organize his/her work load and establish priorities. Does the individual properly plan out work including coordinating with others as required?								
3. **PRODUCTIVITY – QUANTITY:** Consider the individual's overall productivity in terms of quantity of work produced on a timely basis. Consider the individual's ability to meet or exceed job requirements.								
4. **PRODUCTIVITY – QUALITY:** Consider the individual's ability to produce neat, accurate work on a timely basis. Does the employee possess acute attention to detail when it is an essential factor?								
5. **INITIATIVE:** Consider the employee's ability to work without constant supervision and his or her ability to begin and/or follow through with assignments.								
6. **TECHNICAL OR JOB COMPETENCY:** Consider the individual's overall technical knowledge and job competency, and the employee's ability to put this knowledge into practical application.								
7. **ATTENDANCE:** Consider the individual's attendance record. Does the individual attempt to adhere to policies concerning lunch hours and work breaks? Does the individual notify his/her supervisor of absenteeism or tardiness?								

Figure 5.13 → Sample Performance Appraisal —Cont'd

OVERALL PERFORMANCE SUMMARY

Indicate the performance level that most closely reflects how the employee's overall performance measured up to what should normally be expected from an employee with similar experience at this level.

☐ Unsatisfactory	☐ Acceptable	☐ Competent	☐ Competent Plus	☐ Outstanding
Doesn't meet objectives, falls short of required performance; consider trial (probationary period), transfer to a more suitable job or termination.	Usually meets objectives; areas for improvement noted in appraisal, level of performance is less than expected but acceptable.	Consistently meets objectives; full utilization of ability and experience to produce the desired results expected from a qualified employee.	Consistently meets objectives and actively contributes to achievement of overall company goals. Superior performance in some aspects of job. Performance above the competent level.	Consistently exceeds objectives, actively develops teamwork and cooperation, seeks new and better ways to accomplish tasks, extremely capable and versatile in adjusting priorities to current needs; an effective communicator.

PRINCIPAL JOB DUTIES DURING REVIEW PERIOD. Briefly describe the principal work assigned to the individual. List and number specific job objectives and goals established.

_____
_____
_____
_____
_____
_____
_____
_____
_____
_____
_____
_____
_____
_____
_____
_____
_____
_____

Signatures and Dates:

Appraised by:	Director, Human Resources
Department Head	Employee

Continued

Figure 5.13 → Sample Performance Appraisal —Cont'd

JOB PERFORMANCE STRENGTHS (Summarize the employee's most significant strengths.)

_____
_____
_____
_____
_____
_____

IMPROVEMENT NEEDS (Summarize the employee's most significant areas needing improvement.)

_____
_____
_____
_____
_____

DEVELOPMENT PLANS (Summarize the activities such as special coaching, special assignments, off-the-job training, etc., that are planned to help this employee improve job performance during the next appraisal period.)

_____
_____
_____
_____
_____
_____
_____
_____

ADDITIONAL COMMENTS:

_____
_____
_____

EMPLOYEE COMMENTS:

_____
_____
_____

Source: Midcoast Aviation

Promoting Employees

Promoting employees from within is ideal, if at all possible, because staff will develop greater loyalty to a company that clearly indicates it can use them as they mature and acquire more skills. By contrast, an ambitious and capable employee that sees no internal career path will likely start looking for jobs outside the company. Promotions need to be conducted in an equitable manner, however. If a new position is to be created, or if someone leaves or retires, the manager will get best results by posting the opening and giving all interested parties a chance to be considered. In a small company, this may mean interviewing everyone. In a large company, some preliminary screening is necessary and appropriate; however, if only a few applicants are going to be screened out, it may be better for company morale to interview them all. An inexperienced employee who has little or no idea why he or she is not qualified for the opening can become a problem if his or her aspirations are not acknowledged. The interview can be an opportunity for the manager to hear directly from the employee what their aspirations are, to coach them as to how best to prepare themselves for advancement, and to steer the person on to the career path most likely to be best for them and the company.

Compensation Systems

Job Evaluation

Using the task information described in the Comparable Worth section on pages 113–114, and some criteria about responsibility levels, one may evaluate jobs under review in relation to each another. One method for doing this is simple ranking, which ranks and groups the jobs by their value to the firm. Then the groups of jobs are arranged in a series of pay levels.

Ranking the jobs in order of their value to the firm takes into account only the internal situation. The salary that people in these jobs can command elsewhere must then be investigated. Possible sources of such data include:

> Trade and professional groups such as NATA and trade publications such as Aviation Maintenance Technician (AMT), which conduct annual salary surveys;
> Chambers of commerce;
> Major firms in the area;
> U.S. Bureau of Labor Statistics;
> American Management Association;
> Employment agencies; and
> Newspaper want ads.

In analyzing pay in a particular area and comparing it to the company's jobs, be sure to compare actual job descriptions and not just job titles. Based upon the average pay for each type of position developed from a market survey, a midpoint pay level for the company in question can be constructed for each position. Commonly, the low point in each range will be 85 percent of the midpoint and the high will be 115 percent. Such a range will enable employees to join the firm at the bottom of the pay range for that position and receive 30 percent in raises without promotion to a higher job category. A more fine-grained breakdown of each pay range may also be suitable.

Fringe Benefits

Most full-time positions carry with them an array of non-salary benefits, such as paid sick leave and vacation time, paid holidays, medical insurance, life insurance, pension plans, profit-sharing plans, and other supplementary compensation. The primary advantages of these are:

1. Time is specifically allotted for medical and personal needs.
2. In the event of the employee's death or disability, the dependents are provided for in a planned manner.
3. Group purchasing enables the employer to buy more benefits for the same price than employees could obtain alone, particularly with regard to insurance.
4. Good fringe benefits are a major means of employee retention in tight labor markets.

Most companies offer certain items on a mandatory basis, such as a minimum level of life insurance, and then provide optional additional benefits such as disability insurance. There may be choices of medical plans with varying levels of coverage and varying levels of employee participation in payment. The trend in this area is the "cafeteria" approach, in which the company determines a fixed dollar amount or percentage of salary to be used for fringe benefits, and each employee chooses precisely and only that package of services that he or she really wants. As more and more households involve both a husband and wife working for different employers with different benefit options, this ability to custom-design one's benefits is a great advantage.

For most businesses, the typical scale of fringe benefits is in the range of 30–37 percent of basic pay. Benefits—and the costs of providing them— have been rising faster than wages and salaries in the past few decades, so that means to control these costs, especially health care premiums, are clearly important. The flexible benefits approach can give employees more of what each really needs without increasing total costs as much to the company.

The Patient Protection and Affordable Care Act was signed into law on March 23, 2010, by President Obama, with the intent of reducing the number of persons without health insurance coverage by expanding public and private coverage and reducing healthcare costs. The Act established penalties for employers with 50 or more employees that do not offer health insurance to full time workers, effective in 2015. As a result, there has been some concern that businesses with fewer than 50 employees would reduce or drop healthcare coverage options. The provision also created some incentive for businesses to hire more part-time workers, or to limit the number of full-time employees they hire. The full effect of this law has yet to be realized, but a careful investigation of these issues by small aviation businesses is certainly warranted.

Administration of the Total Compensation Plan

Many employees are unaware of the total dollar value of their salary plus benefits. When hiring new staff, it is desirable to be able to cite lower and upper limits for the value of the total compensation package. Administration of the compensation plan will probably also involve periodic activity to stay current with what competing businesses are doing. It should involve a clear and consistent policy about raises, promotions, merit increases, bonuses, and the like. The plan needs to give the manager the flexibility to take quick steps to prevent the loss of a key employee without causing problems among the other staff. It must provide the freedom to bring in specialized talent at the going rate while not slighting long-standing, loyal employees. It must allow a visible career ladder within the firm so that staff can envision their long-range futures. If pay levels and changes are soundly-based on market competition, clearly delineated, and fairly administered, it will go a long way toward minimizing trouble over what can otherwise be an ongoing source of employee discontent.

Disciplinary Problems
Conflict Resolution

Because most businesses involve operating under time pressure with a diverse group of personalities, job situations may lead periodically to disagreements on how something should be, or should have been, done. An autocratic manager or boss who rules, "In the future we'll do it my way, and that's the end of the matter," is losing in several ways:

1. Resentful employees may consciously or unconsciously seek other ways to undercut their boss's authority. Morale will decline.
2. Good suggestions from employees on how to improve the company's procedures will not be heard, and this will mean not only the loss of a good solution today, but also the stifling of future problem-solving ideas and activities.
3. Junior staff members are not given either recognition or the ability to grow, thereby heightening the chance of future inadequacies.

If, instead of utilizing an autocratic approach to disagreements, the manager is willing to be more open-minded, the guidelines in Figure 5.14 can be helpful in achieving a satisfactory solution.

Administering Discipline

Positive discipline through motivation and self-control is what makes most companies work well most of the time. Negative discipline—punishment of some type—may only result in behavior adequate to avoid punishment, not in real correction of the problem. However, even under the best of conditions with trained and loyal employees and excellent leadership, someone is bound to violate the rules now and then, so that there are serious negative consequences. When this happens, there is a need for some generally accepted principles to guide management. In administering discipline, two conflicting goals—protecting the rights of the individual to be disciplined, and preserving the interests of the organization as a whole— must be reconciled. Within this context the following guidelines are generally considered desirable:

1. Provide definite policies covering discipline.
2. Establish reasonable rules and standards.

Figure 5.14 → 21 Steps to Resolving Disputes

A. When Something Happens That Upsets You/Makes You Angry:

1. Ask the person involved for a specific time to talk over something; tell him/her what you want to discuss.
2. Set a mutually agreeable time and place.
3. Prepare. Make notes or run through in your mind what seems to have happened and why you're angry about it.
4. Do it soon. Don't put the other person in suspense overnight or over a weekend, unless it can't be helped.

B. Holding the Discussion:

5. Do it in private.
6. Allow no interruptions.
7. Finish it. Don't leave until both feel that a clear resolution has been reached.
8. Acknowledge that the other may have had a good reason (to him/her) for doing what he/she did, regardless of how stupid it seems to you. Ask for clarification.
9. Listen, interpret, and clarify.
10. Stick to the issue. When discussing what A did, don't get sidetracked by discussions of the times B has done the same thing.
11. Use "I" statements; be accountable for your own feelings.
12. Allow your own anger.
13. Expect and acknowledge the other's anger but don't take responsibility for it.
14. Don't let the other person's anger make you angrier; you are not the target.
15. Stick to one issue at a time. If necessary, set another time and place to discuss other issues or agree to continue then and there.
16. Don't gunnysack ("twice is always").
17. Don't walk away.
18. Don't interrupt.

C. Concluding the Discussion:

19. Seek completion and clarity on the issue; mutual understanding of how to deal with it.
20. Offer reassurance and reaffirmation of the other person.
21. Let it go. Don't bring it up again or blame others. What's done and resolved is over. Be friends and colleagues again.

---

3. Communicate rules to all concerned.
4. Investigate each case thoroughly.
5. Ensure consistency.
6. Provide a logical sequence of progressive penalties.
7. Establish a right of appeal.
8. Document all disciplinary matters in writing, with dates and names.

The "Red-Hot-Stove" Rule

A practical approach to discipline is the "red-hot-stove" approach in which the consequences of a misdemeanor are immediate, consistent, and impersonal. The act and the discipline seem virtually like one event. This involves:

1. Having advance warning—knowledge of the rule and the consequences of breaking it.
2. Immediate and punitive results from the violation.

3. The same results for everyone, no matter what their status.
4. Understanding that the result is not because of who you are, but what you did.

This process may be difficult to apply in every case, but persistence will pay off, at least in part because of its deterrent effect.

The Troubled Worker

Stress and fragmentation of our lives occur not just in the workplace, but also in the rest of life. The above descriptions of how to handle normal disagreements and infractions may not be applicable when the worker in question is dealing with some kind of nonwork trauma. This could be death, divorce, serious mental or physical illness of oneself or a loved one, problems with children or stepchildren, substance abuse, domestic violence, financial worries, or a number of other issues. While the truly professional worker will attempt not to bring these problems to the workplace, sometimes the situation is so severe that effects will show at work.

The supervisor's or manager's role is very sensitive in such situations. Discussions with the employee must be restricted to the problem's effect on job performance. It is not the supervisor's business to become a diagnostician or a counselor. If the issue is not resolved satisfactorily through a supervisor and employee discussion, it may be necessary to seek outside help. Many companies are now contracting with "Employee Assistance Programs" or EAPs, where the troubled worker can obtain expert counsel and, if needed, referral to a detoxification program or other treatment facility. As in all cases of disciplinary action, careful documentation of the steps taken, the commitments made, and their schedule must be retained. This is vital if the individual's employment must later be terminated, because it will provide the necessary documented evidence of failure to perform.

Particularly in the case of drug and alcohol abuse, the supervisor must carefully monitor behavior, and deal with the deterioration in work performance before it poses a threat to the safety or productivity of other workers, or a risk to the company through poor work.

In 1988 the federal government adopted requirements for drug testing of certain personnel in aviation. Aviation employees with safety-or security-related responsibilities are subject to drug testing. Part 135 companies of any size are required to develop and submit a plan to FAA that includes random drug testing. It also requires pre-employment and post-accident testing. While employers are not required to offer rehabilitation, they are encouraged to consider such plans in establishing the total compensation plan. Employers must designate a physician as the Medical Review Officer for administration of the drug-testing plan.

Separation

Whether voluntary or involuntary, there should be some standard procedures for termination. The departing employee may become a vendor, or a source of referrals to the business, or may even at some point be rehired. A person who does not fit the company's present needs may still have desirable skills in changed circumstances. In all cases, an exit interview is recommended to seek information from the outgoing employee about concerns and issues, and to provide management with ideas for changes to benefit the remaining staff. A second purpose of an exit interview is to stimulate the maximum self-esteem and positive feeling about the company in the mind of the departing employee. A sample form for exit interviews is presented in Figure 5.15.

If a company makes the effort to be receptive, even when the parting is not totally amicable, information received from departing employees can be very useful in analyzing problem areas and in providing for organizational improvement. The key to successful exit interviewing is to encourage the employee to talk as freely as possible. Establish yourself as a sympathetic listening post so that his true feelings and opinions may be obtained. The questions in Figure 5.15 covering basic areas may be used to set the interview pattern. Other topics, appropriate to individual cases, should be explored as indicated. The interviewee should be provided with information regarding what will be done with the exit interview form afterwards—will it be kept on file for management actions in response, saved, or kept in the employee's record?

Personnel Policy Manual

This chapter has touched on numerous aspects of personnel management, particularly as it affects the small FBO business. The information presented assumes that the company has no personnel manager, and that the company manager and various supervisors must carry out personnel activities as part of their jobs. Since consistency is such a major feature of many personnel issues, there is a need for a written manual. The manual will likely contain at least the following:

> Purpose;
> Authority of the manual;
> How to use the manual;
> Responsibilities of department heads, supervisors, and employees;
> Employee relations policy;
> Recruiting and selection policy;
> Training and development policy;

Figure 5.15 → Exit Interview Form to Be Used with the Departure of Each Employee

Interview Date _____

Employee's name_____
Job _____ Employment date_____
Department _____
Reason for Separation _____
Departure Date _____

Reason for termination? _____
Do you have another job? _____ Where? _____ Pay? _____
Did you like the work you were doing? _____
How were the working conditions? _____
When you first started here was your job fully explained? _____
Were you introduced to co-workers?_____
Did you enjoy pleasant relationships with co-workers?_____
_____ Exceptions? _____
Did you like your supervisor?_____
Did your supervisor seem to know his/her job?_____
How well did your supervisor handle gripes or complaints? _____

Was it easy to communicate with him/her?_____

Do you feel your pay was fair?_____

Were benefits satisfactory? _____

Was there sufficient opportunity for advancement? _____

Do you know how you stood as far as work performance was concerned?_____

What suggestions do you have for improving policies, procedures, work situations, etc.? _____
_____

What are your plans for the future? _____

Others? _____

Explain benefits and future relationships with company. _____
Comments:_____
_____

Interviewer _____

Source: Dr. J. D. Richardson and Julie F. Rodwell

> Working schedules and hours;
> Compensation policies and procedures; pay and fringes;
> Promotions, transfers and layoffs;
> Attendance, punctuality and absenteeism;
> Safety and security programs;
> Operational rules;
> Profit awareness;
> Complaint and grievance procedures;
> Communications;
> Labor relations; and
> Personnel forms, which may be URLs to web-based forms.

Manual Style

The manual can be anything from a formal, legalistic compendium of all current policies (perhaps in a loose-leaf binder that makes updates easier to incorporate), to a small, perhaps humorous brochure. Many companies are abolishing the old hard-copy three ring binder and replacing it with a company intranet site for employees. In many companies it may be appropriate to have both—a reference site with all the detail, and something abbreviated for each employee. New employees should be required to study the manual, and ask questions. The owner/manager also needs to decide whether he/she alone will update the policies, seek advice from supervisors, or assign the full responsibility for recommending updates to someone else. It is important to clearly maintain revision history and dates of revisions, to ensure that users are referencing the current version of the manual.

Personnel Records

A number of record sheets will accompany the personnel manual. A typical list of needed sheets is presented in Figure 5.16. Examples of some of these have been presented earlier in the chapter.

Employee Organizations

A significant percentage (10.5% in 2018)[21] of American labor is unionized. Although unionism is not as prevalent in general aviation operations as it is among the major air carriers, which range from 16% to 85% unionized, it ranges around 5 percent in most parts of the country. Further, according to the BLS, almost one-half of all aircraft mechanics, including those employed by some major airlines, are covered by union agreements. The principal unions are the International Association of Machinists and Aerospace Workers and the Transport Workers Union of America.

The International Brotherhood of Teamsters represents some mechanics. The financing and power behind the nation's major unions means that if unionization begins, the process may conclude with the establishment of union representation. In recent years some traditionally non-union areas such as office workers have become increasingly unionized. The usual reasons for unionization can often be addressed by fair and reasonable personnel policies. Reasons for unionization may include:

> Changes in management;
> Pay system problems;
> Elimination of promotional opportunities;
> Sudden work-force reductions;
> Unsatisfactory or more restrictive working conditions;
> Poor or inconsistent supervisory practices;
> Desire for participation in management;
> Desire for better income and working conditions; and
> Desire for control over benefits.

Impact on Management

The existence or introduction of a union has an impact on the entire organization. This impact can include:

> Competition for loyalty of employees;
> Possible challenges to management decisions;
> Use of time at work to discuss union issues;
> Stimulation of more careful personnel practices;
> More rigid working rules;
> Possible threats to flexibility and efficiency;
> Centralization of personnel decisions;
> Introduction of outsiders to the management-employee relationship; and
> Higher labor costs.

Every manager should consider very carefully the reasons that contribute to the formation of employee organizations, the impact that such an organization would have upon the business, and the development of a managerial philosophy and action program in response.

Figure 5.16 → Personnel Forms and Records

Activity	Form
1. Researching or studying various jobs to obtain information necessary in developing the organization.	Job Analysis
2. Clearly describing a given job in terms of its contents and tasks involved.	Job Description
3. Describing the requirements of a job in terms of skill, effort, responsibility needed by the person on the job.	Job Specification
4. Advising the personnel office or the employment agency of the type person you are recruiting for a specific job.	Requisition
5. Recording biographical, training, experience and personal information about a job applicant.	Application Blank
6. Conducting an interview with a job applicant.	Oral Interview Guide
7. Checking job applicant's references and data to insure validity and gain additional information.	Reference Check (a) Telephone (b) Letter
8. Conducting a physical examination of applicant to insure ability to handle job and determine overall physical condition.	Physical Examination
9. Orienting or inducting new employees into the organization.	Orientation Check List
10. Recording all necessary date and activities on each employee.	Employee Personal Record or Folder
11. Recording attendance and results from various training programs.	Training Records
12. Evaluating and recording the performance of employees.	Performance Rating Form
13. Recording disciplinary situations and action taken.	Disciplinary Forms: (a) Warning (b) Layoff (c) Dismissal
14. Providing a means of handling employee complaints and grievances.	Grievance Forms
15. Providing a means of obtaining employee suggestions.	Suggestion Forms
16. Collecting information and date from all personnel leaving the company.	Exit Interview Form

Source: Dr. J. D. Richardson and Julie F. Rodwell

Summary

The success of an aviation business can largely be attributed to its personnel. Human resource programs are instrumental in providing an environment that stimulates people to contribute their best efforts. Programs are designed to provide and service a steady flow of personnel into the organization. This flow is like a pipeline with various control points along its length: identification of needs, recruiting, selecting, orientation, training, development, compensation, evaluation, discipline, promotion, and separation. The areas of communications and motivation are especially critical to the organization's success and should receive special managerial attention.

DISCUSSION TOPICS

1. What is meant by the "pipeline concept"?

2. Identify and briefly discuss five "control points" instrumental to the success of human resource management programs.

3. Identify the key steps involved in the selection of employees.

4. What types of questions are appropriate to ask in an employment interview? What questions/information are inappropriate? Discuss the reasons for identifying questions as appropriate or inappropriate and the possible negative results or liabilities for the company of asking inappropriate questions.

5. What issues should be covered in an exit interview?

6. What are the key elements of orientation, and why are they important?

7. What are some factors that could cause complaints about discrimination in the hiring and managing of personnel?

8. How does Maslow's hierarchy of needs relate to the ways a manager can choose to supervise his or her team?

9. What are three reasons for having a personnel policy manual?

Endnotes

1. Peters, T. J., & Waterman, R. H. (1982). *In search of excellence: lessons from America's best-run companies.* New York: Harper & Row.
2. See *http://www.bls.gov/ooh/*
3. *Boeing Pilot and Technician Outlook 2020–2039*, p. 4. Retrieved December 5, 2020 from *https://www.boeing.com/resources/boeingdotcom/market/assets/downloads/2020_PTO_PDF_Download.pdf*
4. See *https://www.bls.gov/ooh/transportation-and-material-moving/airline-and-commercial-pilots.htm#tab-1*
5. See *https://www.bls.gov/ooh/transportation-and-material-moving/airline-and-commercial-pilots.htm#tab-5*
6. See *https://www.bls.gov/ooh/installation-maintenance-and-repair/aircraft-and-avionics-equipment-mechanics-and-technicians.htm#tab-6*
7. See *https://www.bls.gov/ooh/installation-maintenance-and-repair/aircraft-and-avionics-equipment-mechanics-and-technicians.htm#tab-1*
8. Brandt, S. C. (1982). *Entrepreneuring: the ten commandments for building a growth company.* Reading, MA: Addison-Wesley Pub. Co. ISBN-13: 978-0201103823
9. See *http://www.law.Cornell.edu.*
10. Note 10, op. cit.
11. Note 10, op. cit.
12. Cessna Pilot Training Manual, Wichita, KS, undated.
13. "Jackson Vocational Interest Survey," Sigma Assessment Systems, Port Huron, MI; "Strong Interest Inventory,"
14. Blanchard, K. H., & Johnson, S. (1982). The one minute manager. New York: Morrow.
15. Here is a classic, if perhaps apocryphal example: an old story tells about the problems that Chevrolet had in trying to market the Chevy Nova in Mexico. "No va" in Spanish means "won't go."
16. Maslow, A. H. (1970). *Motivation and personality* (2d ed.). New York: Harper & Row.
17. McGregor, D. (1960). *The human side of enterprise.* New York: McGraw-Hill. Also, Ouchi, W. G. (1981). *Theory Z: how American business can meet the Japanese challenge.* Reading, MA: Addison-Wesley.
18. Peters and Waterman, op. cit.
19. See 19, Ouchi, op. cit.
20. See 1, op. cit.
21. Ref. Pew Research Center (*https://www.pewresearch.org/fact-tank/2019/08/29/facts-about-american-workers/*).

6

Organization and Administration

OBJECTIVES

> Recognize the advantages and disadvantages of various business organizations, including sole proprietorship, partnership, and corporation.

> Understand the difference between "line" and "staff" positions.

> Demonstrate the logic of organizing, and its relationship to other managerial functions.

> Understand the organizational principles of specialization and decentralization.

> Recognize that the exact number of people a manager can effectively supervise depends on a number of underlying variables and situations.

> Draw a distinction between formal and informal organizations.

Introduction

Once the workload of an enterprise grows beyond what a single person can do, a greater degree of organization becomes necessary. Various tasks must be assigned to different people, and their individual efforts must be coordinated. As the business expands, this process leads to departments and divisions, with each having its own particular mission. One should consider the resulting organization as a complex machine. Each part performs a necessary function. The different parts are painstakingly matched and assembled in order to meet the stringent demands placed upon the system. A change in any one of the parts will frequently call for an adjustment in several others.

The organization must be developed to accomplish the job, and a social structure must be built to meet the needs of the people doing the work. It is composed of several parts or elements that must be balanced carefully in order to achieve the desired goals. To accomplish tasks, the manager must utilize both technical and social tools of organizing. The techniques of structure, procedures, and manuals must be appropriately used in a social setting that recognizes the influence of individuals and groups on the ultimate outcome of business efforts.

Goals and Objectives

Basically speaking, when the manager sets about organizing business efforts, there are two major initial considerations: (1) what are the goals and objectives, and (2) what resources are available to achieve these goals and objectives? With these two elements firmly in mind, the manager can prioritize and organize the efforts of the business to pursue the desired goals. The two constantly interact, since resources are always finite.

Care should be exercised, however, that the resources (or apparent resources) do not limit the potential of the business. Certain goals may warrant extending the present resources to a point not previously considered possible.

Determining available resources means simply identifying the financial means available, reviewing the personnel capabilities, and surveying the physical assets of the organization. The business must have the required resources available, find ways to get them, or develop a structure more compatible with the currently-available resources.

Answering the following questions methodically and carefully should provide the manager with an idea of what resources are available:

General

1. Do you have a firm and accurate estimate of the resources required?
2. Do you have an inventory of resources on hand?
3. Do you have a written plan for arranging and relating resources to meet planned objectives?
4. Are additional resources available?
5. Do you have a plan for obtaining additional resources?

Financial Practices

1. Do you have current, accurate financial statements?
2. Do you regularly compare your present financial status with your past record?
3. Do you project your financial needs into the future?

Personnel

1. Do you have a current list of your present personnel indicating qualifications, skills, history, and performance?
2. Do you have an ongoing training and development program for your current personnel?

Physical Assets

1. Do you maintain an accurate listing of buildings, equipment, vehicles, aircraft, and real estate?
2. Does your listing indicate value, physical condition, needed improvements, their costs, and operational capabilities?

Having clearly identified the existing resources of the organization, the manager should determine the capacity for the business to acquire additional resources. Some insight into this capability can be gained by answering the following questions:

1. Is additional capital available in the form of either equity, loans, or credit?
2. Can the labor market provide additional personnel of the type and experience level required?
3. Can additional physical resources in the form of real estate, buildings, or operating areas be acquired?

The next step is to establish goals and objectives. The framework for this activity was presented under the managerial function of planning discussed in Chapter 2, Management Functions. Here it becomes the structure that guides the day-by-day organizing activity of the manager and the total business.

Organizational activity develops a structure that will facilitate and encourage the achievement of these goals.

Selecting Information

The body of human knowledge and the ability to access it quickly and accurately are growing at a phenomenal rate. Having access to almost instantaneous information from around the world is a very positive tool, but it can also have the negative effect of overwhelming one with issues and information. A "Renaissance Man" can be defined as a person who is familiar with, even highly steeped in, all areas of human knowledge. The Renaissance, which ended five hundred years ago, was the last time that such a singular accumulation of knowledge was possible. Today, the volume of information available and its rate of increase are just too great.

In light of this, the business owner and manager must somehow select what information to pursue in more depth, and what to ignore. Some of his or her employees may have a tendency to move superficially from one data set to another, thinking they know what it implies for the business, while other, more detail-oriented employees may tend to get stuck in what has been called "analysis paralysis"—the desire to really know enough before making a recommendation or decision. The astute manager will be on the lookout for these tendencies in both himself and his or her employees and provide corrective direction in either case.

Routine-ize the Routine

Given that so many tasks always seem to be in a state of flux, it is extremely important that managers structure their operations so that as many daily tasks as possible can be handled through the use of standard procedures. Every worker's job should consist of some percentage of routine work following prescribed procedures, and some smaller percentage of non-routine work for which non-standard ideas and solutions must be created. Generally, the higher one's position in the firm, the less routine and more non-standard will be the work, yet even top managers need routines, and if this is not their strong suit, they need knowledgeable and effective subordinates, such as administrative assistants and department heads, who will keep them on track.

A World of Change

It used to be said that nothing is sure in life except death and taxes. The new axiom might encompass death, taxes, and change. New technology is constantly being offered on the market, including new computer software and more powerful machines to run it, along with tablets and "smart phones" having powerful and useful apps; new political issues face the airport, new people get involved or disappear, and new problems and issues arise that the business owner and his or her team must tackle.

For many workers, massive uncertainty and being faced with a new situation are very challenging. It is hard for them to believe that out of the current confusion, a road map for success and decisive actions to achieve it will appear. A useful analogy for the manager to use at such times is this:

> "Imagine that a small group of people is trying to put together a big, complex jigsaw puzzle. At first, someone shakes all the pieces out of the box onto the table, and it's impossible to tell if in fact there are several puzzles in the heap, or just one.
>
> "Then, it gets sorted out into piles and one puzzle is separated off. The group begins to find corner pieces and put them in place, then side pieces, and then pretty soon a few middle pieces get placed. Before long, the picture begins to emerge, and after a while there are just a few pieces left that even a child could place correctly.[1]
>
> "But, remember the lesson of the puzzle—every time a new situation arises, it will probably feel overwhelming and confusing. Some people have a higher comfort level than others with that. Some people relish shaping order out of confusion; others are happier when the tasks are more cut and dried and the framework <is> in place. Eventually every pile of loose pieces gets to that point if people persist."[2]

Legal Structure

The manager has several forms of business organization available. By carefully selecting the appropriate organizational structure, he or she can increase profit, decrease taxes, protect business assets, provide for orderly growth of the business, and balance the resources against the challenges provided by the competitive world of business. The following basic types of business organizations are available:

> The sole proprietorship;
> The partnership, including the limited liability partnership (LLP); and
> The corporation, including the limited liability corporation (LLC).

A flying club might also qualify as a not-for-profit educational organization. There are variations in partnerships and corporations, but we will consider only these three major categories. While no selected form of business organization needs to be the permanent choice, it should be recognized that it may be difficult to develop a new form to meet changing needs while the business still operates under the original form. As the business grows, it will change in its operations and needs, and the financial and tax situations will more than likely also change. Likewise, statutory, legal, and tax developments may change the characteristics and advantages of the various business structures available. In view of the changing business world, the manager should consider carefully, early in the process of organizing the business, the legal structure most suitable for its long-term viability.

Let's consider the major advantages and disadvantages of the three principal forms of legal structure.

Sole Proprietorship

The sole proprietorship is usually defined as a business owned and operated by one person. In community property states, the spouse may also have an ownership interest. To establish such a business, all that is needed is to obtain any required local and state business licenses.

Advantages include:

> Ease of formation and dissolution;
> Sole ownership of profits;
> Control and decision-making vested in one owner;
> Flexibility; and
> Relative freedom from regulation and special taxation.

Disadvantages include:

> Unlimited liability for business debts that may affect all the owner's personal assets (additional problems of liability, such as physical loss or personal injury, may be reduced by proper insurance);
> An unstable business life, highly dependent on the health and welfare of the owner; potential lack of continuity;
> Less available capital in most cases; relative difficulty in obtaining long-term financing;
> Relatively limited viewpoint and size, and a lack of "sounding boards"; and
> Difficulty in building equity in the company for sale or retirement purposes.

Partnership

A partnership is an association of two or more people for business purposes. Written articles of partnership are customarily executed, and one expert in this area suggests each prospective partner write out their desires and goals from participation in the enterprise before any formal negotiation.[3] This will clarify, for example, any conflicting ideas about time to be invested versus compensation taken out, about the level of company facilities such as office space, cars, furniture and so on to be acquired before profits are made; and about how long each party is willing to subsidize operations, and to what degree. Written articles normally address at least:

> Name, purpose, domicile;
> Duration of agreement;
> Nature of partners (general or limited, active or silent);
> Contributions by each (at inception, subsequently);
> Business expenses (how handled);
> Authority of each partner to make decisions;
> Chief Executive Officer and how she/he is appointed;
> Separate debts;
> Books, records, and methods of accounting;
> Division of losses and profits;
> Draws or salaries;
> Rights of a continuing partner;
> Death of a partner (dissolution and final disposition of the partnership);
> Employee management;
> Release of debts;
> Sale or transfer of partnership interest;
> Arbitration;
> Additions, alterations, or modifications of agreement;
> Settlement of disputes;
> Required and prohibited acts; and
> Absence and disability.

It is recommended that partners be mutually subject to intensive background, credit, and character investigation because most new ventures that fail do so because of disagreements among partners.

Advantages of partnership include:

> Ease of organization;
> Pass-through taxation, meaning that all profits "pass through" the business to the partners, who pay taxes on their share of these profits;
> Automatic availability of a sounding board for business choices;
> Minimum capital required;
> More capital available than for a sole proprietorship; and
> Broader management base and provision for continuity.

Disadvantages include:

> Potential for conflicts among partners with divergent personalities, goals, values, or other factors;
> Less flexibility for the individual owner, because of the need for consultation with partners;
> Unlimited liability, as with sole proprietorship, unless formed as a limited liability partnership;
> Potential size limitations;
> Capital restrictions; the firm may be inappropriately bound by actions of just one empowered partner; and
> Difficulty of transferring partnership interest.

Most states have adopted legislation permitting partnerships to be organized as limited liability partnerships. LLPs allow the partnership to retain the benefits of pass-through taxation while limiting the partners' liability for the debts of the business and certain actions (torts) committed by employees.

Corporation

The corporation, under Chapter 1 of the Internal Revenue Code, has a separate legal life from its members; it is "an artificial being, invisible, intangible, and existing only in contemplation of the law."[4] Subchapter C of that chapter provides the basic corporate form, while Subchapter S provides the benefits of pass-through taxation to qualifying corporations. Corporations are usually formed under the authority of state governments. Multistate or out-of-state corporations are more complex, because they must comply not only with various state laws, but also with the requirements of interstate commerce.

Other things being equal, the advantages of corporations include:

> Limitations of the stockholders' liability to the amount invested, with respect to business losses (though not with respect to damage suits, and so on);
> Ownership is readily transferable;
> Separate legal existence, even with the demise of all current owners; can be inherited or sold;
> Relative ease of obtaining capital compared with other organizational forms;
> Tax advantages, including owner's fringe benefits and other considerations; some ability to take compensation as dividends instead of salaries, reducing social security tax liability;
> Permits larger size;
> Structure lends itself to easy expansion;
> Delegated authority from the corporation to its employees and officers; and
> Potentially broader skill base as a larger firm.

Disadvantages include:

> Closer governmental financial regulation;
> Relative cost and complexity to set up;
> Activities limited by charter;
> Business could be open to manipulation of minority stockholders; and
> Double taxation as a Subchapter C corporation on corporate net income and on individual salaries and dividends.

This last feature may be overcome by electing Subchapter S status. The purpose of Subchapter S (26 CFR 1.1361–1379) is to permit a small business corporation to employ pass-through taxation; i.e., to have its income taxed to the shareholders as if it were a partnership. One objective is to overcome the double taxation feature; another is to permit shareholders to offset business losses against their other income.

Some conditions for Subchapter S status are that the corporation have 100 or fewer shareholders, all of whom are individuals or estates, that there be no nonresident alien shareholders, that there be only one class of outstanding stock, that all shareholders consent, and that a specific portion of the corporation's receipts be derived from active business rather than enumerated passive investments. No limit is placed on the size of the corporation's income and assets.

Most states have also adopted legislation permitting corporations to be organized as limited liability corporations. LLCs are similar to Subchapter S corporations, with additional advantages: they are typically easier to organize, there is no limit to the number of owners, and the LLC status cannot be revoked by the IRS, as can the Subchapter S status of a corporation.

Given the choices of structure, many businesses may want to advance gradually to more complex forms as they grow. The benefits of incorporation may be particularly strong for family enterprises.

Principles of Internal Organization

No matter the legal structure of the firm, that structure normally does not impact employees in their day-to-day tasks, and the firm has numerous choices of internal structure to get the daily work done. Most aviation businesses are quite small. Aviation Maintenance Technician Magazine reported in 2002 that the average business had 10 or fewer mechanics. This means that there may be little latitude for research and experimentation, or for choices of organizational structure within the company, and that learning from the industry at large may be an important means of staying current.

The Rational Model vs. Some New Approaches

Adam Smith, in his classic *The Wealth of Nations*, set the stage for the success of the industrial revolution by addressing the issue of division of labor, and therefore, specialization.[5] A twentieth-century prophet in the same vein was Frederick Taylor, who pioneered the idea of motion study; namely, each task has only one best way to do it, and finding this best way will lead to efficiency and profit.[6] The advancement of statistical tools of analysis, together with the ability to process large amounts of quantitative information using computers, seems to be a continuation of this school of thought. However, Taylor's approach completely removes human factors from the discussion and seeks to make each worker as machine-like and quantitatively productive as possible.

The recognition of human motivation as an integral part of the rational model came with the work of Frederick Herzberg, a psychologist who became widely recognized in the field of business management. Herzberg developed the Motivation-Hygiene theory of job satisfaction, which states that employees are influenced both by motivating factors that are important in promoting job satisfaction, such as achievement and recognition, and by "hygiene" factors that help prevent dissatisfaction, such as salary and benefits. It is important for the manager to understand that motivating and hygiene factors operate independently of one another, and that both sets of factors are present in the ideal work environment.[7]

Several studies of Japanese productivity and of highly-successful U.S. firms have led to additional views of business organization.[8] The talents required for custodianship of an existing corporate leviathan may now include creativity, innovation, adaptation to change, and high quality. Leaders in this area included Jack Welch, when he was CEO of GE.[9]

Explicit Corporate Philosophy

Corporate values help shape the corporate culture. One tool of management and organization that has been gaining ground is the formal statement of corporate values and goals. This statement may be referred to as either a mission statement or a vision statement, depending on how broadly it is written and whether it focuses on the present or on the future. Regardless of how it is classified, a company statement that is available to all employees and that explains how the corporation wants to handle service can provide the backdrop and value system for everything that goes on in the company. It can, in addition, help to explain the nuances of the organizational structure and provide guidance where the structure does not provide specifics. Examples of such statements include Nordstrom's generous return policy, Frito-Lay's 99.5 percent reliability of daily calls to all stores, McDonald's stress on hygiene, Hewlett-Packard's emphasis on sufficient profit to finance company growth and provide the resources needed for other corporate objectives, and many others.

Organizational Culture

The effectiveness of an organization is also influenced by the organization's "culture", which affects the way the managerial functions of planning, organizing, staffing, leading, and controlling are carried out. Culture includes many facets, such as how people at various levels dress, how often they come to work early or stay late, how much after-hours socializing there is within the rank and file, to what degree communications flow freely from junior staff to top management (or whether it must flow only through approved channels), and how open the company is to trying employee suggestions—these are but a few aspects of the composite that is corporate culture.

As it relates to organizations, culture is the general pattern of behavior, shared beliefs, and values that members have in common. Culture can be inferred from what people say, do, and think within an organizational setting. It involves the learning and transmitting of knowledge, beliefs, and patterns of behavior over a period of time, which means that an organizational culture is fairly stable and does not change quickly. It often sets the tone for the company and establishes implied rules for the way people should behave.

Well-worded slogans can give a general idea of what a particular company stands for. Some examples include:

Delta describes its internal climate with the slogan "the Delta family feeling."

The Disney Corporation refers to its customers as "guests" and its employees as "characters" (implying they are all considered to be on stage while working).

In similar ways, Maytag wants to be known for its reliability, Ford for quality, and so on.

Managers, especially top managers and owners, create the climate for the enterprise. Their values influence the direction of the organization. Changing a culture may take a long time, perhaps 5 years or longer. It demands changing values, symbols, myths, and behavior. This is important for a new manager or business owner to keep in mind. If cultural change is to occur, top management must persistently and consistently reinforce the new norms over a period of years.

Specialization and Job Rotation

A review of the early history of the Industrial Revolution in both the United States and Europe suggests that great strides were made in producing larger quantities of goods at lower cost, partly because of specialization—the application of Adam Smith's ideas. As the number of an employee's duties became more limited, he or she became an expert in a narrow area.

Evidence from Japan and elsewhere suggests that there may be limits to this approach. Some problems include:

> Monotony and boredom;
> Lack of ability to see or understand the whole enterprise;
> Lack of pride in products, as a person only carries out one small task on each item;
> High absenteeism; and
> High turnover.

By contrast, where entry-level managers are rotated to every key part of the firm and required to learn every job, they relate well to the big picture and to the problems of other departments. When senior management has been through this apprenticeship there is more contact with the technical realities of the job. Likewise, when employees on the line are organized so that they see the total effects and importance of their labors, quality and satisfaction go up. When line employees are asked for help on improving the processes (e.g., quality circles), they often contribute enthusiastically and constructively.

Decentralization and Decisions by Consensus

The rational model indicated that the owner/manager should reserve to herself/himself all major decisions. A newer approach is more oriented toward making all decisions at as low a level as possible; that is, as close as possible to the people who will be putting them into effect. The availability of e-mail and the Internet on almost every employee's desk means that no longer is top management the only level of a company that knows detailed information about the firm and other firms. People have access to more information and many will use it, often almost unaware they are doing so, to bring more expertise to their jobs. In consequence, they may want more of a voice regarding how operations are conducted.

Decentralization does not refer to specialization of the work itself, but to the division of the managerial work and assignment of specific duties to various department heads or executive levels. The key issue becomes: how much of the managerial work, the planning, organizing, directing, and controlling will be done by the president, or manager, and how much should be assigned to other personnel? This allocation of managerial work is one of the most critical and most sensitive aspects of the organizing process. The amount of decentralization reflects many considerations, including importance of work, individual capabilities, and managerial desire to control, vs. to decentralize. In many situations, the executive finds a need to decentralize out of sheer necessity, in order to lighten the workload. It is often discovered that such action results in improved morale, better local coordination, and faster action on important matters.

A manager should look closely at the workload in deciding what part or parts of the overall management function should be transferred to executives at lower levels. As discussed in Chapter 2, he or she should remember that delegating responsibilities involves:

> Assigning duties and the attendant responsibility to accomplish them;
> Granting authority and creating an obligation on the part of the subordinate;
> Getting out of the way, so the assigned work can get done.

These three elements have been compared to the legs of a three-legged stool. All three are required for the whole to exist effectively. The assignment of duties without the necessary authority over resources to accomplish them will not succeed, nor will constant micromanagement.

In deciding how much decentralization to attempt, or where to place authority and obligation in an organizational structure, the manager might consider the following guidelines:

1. Who knows the facts, has the technical ability, and can get them together most readily?
2. Who has the capacity to make sound decisions?
3. Is it necessary that speedy, on-the-spot decisions be made?
4. Is it required that local activity be carefully coordinated with the work of other units?
5. How significant are the decisions in the area under consideration?
6. What is the existing workload for managers?
7. Will morale and initiative be significantly improved by decentralization?

When the manager decentralizes the operation in the process of developing an organizational structure, he or she should recognize the importance of this action and the change that will be required in his or her manner of personal operation. Decentralization needs to happen in an equitable way so that, for example, all department heads have about the same authority and none is seen as being the boss's favorite.

Experience has shown that planning is the most critical of the managerial functions if decentralization is to be effective. The process of identifying problems and deciding the correct action to take is something the manager is loath to release to others; it is something he or she is prone to supervise closely. The functions of organizing, directing, and controlling are very important, but they tend to depend upon how the planning duties have been allocated.

Management by Walking Around

A practice first identified by David Packard, cofounder of Hewlett-Packard, "management by walking around" means that the owner/manager gains a direct, unscheduled view of what is going on, including stopping to ask junior staff what they are doing and what the obstacles are to doing it better. Great tact, of course, is required in order not to undermine the authority of the immediate supervisors or cause them unwarranted anxiety about how their department is perceived to be run. The chief executive should take up any observed problems privately with supervisors and managers, and be sure to give them a chance to explain why they took their chosen course. In a sizable tight line/staff organization, management by walking around is the only way the owner/manager can keep in touch with what is happening on the floor.

Rather than relying only on reports from the five to ten department heads and aides who report directly, he or she makes personal, direct observations, as well.

Management by Results

The ultimate result of any business is profit, and the intermediate results are quality and customer satisfaction. Management by results is a variation of management by objectives that explicitly gives the people responsible the necessary latitude to decide for themselves what techniques to use to obtain desired results (within the ethics and quality standards of the firm). If accompanied by a few specific performance measures, this approach can work well.

Span of Control

Span of control is a concept used in organizational activity that is based largely on a theory of human limitations. Ideally, a particular manager in a specific situation can directly and effectively supervise a limited number of subordinates. There is considerable controversy over the nature and implications of the span-of-control concept.

The "correct" span for one manager in one situation at a particular time may vary widely from that of another manager, situation, and/or time. Ralph Davis, in his classic 1951 study, suggested that if a manager is supervising subordinate managers and the work is difficult, then the span should be limited to approximately three to nine.[10] If one is supervising operative employees, acceptable spans are from ten to thirty.

A practical approach may be to realize that the correct span of control is affected by a number of factors and to try to organize the business structure with these factors in mind. In following this practice one should consider:

> The complexity of the work being controlled;
> The degree of similarity to other work;
> The degree of interdependency with other work;
> The stability of the organization and situation;
> The degree of standardization;
> The caliber of the manager, including the span of attention, energy, and personality;
> The availability of the CEO, including demands for outside appearances and public affairs activities, and any other business ventures using up his or her time; and
> The caliber of the subordinates and their motivation.

The specific determination of the span of control must be decided on an individual basis by remembering that the outcome determines the supervisory relationship between superior and subordinate and the shape of the organizational structure. The manager must remember, too, the basic human needs and the primary needs of the organization, for an effectively coordinated and controlled effort.

With the current nearly-universal availability of e-mail and the Internet, employees have access to huge amounts of information as quickly as management does, and the old concepts of "need to know" sharing of information are disappearing rapidly. This consideration influences the span of control.

Another consideration for a small FBO business is that if the organization chart is a many-leveled pyramid, there will be some layers of positions that are purely overhead, such as those of assistant vice presidents, and so forth. The majority of smaller companies cannot afford too many such positions, meaning that the operating managers, such as the chief pilot and maintenance manager, are most likely to report directly to the owner/manager.

Effective Work Groups and Teams

As there are individual personalities and differences, so there are group personalities and differences. Organizational activities tend to create and influence the development of work groups. By knowing something about groups, their characteristics, and their behavior, the manager can improve his or her management. She/he can develop more effective work groups, and must accept the work group as an integral component of the total organization.

Rensis Likert has said, "An organization will function best when its personnel function not as individuals but as members of highly effective work groups with high performance goals."[11] Because of the group's importance to the organization, we will attempt to understand how it works, how its structures and procedures can influence business results, and how it can be used, as well as misused, in advancing the organization's interests. A starting point is the determination of the characteristics of the group. By considering these characteristics, a manager can plan the organizational activity to achieve goals or measure progress from the past. The key characteristics of a group and some important concepts or guidelines from the manager's point of view are shown in Figure 6.1. The manager needs to recognize these characteristics and utilize the pertinent concepts to his or her advantage in order to develop the organizational structure.

Staff Support

In designing an organizational structure, most managers utilize some sort of staff. The staff function is normally separated from the primary chain of command and from the line functions, to produce economy and effectiveness of operation.

The concept of *staff* refers to a unit not directly involved in the end product or salable items of the organization; that is, with the support functions. The CEO

Figure 6.1 → Key Characteristics of a Group

Characteristics	Concepts
Group Goals	Group members are more dedicated to goals they help establish than to those imposed upon them by others.
Structure	There is both formal and informal structure in small groups.
Size	Should be large enough to obtain sufficient ideas, yet small enough not to hamper free communication or inhibit participation.
Leadership	Indications are that every group requires some "central focus."
Participation	Group size, power and status structure and style of leadership are determinants of the nature and extent of participation.
Cohesiveness	Determined largely by the degree to which group goals help satisfy individual needs. If cohesiveness is obtained for official, task-oriented groups, the organization stands to gain through increased cooperation and motivation.
Norms	There is a tendency for the behavior of group members to coincide with group norms.
Agreement	Pressures toward group norms tend to produce agreement. Most business groups prefer to work toward a consensus, meaning acceptance without necessarily fully approving personally.

primarily fulfills a staff function. Serving in the normal fashion, staff units typically:

1. Give advice—the personnel office may offer advice to the manager of the maintenance department on the interpretation of selection screening devices used by the company.
2. Perform services—the accounting department, in developing an analysis of the costs involved in conducting flight instruction, is providing a service for the flight department.
3. Provide information—in many instances the staff provides information on personnel, on employment data, on tax questions, and on many areas in the technical competence of the specialist.

It should be recognized that the distinction between the two terms, line and staff, represents an oversimplification. For one thing, the real relationship in terms of authority, influence, and control is much more complicated. For another, in small organizations most managers may fulfill both line and staff functions. Instead of the departmentalization that is seemingly emphasized by the concept, the real focus should be on the high degree of interdependence required if the major systems in the organization are to function at maximum effectiveness.

Without cooperation and coordination in the pursuit of joint goals, conflict can emerge in line-staff relationships. Line functions may believe they make money for the organization by themselves; staff functions may feel under-recognized and know that they are essential to the bottom line, too. Lack of resolution of these innate differences may lead to many operational problems and ultimately to the need for managerial corrective decisions. Astute managers are aware of the inherent differences in the two functions and seek to keep the two working harmoniously together.

Human Factors

A major element in organizing a business is the recognition of the human aspects of organization. A good starting point is the realization that people work to satisfy needs and that the design of the organizational structure makes a tremendous impact on the achievement of those needs. The typical hierarchy of needs met from working is discussed more fully in Chapter 5, Human Resources.

Our primary concern at this point is meeting human needs through the organization, since the structure of a company defines an environment of formal rules, job descriptions, and communication networks during working hours. This environment can satisfy needs, or block them. It can help develop good attitudes or bad attitudes, and it can partly determine what people think and learn. Company structure becomes a highly important element in getting results. It can contribute to or detract from the satisfaction of human needs. The following are a number of ways that organization design can contribute to human needs:

Small units. By assigning workers to small groups of three to ten, social satisfactions are greater.

Non-isolated jobs. One should not break down work into such extremely specialized and independent parts that a person lacks the opportunity to interact with fellow workers.

Job enlargement. Narrow specialization, or division of the work into extremely small components, may diminish worker satisfaction and affect the employee's opportunity for growth. By enlarging the job, many companies have realized benefits, including increased worker satisfaction.

Job enrichment. The concept of job enrichment implies increasing the challenges of a particular job to allow employees to use a greater range of their skills and abilities. This may be contrasted with job enlargement, which simply seeks to increase the number of tasks performed.

Broad responsibilities. A very narrow area of responsibility tends to create a situation of always giving and never receiving. To satisfy social needs, a person's relationships with others should be reciprocal.

Place or status. The place or status of the individual in the organization may create pride in the position and even in the title itself.

Decentralization. Increasing the freedom of action of the subordinate and the opportunity to satisfy the need for self-expression are two benefits of decentralization.

New Approaches to Organization

Organization is a management tool—a most powerful tool—designed to fulfill specific needs. As the needs change, new ways are required to satisfy the changing conditions. There is an interaction between the manager's knowledge and skill level, the organization's existing situation, and the impact of the environment and its changing demands. The combination of these elements results in the identification of an existing organizational situation.

As we have discussed, change is inevitable in the business world, so the manager must have a genuine concern for the application of new approaches to organization that will make it easier to meet the demands of change. In one situation, it may mean simply the introduction of a formal organizational chart and structure in order to achieve a degree of stability. In another, it may mean identifying the need to relate more strongly to the customer. In still another way, it may require being aware of the special needs resulting from complex problems, special environmental issues, or technological changes. Each of these situations suggests the need for a new approach to organizational design. Some of the new directions that have been suggested include:

1. Hiring "idea" people to promote innovative ways of approaching and developing answers to problems.
2. Hiring individuals who are manifestly "different" from those normally hired in order to ensure fresh ideas.
3. Focusing management control on the outcome of work, rather than the way work is done.
4. Increasing rotation among jobs.
5. Increasing employee participation in decision-making.
6. Encouraging professional association involvement so that people identify with a group outside the organization.
7. Using project teams whenever feasible.
8. Including formal consumer representation.
9. Decreasing the number of supervisory levels or eliminating the superior and subordinate relationship.
10. Implementing individual profit centers.
11. Providing compensation based upon results achieved.
12. Allowing much greater freedom of access to information.
13. Encouraging mobility of individuals.
14. Increasing individual rights.
15. Supporting continued education.

A prototype organization for the business of the future has been suggested, as well. It may not be completely realistic, but it should stimulate additional thought and action in developing the new form needed for future business organizations. This proposal has four major features:

1. Hierarchical and bureaucratic structure is reduced to an absolute minimum.
2. Most middle-level jobs are conceived of as temporary.
3. Individual employees are assigned to competence centers or profit centers.
4. Career development is emphasized by the introduction of many varied programs.

Communications Technology and Applications for Internal Organization

For some years, forward-looking companies have shifted to utilizing telephone conference calls, increasingly sophisticated telephone systems, smart phones with Internet and e-mail capability, and other tools to enable internal communications to be more effective. In addition, the use of applications such as Zoom, Go-To-Meeting, Microsoft Teams, and other applications has transformed the ability to conduct meetings, with attendees physically located in a number of locations (e.g., while working from home, traveling on company business, or located in distributed company locations), rather than requiring those participants to travel to a specific company conference room. These applications provided critical functionality during the COVID-19 pandemic in 2020. The software systems/applications also allow for real-time shared viewing and editing of documents and drawings as well as "archiving" of real-time speeches, presentations, and discussions for later use in training. Larger companies with more than one outlet are turning to these approaches to save money and time and to meet the needs of their clients. While this topic is covered here, it is also highly pertinent to marketing and may provide new means of meeting customer needs, as well. However, face-to-face contact is still needed to build the relationship and work out any difficulties, and the requisite air travel may not be affected as much as

FBOs may fear because, as is true with all new technologies, this enables people to do more, not just do better what they were already doing.

Internal company websites, or intranets, can also help the organization improve its internal communications process. Intranets allow the instant widespread dissemination of policies and procedures documentation to company employees, and help resolve problems related to the revision process for such policy manuals that exist in hard copy form.

Internal Structure Design

When developing an internal organizational structure for a business, the manager should consider such factors as departmentalization, decentralization, delegation, specialization, managerial resources, staff capabilities, and the specific types of business activity involved. Managers should use each of these concepts in developing an effective total organizational structure. The determination of the overall structure is based upon four major considerations:

1. What balance and emphasis should be given to the various departments?
2. How can the span of control or supervision of each manager be effectively utilized?
3. How can dynamic change be provided for?
4. How can this organization be integrated into the other phases of planning, directing, and controlling?

The answers to these questions provide direct guidance in identifying the desired organizational structure.

Formal Internal Structure

In developing the formal structure of the organization, the manager has three basic formats to consider: functional, line, and line and staff. Since many aviation businesses need to utilize elements of all three business structures, it is recommended that the manager understand the terms and be able to use all three concepts in organizing a business.

Functional or Matrix Management

Opinions regarding the value of the functional or matrix form of organization come and go. An example of the functional form is shown in Figure 6.2. The functional organization has a supervisor in charge of each function or project, rather than in charge of specific employees. Each of these supervisors is a specialist and makes decisions for that functional area. The suitability of the matrix or functional organization form to an aviation service business is limited, because many of the line personnel must be specialized and certified (e.g., as pilots or as maintenance technicians); thus, they cannot do another person's job without dual certification. However, just as the manager of a small business wears many hats, the employees do as well, and, in that sense, an FBO may be operating on a functional basis regardless of formal organizational structure. As the owner/manager hires an assistant or supervisor, this person is normally placed in charge of a specific function and all employees receive guidance from him or her in that one area. As additional supervisors are added with functional specialties, the employee must move from

Figure 6.2 → Functional or Matrix Organization

EMPLOYEE	FUNCTIONAL AREA 1 Led by Matt	FUNCTIONAL AREA 2 Led by Judy	FUNCTIONAL AREA 3 Led by Jack	FUNCTIONAL AREA 4 Led by Jean
Mary	O		O	O
José		O	O	
John	O			O
June	O		O	
Spike	O		O	O

O = assists as needed, reports to that functional supervisor for assigned tasks supervisor to supervisor (function to function) as he or she performs different physical tasks.

This is one source of difficulty with the functional organization; employees can become confused, conflicts can develop, supervisors may battle for employees' availability, and problems can arise in evaluating quality of work and performance on the job. On the other hand, *temporary* functional assignments into special-purpose task forces, which have a leader, a mission, and a set amount of time to do the task and disband, can have great value. An example of this might be a flight instructor who is tasked with delivering an airplane to a customer for the sales department.

Line Organization

The second type of formal organization structure is the pure line organization. In this type of structure, the supervisor is in charge of a specific operational unit and of the time of all people assigned to work in that unit. This is frequently thought of as the military form of organization. The line organization has one person in charge of another person, who in turn is in charge of other persons, although the number of layers can be minimized. (See Figure 6.3)

Line and Staff Organization

For companies that have grown beyond the direct supervision stage, the line and staff organization becomes the most feasible form of structure. It differs from the straight line because a staff element has been added to the structure. With regard to actual operations, the structure operates similarly to the line structure. However, the staff element is there to advise the line manager and the CEO on the technical features of that staff specialty, thus assisting that manager in getting the job done. The staff unit is normally designed to provide advice, information, and services to line managers and the CEO. Figure 6.4 suggests a line and staff structure for an aviation business. In this type of structure, the supervisors in sales, instruction, service, and flight have the responsibility for getting the job done in their individual areas, and the staff personnel in administration and finance have the task of advising and assisting in their areas of specialty. For example, accounting might advise on the necessary records and budget guidelines, or administration might advise on OSHA requirements and coordinate report deadlines.

The aviation manager should be aware of all three types of organizational structure and recognize that there are elements of each in every business. The most likely pattern when a business is starting will be functional, then as the business develops the organization will adopt a line structure, and finally, with size, it will assume the characteristics of a line and staff structure.

Informal Internal Structures

The formal organization consists of the official, authorized relationships prescribed by management, usually as depicted by company and department organization charts. However, informal internal structures are also part of every organization. Informal organization consists of the myriad relationships, unofficial

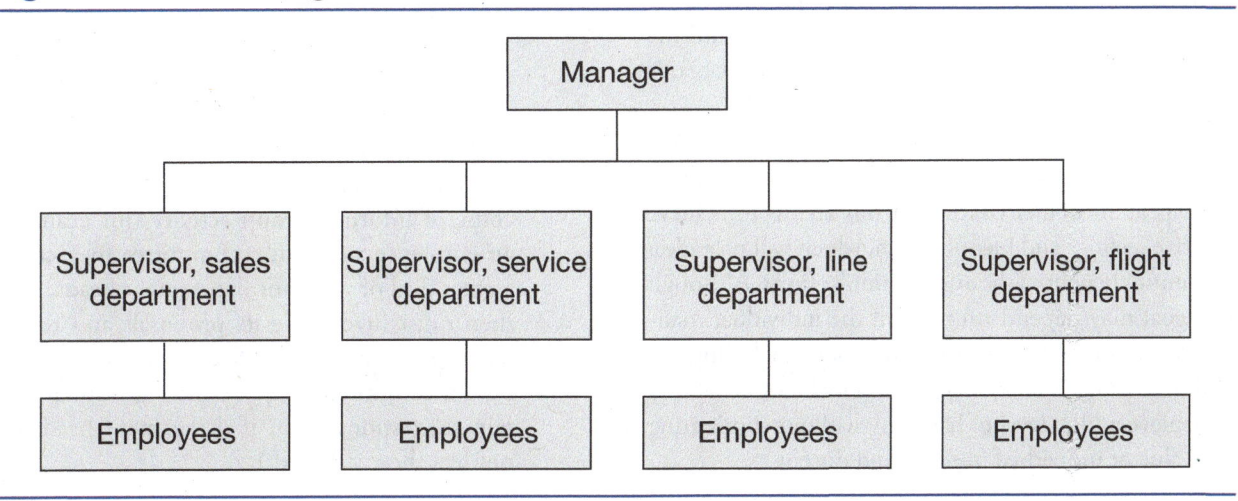

Figure 6.3 → Line Organization

Figure 6.4 ✈ Line-Staff Organization

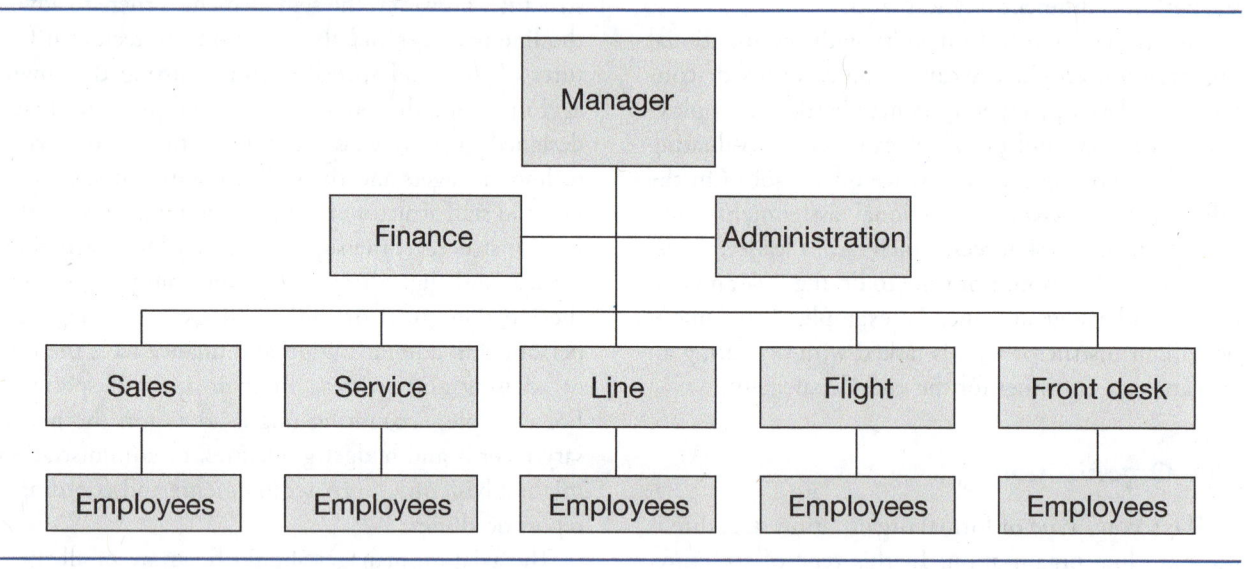

```
                        Manager
                           |
           ┌───────────────┴───────────────┐
        Finance                      Administration
           |
  ┌────────┼────────┬────────┬────────┐
Sales   Service   Line    Flight   Front desk
  |        |        |        |        |
Employees Employees Employees Employees Employees
```

and unauthorized, created by the many individual personalities and groups within the formal structure. These relationships, both good and bad, spring up spontaneously and continually. There is no choice whether to have an informal organization; it is a fact of organizational life. Since it exists regardless of the manager's approval, it is important for management to capitalize upon its advantages and work around its disadvantages. There may be many pockets of separate organization. A healthy company will, however, seek to encourage inclusiveness, rather than cliques.

Managers who do not understand group behavior and attempt to break up informal groups are frequently frustrated when new alignments form and present continuing problems. The successful manager understands the nature of group forces, is perceptive to group needs, understands why employees need to talk together about various aspects of work, and can successfully blend the informal groups' goals with those of the formal organization. A perceptive, yet light touch is needed.

The informal structure is the source of many positive values that can contribute to the organization. Management's chief concern in this area is to achieve positive feelings and identification, which will provide a maximum benefit to the organization's efforts. Although this goal may depend strongly on the individual manager's skillful use of organizational tools (including the ability to direct and control), it also includes the ability to achieve a high level of harmony without eliminating the value of individual freedom and dissent.

Advantages of informal structure include:

1. It fills in gaps and deficiencies in the formal structure.
2. It facilitates the majority of the organization's work.
3. It lengthens the effective managerial span of control.
4. It compensates for the violation of formal organizational principles.
5. It provides an additional channel of communications.
6. It stimulates better management.

Disadvantages of informal structure include:

1. Informal activities may appear to work counter to organizational goals; for example, if a subgroup of employees decides to become more vocal about working conditions. Nevertheless, the informal structure is an important way for management to learn early about worker concerns and desires.
2. There is less ability to predict and control outcome of informal group activity (for example, an employee suggestion committee may create a great deal of work for department heads, who then must investigate its proposals and report back on what may be feasible to implement; such a committee may feel strongly about certain suggestions, even if department heads do not feel they will work).

3. Since, virtually by definition, individuals self-appoint to particular informal groups, there is less interchangeability of individuals within groups, thus reducing managerial alternatives and influence.
4. There can be high costs (for example, in terms of employee time) of maintaining a positive, informal set of relationships.
5. If management is unaware of informal relationships, negative or positive, assignments may be made that create potential problems, either that the assignees get along too well, or don't get along at all, with the result that the assigned task may not get done on time.

The informal organization can be identified and charted using sociometric techniques. The most common applications of this procedure develop: (1) a chart depicting actual contacts between particular people, (2) a chart reflecting the feelings of people toward each other, or (3) the influence-of-power chart, which depicts the level of influence of each member in the group. Each of these techniques provides the means for gaining some insight into the informal structure and provides additional information to supplement the organizational efforts of the formal structure. Keep in mind, however, that if the effort is made to explore informal structure to this degree, first, such exploration may cause resentment, and second, relationships are constantly changing so the outcome may not be very useful. A simpler approach may be to simply observe and take steps such as requiring the listing of relatives already employed by the company on job application forms. A new manager will do well to find a trustworthy peer or mentor who will assist him or her in understanding the "land mines" and "sacred cows" of the informal organization.

The informal group may select a leader who reflects the views of the group toward the formal structure. Even if this leader is removed as a troublemaker or malcontent, the difficulty is not solved, because the group will most likely simply select another representative who, in turn, will present the same views, perhaps even less helpfully. The manager can work with the group leader to obtain full cooperation from the group. This does not mean that he or she should channel orders through the leader or attempt to make that individual a member of the management team. When this is attempted the usual result is the identification of a new informal leader. The manager can, however, devote a little extra attention to winning the support and approval of the group leader when the need arises, or perhaps on a continuing basis.

The informal organization can be used to supplement the formal channels of communication. The informal structure has been recognized as being fast to form, but also being susceptible to distortion of information. It can, however, be utilized to gain input on group discontent or pending problems, and thereby to be of tremendous help to managers. In most instances, it is considered wise to utilize the informal communications network and to deal with the informal leader in a delicate and discreet manner.

In a large organization in which rapid change is occurring, many rumors may develop and circulate through the informal networks of the business. This is of some concern to the manager, because rumors may develop that are based on partial information or misinformation, and may become distorted as they circulate. One technique that management can apply in this situation is a "rumor board." For this, an erasable board is placed in a convenient, yet non-customer area, such as the lunchroom, and employees are invited to write rumors they have heard on the left side of the board. Management checks the board daily and writes either answers, or that no answer is available yet and a date when it will be, and initials this input. A clerical person is assigned to record both rumors and answers as the board gets full, and they can be put on the company's employee intranet site for further reference. Alternatively, an electronic rumor board can be set up on the employee website.

Quality Circles

Quality circles are a feature borrowed from the Japanese, although it was an American who introduced the Japanese to quality planning.[12] A circle of workers is jointly responsible for a finite, complete product and—in addition to being on the production team together—meets regularly to iron out problems and make improvements. Various circles doing the same thing may become quite competitive. Posting weekly performance can be an incentive toward high results.

Task Forces

As mentioned above, task forces may be short-lived and interdepartmental. Many companies use them to brainstorm a problem on an intensive basis, and some firms use an annual or biannual retreat to do much the same thing.

Social Structures

Some firms deliberately encourage and even subsidize sports teams and other clubs, with the intent of mixing all types of employees together in informal settings. A superior's usual aloofness can be set aside for a period. Parties and picnics accomplish some of the same things, depending on how they are run. Management needs to gauge whether employees (and their families, if included) really enjoy these functions, or are just attending for fear of not being thought a team player if they stay away. For example, not everyone is athletic and some employees may not wish to lose face by demonstrating their ineptitude at ball games, fearing their lowered image will then be carried over to Monday morning. Some employees simply feel that their work week is already long enough and they don't want to spend scarce free time on work-related events. In response to these concerns, some companies have started polling employees as to what they would rather do. Public celebration of work achievements is sometimes part of these functions, or the public recognition may be reserved for other times.

Other Networks

Special-interest professional groups within the organization, attended without regard to one's position in the hierarchy, can give senior staff members the exposure that they might not otherwise get to junior staff members, and vice versa. This important source of networking and interaction between senior and junior staff can also help convey directly the corporate values and culture to all levels of employment. Remoteness of senior officials is not necessarily an advantage to the company.

External Pressures on Choices of Structure

In designing the structure of an aviation business, consideration is given, consciously or unconsciously, to the pressures of external elements; pressures that suggest or dictate certain patterns in the organizational structure of the business.

Industry Norms

The aviation industry as a whole tends to promote an organizational design that is somewhat homogeneous throughout the industry. Hangar design, ramp location, services requested by the customer, and competitor activity all exert pressure on the manager to conform to the general industry organizational structure.

The companies supplying the major aircraft product lines (e.g., Piper Aircraft, Beechcraft, Cessna, Gulfstream, Embraer, Bombardier, Dassault) provide a degree of pressure through franchises or other agreements, which result in the development of a structure suggested by the company and similar to others in the organization. These standard structures have generally been found to be successful and have been promoted in order to achieve standardized operations and successful businesses.

Government and Regulatory Influences on Organizational Structure

Governmental activity at all levels plays an important part in structuring aviation businesses. Federal, state, and local regulations require reports and procedures that can cause the manager to structure the organization in a certain fashion. For example, OSHA requirements have caused many managers to write procedures, appoint safety directors, and in some instances reorganize the business. For those managers involved with FAR Part 135 operations, this type of business activity requires a certain organizational structure as well as operating manuals and routine periodic checks that create certain organizational structures. FAA, VA, and state educational agencies have created requirements for flight schools that have literally determined organizational structure for aviation businesses engaged in instruction. Aviation maintenance requirements and practices, as stipulated by the FAA and original equipment manufacturers, very effectively contribute to the structure of aviation maintenance activities.

Practical Applications Guidelines

Thus far, the primary focus has been on basic concepts and principles and planning the organizational effort. We now shift the focus to the practical task of dividing the business into functional areas and assigning these separate functions to specific individuals. Another three-legged stool analogy the manager should keep in mind incorporates the three legs that support a successful business of any type:

1. Delivering outstanding services and products; producing what you are in business to produce.
2. Marketing and sales to make sure people know you have great products and services.
3. Managing the finances and administration, and the paper trails.

Most business owners start their companies because they love what they do and believe they can do it better than others. Their problems arise when they discover they are not very good at, or don't enjoy, the other two legs of the stool. Usually, neglect of marketing and/or finance are what cause businesses to fail, even when products and services are excellent.

In organizing, the manager is constantly balancing two factors: the desired structure and the available personnel. Does she/he design the structure and then attempt to find or train personnel to do the work, or does she/he utilize the available human resources and work with the resulting organization? The manager will undoubtedly need to work with a combination of these two approaches, although the successful manager generally sets organizational goals and then works to build the structure that will achieve them.

In a practical sense, organizing is the basic process a manager uses to unite the efforts of different people to achieve company goals. There are two elements present in the process of organizing: (1) dividing work into separate jobs, and (2) ensuring the separate areas of work are combined into a total team effort. Let's now consider the various steps that become part of this practical organizing activity.

First, identify the basic or recurring pattern of work that takes place in the business. The best approach is to group work on the basis of geographical areas, products, customers, or functions. Many aviation organizations can, and do, utilize all four groupings. This is evidenced by companies who have regional salespeople, who distinguish employees working on Piper aircraft from those working on Beechcraft aircraft, who classify customers by income level and assign salespeople to each major category, and who group work on a functional basis such as service, sales, or charter. This first step becomes an effort in understanding the activities that will take place within the business.

The second step is to describe the relationships that will exist between the various groups within the business. This can be done through a chart on paper, a written description of the working relationships that exist, or both. Typically, this is achieved through an organization chart and organization manual, which may be communicated company-wide by dissemination through the intranet.

Define precisely what each unit will do as a portion of the overall organization. In identifying units, the basic principles of specialization, decentralization, span of control, work groups, staff activity, and human factors should be kept in mind.

Assign business activities to specific individuals and units of the organization. This may be done casually in conversation or formally by charts, job descriptions, operations manuals, meetings, or joint understanding.

Provide for the means of tying the organization together and for developing integrated, cooperative action in the day-by-day transaction of business. Normally, this is accomplished through communications, procedures, rules, and controls.

In summary, the organizing process involves:

> Identifying the kinds of work that take place;
> Describing the relationships among work groups;
> Defining precisely what each group will do;
> Assigning activities to specific individuals or groups; and
> Developing integrative action among work groups.

Problems

The practical activity as outlined here represents the minimum that should be included in the organizational process. Realistically, the manager should recognize that there are many problems inherent in the process and should try to avoid these difficulties. Some of the problems frequently experienced are:

> Lack of clearly-defined duties and resulting friction;
> Expansion of activities beyond assigned areas by overzealous managers;
> Desire to create a situation in which the unit reports high up the administration chain;

> Failure to adapt to necessary changes; and
> Failure to integrate with the other functions of planning, directing, and controlling.

The Organization Manual

One of the most effective tools for developing a business structure is the organization manual, management manual, or handbook. This particular manual does not follow a specific format or content outline, but rather provides a flexible technique for meeting individual organizational needs. As usually visualized, this document is a centralized source for policies, organizational structure, relationships, responsibilities, rules, and procedures. Properly designed, implemented, and maintained, it can be an extremely valuable tool in organizing the business. It becomes a guide to day-to-day operations for individuals in the organization and a source for training new employees. Primarily, it is a means for creating and maintaining the desired organizational structure and activity. It should be communicated to all employees as required through the use of the corporate virtual private network or intranet.

Manual Outline

Although the contents of an organization manual should vary in order to meet the needs of each organization, usually one finds the following subjects covered:

1. Introduction, philosophy
2. Organization structure
   a. Authority
   b. Responsibility
3. Departmental organization
   a. Activities
4. Staff activities
   a. Personnel
   b. Financial

Summary

Organizing the business is a continuing challenge for the aviation manager. Both the external legal structure and the internal practical structure must be established to support the goals and activities of the company. The choice of ownership structure should be made consciously and reviewed periodically. Internal organization should recognize the needs of the various company departments, as well as the informal structures that help a company function. External pressures and industry norms may need to be considered, as well.

Endnotes

1. Many children are much better at puzzles than adults and there can be great value in inviting children's opinions and advice at times of business confusion. Children tend to have more open and agile minds.
2. Community Facilitation & Development, Ashland, OR.
3. Brown, D. (1980). *The Entrepreneur's Guide*. New York: Ballantine.
4. Op. cit.
5. Smith, A. (Author), E. Cannan (Editor), G. Stigler (Preface) (1976). An Enquiry Into the Nature and Causes of the Wealth of Nations, edited and with an Introduction, Notes, Marginal Summary, and Index by Edwin Cannan. Chicago: The University of Chicago Press.
6. Taylor, F. W. (1967). *The Principles of Scientific Management*. New York: Norton.
7. Herzberg, F. I. (1959). *The Motivation to Work*. New York: Wiley.
8. The reader is recommended to review, for a more detailed study, several "classic" texts, including: Ouchi, W.G. (1981). *Theory Z: How American Business Can Meet the Japanese Challenge*. New York: Avon Books. Pascale, R.T. & Athos, A. G. (1981). *The Art of Japanese Management: Applications for American Executives*. New York: Warner Books. Peters, T. J. & Waterman, R. H. (1982). *In Search of Excellence—Lessons from America's Best-Run Companies*. New York: Warner Books. Weiss, A. (1984). *Simple Truths of Japanese Manufacturing*. Harvard Business Review, 84, July-August 1984, 119–125.
9. For example: Byrne, J.A. & Welch, J. F. (2001). *Jack: Straight From the Gut*. New York: Warner Books. Johnson, C.R. (1999). *CEO Logic: How to Think and Act Like a Chief Executive*. Franklin Lakes, NJ: Career Press. McCoy, C.W., Jr. (2002). *Why Didn't I Think of That? Think the Unthinkable and Achieve Creative Greatness*. Saddle River, NJ: Prentice Hall Press. Slater, R. (1999). *Jack Welch & The G. E. Way: Management Insights and Leadership Secrets of the Legendary CEO*. New York: McGraw-Hill.
10. See *https://ucl-new-primo.hosted.exlibrisgroup.com/primo-explore/fulldisplay?docid=TN_cdi_crossref_primary_10_1177_000271625228100140&context=PC&vid=UCL_VU2&lang=en_US&search_scope=CSCOP_UCL&adaptor=primo_central_multiple_fe&tab=local&query=any,contains,Ralph%20Davis,%201951&offset=0* for access to the full text.
11. Likert, R. (1961). *New Patterns of Management*. New York: McGraw-Hill.
12. See the W. Edwards Deming web site at http://www.deming.org.

DISCUSSION TOPICS

1. How should a manager use the organizational structure to address a) change and b) routine tasks?
2. What are the key advantages and disadvantages of the three legal structures? In starting a new FBO, which one would you choose, and why?
3. Describe the formal and informal organizations. How can they work in harmony?
4. Discuss the organizational principles of specialization and decentralization.
5. What is span of control and how can the business choose appropriate spans?
6. Why is an organizational manual useful? Name three elements it should include.

7

Management Information Systems

OBJECTIVES

› Understand the necessity of a business information system.

› Understand how to effectively manage a business and monitor its profitability using information systems, financial reports, and ratio analysis.

› Relate the requirements of an effective management information system to an aviation business.

› Recognize the problems in maintaining an effective management information system for an aviation business.

› Demonstrate the origins, pathways, and use of key information system elements for an aviation business.

› Describe the factors to consider in designing an effective records system and upgrading computers for an aviation business.

Introduction

Computer-based or manual management information systems transform data into information useful for decision-making. This chapter discusses the data types needed, the data formats and flows involved, and the data analysis and applications utilized in a typical fixed base operator business. Finally, this chapter covers the selection of tools for data processing and explores several computer-based management systems commonly used by FBOs.

Industry Changes—Franchises and Computerization

In the 21st century, significant factors affecting FBO management information systems include the following:

1. The availability of special-purpose computer hardware and software designed to keep track of every aspect of an FBO's business means that the competitive company will be computerized in virtually every respect. Leading vendors of industry software are FBO Manager, FBO Director, MyFBO.com, Corridor, and TotalFBO; generic systems suitable for any business are QuickBooks, Oracle, and Sage.

2. Not only enabling traditional tasks to be automated and performed in a more precise and timely manner, computers also accomplish a greater volume and variety of tasks. While expectations for quantity and quality of work increase, the introduction of a computer system to an office does not necessarily save human

labor; rather, people will likely do more work as they learn to use both the new system and all of the new applications that are available.

3. Despite the availability of more sophisticated information management systems, "garbage in" still produces "garbage out." A haphazard approach to managing information produces muddled results even if one obtains well-designed forms, a state-of-the-art computer system, and the most expert advice.

4. The rising number of FBOs with multiple locations, such as franchises and chains, means that aviation businesses are sharing a centralized approach to information procedures, including standardized forms and customized computer systems. FBO software vendors have turned to the Internet to address the unique needs of FBO franchises.

System Purposes

A manager should ensure that good records are kept on a regular basis in any small business in order to:

> Ensure financial survival,
> Enable performance improvement,
> Create special reports,
> Comply with taxes and legal obligations, and
> Provide a reality check for business owners.

Financial Survival

Knowing sales volume, whether payments have been received, the cash on hand, the cost of opening the doors each day, and the break-even level of activity are basics for the survival of any business. Even a small business will involve more information than the manager can carry in his or her head. It is particularly easy to overlook upcoming expenses or misjudge incoming sales. Any plans for expansion will also require knowledge of unit costs, profit margins, and the competitive landscape. Most importantly, a small or new business cannot wait until the end of its fiscal year to evaluate its financial performance. Monthly status reports are essential, and in some cases, weekly or daily posting of results may be appropriate.

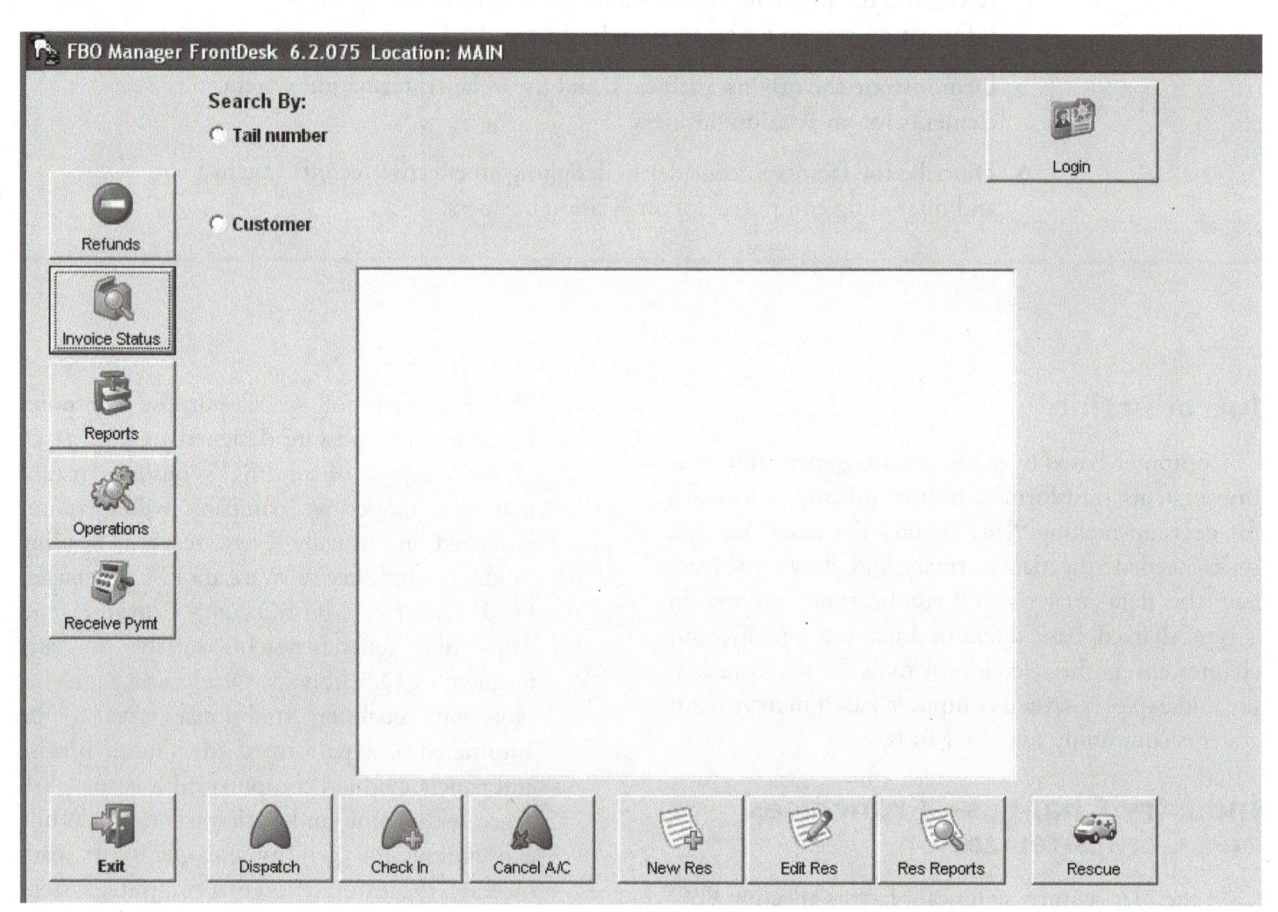

Main screen from FBO Manager Front Desk (using touch screen interface). Courtesy of FBO Manager.

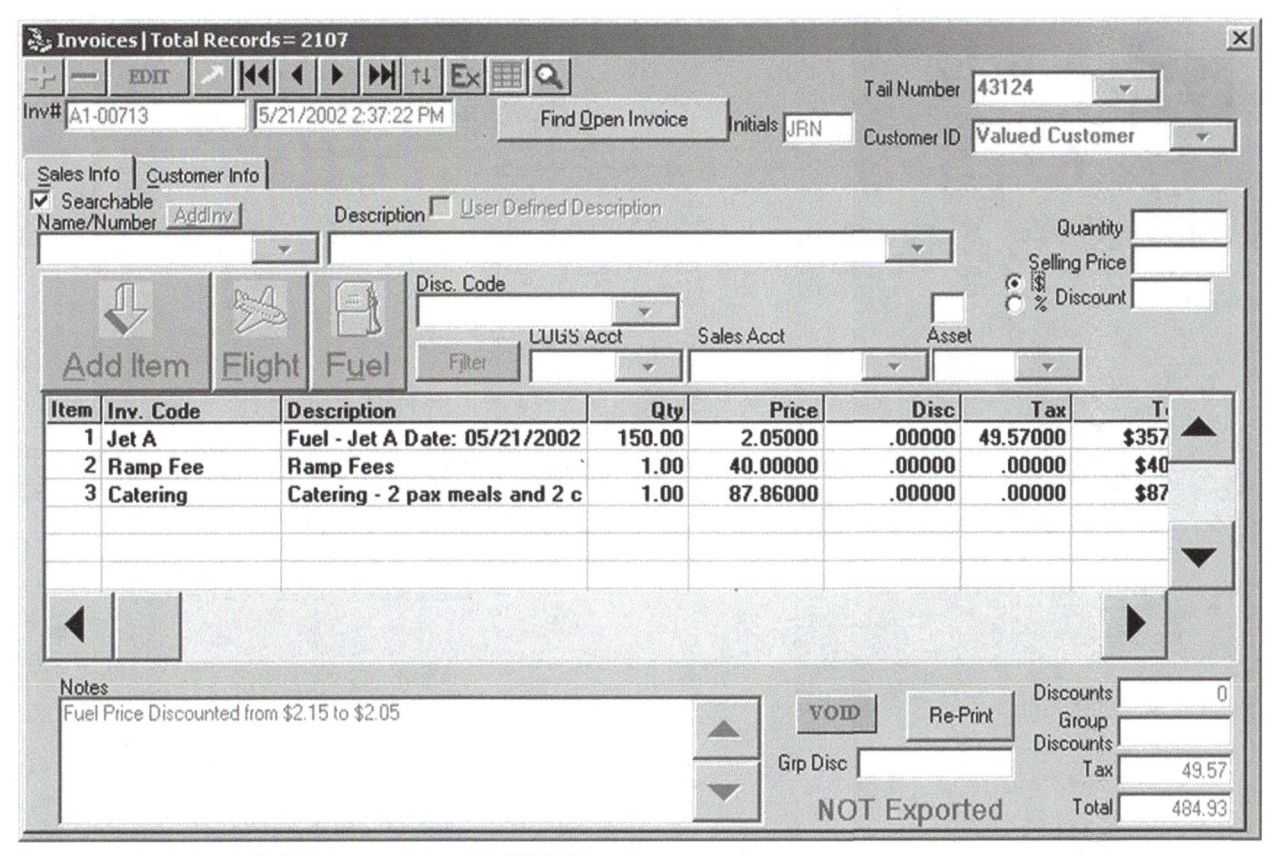

Invoice screen shown in FBO Manager. Courtesy of FBO Manager.

It is recommended that the operator create an income statement for each section of the business: maintenance, flight school, pilot shop, charter, concierge services, inventory, and so forth. The operator will also need to have an income statement for each human or equipment "asset" within each profit center of the business. For example, what are the gross and net of each instructor, each mechanic, or each aircraft on the flightline? An operator needs this information to fully evaluate the profit centers and identify the money-losers within the company, so that corrective actions can be taken. Additionally, if cash flow becomes a problem, the operator should know which equipment assets can be sold quickly to generate capital. Although the cost of accounting software may seem high, the operator must have access to this information to profitably manage the business.

Performance Monitoring and Improvement

Good information systems permit the examination of the whole business and also subareas, such as projects, sales territories, individual employees, profit centers, and cost centers. This data can then lead the manager to take correct or prevent any problems. Accurate comparative data on employee performance such as productivity, hours worked, and absences can be important in establishing a clear and fair case for promotion or termination. The earlier it is discovered that a certain area is underperforming, the sooner and easier it is to address problems. Good management entails almost constant course corrections in all areas of the business.

Special Reports

Apart from regularly scheduled postings, good records can quickly and accurately generate reports on specific subjects as needs arise. For example, a special report might be five-year trends in parts costs compared with maintenance labor costs to determine a policy on sale of parts to self-repairers. The analysis would examine whether the profit is greater on sale of parts or on labor and how the two are likely to relate in the future, based on past trends. Another special analysis might look at the allocation of time of a certain group of employees to particular tasks. Jobs might be reorganized to use less costly personnel

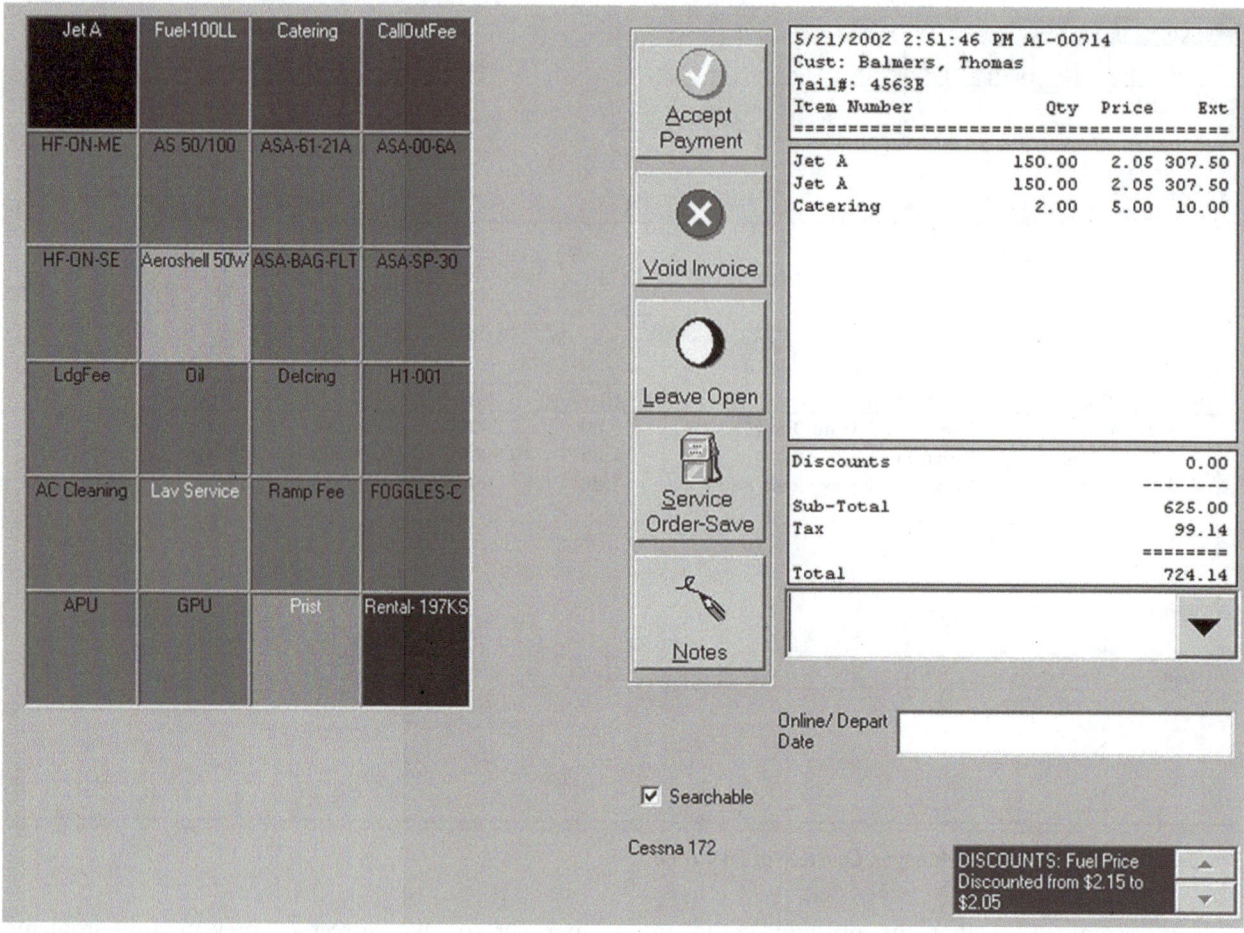

Sales screen providing for user-defined layout of items and colors. Courtesy of FBO Manager.

for less profitable areas of activity. A look at seasonal and daily peaks might change marketing and work shift strategies, and so on. The manager's ability to have special reports prepared depends on amount and detail of the readily available data in the firm and how efficiently that data may be analyzed. Much more can be found from data kept for accounting purposes than is generally intended, and computerization makes it relatively easy to create such reports. For example, Figure 7.1 shows fuel sales ranked by customer volume over a three-year period. Although jet fuel costs may be different now, this chart clearly provides a wealth of useful information.

Taxes and Legal Obligations

A good data system will not only ensure that required reports and payments are made on time, but will also flag upcoming events so that they can be anticipated.

Checks on Reality

Good paper records of inventory, cash on-hand, accounts payable, accounts receivable, and sales volume can protect against employee neglect or theft. A company in which a manager regularly checks records against stockroom and office observations will establish an atmosphere in which pilfering or "cooking" the books is less likely. If an employee or customer is stealing, good records are essential for discovery, prosecution, and insurance claims.

System Processes

Integrated Flow

Data systems all begin with the automated or manual recording of information at its point of origin (usually the point of sale). The flow from that point on should be fully computerized. The data should flow freely from user to user, with repetitive work and

Figure 7.1 → Fuel Sales by Type by Customer

Fuel Sales by Type by Customer for 11/01/1998 to 12/31/2001
Jet A

Customer	Quantity	Taxes	Sub-Total	Total	Eff Price
Moodyman, James	34600.00	$13,072.28	$74,390.00	$87,462.28	$2.53
Mattingly, Robert W	10703.40	$2,572.29	$22,542.71	$25,115.00	$2.35
Boyd gaming	6806.70	$671.27	$14,549.31	$15,220.58	$2.24
Valued Customer	6777.90	$1,079.50	$14,557.54	$15,637.04	$2.31
Balmers, Thomas	6181.50	$773.46	$12,934.62	$13,708.08	$2.22
Beaudoines, Jay	3200.10	$54.39	$6,424.17	$6,478.56	$2.02
7331	3070.00	$0.00	$6,166.75	$6,166.75	$2.01
McAlister, Stuart	2969.80	$809.81	$6,988.07	$7,797.88	$2.63
Williamson, John	2268.90	$826.74	$4,748.45	$5,575.19	$2.46
McClures, Steve	2019.00	$578.28	$4,290.85	$4,869.13	$2.41
Stevens, Samantha	1720.00	$410.78	$3,626.00	$4,036.78	$2.35
A.VFuel Test	1599.20	$416.80	$3,398.28	$3,815.08	$2.39
Stuart-Macdonald	1548.00	$517.27	$3,328.20	$3,845.47	$2.48
Baltzer, Jacob	1520.00	$533.33	$3,268.00	$3,801.33	$2.50
Allens, David	1241.80	$364.27	$2,300.77	$2,665.04	$2.15
Del Monte Aviation	1103.00	$395.40	$2,261.15	$2,656.55	$2.41
Claus, Santa	1074.60	$75.80	$2,270.34	$2,346.14	$2.18
Winkler, Henry	500.00	$75.78	$1,075.00	$1,150.78	$2.30
Mouse, Mickey	350.00	$56.83	$752.50	$809.33	$2.31
Miller, Lee L	310.00	$117.48	$666.50	$783.98	$2.53
AvFuel	300.00	$0.00	$645.00	$645.00	$2.15
Bees, Wade	300.00	$37.90	$645.00	$682.90	$2.28
Executive Jet	252.00	$0.00	$541.80	$541.80	$2.15
Pedro	250.00	$94.73	$537.50	$632.23	$2.53
Dawsonest, Ken	200.00	$75.78	$430.00	$505.78	$2.53
415WM	200.00	$0.00	$430.00	$430.00	$2.15
Rowers, Dan	150.20	$0.00	$322.93	$322.93	$2.15
Calverts, Dennis	150.00	$0.00	$322.50	$322.50	$2.15
Mullen, Sandi	136.60	$51.76	$293.69	$345.45	$2.53
Nelson, John	122.00	$0.00	$262.30	$262.30	$2.15
Gilligan, Jim	116.00	$0.00	$249.40	$249.40	$2.15
Deerport Aviation	100.00	$0.00	$155.00	$155.00	$1.55
Whitman, Chip A	100.00	$37.90	$215.00	$252.90	$2.53
Chambers, Art	74.00	$0.00	$159.10	$159.10	$2.15
Haspels, Ralph O	60.00	$22.74	$129.00	$151.74	$2.53
Gibbsons, David	49.66	$0.00	$106.77	$106.77	$2.15
Taylorama, Greg	49.00	$0.00	$105.35	$105.35	$2.15
Saelzle, Peter	5.00	$0.00	$10.75	$10.75	$2.15
Kellam, David S	-6546.00	$85.64	($14,073.90)	($13,988.26)	$2.14
TOTAL	85632.36	$23,808.21	$182,026.40	$205,834.61	

Courtesy of FBO Manager.

human intervention minimized. This means that the initial data should be collected in a format that suits the needs of its final users and perhaps in more detail than is needed for intermediate purposes.

Requirements of an Effective System

In all the diverse areas where information should be collected, there are a number of common requirements. The management information systems set up to address these requirements must be timely, must aid in the allocation of resources, and must improve decision-making by assisting in the selection of alternatives. They should be adequate to fit the needs of the particular business without being over-designed, and should address the total operation of the company, not just accounting and financial records. Understandable and operational, these requirements should adhere to the existing organizational structure but also be flexible. Most importantly, they should flag problems or deviations from the norm and indicate corrective action.

First Steps

In creating the business' information systems, the manager should develop a master plan that covers the basic requirements and considers any additional needs. One may develop a master plan by starting with the desired output and working backwards to identify necessary inputs. First steps include:

1. List long-term objectives for the system.
2. Establish the required reports and data.
3. Define the quantities and qualities of information currently being collected and evaluate its adequacy using chronological or input-output process flows.
4. Make short-term system improvements that are consistent with long-term objectives.
5. Establish the schedule, budget, and responsibility for the long-term plan.
6. Implement and execute the plan.

The long-term objectives for the data system may be drawn from the overall company business plan (see Chapter 2, Management Functions). Specific data requirements will need to consider:

> Data source,
> Frequency of data collection,
> Frequency of data transmission (batching or real-time),
> Time period for complete information cycle,
> Time delay between an event and its report to a manager in useful form,
> Format of data before and after tabulation or analysis, and
> Adequacy of existing data (do current systems sufficiently guide management?)

Aviation Management Information Systems

Human Resources

Many personnel issues arise not on a regular basis, but in relation to changes in personnel and personnel requirements. Some personnel issues, however, that do occur on a regular basis are time sheets, punctuality and absence records, sick time, and leave time. These records will likely be kept by the employee on a daily basis and submitted on a weekly or biweekly basis for payroll processing. For employees wearing more than one hat or belonging to more than one profit center, time records will report not just time on the job but time by project. It can also be useful to track the manager's own time among activities such as marketing, auditing, public relations, administration, other overhead, and direct revenue-generating tasks.

Figure 7.2 shows an array of potential human resource data requirements, sources, and controls.

Financial

The cash data needs of the organization are shown in Figure 7.3.

Material

"Material" covers all tangible assets of the business such as buildings, aircraft, cars, inventory, shop machines, and office equipment. Information is needed on the condition, contribution, and replacement of these assets, as well as flags for scheduled preventive maintenance. Figure 7.4 lists such assets, their information sources, and controls. Figure 7.5 shows a sample physical inventory worksheet for monitoring the parts and supplies in the maintenance shop.

Figure 7.2 → Human Resource Data Needs

Activity	Source of Information	Control Measure
1. Identifying human resource requirements	Inventory of company's human resources Job requirements	Company goals, objectives Organization requirements
2. Recruiting	Examination of practices Review of results	Turnover goals Stability desired Training needed
3. Selection	Personnel records Performance results	Organization requirements Position requirements
4. Training	Training records	Training requirements Individual needs Defective rate
5. Using human resources	Personnel records Time cards Job orders	Turnover rate Tardiness rate Labor productivity Organization requirements

Figure 7.3 → Cash Data Needs of an Organization

Activity	Source of Information	Control Measure
1. Obtaining adequate funds	Organization requirements Financial statement analysis	Goals Comparative data
2. Accounting for funds	Accounting record Routine reports	Accounting standards Comparative data
3. Utilizing funds	Financial analysis	Organization goals Historical data Industry data

Figure 7.4 → Summary of Business Assets

Activity	Source of Information	Control Measure
1. Aircraft	Physical inspection Profit records Customer indications	Organization goals Historical records Industry data
2. Inventory	Routine inventory records Profit records Customer indications	Goals Historical records Industry data
3. Physical plant	Physical inspection Preventive maintenance records Engineer reports Customer indications	Company goals Historical records Comparative data
4. Equipment	Physical inspections Preventive maintenance records Customer indications	Company goals Historical data Manufacturer requirements

Figure 7.5 → **Sample Physical Inventory Worksheet**

10/27/2009
5:08PM

Horizon Business Concepts, Inc.
Physical Inventory Worksheet

Page: 1

Type	Part Number	Description	Unit Cost	Bin/Location	Dept	Current Stock	Actual Count
RP	099001-041	EXHUST SYST ASSY	998.00	SH-B4	SP	2.0	
RP	102280-1	CANOPY FAIRING AFT	87.50	SH-B3	SP	2.0	
RP	102344-19	MOLDING WINDO LH	68.00	SH-B11	SP	1.0	
RP	102428-1	BEARING RETAINER	9.00	SH-C4	SP	10.0	
RP	1182-B3	SPARK PLUG	0.65	SH-C18	SP	10.0	
RP	1182J3	SPARK PLUG	0.72	SH-C18	SP	24.0	
RP	123456789ABC	FILTER, ELECTRONIC	37.50	SH-F11	SP	3.0	
RP	152 SPINNER	SENSENICH PROP MOD	185.00	SH-D19	SP	2.0	
RP	1C172BTM7359	PROPELLER ALUM SPACER	1,477.00	SH-D19	SP	0.0	
RP	202033-1	WEIGHT	36.00	SH-D18	SP	-4.0	
RP	2231	Carb Spring	0.34	SH-C22	SP	49.0	
RP	23450	Gumout Carb Cleaner	0.15	PL2	SP	144.0	
RP	301025-501	STABILIZER R/H	649.00	SH-B4	SP	2.0	
RP	401121-706	LINE	37.49	SH-Z11	SP	1.0	
RP	406006-501	AIRBOX	22.00	SH-B1	SP	19.0	
RP	453331	Rotor	0.31	SH-B15	SP	144.0	
RP	456678	Cessna Door Lock	6.00	SH-D19	SP	9.0	
RP	479729	FILTER NYLON	30.00	SH-Z11	SP	9.0	
RP	503006-2	THERMOID TUBE	9.95	AV4	SP	4.0	
RP	5102270-3	SPACER	0.25	SH-C12	SP	14.0	
RP	5102329-3	SEAT ATTACH BRKT LH OUTBO	37.00	SH-B2	SP	1.0	
RP	5102361-501	RETAINER	37.00	SH-C11	SP	2.0	
RP	5201174-5	FAIRING	864.00	SH-B5	SP	1.0	
RP	5302039-1	BRACKET	17.00	SH-B3	SP	10.0	
RP	5406001-1	DUCT	15.00	SH-B1	SP	1.0	
RP	5503006-502	BUTTERFLY	19.00	SH-C16	SP	10.0	
RP	5607003-502	ACTUATOR ASSY	14.00	SH-C16	SP	2.0	
RP	5803007-108	PLACARD FLASHING BEACON W	0.99	SH-C19	SP	10.0	
RP	5804018-2	SPACER	13.75	SH-C22	SP	10.0	
RP	600X6 TUBE G/Y	TUBE, GOODYEAR	17.95	SH-D2	SP	2.0	
RP	607004-4	COLLAR OUTBOARD	8.65	SH-A3	SP	3.0	
RP	641431	RING, FOURTH GROOVE	3.33	SH-A3	SP	3.0	
RP	701063-2	BRACKET	277.00	SH-A4	SP	1.0	
RP	7ZCKS6	Sensensich Prop Bolt Kit	14.79	SH-A5	SP	1.0	
RP	78-23444	Sandpaper, Wet/dry	0.08	SH-Z9	SP	50.0	
RP	78A110	LOCK CLIP	0.25	SH-A5	SP	10.0	
RP	7CS10425-5	BRACKET ASSY	79.00	SH-A3	SP	2.0	
RP	7F10401-13	COVER	169.00	SH-B9	SP	1.0	
RP	7L10238-13	NUT	3.35	SH-Z19	SP	1.0	
RP	7P10616-17	BRACKET	0.00	SH-Z11	SP	0.0	
RP	803007-13	PLACARD,CANOPY LOCK	0.00	SH-C19	SP	0.0	
RP	804019-509	FAIRING	0.00	SH-B3	SP	0.0	
RP	8357672	Hubbell Surge Fitting	0.16	SH-A11	SP	52.0	
RP	85-43-201-38	SEAL- WASHER	0.00	SH-D22	SP	-1.0	
RP	A3236SS012-27	DIMPLE WASHER	0.00	SH-Z14	SP	0.0	
RP	AES1300-1	BOLT	0.00	SH-Z14	SP	-4.0	
RP	AN175-14	BOLT	0.00	SH-Z15	SP	0.0	
RP	AN500-416-6	SCREW	0.00	SH-Z15	SP	0.0	
RP	ANT-34A	BOLT	0.00	SH-Z16	SP	0.0	

Counted By: _____   Date: _____

174 Essentials of Aviation Management

Type	Part Number	Description	Unit Cost	Bin/Location	Dept	Current Stock	Actual Count
RP	AN960-516L	WASHER	0.00	SH-Z16	SP	0.0	
RP	C21-9159	COMPASS FWD GASKET	0.00	SH-D23	SP	0.0	
RP	CH27-1477	Fuel Filter	17.00	SH-F11	SP	16.0	
RP	CH48109	Oil Filter, Spin On	8.78	SH-F11	SP	11.0	
RP	CH48110	Oil Filter, Spin On	9.11	SH-F11	SP	12.0	
RP	DC12053OZ	GLASS PRIMER,RTV732	0.00	PL2	SP	0.0	
RP	GAES102228-10	SCREW	0.00	SH-Z17	SP	0.0	
RP	GX147-2153	Ground Cable	0.00	SH-C22	SP	0.0	
RP	KCP2 1/2	FIRE EXT	41.00	SH-E10	SP	1.0	
RP	LW16511	CAMSHAFT 0320-H2AD	0.00	SH-R15	SP	0.0	
RP	M647	SPARK PLUG GASKET	0.00	SH-C17	SP	0.0	
RP	MS20364-832C	NUT	0.00	SH-Z18	SP	0.0	
RP	MS24665-360	COTTER PIN	0.19	SH-Z20	SP	-4.0	
RP	MS35333-38	WASHER LOCK	0.00	SH-Z21	SP	0.0	
RP	NAS451-56	7/8" STEEL HOLE PLUG	0.00	SH-Z22	SP	0.0	
RP	PA1912	Brake Line - 19x1/2"	7.00	SH-C19	SP	12.0	
RP	PX145-19	Tire	17.00	SH-R5	SP	-1.0	
RP	REM-40	SPARK PLUG	0.00	SH-C17	SP	0.0	
RP	REM38	SPARK PLUG	0.00	SH-C17	SP	0.0	
RP	REM40E	SPARK PLUG	0.00	PL3	SP	0.0	
RP	S-220	PLASTIC REPAIR KIT	0.00	SH-Z22	SP	0.0	
RP	SK130-1	THROTTLE CABLE CLAMP MOD	0.00	SH-B1	SP	0.0	
RP	ST-43189	C-172 Strut	0.00	SH-R5	SP	0.0	
RP	T17-50	Cessna Tire-10"	59.00	SH-R5	SP	29.0	
RP	Z1974126	Battery, 12 Volt	15.00	SH-A13	SP	8.0	

Counted By: _____ Date: _____

Courtesy of TotalFBO Accounting and Business Management Software by Horizon Business Concepts, Inc.

Aviation Operations

This topic is crucial because it represents the areas where the business makes its money. Some of the information needed in these areas originates elsewhere in the system, and must be converted or transferred for managerial decisions about operations. Figure 7.6 summarizes these aviation operations data areas.

Legal and Tax Information

The accounting system yields data on the various taxes due and the schedules for payment. Legal information may be more sporadically generated but is vitally important and must be well organized and addressed.

Market Information

Information regarding industry trends and competing firms' performance comes from many sources, as detailed in Chapter 3, Marketing. The manager must decide how to monitor and assemble this information so that it is readily accessible.

Technological Information

Aviation technology is constantly changing, and decisions must be made about whether to use new manufacturers, open new product lines, or adapt new methods. The success or failure of each test must be documented, as should requests from customers for new goods and services.

Figure 7.6 → Aviation Operations Data

Activity	Source of Information	Control Measure
1. Line operations	Gas/oil sales records Related sales Customer indications Personnel reports Profit objectives	Volume goals Profit records Historical data
2. Maintenance	Volume of work Customer indications Work redone Labor allocation FAA inspection	Volume goals Project objectives Historical data Industry data
3. Instruction	Volume of work Customer indications Profitability Instructor and aircraft utilization records FAA inspection	Volume goals Project objectives Historical data Industry data
4. Flight services	Work volume records Customer reports FAA inspections Profit records Utilization	Volume goals Profit objectives Historical data Industry data
5. Aircraft parts sales	Sales volume records Profitability Customer indication Sales reports	Unit goals Historical data Industry data
6. Miscellaneous (Rental, parking, and so on)	Specific activity reports Profit analysis	Predetermined goals Historical data

Analyzing Business Activity

Sometimes described as financial analysis, this task is a combination of checking over the accounting data and critically reviewing the operating procedures and accomplishments. Accounting data provide the measures of actual activity and management expertise, as expressed through the business plan. These must be used to set up performance standards to measure the actual results.

General Procedure

Sources of performance standards to apply to the business can come from:

> Results from a previous operating period;
> Contrasting two previous operating periods;
> Industry averages, from the National Air Transportation Association (NATA) or other organizations;
> Competing businesses, as obtainable; and
> Similar businesses, as reported in trade press articles or in advertising.

Analyzing the Business as a Whole

The purpose of this activity is threefold—to determine whether goals and objectives are being met, to identify areas with potential for improvement, and to suggest possible corrective actions.

Specific tasks will include:

> Comparison of financial statements with budget and financial projections,
> Comparison of current financial statement with previous years' statements, and
> Comparison of financial statement with those of similar businesses, if possible.

Specific measures should include:

> Net income,
> Capital structure,
> Profit margins,
> Return on investment,
> Sales by each department, and
> Community considerations.

After an examination of the income statement or balance sheet, it is necessary to inspect the individual components. Data should also be compared to expectations, historical results, industry standards, and business norms. The help of an accountant or banker may be appropriate.

Secondly, the income statement should be examined using a replicable and relative method, such as expressing income in each department as a percentage of the total. Gross and net profit by department should also be calculated, and profit centers ranked by return.

The final step in the analysis of the overall business through financial statements involves the use of ratios. Ratios are developed to show the relationship of key values given on the statements. They may be classified according to their source data as balance sheet, income, and mixed ratios, or they may be classified according to the purpose of the ratio comparison:

> Liquidity ratios,
> Leverage ratios,
> Profitability measures, and
> Managerial efficiency measures.

Prior to examining key ratios in each of these areas, certain basics regarding the use of this technique should be considered. A ratio is most useful when compared to previous values, to industry averages, and to generally-accepted standards. Ratios based upon poorly-prepared statements are practically worthless and may be misleading to the unsuspecting analyst. Finally, ratios indicate business situations in a general sense and identified trends require further investigation in order to fully understand the situation and its underlying financial drivers.

Liquidity Ratios. These ratios describing liquidity of assets are helpful in analyzing the ability of a business to meet its obligations over the short-term and to move rapidly in seizing unexpected business opportunities.

Current Ratio. The current ratio is defined as current assets divided by current liabilities. A ratio over unity is desirable, indicating that the company can pay its obligations in full over the short-term. The value of "for-sale aircraft in inventory" will greatly influence this ratio. A more stringent version of the current ratio, termed the quick ratio, excludes inventory, prepaid expenses, and other illiquid accounts from current assets.

Tangible Net Worth. Computed as total assets less total liabilities and intangible assets, tangible net worth estimates a company's liquidation value. Tangible net worth is the value of a company excluding "goodwill", patents, copyrights, or similar intangible property.

Net Working Capital. Computed as current assets less current liabilities, net working capital is a measure of short-term liquidity and operational efficiency. A substantial positive value indicates greater potential to invest and grow; a negative value may indicate trouble paying creditors and funding current operations.

Ratios of managerial efficiency. Ratios in this category are designed to analyze the efficiency with which managers handle the assets of the organization. Specific ratios cover the various assets of a business, namely accounts receivable and inventory.

Receivables. The average collection period for accounts receivables measures management's efficiency in handling the assets that are tied up in outstanding accounts. Minimizing the time between sales and reception of payment is desired as the soundness of a company's credit policies are indicated by this relationship. A collection period of between 30 and 45 days is tolerable; if collections exceed 45 days, customers may be taking advantage of overly lenient credit policies.

Inventories. Inventory turnover, which is the rate at which a company replaces inventory, reflects the efficiency of management in stocking products for sales. The inventory turnover ratio is computed by dividing costs of goods sold by average inventory for a period. The desirable value must be identified by experience, and is specific to the nature of the product. A high ratio indicates inability to fulfill sales while a low ratio indicates inefficiency in generating sales.

Leverage Ratios. Ratios in this area reflect the balance of borrowed funds and ownership funds. They provide an indication of management's ability to "lever" funds into additional assets by way of more borrowing or investment. The existing leverage provides a measure of the support and faith lenders have in the organization and the existing management.

Debt to Earnings Ratio. Debt may be divided by earnings before interest, tax, depreciation, and amortization (EBITDA). This ratio measures the amount of income generated and available to alleviate incurred debt. Higher leverage, though riskier, allows a greater return on investment. It reflects the trust that creditors place in the management of the organization as well as the management philosophy.

Debt to Equity Ratio. This ratio reflects the mix of loans and ownership used to finance a company. A value greater than unity indicates that the equity of the creditor exceeds that of the owner. High values are generally undesirable as they indicate that the business has aggressively financed its growth with debt, leading to significant interest expense.

Assets to Equity Ratio. Also called the equity multiplier, this ratio is a risk indicator that reflects again the leverage in the amount of assets financed by stockholders' equity.

Profitability Ratios. Ratios in this area provide a measure of the profitability of operations. Earnings or profits are compared with key indicators of investment and ownership to reflect the profitability of such activity.

EBITDA. Computed by adding interest, taxes, depreciation, and amortization to net income, EBITDA is an overall assessment of profitability.

Net Profits to Net Worth Ratio. With this ratio, the owners are provided with a measure of the earning power of their investment. The desirable return will depend upon the alternatives available, the risk tolerance of the investor, and industry averages. Considering the risk element in the aviation business and relatively secure alternatives, the owner's return should be approximately 10–20 percent.

Net Profit to Sales Ratio. The profit for a period compared to the sales for the same period offers an indication of cost efficiency in producing the goods and services sold.

Special-purpose ratios. A large number of items in the financial statements can be evaluated through ratio analysis. Individual expense items can be related to sales. Income categories can be compared with total sales. Ratios of this type are useful in period-to-period comparative analysis, peer-to-peer industry analysis, and for budgetary control work.

Management Audit

The management audit is the second procedure, in addition to financial ratio analysis, that may be used by the manager in analyzing the overall business. This process uses a number of the same tools and techniques employed in financial analysis, but the general concept is different. What is a management audit? By definition, it is a systematic checklist approach to the analysis of a business, its functions, operations and decisions. By carefully reviewing each of the functional and operational areas of a business, and comparing existing conditions with previously identified goals and required or desired standards, some measure of strengths and weaknesses can be obtained and corrective action suggested. By examining the entire business in this manner, an auditor is really examining management, hence the name "management audit."

Appendix V (see the book's website) contains an Aviation Management Audit. It is designed as a self-audit, one that can be conducted by the management team. To do this as it should be done requires a thoughtful, objective point of view. Of course, the audit can be expanded and made much more beneficial when administered by an impartial third party such as an aviation management consultant. By having such an outsider conduct the audit, a higher degree of objectivity can be achieved and more penetrating questions can be developed. The participants can also use the audit in a seminar setting to achieve a higher level of management awareness and an exchange of information. A typical seminar has a small group of 18 to 25 aviation managers in attendance. Each has received an audit manual and has been requested to complete the self-audit prior to arrival. The seminar reviews the audit manual with individual replies, group comparison, and leader comments and analysis of each of the items. Further benefit could be received by the tabulation of accumulated audit responses and perhaps even the preparation of nationwide norms and standards.

The management audit is not an accounting or financial audit, but a review of the overall business as well as of the individual departments. A thorough audit provides management with invaluable information and advice regarding the organization; consequently, it is considered a valuable supplement to the regular information system portrayed throughout this chapter. Conducted at regular intervals, an audit promotes change to mitigate the effects of risks that impede the business from achieving financial and operational objectives.

Analyzing Departmental Activity

Introduction

The normal tools used in analyzing departmental activity are:

> Plans and objectives,
> Budgets and schedules,
> Operating controls, and
> Financial records.

Although each is a separate technique, they combine as a sequence that forms a method of analyzing department activity. The process is described so that the manager may consider the major elements of each step.

Plans and objectives. Identifying plans and objectives represents the first step in conducting a thorough review of departments or profit centers. Without a definite business objective, it becomes difficult to determine whether the activity of that portion of the business is successful. To paraphrase the Cheshire cat from *Alice in Wonderland,* "If you don't know where you're going it doesn't matter which way you go, and you won't know if you get there." The following are questions that should be asked in this first step:

1. What are the objectives of the department?
2. Are the objectives definite, clear-cut, and, most importantly, measurable?
3. Are they attainable?

4. Are the objectives understood and accepted by those involved in their attainment?
5. Are the department objectives compatible with and supportive of overall company objectives?

By using these questions, a manager should be able to develop an initial set of standards to measure business success or failure.

Budget. Practically speaking, a budget is a quantitative expression of the plan or objectives of the department. As such, it becomes an aid in coordination and implementation. It is indispensable in grappling with uncertainties, and its benefits exceed its costs. Budgets are useful vehicles for progress and improvement.

Operating controls. These controls are used as guidelines to prevent excessive spending, to provide ample supplies, to monitor personnel and aircraft utilization, etc. For the parts department, the control may be inventory limits; for aircraft sales, there may be a unit dollar limit; for aircraft service, there may be a mechanic time limit; for credit, there may be a series of collection period limits; for flight instruction, there may be a set number of flight hours per instructor; and for passenger activity, there may be a mandatory procedure for briefing and debriefing with each flight.

A series of general questions should help in the examination of a department's operations:

1. What is the true worth of departmental assets? List them.
2. What is the extent and character of departmental liabilities? List them.
3. Does the ability exist to earn a fair return?
4. What is the capacity to withstand setbacks?
5. What is departmental labor efficiency?
6. How efficient is the utilization of equipment and allocation of labor?
7. Is departmental and company-wide space utilized properly?
8. Are the departmental records maintained accurately?
9. Do the activity indicators reflect constant action (e.g., sales calls; inactive student follow-up)?
10. Are salaries adequate? Too high? Too low?
11. Do expenses appear in line? Do they match budgeted figures?
12. Is the department inventory maintained accurately? Is it adequate? Excessive?
13. Are the department-generated receivables too high? Too low? What is the bad debt expense?
14. What is an analysis of departmental income by key categories?
15. What is break-even for the department? Determine key indicators and drivers.

Financial records. The financial records for this purpose are the income statement and various supplements used to amplify or explain indications of the statement. Figure 7.7 illustrates the general ledger departmental income and expense report provided by TotalFBO. Using this report as a guide, a manager may conduct a review of a department's activity. The manager should examine each subject category on the statement line by line to determine whether departmental objectives have been achieved, operations are efficient, and problems exist.

As an illustration, suppose a review of several lines on the income statement is performed. The following analysis shows possible and practical comparisons to make:

Total Sales. Compare sales volumes with budgets on different time basis. Analyze the various components of the total sales figure (new, used, geographic, demographic) to determine relationships between sales and marketing actions, suggesting future advertisement activities.

Cost of Sales. Consider the efficiency of buying practices. Identify possible weaknesses and shortages to take preventative actions.

Consider expenses associated with creating revenue streams. Compare wage expense with budget projections, historical data, and other businesses. Examine job requirements and consider impact on incumbents in the job. Ensure compliance with federal and state regulations.

Gross profit. Compare with budgeted figure and with data for similar periods of operation. A large deviance between observation and expectation warrants further examination and analysis.

Net Income for the Period—Profit or Loss

In like manner, the manager should first determine the general situation and then identify problems that may exist within each line. In conducting this type of

Chapter 7 Management Information Systems 181

Figure 7.7 ✈ General Ledger Standard Income Report

Run: 10/27/2009
5:09PM

Horizon Business Concepts, Inc.
General Ledger Standard Income Report
For The Period: 1/01/2009 Through: 5/31/2009
All Departments Consolidated

Page: 1 of 2

Income

		5/31/2009 MTD		5/31/2009 QTD		Year-To-Date	
4000	Parts Sales	1,578.26	6.6 %	3,651.25	9.4 %	15,059.44	8.0 %
4010	Fuel Sales	2,982.75	12.5 %	5,670.57	14.6 %	26,995.81	14.3 %
4011	Minimum Fueling Surcharge	17.22	0.1 %	47.15	0.1 %	72.92	0.0 %
4015	Into-Plane Fees	816.66	3.4 %	1,615.48	4.2 %	2,571.17	1.4 %
4020	Oil Sales	6.63	0.0 %	6.63	0.0 %	26.43	0.0 %
4050	Miscellaneous Shop Supplies Sales	126.98	0.5 %	322.21	0.8 %	355.43	0.2 %
4060	Solo Aircraft Rental Revenue	260.64	1.1 %	378.72	1.0 %	26,036.80	13.8 %
4070	Dual Aircraft Rental Revenue	658.60	2.8 %	1,214.80	3.1 %	24,225.50	12.8 %
4080	Charter Revenue	12,085.80	50.7 %	14,683.93	37.8 %	21,128.01	11.2 %
4081	Charter Income - Aircraft Standby	1,190.00	5.0 %	1,830.00	4.7 %	2,545.00	1.4 %
4082	Charter Revenue - Pilot Standby	150.00	0.6 %	500.00	1.3 %	740.00	0.4 %
4083	Charter Revenue - Extra Charges	0.00	0.0 %	0.00	0.0 %	35.00	0.0 %
4085	Charter Revenue - Catering	260.00	1.1 %	475.00	1.2 %	665.00	0.4 %
4086	Charter Revenue - Other Expenses Charged	300.00	1.3 %	475.00	1.2 %	620.00	0.3 %
4090	Headset Rental Revenue	39.50	0.2 %	68.50	0.2 %	130.00	0.1 %
4100	Flight Store Revenue	7.50	0.0 %	31.50	0.1 %	48.35	0.0 %
4130	Hangar And/or Tie Down Revenue	385.00	1.6 %	770.00	2.0 %	1,155.00	0.6 %
4140	Instruction - Primary Revenue	226.10	1.0 %	492.10	1.3 %	16,107.80	8.5 %
4150	Instruction - Advanced Revenue	78.40	0.3 %	78.40	0.2 %	4,123.60	2.2 %
4160	Ground School Revenue	111.15	0.5 %	214.65	0.6 %	288.75	0.2 %
4170	Sublet Work Revenue	2,000.00	8.4 %	3,250.00	8.4 %	3,250.00	1.7 %
4175	Outside Repairs Revenue	(2,000.00)	-8.4 %	(1,925.00)	-5.0 %	(1,925.00)	-1.0 %
4180	Shop Labor Revenue	2,232.50	9.4 %	3,740.00	9.6 %	27,628.50	14.6 %
4200	Off-Site Fuel Credit Revenue	(138.35)	-0.6 %	(138.35)	-0.4 %	(173.44)	-0.1 %
4230	Flight Package Sales	0.00	0.0 %	0.00	0.0 %	15,825.00	8.4 %
4900	Finance Charge Revenue	177.44	0.7 %	365.93	0.9 %	384.11	0.2 %
4930	Miscellaneous Revenue	50.00	0.2 %	50.00	0.1 %	100.00	0.1 %
4950	Freight Revenue	232.75	1.0 %	1,031.75	2.7 %	1,061.75	0.6 %
	Total Income:	**23,835.53**	**100.0 %**	**38,900.22**	**100.0 %**	**189,080.93**	**100.0 %**

Cost Of Sales

		5/31/2009 MTD		5/31/2009 QTD		Year-To-Date	
5000	Parts Cost Of Goods Sold	1,006.73	4.2 %	2,772.46	7.1 %	11,178.47	5.9 %
5010	Fuel Cost Of Goods Sold	1,939.17	8.1 %	3,692.61	9.5 %	18,777.42	9.9 %
5020	Oil Cost Of Goods Sold	1.98	0.0 %	1.98	0.0 %	3.11	0.0 %
5040	Purchases Discounts	(9.14)	-0.0 %	(9.14)	-0.0 %	(9.14)	0.0 %
5050	Charter Costs	2,057.93	8.6 %	4,011.10	10.3 %	5,352.06	2.8 %
5070	Fuel & Oil Expense - Flight School Acft	1,005.45	4.2 %	1,800.50	4.6 %	2,630.99	1.4 %
5080	Ramp Vehicle Fuel/Oil Cost	0.00	0.0 %	28.89	0.1 %	28.89	0.0 %
	Total Cost Of Sales:	**6,002.12**	**25.2 %**	**12,298.40**	**31.6 %**	**37,961.80**	**20.1 %**
	Gross Profit:	**17,833.41**	**74.8 %**	**26,601.82**	**68.4 %**	**151,119.13**	**79.9 %**

Expense

		5/31/2009 MTD		5/31/2009 QTD		Year-To-Date	
6160	Advertising Expense	0.00	0.0 %	87.50	0.2 %	87.50	0.1 %
6170	Freight Expense	460.90	1.9 %	1,578.65	4.1 %	2,110.55	1.1 %
7100	Administrative Salaries Expense	7,025.24	29.5 %	14,050.48	36.1 %	44,478.72	23.5 %
7110	Salaries/Wages Expense	17,737.02	74.4 %	34,280.91	88.1 %	91,329.53	48.3 %

Continued

Figure 7.7 → General Ledger Standard Income Report —Cont'd

Run: 10/27/2009
5:09PM

Horizon Business Concepts, Inc.
General Ledger Standard Income Report
For The Period: 1/01/2009 Through: 5/31/2009
All Departments Consolidated

Page: 2 of 2

		5/31/2009 MTD		5/31/2009 QTD		Year-To-Date	
7130	Payroll/Employee Benefit Expenses	1,998.28	8.4%	3,995.10	10.3%	5,992.91	3.2%
7140	Taxes-FICA	1,400.52	5.9%	2,772.64	7.1%	3,982.65	2.1%
7150	Taxes-Medicare	327.56	1.4%	648.45	1.7%	931.44	0.5%
7160	Taxes-FUTA	144.84	0.6%	321.36	0.8%	482.46	0.3%
7170	Taxes-SUTA	523.19	2.2%	1,215.73	3.1%	1,859.70	1.0%
7200	Building/Office Rent Expense	1,785.00	7.5%	3,570.00	9.2%	3,570.00	1.9%
7240	Office Supplies Expense	0.00	0.0%	0.00	0.0%	1,284.00	0.7%
7250	Taxes-Fuel	0.00	0.0%	1.75	0.0%	1.75	0.0%
7260	Taxes-Sales	0.00	0.0%	35.27	0.1%	35.27	0.0%
7360	Flight School Supplies	0.00	0.0%	6.25	0.0%	6.25	0.0%
7370	Insurance-General	0.00	0.0%	0.00	0.0%	2,743.00	1.5%
7380	Insurance-Auto	0.00	0.0%	0.00	0.0%	500.00	0.3%
7390	Insurance-Aircraft	2,175.00	9.1%	4,350.00	11.2%	10,025.00	5.3%
7420	Maint-Aircraft	178.12	0.8%	817.71	2.1%	957.37	0.5%
8520	Interest Expense	445.85	1.9%	556.68	1.4%	556.68	0.3%
8530	Credit Card Fees Expense	144.63	0.6%	200.37	0.5%	265.93	0.1%
	Total Expense:	34,346.15	144.1%	68,488.85	176.1%	171,200.71	90.5%
	Profit Before Other:	(16,512.74)	-69.3%	(41,887.03)	-107.7%	(20,081.58)	-10.6%

Other Income

	5/31/2009 MTD		5/31/2009 QTD		Year-To-Date	
Total Other Income:	0.00	0.0%	0.00	0.0%	0.00	0.0%

Other Expense

	5/31/2009 MTD		5/31/2009 QTD		Year-To-Date	
Total Other Expense:	0.00	0.0%	0.00	0.0%	0.00	0.0%
Net Income:	(16,512.74)	-69.3%	(41,887.03)	-107.7%	(20,081.58)	-10.6%

Notes: Report Period: 1/01/2009 - 5/31/2009
All Departments Consolidated.

Courtesy of TotalFBO Accounting and Business Management Software by Horizon Business Concepts, Inc.

---

analysis, he or she must remember that the income statement is:

> The result of facts,
> The result of managerial guidance,
> The result of personal judgments and evaluations.
> The result of multiple evaluations

Analysis supplements. Special supplements to the financial statement are tremendously valuable in examining the various operational activities and developing insight into possible cause-and-effect relationships.

Tables, charts, graphs, and worksheets can be prepared for a variety of departmental activities and will present specific, detailed data for the desired business activity. The supplements may be practically applied to the activities of service, sales, parts, flight, passengers, line, or other departmental activities. Typical supplements of interest to managers include:

> Accumulated aircraft sales per salesperson;
> Monthly parts sales compared to projections;
> Aircraft utilization by aircraft and by type or purpose of flight;
> Charter sales by month;

- Student completion rates;
- Instruction hours per month;
- Daily, weekly, and monthly fuel sales;
- Fuel inventory records;
- Transient aircraft count;
- Passenger analysis; and
- Employee training records.

Post-analysis action. Such action may be corrective, reinforcing, or sometimes even innovative. The general objective is to improve the efficiency and profitability of departmental operations and ensure that they contribute to the achievement of overall organization objectives.

Taking Action

The analysis of information system data should point toward some selected action. Normally there are two alternatives: (1) to revise or adjust the basic plan of the organization for the specific area under consideration, or (2) to adjust the business activity that is taking place. Of course, there is a third alternative available—do nothing and hope things will improve. Chances are that they will worsen if ignored, so this alternative is not realistic. Typically, the action taken is to make some adjustment in the current business activity.

Selling might be accelerated, hours changed, layout altered, personnel added, or a similar activity undertaken to help meet an objective of the organization. In some situations, the objective or plan of business must be revised to meet new circumstances; however, the manager normally makes some change in the business activity to achieve the selected goals. Figure 7.8 lists typical actions taken by an aviation manager after completing an analysis and finding that:

1. Aircraft revenue passenger seats are found to have 50 percent utilization.
2. Gross profits relative to total sales is shown to be 30 percent.
3. The return on investment (before taxes) is reported as 6 percent.

In each of the areas, a careful analysis must be performed, and subsequent decisions must be made on appropriate action. A later section of this chapter is devoted to techniques used in analyzing both the overall business and individual departments. Special attention should be given to this as it is an important

Figure 7.8 → Management Analysis and Action

Analysis	Action Taken
Instruction hours low	Promote additional students by telephone
Fuel sales low	Revise target volume of sales
Maintenance profits low	Reduce overhead
Parts sales low	Direct mail or e-mail advertising campaign
Accounts receivable high	Special collection letter(s)
Weak liquidity position	Obtain additional credit reserves

expansion of the information system utilization discussion.

The entire cycle of events involved in the use of the information system by a manager is depicted in Figure 7.9.

Controlling the information flow in an aviation activity is no easy task. It implies much more than just having information. The manager controls the availability of information, its analysis and the action taken as a result.

Records and Record-Keeping

Although records and record-keeping are obviously important, many aviation firms fail to perform these tasks adequately. Record-keeping is almost a nuisance, and hiring personnel to do it generates expensive overhead. For the manager who yearns to experience the thrill of flying, record-keeping becomes an unpleasant task. Especially for the individual who lacks appreciation of its benefits or an understanding of its legal and tax underpinnings. Keeping immaculate records is paramount to efficient and effective busines operations at all levels of management.

Good records are essential. Bankrupt aviation businesses typically have failed to record adequate information. Conversely, successful businesses have effective record-keeping systems. Good records, as discussed in the earlier section on the components of an overall management information system, are necessary to achieve better control over operating results. They are also desirable in supporting credit applications to bankers, filing tax returns with the

Figure 7.9 → Information Cycle

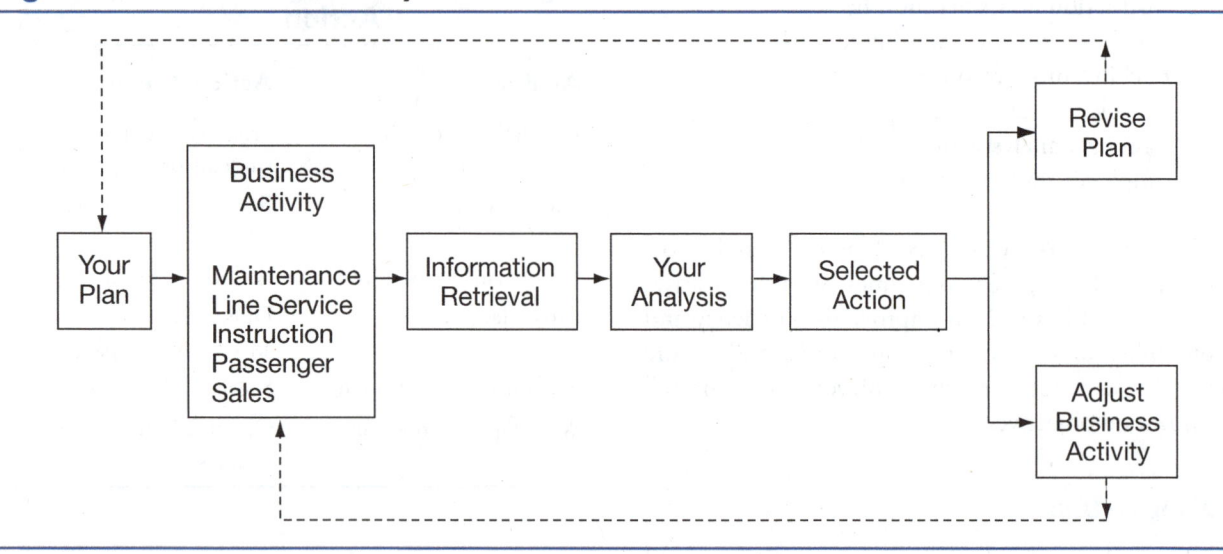

Your Plan → Business Activity (Maintenance, Line Service, Instruction, Passenger Sales) → Information Retrieval → Your Analysis → Selected Action → Revise Plan / Adjust Business Activity

---

government, and compiling reports for regulatory agencies. Internally, a good record system allows for the early detection of employee fraud, material waste errors, spoilage, and other losses requiring prompt correction. It may also pinpoint employee skill deficiencies as well as internal organizational problems.

Records Design

The amateur seldom designs information systems effectively. In setting up a system, careful attention must be given to the achievement of the objectives mentioned earlier in the chapter. System requirements associated with the three areas of managerial concern—people, money, and material—should be identified and included in the design of the system. The design of efficient, economical forms is very important. They should be prepared so that completion time is minimized while necessary information is provided. Each form should clearly state its purpose in its heading. The form size should be suitable for both filing, posting, and sharing systems. Forms should be kept simple yet still capable of collecting comprehensive information. Recurring infomation should be already entered on the form. All prospective users should check any proposed forms to ensure that they contain the desired data types and fields.

Most, if not all, forms used today by aviation businesses are computerized—either generated on the screen or preprinted, completed, and scanned on the computer. In addition to material, software vendors can frequently provide valuable assistance in analyzing a system and its needs. They can adapt and work up special forms to fit individual company needs. Many of these forms are practical source documents that may guide personnel in performing operational procedures as well as provide financial data.

Correspondence

Correspondence or letter writing used to be the most prevalent form of office paperwork. It has been replaced for most purposes by e-mail. E-mail has the advantage of speed, range, and being able to connect the correspondent with several people at one time. However, e-mail can be inappropriately forwarded to other parties, subpoenaed even when the computers involved have deleted the message, and is in general less professional and more casual than a letter. Regardless, postal delivery of bills and statements from businesses to consumers is being eroded by electronic alternatives, with such mail volumes decreasing steadily since the start of th 21st century. Given the average cost of a letter in time and money—reading, drafting, typing, editing, and mailing—it is apparent that correspondence should be an economic concern to the aviation manager.

Two basic approaches may be taken to improve the correspondence handled by a front office: (1) training the letter writers and (2) lowering correspondence costs. Letter writing, like other skills, can be improved by training and practice. Training can increase the readability of correspondence, reduce the length of letters, reduce the number of rewrites, and increase the number of letters per writer. There are many sources for this kind of training.

Reducing the costs of correspondence primarily involves improving the rate of production. These improvements can be achieved by the following methods:

Reducing planning time. Planning the correspondence frequently takes more time than writing the letter itself. Writing letters that are short and easy to read using templates, guides, and other prepared kinds of correspondence can substantially reduce the planning time.

Reduce reading time. Correspondence that is poorly conceived and written is frequently difficult to understand and takes longer to read. By improving the writing and content of letters, the time commitment required of recipients in both reading the letter and clarifying its message via more correspondence is reduced.

Reducing writing time. The time taken to prepare a letter of a fixed length will vary by its preparation and purpose. The average Composition time is 59 minutes for a handwritten letter, 30 minutes for a dictated letter, and 3.5 minutes for a form letter. To reduce writing time, the following guidelines are useful:

> Greater use of templates;
> Greater use of voice recognition software or computer-aided design diagrams;
> Fewer rewrites and fewer required approvals;
> More judicious use of telephones, including effective use of voicemail to impart information;
> Greater use of the informal reply;
> Greater use of routing slips; and
> Complete elimination of hand-drafted letters; rather, all employees should write correspondence on a computer even if the final deliverable is a physical letter.

Reducing reviewing time. The average letter receives three reviews prior to being mailed. More care in preparation or more adherence to guide letters may reduce these iterations or even eliminate reviews completely. Reviews for inspection by the author or approval by top officials can be reduced in this manner. Delegating signature authority, routing courtesy copies, and eliminating perfunctory approvals will all reduce reviewing time.

Reducing word processing time. By writing shorter letters, using form and guide letters, following a simplified format, and using window envelopes, word-processing time can be noticeably reduced.

Reducing delivery time. The time from preparation to delivery can be reduced by cutting the number of reviews, lowering the signature level, improving accuracy in addressing, and utilizing priority and express mail services as appropriate and cost-effective.

Reducing filing time. Filing time can be cut substantially by using good judgment as to which information is most pertinent. For those items that require filing, only the necessary copies should be prepared and filed. The use of a subject line or filing key by the author will greatly reduce the filing time of the clerk or office manager. Another quick way to file is to keep one set of correspondence in a chronological file and another set in a subject file. Files should be purged at least annually and only key legal and financial records saved. Electronic document archiving is also an effective alternative to manual filing systems.

Using supplies and equipment effectively. Office productivity can be improved by the efficient use of computers, scanners, printers, copiers, and mailing supplies. Information available by computer can be stored on the computer, for example through website bookmarking, and rarely needs to be printed and filed.

It should be recognized that most business activity is accomplished through words, and many of them are written or typed. The better the manager writes, the better the job is done and the more successful the organization's administration. The goal of correspondence management is the achievement of these objectives.

Records Management

Records have been described as the working tools of management, the memory of the organization, and the source of many kinds of valuable information vital in decision-making processes. The tremendous increase in the volume of business records in the past few years has created a challenge to most organizations—a challenge that grows with each passing year and with organizational growth. The reason for this growth is clear. Rapidly developing technology, a changing economy, the use of photocopying and scanning, and the increasingly complex aviation business all contribute to the growth of records; such problems in records management are commonly referred to as "red tape," "information explosion," "flood of forms," and the "paper-words jungle." The challenge to management of aviation businesses is to bring the tidal wave of records under control and to create a system that serves the needs of the company as efficiently and economically as possible.

Record identification. For most aviation businesses, the following records are normally critical to company operations:

> Correspondence, including letters, e-mails, and memos;
> Business forms;
> Reports and summaries;
> Maintenance records for company and customer aircraft—perhaps the most crucial records required of an FBO;
> Standard organizational instructions and procedures; and
> Handbooks and manuals.

One can also divide records into administrative and operational records. The administrative records, which make up 10 to 15 percent of the total, include rules and regulations, policy and procedures manuals, financial and planning records, articles and by-laws, and agendas and minutes. The operational records, which make up the remaining 85 to 90 percent, are made up of purchase orders and requisitions, claims, bills of lading, personnel records, construction records, invoices, and maps and blueprints.

All records follow a similar path or cycle in their existence. They are (1) created, (2) classified, (3) stored, (4) retrieved when needed, and (5) returned to storage or (6) destroyed. Each of these steps must be understood and controlled as part of an active records management programs.

Records management program. There are many approaches that could be taken in the identification of the various components of records management. On a practical basis, records management encompasses all those activities dealing with the creation, maintenance, and disposition of records. The following are parts of a well-rounded records-management program:

> Control of reports,
> Control of administrative directives,
> Control of paper forms,
> Control of paperwork procedures,
> Records protection,
> Control of computer and network access,
> Mail service controls,
> Files control,
> Records retirement or disposition by shredding or recycling,
> Storage and archiving of records, and
> Reviewing requests for space and equipment for records.

Report controls becomes a major concern to the manager when he or she views the number of internal and external reports required in the operation of an aviation business. The necessity for a reports-control system to monitor important areas and to ensure that all external report commitments are met on schedule can become critical to the success of a business. Internal reports should also be monitored and controlled with the same concern, since the organization originates many of these reports. It is necessary to avoid duplication and overlapping of data and to cull unnecessary reports.

The length that records must be kept varies with the content. Land acquisition records should be kept permanently as should legal documents establishing the company. Tax records should be disposed of after the statute of limitations on audits. Annual reports should be kept permanently, as should items of potential historic interest about the company. The Tort Reform Act now provides for an 18-year limit on product liability suits since 1994; however, the period for which aircraft purchase and aircraft maintenance records need to be kept is still the lifetime for each item.

Administrative procedure controls is another problem that rises as the organization grows, stimulated by the complexity of aviation. To control this area effectively, it is necessary that guidelines be adequately developed consistently implemented for the organization. The content of this framework will vary depending upon the needs of the organization, diversity of activity number of personnel, geographic locations, and management pattern. It may range from a simple system of working procedures and informal memoranda to a more complicated structure with a formal manual that specifies the complete system of procedures and directives and their administration.

The control of forms used in an aviation business becomes an extremely important part of a records management program because of the many requirements for information and records pertaining to passenger statistics, student activity, maintenance progress, and fuel allocations as well as internal departmental activities and analysis. They are necessary to transmit information and instructions and to record data. Forms are carefully designed papers that perform, simplify, and standardize office work. They accumulate and transmit information for reference or decision-making purposes. They may be classified in various ways, but a typical system classifies them according to their business functions such as purchase forms, sales forms, correspondence forms, and accounting forms. There are three objectives in forms control: (1) to eliminate as much unneeded information and as many business forms and records as feasible, (2) to combine as many of the business forms as possible, and (3) to simplify the forms in content, arrangement, and method of preparation. With these objectives in mind, the manager can utilize the following three principles in the standardization and control of forms:

1. A form should exist only when there is a need for it.
2. The size, quality, and color of paper used in all forms should be standardized to reduce costs and confusion.
3. The design, use, and replacement of business forms should be centrally controlled.

One effective method of attaining the desired objectives and achieving efficient and economical forms is through the use of a design checklist. Such a checklist can be used in reviewing existing forms or in developing new forms.

The control of paperwork procedures is another key element in a complete management program. With rising costs in every portion of the business, the manager must control the use of personnel involved in handling paperwork. Clear-cut procedures allowing maximum productivity of the personnel will result in the best possible results. A logical flow diagram of the various steps involved in the use of paperwork will help the manager visualize the complete procedure and control the process. For larger organizations with many personnel engaged in paperwork, work standards can be utilized successfully. In some aviation businesses, a program of work measurement and evaluation is beneficial.

With the relatively low cost of today's computers and the myriad of software programs available to assist businesses in accounting functions, standardization of forms is easily performed via computer. Computers actually reduce the amount of paperwork generated through the utilization of e-mail and other electronic communication means.

Mail-service control should be reflected through a system that is tailored to meet the particular needs of each organization. Accuracy, speed, and economy are the three key criteria. Mail-service procedures will be quite different in various organizations, depending upon size and type of business activities. Since mail is one of the communication links of an aviation business with its customers and suppliers, the mail function should be equipped, organized, staffed, and supervised to do the job effectively. Management should provide for practical organization, up-to-date equipment, written procedures, planned layout, informed personnel, and controlled costs. Each of these areas needs to be carefully reviewed by management, and action taken to ensure that the desired level of mail service is being achieved. The responsibility for mail service must be assigned and fixed. The proper equipment to expedite the handling of mail should be identified and obtained. The procedures for handling internal as well as external communications should be planned and written for reference purposes. The layout or arrangement of a mailroom varies depending upon the amount of activity, but the primary emphasis is on efficiency. Mail-service personnel should be trained to understand and handle their mail service for the company. They should be knowledgeable about company policies, mail procedures, and postal regulations, as well as dependable, accurate and thorough. The costs of mail service must constantly be reviewed. There are many specific ways to cut costs without reducing service. Alert and trained personnel using the proper equipment in an efficient layout can identify these opportunities and take advantage of them. In today's world of high-speed communications, virtually all aviation businesses find a scanner or fax machine indispensable, and it is wise to have a separate phone line for a fax machine unless the business is very small.

File control includes the identification and development of an appropriate filing system, the acquisition of supplies and equipment, the training of personnel, and the development of working procedures that will result in the filing activity needed by the organization. A most crucial step is to ensure protection of active files through a daily process of duplication or storage. The operational adequacy of the filing system can be examined through the use of a file audit. Digital or physical, a file audit is a systematic examination of documentation in day-to-day operations. The following checklist illustrates the technique of auditing the filing system and the records-management activity:

1. Are files under central control?
2. Is there a records manual? Is it up to date and adequate?
3. Are files neatly maintained?
4. Is filing kept up to date?
5. Are personnel adequately trained?
6. Are all records marked with a retention period?
7. Are obsolete records destroyed?
8. Are records adequately protected?
9. Has provision been made for expansion?
10. Do different files duplicate records? Is this necessary?
11. Is employee handwriting of records legible?
12. Has scanning or microfilming been examined to determine its feasibility?
13. Is provision made for the transfer or removal of inactive records?
14. Is there an inventory of all records?
15. Have records been classified in terms of being vital, required, routine, and so on?
16. Are control reports required and submitted?
17. Is there an ongoing program of evaluation and follow-up?
18. How much does records maintenance cost?

Through this kind of critical examination and its implied follow-up activities, the files of an organization can be developed and maintained at the desired level of capability. As with other administrative or support functions, the level of sophistication will vary with the size and complexity of the business. The full file audit can be of tremendous value to any organization.

Records retirement or retention is another extremely important element of administration. The volume of data required to manage an aviation activity today has reached massive proportions. This has resulted in stuffed files and the necessity for a planned program for record retirement. Various records are kept or disposed of according to different requirements such as the Statutes of Limitation and the stipulations of different regulatory agencies. To assist in evaluating records for retirement, the manager might consider the following:

> Value for administrative use;
> Value for legal use;
> Value for policy use;
> Value for fiscal use;
> Value for operating use;
> Value for research;
> Value for historical use;
> Supporting value (how may the record support other records of importance?);
> Document duplication (how many copies of the record are available?);
> Content duplication (how is the information represented in some other record?); and
> Volume of file (the value of all records should be weighed against the volume of the file in terms of equipment, file size, floor space, and resultant storage cost).

The National Fire Protection Association has developed a record classification system that should be helpful to aviation businesses in developing a retirement/retention program.[1]

Vital. Those records that underlie the organization and give direct evidence of legal status, ownership, assets, and liabilities. The loss of them would seriously affect continuing a business.

Important. Those records that are essential to the business and could be reproduced from original sources only at great expense.

Useful. Those records that have routine significance. The loss of them would cause temporary inconvenience but they are not essential to continuing a business.

Nonessential. Those records of no present or future value that should be destroyed. As a guide to everyday activity, many organizations have developed a schedule covering the retention period of their records. This schedule should be adjusted to meet the needs of individual businesses and the changing requirements of federal, state, and local laws. When records have served their purpose, shredding or burning may dispose of them. These techniques are useful when the contents are confidential and mutilation is necessary. It is good management to account for the destruction of all important papers and to have a system that provides for the retirement of the others. As mentioned earlier, electronic archiving plays an important part in the records retirement and retention program. Although the emphasis is normally placed upon the immediate dollar savings resulting from reduced storage requirements, the importance of preserving business experiences and history should not be overlooked. Information regarding motives and reasons for making decisions can become useful in later years.

Records storage can be considered in two forms: operational storage and inactive storage. The key governing factors in maintaining a proper balance in storage are availability, cost, and organization. Operational storage, as mentioned earlier under files control, is concerned primarily with material being used daily, weekly, or on some regular basis. Of course, records that are not valuable should be destroyed before they reach the files for storage. A retention guide will greatly assist an organization in identifying what to retain in storage and for how long. Most organizations retain far more paperwork than is necessary. The National Records Management Council has estimated that 95 percent of all corporate paperwork over a year old is never used as a reference. Carefully screening records according to a retention list, and transferring only those necessary valuable records to a low-cost storage location can realize considerable savings. Storage facility requirements vary considerably. Generally they should offer protection from theft, fire, dust, dirt, moisture, and vermin, and should provide reasonable accessibility. A classification and retrieval system is essential—there is no value in having records stored if the right ones can't be found when needed.

A dynamic organization will undoubtedly experience a continuous need for additional filing space and equipment. Requests for such additional capability should be considered carefully. When the file cabinet or computer storage fills up, it may be time to get rid of some records through archival and elimination, rather than buying additional equipment.

Communications

Success in an aviation business depends increasingly upon communications. Managers need to study the techniques and devices that make communications as successful as possible. Communication problems are troublesome and businesses should consequently give them careful attention.

The first step in developing a satisfactory communication plan for an aviation business is to analyze and identify both the internal and external needs of the organization. Careful consideration should be given to such factors as types of communications, internal layout, types of employees, work-flow, customer distribution, and economic concerns. Such an analysis will consider the following areas:

> The number and kinds of communications that are transacted internally and externally;
> Frequency, duration, and timing of internal and external communications;
> The relationship between cost and service level;
> The importance of speed;
> Whether the responsibility for communication is fixed;
> The importance of exact understanding and accuracy for some or all communications;
> Whether there is a need for direct communication hookups to branches, suppliers, or customers;
> What the key internal communication system needs are; and,
> The communications ability of the organization's personnel.

The second step is acquiring and installing the desired communications equipment and implementing techniques to meet the internal and external needs of the organization. Electronic and regular company bulletin boards, correspondence, newsletters, booklets, meetings, suggestion systems, attitude surveys, e-mail systems, intercom systems, messenger systems, pagers, walkie-talkies and closed-circuit TVs are examples of primary internal communications equipment. The external methods of communication include electronic and regular mail service, telecopiers, facsimile machines, and telephones. Each of these areas contains a wide variety of possibilities and each should be considered carefully to ensure that organizational needs are being met as completely and economically as possible. Communications equipment sales personnel or service representatives can greatly assist in identifying needs and matching these needs with available equipment.

The third step in creating an adequate communication network is training personnel in the communication process and in how to use available communication equipment. Training in communication is a continuous effort that should include all personnel and should focus on eliminating problems that plague oral and written communications. A great deal can be accomplished through the use of lectures, training sessions, practice, demonstrations, bulletins, illustrative literature, and instruction sheets.

Duplicating Information

Office copying machines have improved a great deal in recent years, and the price has decreased. Considerations in regard to photocopying equipment include whether to rent or buy and whether to procure a grayscale or color copier.

There are machines of every level of power and capacity. To save staff time and ensure good quality, offset printing may be favorable for jobs with large volumes or high complexities.

It is typically more economical to have a local printing company handle very complex or voluminous print jobs, especially color printing. When determining the most economical option, be sure to factor ink, toner, paper, and staff costs into the do-it-yourself option.

Aviation Accounting

Accounting is simply maintaining a record of business activities. It is the systematic recording of the financial transactions and facts of a business so that the status of the business may be periodically determined. Results of its operations must be presented correctly and understandably on income statements and balance sheets. A businessperson needs to know very frequently how the business is financially performing. Which lines of aviation activity are the most or least profitable? What are the cash needs? What are the working capital requirements? The business owner can get reliable information like this only if there is a good accounting system, which is a key component of the Management Information System (MIS). The manager or owner of an aviation business is normally very knowledgeable about the technical aspects of the business. However, he or she may not have a background in accounting and is an aviation enthusiast first and a businessperson second. As a result, the typical manager avoids accounting and continually delegates it to others. Ultimately, he or she becomes unable to use it as a managerial tool. Because of this tendency, the following material points out the relative simplicity of aviation accounting, which can lead toward a more successful business.

Accounting Flow

The manager accounts for business activities with financial statements such as balance sheets and income statements. Essentially, a balance sheet shows what the business owns (assets), what it owes (liabilities), and the investment of the owners in the business (net equity) at a certain point in time. The income statement, on the other hand, is a summary of business operations for a certain period, usually between two balance sheet dates. It indicates the financial results of the company's operations for the selected period of time. In very general terms, the balance sheet is a snapshot of existing conditions, and the income statement explains how the business arrived at those conditions. The first step in generating financial statements is the completion of the source documents—a record of individual business transactions. These transactions include sales, purchases, payroll, and obligations such as depreciation, insurance, and taxes. From the source documents the accounting information flows through the system and becomes input to the financial statements. Figure 7.10 graphically depicts this flow of information through the accounting system. The manager should be familiar with this flow of information and the various records involved. In order to more fully understand the flow of accounting information, each step in the process is examined.

Source Documents

For the typical aviation activity, the normal source documents are sales invoices, journals, general ledger, and financial statements.

Figure 7.10 → Information System Activity Flow Chart

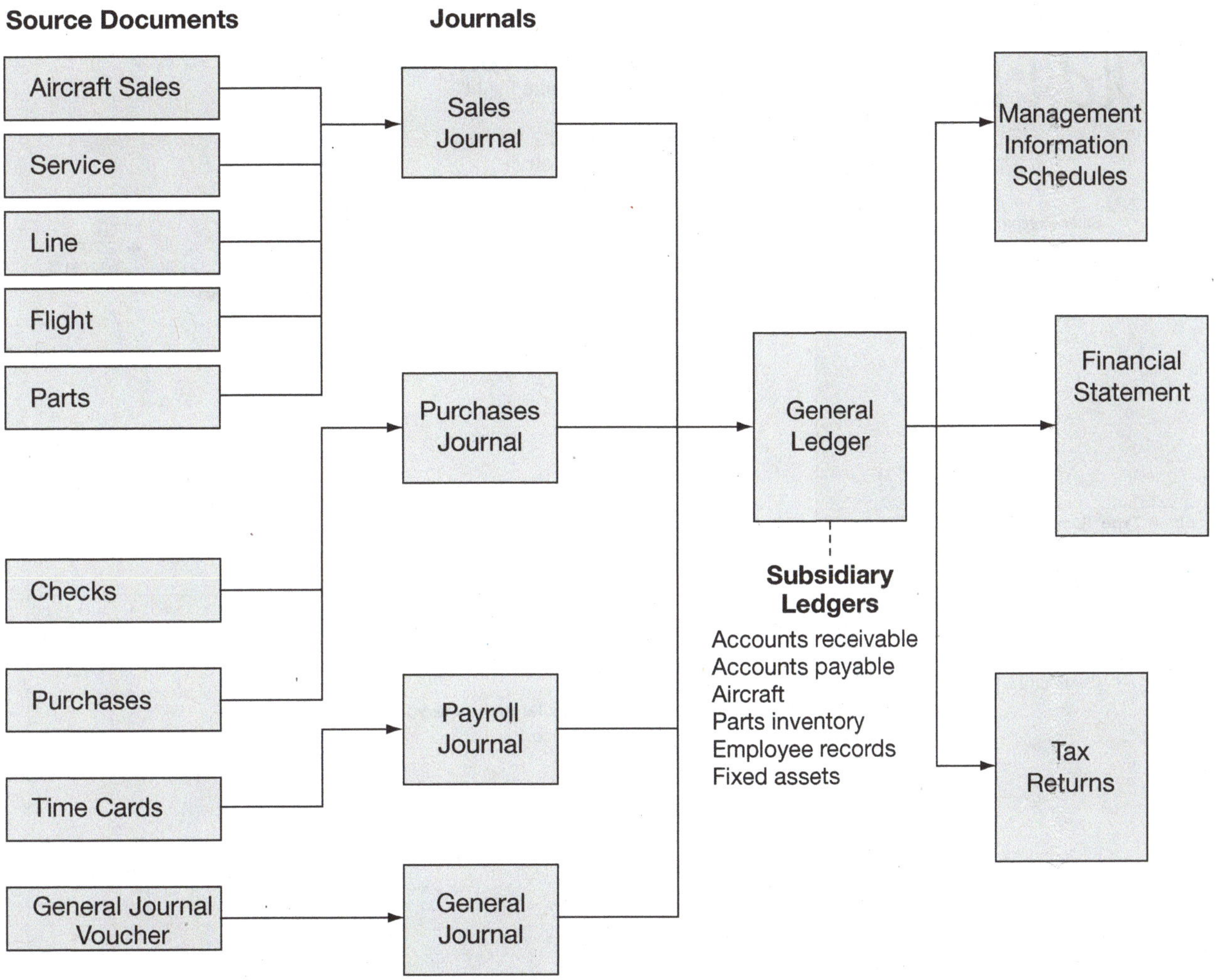

Source Documents: Aircraft Sales, Service, Line, Flight, Parts, Checks, Purchases, Time Cards, General Journal Voucher

Journals: Sales Journal, Purchases Journal, Payroll Journal, General Journal

General Ledger

Subsidiary Ledgers
- Accounts receivable
- Accounts payable
- Aircraft
- Parts inventory
- Employee records
- Fixed assets

Outputs: Management Information Schedules, Financial Statement, Tax Returns

---

Sales invoices will be from the following areas:

> Aircraft,
> Charter,
> Passenger,
> Flight instruction,
> Service,
> Line activity,
> Parts,
> Rental,
> Cash receipts,
> Checks,
> Vendor invoices,
> Time cards, and
> General journal vouchers.

Figure 7.11 shows a sample computerized invoice.

Appendix VI (see the book's website) illustrates typical forms used as source documents for these kinds of business activities. Today these will all be computer-generated. The development of individual forms to meet specific business requirements should be done with great care. Forms control is an important responsibility of the central office.

Journals

Information from the individual source documents is entered into the journals—the first accounting record of the transaction. These books are frequently called the "books of original entry" and they represent a chronological record of all transactions conducted by the business. A journal is simply a financial diary.

Figure 7.11 ✈ Sample Computerized Invoice

totalFBO®

Horizon Business Concepts, Inc.
"Where Your Dreams Take Flight"
address
city, ST 99999
phone numbers, etc
and another line

Sales Person: T-Support

Invoice: TU-01025

4/12/2009
3.32PM

Sold To: HBC - Flight School Fuel/Oil

Ship To: HBC - Flight School Fuel/Oil

Line#	Type	Item/Description	Cr?	Aircraft	Quantity	Units	List Price	Disc	Unit Price	Extended
1	Fuel	100LL			40.0	Gallon	2.000	2	1.950	78.00
		Avgas			Meter Start: 3986.5		Stop: 4026.5			

Taxes Included in Subtotal:
Fuel FET: 7.76
Local Flowage Fee: 2.80
State Sales Tax: 5.40

Subtotal	78.00
Sales Taxes/Fees	5.40
Total Due	83.40

Thank you for your business, or whatever trailer message you'd like
at the end of each invoice.

Printed: 10/28/2009

Courtesy of Total FBO Accounting and Business Management Software by Horizon Business Concepts, Inc.

A typical aviation accounting system has the following journals:

> Cash receipts and sales journal,
> Purchase and cash disbursements journal,
> General journal, and
> Payroll journal.

In order to ensure the flow of accounting information, administrative procedures must be established that ensure the accurate completion of all forms and the timely submission of all completed forms to the accounting office. At the end of an accounting period, the column totals of all journals are balanced and posted to the general ledger. The use of a computer is essential to automatically and rapidly take care of this flow.

General Ledger

The ledger is a computer file made up of the accounts identified in the business. A separate account is maintained for each type of asset, each type of liability, and each element of equity. The data transferred from the journals is entered into the appropriate ledger account. The balances in the ledger accounts are then reflected in the financial statement when it is prepared. Any transactions during the operating period result in an increase or decrease in assets, liabilities, and owner's equity.

Financial Statements

As previously mentioned, business activity is summarized and presented in the two financial statements—balance sheet and income statement. Figures 4.4 and 4.5 in Chapter 4, Profits, Cash Flow, and Financing illustrate these two reports. Accounting records and books of a business are not directly involved in decision-making; however, they are used by the finance manager or outside accountant to prepare financial statements and special supporting documents that are used by the owner/manager in decision-making. The preparation and handling of data, documents, and records has a tremendous influence on the outcome as represented by the financial statements; thus, it is essential that the owner or manager has an extensive working knowledge of the system and be involved in some of the decisions in designing and administering it.

Accounting Activity Flow Chart

The complete flow of accounting information from the source documents to the financial statements was depicted in Figure 7.10, and is broken down into individual areas in Appendix VI (see the book's website). This illustration suggests the basic outline of an aviation business accounting system and is not intended to be all-inclusive. Source documents can be expanded or contracted, journals realigned, or ledgers developed to meet specific business needs. Although the balance sheet and the income statement are considered the two major products of the accounting system, other system products are extremely valuable to the manager. These materials may include income tax returns and information schedules used by the manager in making operational decisions on a continuing basis.

A major managerial function is organizing, usually encompassing the development of an organizational structure and the allocation of functions and duties to units or individuals. The manager's primary objective is to establish work teams that can function efficiently and profitably. As a part of this effort, it is necessary to develop a flow of information pertaining to each work team or activity center in order to ascertain the actual business results. This information flow is needed to measure progress and provide direction for future management activity. A clear picture of how records are processed (who does what, when and where?) is necessary for training personnel in the process. Variations undoubtedly exist from business to business, but the general framework should be like the one presented in Appendix VI. Management must ensure a logical flow of information, provide for necessary personnel training in the use of the system, and maintain a policy of checks and controls to guard against a loss of resources through inefficiencies or theft.

Accounting System

Each account is given a number that serves as its identifier and as the means of guiding and controlling the flow of information through the accounting system. At a minimum, the accounting records should provide information on:

> Assets, including real estate, equipment, inventory, receivables, and cash;
> Liabilities to banks, suppliers, employees and others, including income taxes due;

- Owner's equity in the firm; and
- Sales, expenses and profit for the accounting period.

In developing a system to meet these requirements and the needs of the organization, the aviation manager can take one of several approaches. With the help of a local accounting firm and appropriate software support, he or she can (1) develop his or her own system, (2) turn to one of the many large accounting firms that have an aviation accounting system available, (3) solicit the help of large printing houses that have developed complete systems, or (4) obtain assistance from one of the aircraft manufacturers or specialty vendors that has developed complete systems for its dealers.

In view of the unique issues encountered, it would be better as soon as fiscally possible to acquire a system designed specifically for an aviation business and to avoid the problems of developing one from the very beginning. There are now several primary vendors of aviation software; however, this is a rapidly changing area. As already discussed, one must be careful to select the appropriate vendor.

The TotalFBO system is an excellent example of a financial information system designed specifically for the aviation business community. This system is made up of the following components:

- Financial statements,
- Chart of accounts,
- Source documents,
- Accounting system,
- Accounts receivable portion,
- Accounts payable portion,
- Payroll section, and
- General ledger.

All the components are fully-integrated, with each keyed to the chart of accounts. The statements, source documents, and the accounting system are preprinted with the appropriate account numbers to facilitate efficient operations. It is a compact structure with the capacity for flexibility and growth. Source documents comprise an integral part of the accounting system. They are designed specifically for aviation businesses and are coded with account numbers to facilitate their use and reduce transfer errors.

Profit Center Accounting

The typical aviation business is engaged in many types of activities: selling aircraft, providing service, giving flight instruction, providing air transportation, selling parts, pumping fuel, renting aircraft, leasing space, as well as other related enterprises such as car rentals. In order to properly operate each of these activities, the manager needs to have current information on the profitability of each. He or she needs an accounting system that will properly allocate the revenues and expenses of business activities to the corresponding department, allowing the worth of individual profit centers to be judged. The system will also allocate company-wide overhead on a percentage basis to each profit center.

In order to develop and utilize a profit center accounting system, it is necessary to understand and implement the following principles:

Responsibility-based accounting recognizes various activities in an organization and traces costs, revenues, assets, and liabilities to the individual managers who are primarily responsible for making decisions for that activity. Ideally, revenues and costs are each recorded and traced to the one individual in the organization who is responsible for the driving business activity.

Controllable costs are critical to responsibility accounting. These costs are directly influenced by a manager within a given time span. Thus, one must identify the manager of a specific activity and delineate the time span under consideration. It is difficult, however, to determine the degree of control a manager has over a cost, labelling the item as controllable or uncontrollable. There is the additional problem of trying to clearly assign responsibility for a business activity to just one person. Perhaps several individuals exert influence over a cost center. The question is: "Who is the one person in the organization with the most decision-making power over the item in question?"

Uncontrollable costs are generally excluded from a manager's individual performance report. However, each manager may be assigned some of these costs to become more aware of the operational problems of the total organization.

Cost allocation is an inescapable problem in nearly every organization. How should the costs of an aviation business be split among the flight department, line department, service department, and so on? How should staff costs, computer, advertising, and equipment costs be allocated? These are tough questions and the answers are not always clear. Insight into unclear cost allocation issues is necessary to better assess department performance and maximize the overall business profit.

Three major aspects of cost allocation include:

1. Delineating the cost center, namely the department, product, or process.
2. Tabulating the costs that relate to the cost center, including material, labor, and overhead.
3. Choosing a method for specifically and consistently relating costs to cost centers. This normally means selecting an allocation basis for the individual costs.

The cost centers of primary concern are those departments or work areas that have been identified by the accounting system.

Allocating and summing the costs to be analyzed are a basic concern of the manager and a primary function of the system. The information system must identify the key costs and provide the mechanism for accumulating and assigning them to the manager. The actual method of relating costs to business areas will vary, depending upon the criteria selected for making decisions. Possible criteria include:

> Physical identification,
> Services used,
> Facilities provided,
> Benefits received,
> Ability to bear costs, and
> Fairness or equity.

Allocation bases for overhead. There is a tendency to aggregate overhead costs in pools and use one allocation base for each pool. The following bases are widely used:

> Physical units produced,
> Direct labor hours,
> Machine hours,
> Direct labor costs, and
> Direct materials.

The major problem in choosing a proper base is relating overhead to its most closely related cause. However, the driver that is easiest and cheapest to apply is often selected.

In allocating costs to properly achieve objectives, the administration must select an allocation system that influences employees and managers to take the correct action. The best system measures cause-and-effect relationships of business activity. The fully allocated versus partially allocated cost question can never be answered with one solution for all situations. There are many variables that influence the decision to fully allocate all costs. Among those frequently considered are:

> The size of the line organization,
> The sophistication of the information system,
> The relative cost of the allocation,
> The importance of cause-and-effect relationships,
> Managerial awareness of the concept, and
> Anticipated personnel reactions.

Existing cost allocation systems in many businesses are crude; however, tremendous improvement can be realized through the intelligent use of averages, the development of an optimal system, and the tendency toward full allocation of costs.

A suggested procedure for allocating costs is contained in the following steps:

1. Use the income statement for the business as the starting point.
2. Determine initial policy on allocation, that is, full or selective.
3. Review each item in the total income and expense columns.
4. Determine allocation unit and method of allocation.
5. Extend the selected allocations for several financial periods.
6. Analyze and compare the consequences.
7. Ask the question: Does this system provide me with the necessary profit-center evaluative data?
8. Consult with managers of the individual departments on the appropriateness of the selected system.
9. Identify additional requirements.
10. Resolve any difficulties.
11. Implement the allocation system.
12. Communicate and educate.
13. Evaluate the system after a period of operation.
14. Modify as required.

Contribution concept. The term contribution or contribution margin is an expression used in accounting but frequently not understood by managers. It is important to the successful application of the profit-center concept and ultimately to an analysis of departmental and overall business activity.

The contribution margin is considered the excess of income over the variable expenses. It can be expressed as a total, as an amount per unit, or as a percentage. To more easily present the concept, one may develop a graphical presentation of contribution margin. The three components used in the graph are variable expenses, fixed expenses, and income. The horizontal axis of the graph represents the number of units handled or the volume of services delivered. The vertical axis is graduated in dollars and will be used for dollars of expense or income. The horizontal axis measures the number of goods produced or the volume of services performed. The vertical axis measures the amount of revenue or expense, in dollars. The first step is to add the variable expenses, the costs that increase directly with increasing volume. Next, add the fixed expenses, which are the costs that remain constant over changes in volume. Finally, add the income or revenue as reflected by varying sales volume. The combination provides the cost/volume/profit chart—better known as the break-even chart. Figure 7.12 is an expanded version of the chart. It identifies the components just described as well as the contribution margin. This is the same chart used in Chapter 4, Profit, Cash Flow and Financing (see Figure 4.3). The contribution margin is defined as the excess of income over variable expenses; we can see from the graph that, below the break-even point, there is still a contribution (to the organization) of dollars toward the fixed costs incurred by that department or activity. The manager's goal is to sell enough units so that each department or activity realizes a profit. But in the event a profit is not realized, he or she should recognize the "sunk cost" aspects of fixed expenses and appreciate that some portion of those costs are being offset by that business income. More importantly, if any component of business income is eliminated, the remaining business units have to shoulder those fixed expenses, effectively raising their break-even points.

Information System Tools

Introduction

No small business of the 21st century is complete without a computer system, most likely a desktop networked small computer system.

Figure 7.12 ✈ **Contribution Margin as Graphically Illustrated in a Break-Even Chart**

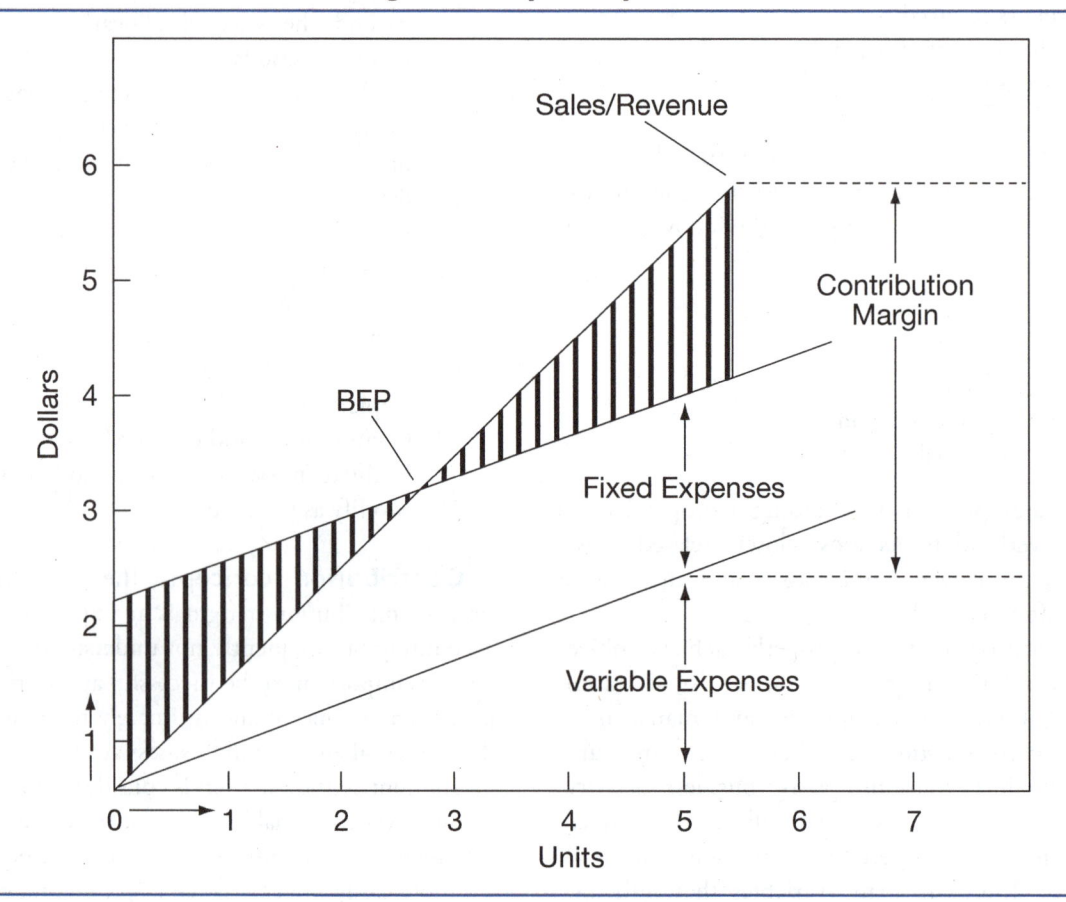

Computers are to manual accounting systems as the automobile is to the horse. The norm today is to expect to upgrade every few years as hardware and software improve; the staff must be computer-literate and change must be expected and embraced.

A manager should not expect to reduce personnel or other administrative costs with a computer upgrade. He or she will find however, that after upgrading software or hardware, existing functions can be accomplished quicker and more precisely. In addition, new means of data storage, analysis, and visualization become available. This opens up new horizons for better managerial control and faster reaction to problems.

In the 21st century, the typical office purchases desktop computers for word processing, data analysis, and other sophisticated functions.

There are many choices of computer systems, and these are discussed below.

Desktop Computers

Handling all functions in-house through networked desktop computers is becoming much more common as the capabilities of these systems increase. Advantages include:

1. The system belongs to the business, which means a greater degree of confidentiality is possible.
2. It can be a multi-user system—everyone in the office who needs access can connect in through a terminal.
3. Because of this feature, every part of the business using a terminal can access the same data. For example, the repair shop and front desk can simultaneously use it for a cash register and inventory control program, respectively, and daily reports on sales and inventory can be created and distributed.

On the other hand, if a computer network is down, operations may be severely hindered.

A desktop computer can conduct all the accounting functions that might be done manually by the bookkeeper or sent out to a service bureau. In addition, the following can be done with ease and speed:

1. Drafting letters, individually typed, with salutation and as many references to individuals or projects as necessary
2. Sorting, for example, of mailing lists by zip code, date of sales, aircraft owner status, and any number of other criteria.
3. Alphabetizing lists.
4. Printing invoices.
5. Printing shipping documents.
6. Retrieving and displaying of customers' account and payment status.
7. Composing contracts or leases with standard text and customizable insertions.
8. Conditional printing, causing a paragraph to be inserted in a document only if certain conditions are met.
9. Cash register functions.
10. Automatic inventory tally and reorder reminders.
11. Budget projections including any number of "what-ifs."
12. Break-even analysis.
13. Special reports on any data collected on the computer; for example, productivity, allocation, and cost trends.
14. Forecasts using various regressions.
15. Checking of spelling and grammar.
16. Tracking of projects and programs.

Selecting the Right Functionality

It is important that the operator choose a software package that has three main functions. The first is the production of complete financial statements, including but not limited to income statements, cash flow analysis, accounts receivable, general ledger, flight and charter accounts receivable, payroll, inventory analysis, checkbook, aircraft leaseback reports, and fuel consumption reports. Before choosing a software package, the operator should list the requirements for each type of statement needed, as well as the frequency needed. One should verify that the system can generate each financial statement currently required, and be expandable as the business grows. An operator does not want to spend thousands of dollars on a system and still need an outside bookkeeper to generate financial statements, a duplication of time and effort.

The second area the operator should be aware of is hardware and software requirements and their costs. The operator must consider the cost of computers, computer programs, and any other devices. The dedicated software packages determine the cost of the software by the number of terminals. The cost of the hardware may average around $1,000 to $2,000 per computer. The cost of networking is dependent on the company's physical layout. The cost of the software is also dependent upon the number of computers networked together.

The third area the operator must consider is customer technical support and software updates. Technical support usually has an annual fee. The operator should consider the quality and quantity of available technical support, the length of time the required to interface with technical support, and the frequency of program updates. Updates may or may not be included in the technical support contract; purchasing software and hardware is not a one-time event.

Computer Selection Considerations

1. Decide which functions are most economically automated and pick the software capable of performing those tasks.
2. Select the hardware compatible with any current and future software. Specialized aviation software, as well as any proposed derivative, must function on the hardware.
3. Random Access Memory (RAM), one form of computer memory, works with the Central Processing Unit (CPU) to manage open programs. New computers should offer a minimum of 16 GB of RAM.
4. Consider the clock speed of the CPU. A new system should offer processing speeds of around 2.5 GHz.
5. The internal data storage capacity of the computer should be around 256 GB for average users.
6. The computer should also have a USB (or USB3) port, solid-state drive, and the latest virus protection program.

FBO Management Software Choices

Generic Business Software. When an aviation business is beginning operation, it may choose to utilize common software such as Intuit QuickBooks or Sage 50. The company may hire a person to customize the software to its particular needs; for example, QuickBooks may be used to calculate running balances on customers' flight and maintenance accounts. Usually, the FBO hires a specialist who is already familiar with its operations. This can be a difficult process because it is rare to find someone who is sufficiently familiar familiar with generic accounting programs as well as aviation business operations.

There are potential disadvantages to using generic software. As the business grows, the software applications become more complex and consequential within the company. One example of a common issue not addressed by generic accounting software is co-mingled fuel, which occurs when the fuel an FBO purchases for resale is stored in the same facilities as fuel purchased by another company for its own consumption.

FBO-Specific Software. Aviation-specificity has emerged as a key design feature and requisite for today's FBO accounting software products. This software goes well beyond bookkeeping to include such areas as customer service and inventory control.

TotalFBO. A powerful and popular software package is TotalFBO by Horizon Business Concepts, Inc. The complete system includes flight school, maintenance, inventory, scheduleing, charter, fuel, payroll, and more. The company has full-time customer support available for an annual fee. Almost any report needed by the operator can be generated in real-time, and the user has the capability to design macros for custom needs. The advantage is an integrated, networked system for the entire company. A main disadvantage is initial setup and training time for the staff; however, the operator can spread out this lost time by setting up sections of the company separately. This system can generate general ledgers, income statements, cash flow analysis, bank deposits, and payroll, among others. Horizon notes:

> "Since TotalFBO™ is so broadly based, almost any type of aviation business can make use of its features." Included are features for Flight Schools, Repair Stations, Airport Managers, Engine/Accessory Overhaulers, Charter Operators, as well as FBOs. The program is modular so that users don't get overburdened with items that are not needed for a particular operation."

Features available in the various modules of TotalFBO include:

> Account integration for speed and accuracy;
> Full GAAP accounting functions;
> Accurate invoice pricing by customer;
> Scheduling of flights and maintenance;
> Quoting of charters adjusted for winds and aircraft performance;
> Creation of many, specialized reports;
> Checks of pilot currency;

Front desk clerk runs customer's credit card after processing her account through TotalFBO.

> Tracking of Part 135 crewmember duty logs;
> Aircraft maintenance projections;
> FAA forms: 337, 8130-3, 8710;
> Weight & balance tracking;
> Bar code input & printing;
> Core tracking
> Integrated employee timeclock;
> Front counter super screen;
> Checks of mechanics' currency;
> Customizable instruction syllabus;
> Complete flight manifest;
> Full departmental accounting;
> Fuel truck interface;
> Automatic credit-card processing;
> TouchScreen invoicing;
> Enterprise-wide data sharing; and
> Emailing of documents in PDF format.

FBO Manager by Cornerstone Logic is similar to TotalFBO in terms of capability and support for flight school management, charter flight scheduling, fuel sales, and customer tracking. The primary distinction is that FBO Manager links to the customer's own choice of accounting software to create the financial and activity reports. FBO Manager typically requires minimal setup time but may require the additional services of a bookkeeper.

The **Corridor** family of aviation service software, offered by CAMP Systems, includes multiple modules for different aviation business activities. Its FBO and MRO software packages are fully customizable and compatible with generic accounting software. Supplemental versions are Corridor Go and Corridor Analytics, which connect mobile devices and business intelligence dashboards, respectively.

MyFBO.com is an online multiple locations FBO accounting software product that does not require the customer to maintain a dedicated server. In addition to billing and payment tracking, its capabilities include leasing and contract accounting, credit card processing, fuel farm management, ground service reservation, flight dispatching and inventory management. Due to its relatively low cost, easy implementation and user-friendly interface, MyFBO.com is ideal for small to mid-size FBOs.

FBOperational Fuel Management System is a products of PRG Aviation Systems. The Entry Level System (ELS) version, designed for small to mid-size FBOs, provides basic sales tracking and accounts receivable management. Additionally, its reporting functions can generate invoices and post data to the user's general ledger software for financial statement purposes. FBOperational Fuel Management System, for larger FBOs, incorporates advanced capabilities such as price-to-sales ratio predictions and daily order tracking. These advanced capabilities help to control costs, maximize profit and determine staffing requirements. Corporate flight departments, fractional ownership companies, airlines, and other businesses with relatively large customer bases are the typical users of FBOperational.

Business Aviation Software Engine (BASE) system, a product of Wellington-Royce Corporation, comes equipped with over 22 applications. Three of these applications are dedicated to basic accounting functions: accounts payable, accounts receivable and general ledger. The remaining applications such as the customer relationship management and aircraft management modules are optional and can easily be incorporated as a customer's business grows. BASE provides larger full-service FBOs with complete sales and analysis accounting functions; however, the software requires the installation and maintenance of a dedicated server.

The primary disadvantage to these packages is they are designed for a very specific and complex market. A limited number of potential customers translates into high costs and low revenues The primary advantage is that the software packages are highly tailored to the industry and are being offered by vendors that understand the nature of an aviation business. Improvements are continually being made as customers make suggestions and develop new needs.

Computer Service Bureaus

Service Bureaus are shrinking in scope as businesses have shifted to more powerful in-house computer systems. They commonly now handle only payroll and perhaps benefits. Depending on the areas a manager chooses to outsource, a service bureau will obtain a business' raw data, including:

> Checks;
> Sales slips;
> Receipts;
> Journal entries; and
> Payroll data: names, hourly rates, hours worked, withholdings, etc.

The bureau will then use its equipment and labor to produce reports and other documents such as:

> Paychecks;
> Cash receipts journals;
> Check registers;
> Payroll registers;
> Sales journals;
> General journals;
> General ledgers;
> Receivables ledgers;
> Payables ledgers;
> Property ledgers;
> Balance sheets;
> Income statements;
> Aged accounts-receivable listings, with customers' statements;
> Inventory status reports;
> Payroll reports, including those to be filed with government agencies;
> Budget reports; and
> Operating ratios.

The bureau will work with the business to establish a virtually seamless process, so that it is difficult to tell where the company computer work left off and the outside bureau stepped in for the contracted functions.

If these services are performed by a bank, which is an option, the bank will also pay invoices, issue checks, and deduct the funds from the business' account. The company's personnel do not need to know anything about computers, though they will still need basic accounting knowledge to verify the accuracy of the outsourced work. Unfortunately, outsourcing these services can be expensive. Also, the necessary turnaround time means that instant results are rarely available.

Other Devices

Other electronics available to managers today include ultraportable laptop computers, cellular telephones, and tablet devices, which all help save time and improve organization.

Business Security

Confidentiality and Control of Information

All businesses need to restrict some information to the manager only. Locked file cabinets or a safe are possibilities. Computers have integrated security provisions. For sensitive information, the user must provide a user identification code or file password. Computer "hackers" are forever finding ways to beat such systems, and computer manufacturers are forever making systems more elaborate.

There is concern about losing records through computer or operator failure. Financial records such as monthly statements should be printed on paper. All computer work should have a back-up file, which is created automatically by some systems. For critical data, a duplicate disk should be placed in secure storage. All computers should be used to save all basic records, such as register receipts, for at least three years. Power aberrations should be guarded against with a surge protector, which are relatively inexpensive. With these precautions in place, the risk of losing computer data is no more likely than it is for other kinds of office records. One risk from computerization is the presence of viruses. Computers are most likely to be infected with malware they are connected to other systems. New virus problems may appear as quickly as old ones are resolved, so the threat must be continuously monitored.

Types of Losses

In addition to theft by outsiders, every employer needs to watch for the possibility of internal theft by employees. Embezzlement is the fraudulent appropriation of property by a person to whom it has been entrusted. Theft can span many levels of sophistication and size, including:

> Stealing office supplies,
> Stealing parts or material in inventory
> Giving unofficial discounts to friends, family, or other customers

- Clocking in unworked hours,
- Using company time for personal interests,
- Pocketing cash received as payment for sales,
- Lapping—stealing payments and using subsequent payments to cover for receivables,
- Check-kiting—depositing check payments in both a business and personal account, covering the checks before banks clear them,
- Mischaracterizing personal expenses,
- Seeking reimbursement for fake purchases,
- Receiving kickbacks from vendors in which an employee and the vendor split profits from inflated prices of goods or services,
- Submitting false refunds for sold goods or services and stealing the refund payment,
- Submitting false receipts for petty cash,
- Use of company postage, computers, copiers, printers or other equipment for personal interests.

Methods of Combating Losses

No system where delegated authority exists is foolproof. By definition, other people are in positions of trust. Criminal and credit records can be checked before hiring but that will not identify the potential thief. What employers can do includes:

1. Set a scrupulous example.
2. Establish a climate of accountability.
3. Design an accounting system with sufficient internal controls.
4. Separate the duties of employees so that the same person does not both handle and record incoming or outgoing check and cash payments
5. Obtain operating statements regularly, compare them with prior data, and investigate any anomalies.
6. Look for clues—an employee who never leaves their desk or takes vacations, increases in returned goods, increases in bad-debt write-offs, decreases in cash sales, decreases in inventory, unusual changes in overall profitability, increases in expenses, and slow collections. These items are not necessarily indications of theft, and may warrant investigation. Close and constant scrutiny by a manager will act as a deterrent.

Summary

The information system in an aviation business, as in any business, plays a key role in helping management determine whether the business is developing according to plan. As the business grows, information systems will likely need updating. Information should be kept on four subject areas: (1) human resources, (2) financial, (3) material, and (4) aviation operations. Records and record-keeping are necessary components of the information system, and they should be tailored to the needs of the individual business. Appendix VI (see the book's website) illustrates the forms and source documents likely to be needed.

DISCUSSION TOPICS

1. Name three reasons why a business information system is needed. Which is the most important reason, and why?
2. Outline the principal requirements of an effective business information system for an aviation business.
3. What are the major problems in maintaining an effective records system in an aviation business?
4. Outline in a flow diagram the origins, pathways and use of the four key elements of aviation business information.
5. Identify the purpose of each of the following ratios: (a) liquidity, (b) efficiency, (c) leverage, and (d) profitability.
6. Describe at least two examples of the ratios discussed in the preceding question.
7. What is a management audit? Give some reasons why it is not more widely used.

Many computer vendors are now offering systems that can serve all aspects of FBO operations; before embarking on computerization, some analysis of choices and the readiness of the existing system should be considered. Many forms can be produced by computer and potentially save storage space.

The task of analyzing business activity to measure progress should be undertaken regularly. Both financial analysis and a management audit (shown in Appendix V at the textbook website) are useful tools for this; they should be applied to individual profit centers or departments as well as to the business as a whole.

Endnote

1. *See http://www.nfpa.org*

8

Operations: Flight Line and Front Desk

OBJECTIVES

› Recognize differences and similarities between the flight line and the front desk.

› Describe the various functions of the flight line operation.

› Give examples and explain the difference between internal and external customers of a fixed base operation.

› Recognize the areas, issues, and functions supported by a well-developed procedures manual.

› Recognize the factors inherent in making the flight line a profit center for itself and other parts of the operation.

Introduction

The flight line and the front desk are both highly-visible nerve centers of the organization. These two operational activities are extremely important in establishing the desired image of the business and in contributing to the efficient flow of business activity. The flight line is that portion of the organization that deals with the customer in or around the aircraft, on the ramp. The front desk is the heart of the business that deals with the customer who comes into the facility by aircraft, and with those who drive or walk to the business. Operating the two efficiently will contribute tremendously to the success of the business. Efficient and safe operation comes from a clear identification of required functions, positive organization of resources, and thorough training of assigned personnel.

The physical location of the front desk is normally in the lobby or reception area of the business and close to the administrative offices.

The two activities, flight line and front desk, are considered as separate entities in this chapter. In many businesses they are clearly separate, while in others one can find a variety of combinations. To management, it is important that the two be organized according to the physical layout and the overall objectives of the business. In addition to the material in this chapter, job descriptions and necessary skills are presented in Chapter 5, Human Resources.

Customer Service

As the first points of contact for all customers, whether pilots, arriving passengers, or other visitors such as vendors, the flight line and the front desk are the places where customers form their impressions of the company. First and foremost are the employees and how they treat all visitors. Do they have side conversations on the phone or with another employee when someone is waiting for assistance? Or, do they

203

go out of their way to provide assistance when someone seems a bit lost? Morton (2002) notes that even one negative contact can destroy all the positives:

". . . superior personal contact from every employee at the airport is a must. . . . every encounter must be positive. If only one employee fails to greet, smile, and provide sterling service, the whole evolution can prove faulty. . . . Statistics indicate that (the customer experiencing poor service) will tell at least 10 people of this incident."[1]

Staying Up to Date
Technological Change

Technology that enables efficient office procedures continues to improve and change at a rapid pace. Smartphones with mobile apps, Wi-Fi internet access, networked computers, and other office productivity devices have had significant impact on how business is conducted. Customers form a favorable impression from seeing a business that uses state-of-the-art technology, and the manager must be aware of new developments, incorporate new applications of these technologies as they emerge, and budget for their acquisition and the necessary training of staff in their use.

Facility Appearance

The general appearance and upkeep of the facility and its surroundings can say a lot about the business. A lobby or lounge that is not kept clean and appears to be in a general state of disorder does not provide a good first impression. A maintenance shop in which spare parts litter the floor and tools are left out in the open in a haphazard manner does not instill confidence in aircraft owners regarding the quality of work performed in that facility. A prospective new student pilot who is met by individuals in the office who are indifferent or unfriendly, and who notes that the aircraft on the ramp look as if they have not been washed in months, may be reluctant to learn to fly at this operation.

Thus, poor appearance can, and does, have tremendously detrimental effects on potential business. Appearance, signage, and uniforms or dress code all contribute to an image of brisk professionalism and should not be treated lightly or dismissed.

A negative example: Students from an aviation management class at a university in Atlanta, Georgia were asked to select an FBO in the metropolitan Atlanta area and critique it. In submitting their written and oral reports to the class, the majority of students were amazed at the poor customer service attitudes and the general condition of the facilities they visited. One student saw incidents of unsafe fueling, lack of pre-flight inspections, a dissatisfied aircraft owner returning an aircraft radio he had brought in earlier for repair that still did not work, and a student whose lesson had just been cancelled because the flight instructor had elected to take a charter flight instead. All of this occurred during a one-hour visit!

Flight Line

The flight line is that part of the business that marshals and parks arriving aircraft, greets passengers and crewmembers, services aircraft, and marshals departing transient aircraft; services local tenant customers; and supports other departments of the organization. It normally provides temporary parking for transients, fuel and oil, minor maintenance, and flight servicing for larger aircraft. In the process of performing these duties, the line department may operate fuel trucks, ground power units, tugs, "follow-me" vehicles, pre-heaters, aircraft tow bars, vacuum sweepers, pressure washers, snow removal or ramp sweeper equipment, and lavatory service equipment. It may provide fuel trucks, high-pressure air, air conditioning (heating and cooling), oxygen and nitrogen recharging, passenger and cargo access ladders, baggage vehicles, lavatory service equipment, water- and food-servicing equipment, aircraft de-icing, aircraft servicing ladders, catering, ice and coffee, newspapers, and block heaters for piston-engine aircraft.

For local (based) customers, aircraft parking services are generally purchased on a monthly or annual lease. Fuel, oil, and other services are much the same for both transient and based aircraft. In a

Aviation lineman marshals aircraft to the ramp.

fully departmentalized organization, the line department may also provide services to the company's own flight department and the maintenance department by parking, fueling, and washing aircraft, as well as providing services comparable to those provided to customers. Even in smaller organizations, for which departmentalization is not complete, it is beneficial to identify any other parts of the business receiving line services, so that appropriate costs are allocated to the correct department.

It is important to recognize at the outset that line operations are primarily a service function. The activity is not a product sales business, but a service business. The courtesy, efficiency, and flexibility of the services provided are the main selling points and the primary reasons for repeat sales. There is wide variation in the size of line operations at airports around the country, including the number of aircraft passing through or based at the facility, and the volume of fuel pumped. Indeed, in some businesses the flight line activities are the predominant source of income. Regardless of its size and the contribution of the flight line to total company sales, it should be operated as a profit center. This is accomplished through the application to the flight line of managerial functions of planning, organizing, directing, and controlling.

Aviation businesses must be concerned first with the flight crew and passengers as customers. The image created by the initial contact with the business is likely to be lasting and will be extrapolated to the entire operation. Since the flight line is such a critical part of the business, we will consider the key elements that contribute to the initial image: line layout, line operations, training of personnel, record keeping, and profitability.

Line Layout

To the transient pilot (as well as to the based customer) taxiing into an aviation operation, the line location and layout will contribute to that important first impression. Is the ramp visible and easy to identify? Is the flight line operational and easily accessible? Are signs legible? Do ramp markings stand out adequately? What is the general appearance of the line? Is it neat and orderly, or does it look like a graveyard for derelict aircraft? Does it appear modern and up-to-date, or does it feature faded advertisements on the side of aged buildings, misspelled signs, and disorder? Is grass growing in the cracks of the pavement? Are tiedown anchors and lines in good condition? Are parked aircraft spaced so closely that taxiing is difficult or dangerous?

The simple process of reviewing the operation "through the eyes of a visitor" may answer these questions. On the next taxi into the ramp, the layout should be reviewed to determine whether it is:

> Functional, with an obvious commitment to safety;
> Easy to understand and navigate by a transient or complete newcomer;
> Positive in image; and
> Practical to support.

Naturally, the size of the ramp layout varies depending upon the volume of business, the type and size of aircraft that normally access the ramp, and the services offered. Regardless of size, there is a great deal that can be done to develop and maintain a line that is efficient for the volume of business and that presents a neat, clean, attractive image. The physical layout of the flight line or ramp varies significantly with the number and size of aircraft involved. A small airport that serves primarily light training aircraft may have fixed fuel pumps, a hard surface upon which to taxi for service, and tiedown points for the aircraft. A larger facility may have a parking spot adjacent to the main terminal for discharging passengers and improved hard-surfaced tiedown areas for the aircraft. Service is usually provided to the aircraft in either location by fuel trucks. Five basic considerations govern the size and configuration of the ramp:

1. The size of the loading area required for each type of aircraft.
2. Aircraft parking configuration; nose in, angled nose in, nose out, angled nose out.
3. Mix of based and transient aircraft.
4. The number of flight operations conducted.
5. The mix of aircraft type and size (e.g., light piston twin, turboprop twin, light jet, heavy jet, etc.)

Within these ramp requirements, aircraft can be grouped adjacent to the terminal in four basic parking systems (systems 1 and 2 are used primarily for general aviation ramps, and systems 3 and 4 are more generally used for airline operations):

1. Frontal or linear system. Aircraft are parked in a line immediately adjacent to the terminal building.
2. Open-ramp or transporter system. Aircraft are parked in groups away from the terminal building.

3. Finger or pier system. Fingers or protrusions extend out from the terminal building into the ramp area, allowing additional aircraft to be parked using the frontal system.
4. Satellite system. Small buildings are located on the ramp and connected to the terminal by means of a tunnel. Aircraft are parked around each satellite building.

At larger airports and for airline operations, aircraft need to be serviced at their respective gate positions. Fueling may be accomplished by trucks, fuel pits, or hydrant systems. Other servicing may be accomplished through fixed installations; mobile equipment is also widely used. Lighting, ramp marking, blast protection, noise, passenger comfort, security, communications, and safety are concerns at all flight lines, and they become even greater concerns with larger aircraft, larger numbers of aircraft, and more passengers.

Line Operations

The operation of a flight line varies according to the size of the airport. The same basic functions of greeting arriving aircraft, marshaling, and servicing, however, will be accomplished regardless of location. As an example of the sequence of events in a line operation, consider what typically happens at a medium-sized airport.

A visitor's initial contact with a fixed-based operation (FBO) could be via the UNICOM, a direct radio link between the aircraft and the service organization. Designated UNICOM frequencies include 122.7, 122.725, 122.8, 122.975, 123.0, 123.05, and 123.075 at non-towered airports, and 122.95 at towered airports. Many locations (particularly larger FBOs) also have an ARINC (Aeronautical Radio, Inc.) frequency for this purpose. Through the UNICOM or ARINC frequency the incoming aircraft, while still airborne or on the taxiway, can request fueling assistance or parking directions, inquire about ground transportation facilities and service, and obtain other assistance. In general, early communications can greatly facilitate the service stop of an aircraft at a given location.

The practice of many corporate and air taxi flight crews of contacting the FBO by radio prior to arrival on the ramp can help facilitate an initial favorable impression on the visiting travelers and provide the FBO staff with the information needed to provide superior service.

The second point of contact with the line may be a vehicle marked "Follow Me" for the larger operations or the line service representative for smaller ramps. The "Follow-Me" vehicle enables ramp personnel to cover a larger area. The vehicle may have the capacity to provide for immediate transportation of personnel, luggage, or supplies to the hangar or terminal. After assisting in parking the aircraft, the line representative creates a favorable impression by greeting the customer (if the aircraft is owner-flown) or a member of the flight crew (for professionally-flown aircraft), inquiring about needs, and doing as much as possible to facilitate requests. Of course, the basic function of line service personnel is taking orders for fuel, oil, service, and other aircraft-related business needs. This is the primary reason for his or her presence. Delivering these services in a friendly, timely, and helpful manner as well as ensuring that the front desk is made aware of any other needs (such as provision of a rental car, or information about local restaurants, motels, etc.) is the ultimate measure of their success.

The layout, construction, and facilities of the flight line deserve the special attention of the manager. The following considerations should be reviewed periodically from an operations standpoint to ensure the needs of the organization and the customers are being met.

1. Is the line layout operational and safe from the incoming pilot's point of view?
2. Is the design and construction of the line itself adequate with regard to size, parking areas, safety in ingress and egress, pavement markings, and weight-bearing capacity of ground or hard surfaces?
3. Is the designated traffic flow (where applicable) evident to all users?
4. Are the facilities of the line adequate for the level of activity being serviced; are there an adequate number of available and proper-size chocks, and adequate tie-down facilities? Are hangar facilities available for transient aircraft, if requested?
5. Are there adequate service facilities to meet expected customer needs?
6. Is the operational safety of the flight line given special emphasis? Are safe parking areas for fuel trucks identified and utilized? Are adequate fire extinguishers available, and required grounding facilities present and in use during fueling operations?
7. Is the flight line secure, with access restricted to authorized persons, so that transient pilots can conduct their business elsewhere with confidence in the safety and security of their aircraft?

8. Do line personnel receive adequate ongoing training in customer service, security, and safety?

Transportation arrangements for passengers, luggage, and cargo will naturally vary, dependent upon the size of operation, the volume of each, and the ramp parking system. Passengers and flight crews cross the ramp by foot or by vehicle. Smaller airports tend to rely completely on foot passage from the airport to the terminal or hangar. Ramp safety and security for pedestrian traffic between the FBO lobby and the aircraft are very important operational considerations, and line personnel are key first-line observers and enablers of the FBO's ramp safety program. A red carpet at the aircraft is provided by many FBOs. This is especially important for outbound passengers because its function as a doormat helps keep the aircraft clean.

Baggage and cargo handling facilities will likewise vary from motorized ramps, trains, or carts to wheeled luggage racks that the pilot or passenger uses to push luggage to the parking lot. Some FBOs have golf cart-type luggage and personnel carriers or a van or bus to carry passengers and luggage from aircraft parked farther from the facility—particularly in inclement weather. Many passengers prefer to handle their own luggage.

The actual handling and servicing of aircraft by flight line personnel is important to the customer and to the aviation manager. These activities must be accomplished to the customer's satisfaction as well as efficiently and safely. Aircraft handling and servicing can be viewed as several distinct operations. Each requires an organizational structure, personnel training, and a high level of safety. The major flight line activities are:

> Directing the movement of aircraft (marshaling);
> Parking aircraft;
> Tying down and chocking aircraft;
> Towing aircraft;
> Taxiing aircraft;
> Fueling aircraft; and
> Servicing aircraft systems (e.g., crew and passenger oxygen, catering, coffee, water, and lavatory).

Written (preferably in a manual provided to each employee) operational procedures and safety practices are needed in order to prevent injuries to personnel and damage to aircraft, and to create and maintain an efficient and profitable activity that meets the customers' needs. Initial and periodic review of the organization's operational and safety procedures, practices, and expectations for each employee who serves in a safety-related position are important in maintaining a climate of safe and efficient operations. Federal Aviation Administration Advisory Circular number 00–34A, Aircraft Ground Handling and Servicing, contains useful data with generally accepted information and safety practices for many of the flight line activities listed above.[2]

Line Administration

The administration of the flight line is a critical element in the eyes of the customer and in the ultimate success of the operation. The term "administration" is used here to embrace both the internal administration of the flight line operation and the overall coordination of the flight line with the total organization. The overall coordination is frequently the responsibility of the front desk. This important function will be covered in detail in a later portion of this chapter.

As previously noted, the most essential element when an aircraft arrives on the FBO ramp is the impression created by the actions and attitudes of the line personnel. The visual and mental image created by these individuals can do much for the company. The desired image is that of a professionally–competent, friendly, courteous, efficient, and concerned individual who is genuinely interested in the welfare of the customer.

During the customer's stay, one way of enhancing efficiency and accomplishing successful servicing of the customer's aircraft is through the use of a simple order form. Figure 8.1 depicts a typical line service request form, generated by the Total FBO software system. This form serves as a reminder to the service personnel, provides specific written instructions, and may be used to obtain additional information about customers and their needs.

The order form becomes a method of ensuring that the desired service has been performed, and it provides a basic record of what was accomplished. The internal paperwork to record and charge the customer for all services rendered will vary depending upon the method of payment. Cash, charge, or credit card may be used to settle the bill.

One method of handling this transaction is through the use of a Line Invoice Form as illustrated in Figures 8.2 and 8.3. On these computer-generated forms is space for a custom or periodic greeting or message to customers.

The third phase of handling a flight line customer is the "send-off." The departure of a customer, although frequently not as recognized as it should be, is as important as the other phases of the visit and should be specifically recognized by the organization and by the line personnel.

Every effort should be made to ease and facilitate the departure. Rolling out the red carpet, assistance with baggage, last-minute check on the adequacy of the service, help with chocks and tie-downs, start up, taxi assistance (helping the flight crew ensure clearance from other parked aircraft and obstructions on the ramp), and a cheery wave will all influence the customer to return for future business.

Figure 8.1 → Sample Line Service Request Form

Horizon Business Concepts, Inc.
Client Request Report

10/27/2009 — Page: 1

TotalFBO Request ID: 10

Customer Service Request

Aircraft Number: N1234
Aircraft Type: KING AIR
Type of Flight:
Arriving: 2 Pax   Flt #:
Departing: 2 Pax   Flt #:

Arriving: 10/25/2009 02:00PM
Staying: Quick Turn
Arriving From:
Departing To:

Departing: 10/25/2009 06:45PM
☐ Customs Requested

Customer Name: Adams, Ronald R
Address: 5710 S. Union
         Tulsa, OK 74107
Phone: 1-918-446-9683   Fax: Email:

Aircraft Based At:
Home Airport:
Home FBO:
FBO Phone:

	Name	Cell Phone
Pilot		
Copilot		
Crewman 3		
Crewman 4		

Aircraft Services Requested

Fuel?   Top Off   Posted Price: 0.0000   Quoted Price: 0.0000
Fuel Type: JET-A   ☐ Include Prist?

Oil?   Check   Posted Price: 0.00
Oil Type: AS15W50

Parking Requested:
Quick-Turn

Other Facilities:
☐ Conference Room
☐ Courtesy Car

Ramp Services:
☐ Air Conditioning
☐ Air Start
☐ Aircraft Preheat
☑ Aircraft Tug
☐ Aircraft Deicing
☐ Baggage Handling

☑ Aircraft Security
☐ Power Cart
☐ Aircraft Stairs
☐ Aircraft Defrosting
☐ Freight Handling
☐ Passenger Handling

Parking Instructions:

Replenishments:
☑ Coffee Service
☑ Ice Service
☑ Newspaper
☐ Nitrogen
☐ Oxygen
☐ Water

Aircraft Cleaning:
☑ Lavatory Service
☐ Wash Aircraft
☐ Clean Interior
☐ International Trash
☐ Dry Cleaning

☑ Clean Windscreen
☐ Wax Aircraft
☐ Waste Disposal
☐ Dish Washing
☐ Linen Service

Maintenance:
☐ Aircraft
☐ Avionics
☐ Interior
☐ Prop

Scheduled By: George
Confirmed:
By:

Dispatcher:
Request Entered: / / 12:00 AM   Cancelled:
Last Updated: 10/27/2009 5:14 PM   By:

Courtesy of Total FBO Accounting and Business Management Software by Horizon Business Concepts, Inc.

Figure 8.2 → Sample Line Invoice Form

FBO PLUS: INVOICE: SALE

Remit to:	FBO Manager
	1304 Langham Creek Dr.
	Houston, TX
	USA
	Ph: 281-492-9500
	Fax: 281-492-9504

Bill to:	FBO Manager	Ship to:	FBO Manager

Invoice Date	PO Number	Terms	Tail Number	Type	Invoice Number
08/26/2009		Net 30	123SW	INVO AR	A1-00855

Item Number	Inv. Item	Description	Quantity	Unit Price	Unit Discount	Unit Tax/Fees	Total
1	Jet A	Fuel - Jet A Date: 08/26/2009 10:21:51 Equip: Jet 1 Meter: WAC1-Front Tail=123SW	111.000	2.100	.0000	.3360	$270.40
2	Jet A	Fuel - Jet A Date: 08/26/2009 10:21:51 Equip: Jet 1 Meter: WAC1-Rear Tail=123SW	1111.000	2.100	.0000	.3360	$2,706.40
3	Aeroshell 50W	Aeroshell 50W	2.000	3.000	.0000	.1800	$6.36
4	Catering	Catering	5.000	10.990	.0000	.6600	$58.25
5	Ramp Fee	Ramp Fees	1.000	35.000	.0000	.0000	$35.00

Lear Jet
Current Balance: $0.00
These are test notes that you can put on your invoice! Just choose [File]/[Configure Business]/[Configure Business] and click on the invoice notes tab.

You can use this space for payment terms information, liability disclaimers or to put a message of the month.

Good luck with your FBO Manager or Flight School Manager demo!

Discounts	$0.00
Group Disc	
Sub-Total	$2,662.15
FET	$256.62
FL TAX	$157.64
Total	$3,076.41
Cash	$3,076.41
Change Due	**$0.00**

Customer Signature:_____

08/26/2009 10:23

Courtesy of FBO Manager.

210 Essentials of Aviation Management

Figure 8.3 ✈ Sample Line Invoice Form

Horizon Business Concepts, Inc.
"Where Your Dreams Take Flight"
address
city, ST 99999
phone numbers, etc
and another line

Sales Person: George

Invoice: TU-01028   10/25/2009
                               5.14PM

Sold To: Ronald R Adams   Ship To: Ronald R Adams
           5710 S. Union
           Tulsa, OK 74107

Line#	Type	Item/Description	Cr?	Aircraft	Quantity	Units	List Price	Disc	Unit Price	Extended
1	Fuel	JET-A w Prist		N1234	50.0	Gallon	4.950	6	4.670	233.50
		JetA			Meter Start: 0.0		Stop: 25.0			
					Meter Start: 0.0		Stop: 25.0			
		Amount on this line includes 1.00 for Prist								
		Fuel Ticket #: 11234								
2	Oil	AS15W50		N1234	2.0	Quart	8.95		8.95	17.90
		Oil								
3	Svc	4310.00		N1234	1.0	Each	65.00		65.00	65.00
		Lavatory Service - Line								
		Ticket #: 11234								

Taxes Included in Subtotal:			
Fuel FET:	10.95	Subtotal	316.40
Local Flowage Fee:	3.50	Sales Taxes/Fees	18.88
State Sales Tax:	18.88	Total Due	335.28
		Paid-Amex ..2345	335.28

Thank you for your business, or whatever trailer message you'd like
at the end of each invoice.

Printed: 10/27/2009   Page: 1

Courtesy of Total FBO Accounting and Business Management Software by Horizon Business Concepts, Inc.

Training Line Personnel

The flight line is often the place where entry-level personnel and part-timers are employed, because young people tend to seek these jobs in order to be around aircraft, and are prepared to handle the menial nature of some of the work, such as washing and fueling aircraft. However, this is not unskilled work. Washing aircraft involves a gentle hand so that aircraft finish and structures are not damaged, static ports and pitot tubes are covered before washing and uncovered prior to return of the aircraft to service, and wash-water runoff is properly addressed. Fueling requires special operational and safety training. All of these activities require a customer-service orientation. Because it is generally the customer's first contact with the FBO, it is critically important that these junior-level staff members are trained to be skilled, professional service providers.

Several key areas of training are needed, including customer service, and safety and security. Additionally, line service personnel should be trained to deal with unexpected contingencies, and to understand when to obtain assistance from senior employees or management. A number of fuel companies and other organizations offer detailed training manuals dealing with safety issues.[3] An individual FBO should consider its level of training needs and select accordingly from such sources.

Customer Service Training. As previously emphasized, each employee within the organization must realize that he or she plays an important part in customer service. Whether it is the lineperson fueling aircraft, the receptionist, or the FBO owner, customer service must be first and foremost in everyone's mind.

One important concept of customer training is defining the customer. In addition to individuals purchasing products and services from the business, customers can also include vendors, visitors, and other employees (internal customers).

For example, if the chief maintenance inspector has procrastinated in completing the paperwork on an aircraft repair, thereby delaying the airplane's return to service, the flight instructor and student pilot who intended to use it for a flight lesson will be negatively impacted. Not only is the student pilot, usually considered a typical or "external" customer, affected, but the flight instructor who works for the same company has been affected as well. Thus, an employee that the external customer may not even see was responsible for the company delivering poor service.

Employees who are rude to each other, or who do not honor each other's deadlines and time schedules, negatively impact the entire operation.

Whether general aviation is in a down period, with few people actively engaged in flight activity, or the industry is booming and business is brisk, excellent customer service can be the one element ensuring customer satisfaction, which means repeat business.

Safety and Security Procedures and Training. In the post-9/11 world, aviation operations have a legal and moral responsibility to provide for the security of customers and employees, and to act as a first line of defense in the security of the National Airspace System (NAS). Security training should be mandated for all employees, and strict security procedures and guidelines should be provided in the front-desk procedures manual described later in this chapter. These security procedures should include continuous awareness of persons and activities in the FBO by all employees, clearly-defined requirements for ramp access, guidance for responding to various levels of threat, including when and how to contact appropriate law enforcement, and compliance with security requirements mandated by TSA, FAA, and the local airport authority. The Transportation Security Administration (TSA) document "Recommended Security Action Items for Fixed Base Operators" (available at https://www.dot.ny.gov/divisions/operating/opdm/passenger-rail/passenger-rail-repository/ga_fbo.pdf) contains excellent guidance for developing security policy and awareness.

As is true for a growing number of segments of the aviation industry, FBOs should formulate and maintain strict adherence to a Safety Management System (SMS). An SMS provides guidance to all employees regarding the safety procedures and policies that have been developed to help ensure safe operations through commitment by all personnel to embracing the organization's safety culture. FAA guidance for developing and implementing a strong safety management system can be found in Advisory Circular 120–92B, "Safety Management Systems for Aviation Service Providers" (available at https://www.faa.gov/documentLibrary/media/Advisory_Circular/AC_120-92B.pdf). The following excerpts from AC 120–92B outline fundamental concepts in the development and use of a strong safety management system.

Characteristics of an SMS. "An SMS is an organization-wide comprehensive and preventive approach to managing safety. An SMS includes a safety policy, formal methods for identifying hazards and mitigating risk, and promotion of a positive safety culture. An SMS also provides assurance of the overall safety performance of <the> organization. An SMS is intended to be designed and developed by <the organization's> own people and should be integrated into <its> existing operations and business decision-making processes. The SMS will assist <the> organization's leadership, management teams, and employees in making effective and informed safety decisions."

SMS Fundamentals. "SMSs can be a complex topic with many aspects to consider, but the defining characteristic of an SMS is that it is a decision-making system. An SMS does not have to be an extensive, expensive, or sophisticated array of techniques to do what it is supposed to do. Rather, an SMS is built by structuring <the organization's> safety management around four components: safety policy, safety risk management (SRM), safety assurance (SA), and safety promotion."

A strong climate of safety, and an obvious organizational and management commitment to making safety the foundation of all operations, is not only ethically responsible, but is also an excellent business decision. It is strongly recommended that the organization develop an SMS, with input from all employees, that the SMS have the clear and enthusiastic endorsement of senior management, and that all employees are trained in its content, application to daily operations, and their responsibility to comply with its tenets.

Service Array and Profitability

The ability of a flight line to generate profits for the organization varies, depending upon many factors, including size, volume of aircraft traffic, competition, management, operative personnel and the general economy. Regardless of these factors, the flight line should, as already mentioned, be operated as a profit center. Even on the smallest airport, the records should reflect the profit picture for the line operation. During past periods of economic recession, experience has shown that for the large majority of general aviation businesses around the country, the fueling operations of the flight line provide the "bread and butter" income that enables them to survive. Large metropolitan airports have found that it is difficult to operate a general aviation activity without a fuel concession. Where the airport authority or city retained the fueling rights, the airport was frequently plagued by a turnover of general aviation businesses.

Line operations and fuel service can be profitable if properly managed, even though self-service fueling has become more competitive in some areas, as discussed below. Customer courtesies, such as coffee, clean restrooms, a well-equipped flight-planning area, and reading materials help keep customers coming back for full service; thus, the flight line and front desk should always work together to ensure satisfied customers. Service quality seems likely to continue as a primary consideration. This means that effective administration, efficient facility layout, and friendly, professional employees can help assure consumer satisfaction.

Fueling

The fueling activities perceived by the aviation customer are only the tip of the iceberg when the whole process of supplying contamination-free fuel in a safe manner is considered. Owing to a number of changes in aircraft fueling requirements, this topic requires special attention. Numerous publications by the National Fire Protection Association, the American Petroleum Institute, and the Federal Aviation Administration provide guidelines on the safe handling and storage of fuel at airports.[4] At air carrier airports, Federal Aviation Administration requirements found in 14 CFR 139 require that FBO line employees undergo specific training. Supervisors must complete an approved fuel service course

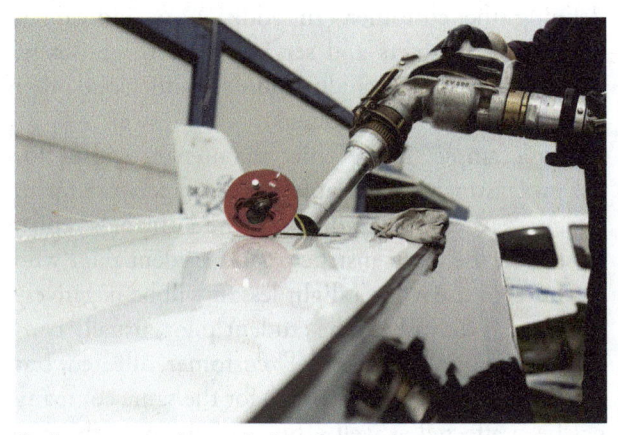

Fuel service can be profitable if properly managed.

and must then, in turn, train other line employees. Certain economic aspects are summarized here.

Trends in Use of Alternative Fuels

The FAA has gradually approved the use, through Supplemental Type Certificates (STCs), of cheaper autogas in increasing numbers of aircraft with reciprocating engines, to replace 80-octane avgas. Szymanski (2002) reviewed the performance and safety issues involved.[5] Some aircraft, such as ultralights, were designed from the outset to use auto gas or "mogas."

The trend toward autogas is good news for piston aircraft owners because of cost savings. But it has been bad news, both for FBOs and for government agencies collecting flowage fees or fuel taxes. The declining demand for 100/130 octane and 100LL fuels (caused by many factors) means that many airports have difficulty in obtaining fuel suitable for light aircraft, thus strengthening the market for alternatives such as auto gas.

If the number of auto gas users increases, there will be an increasing economic effect unless procedures change at airports. The volume of auto gas being brought by car or truck to airplanes and the increasing numbers of pilots transferring it by diverse means from car to aircraft represent a growing potential safety hazard at airports. Some FBO owners are now providing auto gas as a customer alternative. An additional concern is the growing practice of blending ethanol in gasoline designed for delivery to automotive vehicles, because ethanol can cause deterioration and corrosion in aircraft fuel lines, and it attracts water in fuel systems.

There is increasing interest in the certification and utilization of alternative aircraft fuels, including biofuels, to replace 100LL avgas, as well. Several companies are currently working with the FAA to certify such alternative fuels. It is anticipated that once these alternative fuels are certified, they will begin to replace 100/130 and 100LL as primary forms of avgas at FBOs. The precise dynamics of this replacement process, however, will depend on FAA and OEM approval of the fuels for use in aircraft engines, their availability, and pricing.[6]

Self-Fueling

The trend towards self-fueling is not represented only by ultra light operators bringing five gallons at a time to the airport in their automobiles. Increasing numbers of corporations with based aircraft are supplying their own fuel. FAA guidelines and individual public airport requirements in regard to airport minimum standards can be used to ensure that these operations are safely conducted. But the FAA requirements with respect to competition at federally-funded airports mean that as long as standards are met, corporate or other self-fueling must be permitted.

Many FBOs, when polled, identify self-fueling operations at their airports. Self-fueling is to some extent inevitable, yet constitutes losses of revenues and profits, so that FBOs should be vigilant to observe what trends are occurring in self-fueling and whether, with service changes, any of those customers could be attracted back to the FBO.

Some years ago, an NATA Task Force polled 112 corporate self-fuelers asking their reasons for the change, and found reasons that likely still hold true, although proportions may have changed:

> Better price—87%;
> More control, avoid misfueling, invoice errors, lack of detail—65%;
> Lack of timely service—49%;
> Better aircraft security—43%;
> No turbine fuel, had to install own—19%.

Only 42 of the 112 consulted with the FBOs at the field before deciding on self-fueling. Clearly, FBOs need to present a better customer service image to their corporate clients and find out more about their frustrations and plans.

The FBO seeking to persuade an airport-owner landlord to not permit self-fueling may be able to present the best case by demonstrating the company's high level of expertise, fairness in pricing, and quality facilities. Early discussions with each corporate-based aircraft owner—staying close to the customer with regard to needs—are very important. Moreover, a willingness to negotiate lower per-gallon charges for customers who will sign a bulk-purchase fuel contract may prevent those customers from undertaking a self-fueling operation.

Front Desk

The front desk is the nerve center of the FBO. As such, it represents the organization to the customer, to the public, and to the employees. It deals with the individual who comes into the FBO by aircraft as well as the person who drives or walks into the business.

As the hub for all business conducted by the company, it is the reception desk and the public relations center. It coordinates business transactions with all departments and strives to ensure that the customer receives the most efficient and courteous service possible. Its secondary goal is to ensure that all company procedures and guidelines are followed in achieving a successful front-desk operation.

As with many other aspects of the aviation business, the *functions* of a front desk must be provided for by every organization. The very small "mom-and-pop" operation may not have a formal reception desk, but specific individuals find that they serve the same function. The majority of middle-sized airports and most general aviation businesses of any size have a front desk. The very large airport will have a passenger terminal and then each fixed base operation will have its own front desk, usually located at the FBO's separate business location on the airport. Some FBOs operate without a staffed front desk at all, but this approach is likely to harm the business.

Procedures

As the hub of the business operation, the goal of the front desk's staff is to ensure compliance with all company procedures and guidelines. Developing and providing a front-desk procedures manual can achieve this very efficiently. An outline for such a manual is shown in Appendix VII. Effective telephone skills are an important element of front desk customer service. Being polite, friendly, and professional on the telephone requires a certain amount of practice, and the benefits can be tremendous. Knowing how and when to put people on hold, how to take a good message, and how to tactfully deal with an angry or upset customer, are all important skills to master. Companies that choose to use a voicemail menu instead of a live person on the reception telephone may also be harming their business. A 2001 article in Aircraft Maintenance Technology asks:

"How much business are you losing because people give up (on your voicemail menu system) and

The front desk is the nerve center of an FBO. Courtesy of Landmark Aviation.

call your competitor who uses a real, live receptionist that is geared to get the callers connected with the folks that can help them?"[7]

Staying Close to the Customer

An aviation service business should conduct periodic formal or informal customer surveys. These may relate to satisfaction and performance, to new products or services and how they are viewed, or to problems such as noise-sensitive areas. By conducting surveys regularly and by addressing the contribution of FBO operations to the local economy as well as asking about customer needs, the FBO can help assimilate information to assist in ensuring the economic wellbeing of the airport. This is discussed more fully in Chapter 12, Physical Facilities. The FBO might seek to identify key users of the airport as a first step in assessing its contribution to the local or regional economy. Whatever the purpose, based-aircraft owners and other local customers, such as regular charter customers, are usually easy to identify from tie-down lists and so forth. Transient operators, however, may not be so easily identified. For this reason, it is strongly recommended that a detailed transient log or guest book be kept, specifying:

> Company, if applicable;
> Phone number, address of pilot in command;
> Number of persons arriving and departing on the aircraft;
> Originating airport for the current flight;
> Destination airport for the current flight;
> Trip purpose at this airport; and
> Comments on service quality.

The logbook or guest book should be placed prominently at the front desk, and personnel should be trained to ask all transient pilots to complete it. If some pilots do not enter the building, then the line staff should have a similar log for them to fill out when their fuel orders are placed. The log can also provide the manager with day-to-day information on activity and potential problems, allowing a quick follow-up with appropriate staff or customers.

Related Services

Many FBOs obtain part of their revenue from activities other than direct aviation services. When this is the case, the front desk staff will likely be the information providers for, and schedulers and coordinators of, these services. Flight crews and dispatchers routinely contact the Customer Service Representative (CSR) at the front desk for assistance in making hotel and rental car reservations for planned flights with overnight stays. Fuel price negotiation is also a frequent request.

FBOs who offer only aviation services will still find themselves being asked for help in finding ground transportation and places to stay, to eat, and to be entertained. Many merchants give discount coupons and display materials for visitor use; coordination with the local Chamber of Commerce can supply these, and the front desk is an excellent place to offer them.

Flight Planning and Services

Pilot Services. As front desk services become more competitive, many FBOs have extended their range of accommodations for pilots and passengers. Traditionally, an area (or separate room) with tables and chairs for flight-planning purposes is provided. One or more telephones to allow pilots to call the Automated Flight Service Station network for preflight briefings and flight plan filing are generally provided, and it has become the norm to provide internet connectivity and computer equipment, to allow pilots to obtain current weather data and flight planning information, and to file their flight plans. The following section outlines the types of weather services normally utilized by pilots for preflight planning, and many of these are provided by FBOs as a service to their customers.

FAA Weather Services

As noted in the FAA *Aeronautical Information Manual* (AIM), Chapter 7–1–2, the FAA provides a number of weather services for pilots and other aviation professionals to obtain current and forecast weather information that may impact a flight in the National Airspace System (NAS). The following information is excerpted from the AIM, Chapter 7–1–2:[8]

1. "The FAA provides the Flight Service program, which serves the weather needs of pilots through its flight service stations (FSS) (both government and contract via 1-800-WX-BRIEF) and via the Internet, through Leidos Flight Service."
2. "The FAA maintains an extensive surface weather observing program. Airport observations

State-of-the-art flight planning room at Landmark Aviation. Courtesy of Landmark Aviation

(METAR and SPECI) in the U.S. are provided by automated observing systems. Various levels of human oversight of the METAR and SPECI reports and augmentation may be provided at select larger airports by either government or contract personnel qualified to report specified weather elements that cannot be detected by the automated observing system."

3. "Weather and aeronautical information are also available from numerous private industry sources on an individual or contract pay basis. Information on how to obtain this service should be available from local pilot organizations."

4. "Pilots can access Leidos Flight Services via the Internet. Pilots can receive preflight weather data and file domestic VFR and IFR flight plans. The following is the FAA contract vendor: Leidos Flight Service Internet Access: http://www.1800wxbrief.com For customer service: 1-800-WXBRIEF".

Weather Information Systems

The National Weather Service (NWS) provides a spectrum of aviation weather products produced by the Aviation Weather Center (AWC) in Kansas City. Access to this information can be obtained online at http://aviationweather.gov/.

In addition to the weather data (both reported and forecast) provided by the FAA and the National Weather Service, a number of private vendors have developed computerized and online systems for private clients, and increasing numbers of FBOs are offering these services.

Other Pilot and Passenger Services

Many FBOs offer additional facilities and services, such as exercise rooms, quiet and comfortable pilot lounges with TV and workspaces, sleeping rooms for flight crews, meeting and conference rooms, and clerical support services. As business/corporate/private aircraft travel increases, growth in specialized high-quality front-desk services will likely continue, although careful monitoring of these more luxurious facilities is needed to determine if they are really used and valued.

Summary

The flight line and front desk are the first points of contact for aviation customers arriving by air and by ground. The impression created by each is strongly influenced by the physical layout, the level of upkeep, and the knowledge and friendliness of the personnel. Three factors generally create a strong impression: (1) the initial reception, (2) the provision of service, and (3) the sendoff. These characteristics affect the ultimate profitability of the entire business, since the flight line and front desk not only serve customers directly but also refer prospective clients to other parts of the business, such as the maintenance repair facility.

In many aviation service businesses, the true hub or nerve center of the operation is the front desk. The individuals serving here coordinate many customer-related activities as well as other aspects of the business. The functions of the front desk should be based on a well-written and thorough procedures manual that ensures a consistent level of quality regardless of who is on duty, and that contains a rapid reference to key company policies on how to deal with various customer and other transactions.

DISCUSSION TOPICS

1. Distinguish between the flight line and the front desk. What do they have in common?
2. Name six functions of the flight line. Discuss the importance of each to the customer.
3. What should be the principal objectives of a flight line training program? Why?
4. What kinds of employees are typically assigned to the flight line and what problems does this create?
5. What factors will enable the flight line to generate profits for itself and other parts of the business?
6. What areas, issues, and functions can be aided by a procedures manual? By operations checklists?
7. What is meant by saying that the reception desk is the nerve center of the aviation business?
8. What is an SMS? What are the primary components of an SMS? What are the benefits to the organization that has developed and is properly using an SMS?

Endnotes

1. Morton, D. (2002). A philosophy of airport customer service. *Airport Magazine, 14*(2), 44–49.
2. Federal Aviation Administration (1974). Advisory circular 00-34A Aircraft ground handling and servicing. Washington, DC: FAA. Available online at: https://www.faa.gov/regulations_policies/advisory_circulars/index.cfm/go/document.list?omni=ACs&rows=10&startAt=0&q=00-34A&display=current&parentTopicId=
3. See, for example, the American Petroleum Institute http://www.api.org.
4. See API publications at *https://www.techstreet.com/api/pages/home*; also NFPA 407 "Standards for Aircraft Fuel Servicing," at *http://www.nfpa.org/codes-and-standards/document-information-pages?mode=code&code=407*; as well as FAA advisory circulars 150/5230-4B and 00-34A.
5. Szymanski, J. (March, 2002). Octane 101: Autogas vs. avgas. *Aircraft Maintenance Technology*.
6. See, for example, https://www.swiftfuelsavgas.com/
7. deDecker, B. (October, 2001). First impressions count! *Aircraft Maintenance Technology*.
8. Federal Aviation Administration (2020). *Aeronautical Information Manual*. Available at https://www.faa.gov/air_traffic/publications/media/aim_basic_w_chg_1_and_2_7-16-20.pdf.

9

Flight Operations

OBJECTIVES

> Understand the requirements for an air taxi operator to be approved for FAR Part 135 operations.

> Recognize the differences between air taxi and commuter operations.

> Discuss the problems and opportunities inherent in a flight instruction program.

> Describe several opportunities involving actual flight operations to ensure a profitable business.

> Explain the role flight training devices can play in a flight instruction program, including benefits to both the student and the flight school.

Introduction

The term "flight operations" refers to the provision of aircraft and/or personnel for flight services, and the monitoring and oversight of these activities to ensure profitability.

Types of Flights

The reader should refer to the "Taxonomy of General Aviation" shown in Chapter 1, Figure 1.4. The following types of flight operations may be conducted:

Air Transportation
> Charter;
> Air taxi;
> Aircraft rental;
> Aircraft leasing;
> Aircrew and ferry services;
> Air cargo; and
> Air ambulance.

Use of Aircraft for an Activity While Airborne
> Flight instruction;
> Aerial patrol—including powerline, pipeline, forest, fire, highway, border, wildlife management;
> Aerial advertising-banner towing;
> Aerial application, including crop dusting (pest control), seeding, fertilizing;
> Sightseeing services;
> Helicopter operations;
> Gliding and sailplaning;
> Ballooning;
> Parachuting;
> Ultralight operations;
> Aircraft demonstrations;
> Medical evacuation;

- Search and rescue;
- Fish spotting; and
- Aerial photography.

This is only a partial listing of the types of flight operations that may be offered. Each must be considered in terms of what types of client are involved—business and executive, other commercial clients, flight students, or sport and recreational fliers.

Market Trends

Flight operations constitute a key profit center for many FBOs. Numerous services can be offered, some highly compatible with one another and others perhaps less so.

In the late 1970s and 1980s, the U.S. general aviation industry saw a massive reduction in aircraft shipments, and a decline in active aircraft. As a result of the events of September 11, 2001 and other factors, the GA industry has been subject to regulations, procedures and restrictions that can make flight operations difficult. Moreover, the world-wide effects of the COVID-19 pandemic in 2020 created significant and in many cases unprecedented difficulties for the entire aviation industry. With respect to GA airplane shipments during the period from 1994 to 2019, data published by the General Aviation Manufacturers Association (GAMA) show a peak of 4277 units shipped in 2007, and a fairly level trend of annual deliveries in the low- to mid-2000s range.[1] As a result of this variability in aviation operations as a reaction to domestic and world-wide events, the FBO management must remain alert for new flight operation market opportunities. In recent years, FBOs around the country have become involved in such diverse areas as:

- Air ambulance operations;
- Airline aircraft servicing;
- Aerobatics instruction;
- Aircraft management;
- Aerial firefighting;
- Air taxi service;

and many others.

System Issues Affecting Flight Operations

Seven main issues affect general aviation flight operations:

1. Loss of access to certain airports and airspace, due to the terrorism of September 11, 2001.
2. Loss of airports and restricted access to airports before September 11, 2001.
3. Loss of access to air carrier airports in key locations.
4. Changes in airspace usage.
5. Lack of understanding of the economic benefits of general aviation, both as a form of transportation and as a means of performing various functions while airborne.
6. Increasing levels of complaints, from airports throughout the country, directed toward aircraft and airport-area noise levels.
7. A volatile customer base, meaning that business planning, and planning for time and money to spend on the above issues, are difficult to budget.

Many of these issues are discussed in greater depth in Chapter 12, Physical Facilities.

Terrorism and its After-effects. General aviation flight operations were deeply affected by the terrorist attacks of September 11, 2001. Flight schools were shut down. A considerable amount of the national airspace was closed to general aviation flights. For example, a lead article on the NBAA website dated April 25, 2002, complained that "America's businesses still do not have access to DCA"—Reagan National Airport (DCA) handled 60,000 GA operations per year prior to September 11, 2001, of which 90 percent were business aviation.[2] In October of 2005, over four years after prohibiting general aviation aircraft from accessing Reagan National Airport (DCA), the Transportation Security Administration began allowing very limited GA operations at DCA through the DCA Access Standard Security Program (DASSP). This program placed severe restrictions on GA aircraft traveling to/from DCA, among them inspections and passenger screening by the TSA, the requirement to use only DASSP gateway airports and FBOs for departures on the flight leg to DCA, and the requirement to carry an armed security officer on each flight. Operators who participate in the DCA Access Program must also be vetted and be granted a waiver to conduct flight operations at DCA.

Access to Major Airports. The nation's busiest airports and airspace became more congested in the 1990s, and the most restrictive of the impacts of the September 11th events were primarily short-term in nature, with long-term traffic growth projections through 2025 to be roughly what they were prior to these events. Airport/airspace congestion is related to airline traffic growth, fewer air traffic controllers, airline schedules focused on peak periods, and hub-and-spoke airline service development that causes more landings and takeoffs per passenger trip and congestion during peak periods. The terrorist attacks of September 11th reduced airline enplanements and put a temporary hold on aviation growth; however, as the public became more assured about flying again, the upward growth trend, and congestion at major airports, continued, until the unprecedented reduction in air travel as a response to the COVID-19 pandemic in 2020. As a result of growth pressures, various pricing and management strategies have evolved to reduce the access of general aviation aircraft to certain airports and airspace areas in order to reserve space for airline aircraft. Moreover, the continuing use by FAA of Temporary Flight Restrictions (TFRs) has become a daily "fact of life" for general aviation pilots, particularly those who need to operate to/from major airports near large metropolitan areas.

Airport/airspace access is of particular concern for business aviation operations, for which passengers often seek the same major city destinations as are utilized by the airlines. In general, rural and recreational general aviation operations are less severely impacted. The general aviation industry, business aviation operators, and flight operators such as FBOs need to be cognizant of this trend and vocal through the various aviation advocate groups, including AOPA, NBAA, and EAA, to ensure that general aviation continues to have reasonable access to the aviation system. While all airports that receive public funds are intended to be open to all users, the reality of the situation can be more restrictive, particularly at the nation's most crowded airports. The airport "slot" system, whereby airlines compete for available landing slots, leaves little room for non-scheduled flights. NBAA described how one such problem unfolded:

> "The most important battle for airport access, however, would be over the right to operate at Boston's Logan International Airport. In March 1988, the Massachusetts Port Authority (Massport) unveiled its Program for Airport Capacity Efficiency (PACE), which used high fees to discourage use of the airport by smaller aircraft. Within a month, NBAA, and other interested parties filed suit in federal court to block the PACE plan but were unsuccessful. However, by the end of the year, the U.S. Department of Transportation (DOT) ruled that PACE was inconsistent with national transportation policy, and in August 1989, an appeals court found that PACE was contrary to federal law. Fees that had been collected at Logan while the PACE plan was in effect during 1988 were refunded."[3]

Changes in Airspace Usage. A number of other issues affect general aviation access to airports. For example, increased globalization and international GA flying require integration with airspace procedures in other parts of the world. For example, the Domestic Reduced Vertical Separation Minima program was implemented by the FAA in January of 2005 in U.S. airspace between flight levels 290 and 410, inclusive. In general, changes such as these are made to align air traffic procedures with improvements made available through the use of advanced technology, as well as to bring those procedures in compliance with ICAO Standards and Recommended Practices (SARPs) and Procedures for Air Navigation (PANS).

In addition, as a result of the September 11th events, the TSA imposed security rules through the Twelve Five Standard Security Program (TFSSP), including fingerprint criminal history checks for flight crew members operating charter flights in aircraft over 12,500 pounds. More recently, the TSA has sought to amend the TFSSP to cause it to apply to all general aviation aircraft over 12,500 pounds, not simply charter aircraft. Industry organizations successfully raised serious concerns about the proposed Large Aircraft Security Program (LASP). As noted on the NBAA website, "The Transportation Security Administration (TSA) has officially withdrawn a proposed Large Aircraft Security Program (LASP). Introduced in October 2008, LASP would have imposed new, onerous and largely unworkable security regulations on general aviation. The proposal included provisions for aircraft operators requiring criminal history record checks for flight crews, checking of passenger names against No-Fly and Selectee Lists, compliance with the Prohibited Items List for scheduled airlines, as well as biennial auditing of an operator's security program. Certain airports serving these operators would have been required to adopt a security program. NBAA led efforts to rewrite the proposal to reflect the unique nature of general aviation".[4]

Economic Benefits of Business Aviation. In 2001, NBAA and GAMA commissioned a study by the accounting firm Arthur Andersen to examine the likely effects of owning a corporate aircraft on the profits of the company involved. They found significant benefits.[5] More recently, NBAA noted in the *NBAA Business Aviation Fact Book* (page 12) "Nationwide, business aviation employs 1.2 million people, and contributes $150 billion to U.S. economic output".[6] These benefits of Business Aviation will continue to increase as the difficulties of airline transportation continue, with fewer flights, reduced capacity, long lines, tiresome security inspections at airline airports, and heightened concerns for safety and security. Growth in fractional aircraft ownership will also be enhanced by these factors.

Airport Noise. No FBO flight department can succeed over the long haul without a good understanding of airport noise. Three major components of noise all have the same effect on airport neighbors. These three include:

> Aircraft engine noise;
> Noisy flight operations techniques; and
> Proximity of arrival and departure flight paths to noise-sensitive land areas.

The issue of airport noise is by no means new, and it has not been solved. Noise control has thus far come from three techniques—quieter aircraft engines (addressed through FAR Part 36), different flight techniques, primarily on arrivals and departures, and reprioritization of uses of noise-sensitive land. Aircraft engines are now much quieter than in the past and this topic is not addressed in this book. Some state aviation agencies (e.g., Oregon, Massachusetts) had active noise abatement assistance programs for smaller airports as early as the 1970s. The FAA, through its operations ban of noisier Stage 2 aircraft with gross weights over 75,000 pounds, has helped to reduce the major source of airport noise. The primary conflict is institutional, in the sense that while the FAA controls routing and altitude assignment for airborne aircraft—and holds firmly to the need for an interstate airport system that does not have a patchwork quilt of operational restrictions such as night curfews—it is local jurisdictions that control land use around airports.

Most metropolitan airports fifty years ago were well outside the urban boundary, and homes and other noise-sensitive uses have simply encroached on airports as the built-up area has grown. At the same time, annual operations have grown from a few score to, in many cases, hundreds of thousands of flights, and people who bought homes when the airport had negligible activity levels feel justifiably frustrated that their environment has changed so much.

Local jurisdictions may put the needs of housing expansion and economic development higher than airport protection in their land use priorities. Pilots and airport business owners may not live in or even near the immediate jurisdiction of the airport, and airports are often perceived as contributing little to the local economy; thus, conflicts in land use priorities continue to develop and to even be encouraged.

FBOs can no longer afford to stand aside and view this as the problem of the airport manager or other operators. The FBO owner/manager must dedicate time to network with the local chamber of commerce, with elected officials, and with community members, to convey more fully what their airport does, and for whom, and to identify and fix any problems caused by his/her business. These are necessary activities for airport survival. This topic is addressed more fully in Chapter 12, Physical Facilities.

In terms of flight operations, the FBO, both in a flight school operation and in operating the flight department, has a tremendous opportunity for applying good noise abatement flying techniques. As is true for justice, these not only need to be done, but also to be seen to be done. Noise abatement techniques are important for the individual business at a given airport, even if no procedures have been adopted by the airport as a whole. If all students, pilots, tenants, and customers are encouraged by seeing signs, receiving verbal instructions, and being provided with handouts about the need for quiet flight, then when there are noise complaints to which the FBO owner/manager must respond, he or she will be able to describe the detailed efforts the business is making to keep noise to a minimum over sensitive areas. Such efforts might include following written posted procedures for flap settings, thrust settings during takeoff and initial climb, areas to avoid overflying, and so on. These procedures can be tailored to the individual airport in consultation with the airport manager, tenant groups, locally-based pilots, and the air traffic control tower, if the latter exists, and should draw on published materials made available to members by AOPA, NATA, NBAA, and other industry groups. These groups offer fly-friendly techniques involving steeper departure paths, quieter flap settings, use of reduced thrust (safely, of course) to depart the area and while operating below a specified altitude before returning to maximum continuous power for the remainder of the climb segment, and so on. Many airports have adopted such procedures and can share them with airports that have not yet developed their own.[7]

In addition to noise abatement techniques, it is important to ensure that airport management does what is needed in terms of signage to communicate these techniques. Operational controls can include such steps as:

1. Use of a preferential runway system—a runway for which the flight path lies over the least noise-sensitive areas, used in as many surface-wind conditions as possible.
2. Displaced landing thresholds, so that arriving aircraft are at higher altitude over noise-sensitive areas and land farther down the runway, from the approach threshold.
3. Use of longer runways, so that departing aircraft can start their take-off roll further back and gain more altitude before flying over sensitive areas.
4. Turning away early on departure, to avoid over-flying noise sensitive areas; adoption of a unique flight pattern for that airport and each runway, to avoid sensitive areas.
5. Reducing power after a prescribed safe altitude or airspeed has been reached on departure.
6. Limitations on particular types of activity during certain hours of the week, particularly at night.
7. Limitations on take-off or landing noise levels above a certain decibel level (set with guidance from FAA and applied consistently).
8. Reversal of the landing pattern to avoid sensitive areas.

While such operating restrictions may be frustrating to pilots, in some places they may be essential to the continued availability of that airport, and cooperation must be required by the aviation businesses in relation to aircraft they rent and lease out. This can be done by such actions as:

1. Teaching noise-abatement techniques in ground school and during flight instruction.
2. Requiring renters to sign a commitment to follow the FBO's operating procedures.
3. Maintaining a clear and responsive process for receiving and dealing with complaints, even if the problem was not caused by the FBO's aircraft.
4. Maintaining close coordination with the airport owner/landlord on what is happening on the rest of the field, in terms of noise abatement and complaints.

NBAA is one industry organization that has been addressing airport noise since the 1960s. Figure 9.1 shows their recommended approach and departure techniques to minimize noise. More details are available in their full noise abatement document.[8]

AOPA also has noise abatement techniques available, as shown in Figure 9.2.[9]

Figure 9.1 → NBAA Noise Abatement Techniques

Noise Abatement Best Practices For Flight Crews
Pilots should always be mindful of noise impacts at airports. Even the "quietest" modern aircraft may disturb those that live near the airport. Care should be taken to minimize the aircraft's noise profile whenever possible by utilizing noise abatement best practices at all airports, especially during night-time and early-morning hours when aircraft operations may be especially disturbing.

> During the flight-planning process, flight crews should familiarize themselves with the airport's noise abatement policies and any applicable noise abatement procedures (NAPs) for the airport they will be using. These may include:
>   > Preferential runway use
>   > Preferential approach and departure paths
>   > Preferred terminal arrival and departure procedures for noise abatement
>   > Other noise-related policies (maximum noise limits, curfews, usage of reverse thrust, engine run-up policies, etc.)
> Contact the airport's Noise Management or Operations department for more information on local noise policies and procedures.

continued.

Figure 9.1 → Continued

> When available, pilots should utilize their company's recommended departure/arrival NAPs or those recommended by the aircraft manufacturer for their specific aircraft.
> Flight safety and ATC instructions and procedures always have priority over any NAP. NAPs should be executed in the safest manner possible and within all FAA-mandated operating requirements.
> Proper pre-departure and pre-arrival crew briefings are essential to ensuring the safe and effective execution of NAPs.
> When airport or aircraft-specific procedures are unavailable, NBAA provides recommended noise abatement procedures suitable for any aircraft type and airport operating environment (see below).

NBAA-Recommended Noise Abatement Departure Procedure with High-Density Airport Option

1. Climb at maximum practical rate not to exceed V2+20 KIAS (maximum pitch, attitude 20 degrees) to 1,000 feet AAE (800 ft. AAE at high-density-traffic airports) in takeoff configuration at takeoff thrust.
2. Between 800 and 1,000 feet AAE, begin acceleration to final segment speed (VFS or VFTO) and retract flaps. Reduce to a quiet climb power setting while maintaining a rate of climb necessary to comply with IFR departure procedure, otherwise a maximum of 1,000 FPM at an airspeed not to exceed 190 KIAS, until reaching 3,000 feet AAE or 1,500 feet AAE at high-density-traffic airports. If ATC requires level off prior to reaching NADP termination height, power must be reduced so as not to exceed 190 KIAS.
3. Above 3,000 feet AAE (1,500 feet at high-density airports) resume normal climb schedule with gradual application of climb power.
4. Ensure compliance with applicable IFR climb and airspeed requirements at all times.

NBAA-Recommended Approach and Landing Procedure (VFR AND IFR)

1. Inbound flight path should not require more than a 25-degree bank angle to follow noise abatement track.
2. Observe all airspeed limitations and ATC instructions.
3. Initial inbound altitude for noise abatement areas will be a descending path from 2,500 feet AGL or higher. Maintain minimum maneuvering airspeed with gear retracted and minimum approach flap setting.
4. During IMC, extend landing gear at the final approach fix (FAF), or during VMC no more than 4 miles from runway threshold.
5. Final landing flap configuration should be delayed at the pilot's discretion; however, the pilot must achieve a stabilized approach not lower than 500 feet during VMC or 1,000 feet during IMC. The aircraft should be in full landing configuration and at final approach speed by 500 feet AGL to ensure a stable approach.
6. During landing, use minimum reverse thrust consistent with safety for runway conditions and available length.

Source: From NBAA Noise Abatement Program by NBAA. Copyright © 2015 by National Business Aviation Association. Reprinted by permission.

Figure 9.2 → AOPA Guidelines for Noise Abatement

Airport Noise: We can make a difference

Through a concerted effort, and by demonstrating your sensitivity to the concerns expressed by the community as it relates to airport noise, your relationship with those affected by airport noise can be significantly improved. But we must be willing to VOLUNTARILY take the steps necessary to be thoughtful to our fellow community members. Should voluntary efforts not be considered important to the airport, you may find your airport facing local legislation to fix the problem, and this solution isn't always in the best interest of the airport or its users.

Here are some ideas that might be applied voluntarily to improve the noise impact at your local airport:

Pilots—

> Be aware of noise-sensitive areas, particularly residential areas near airports you use, and avoid low flight over these areas.
> Fly traffic patterns tight and high, keeping your airplane in as close to the field as possible.
> In constant-speed-propeller aircraft, do not use high rpm settings in the pattern. Prop noise from high-performance singles and twins increases drastically at high rpm settings.

continued.

Figure 9.2 ✈ Continued

- On takeoff, reduce to climb power as soon as safe and practical.
- Climb after liftoff at best-angle-of-climb speed until crossing the airport boundary, then climb at best rate.
- Depart from the start of the runway, rather than intersections, for the highest possible altitude when leaving the airport vicinity.
- Climb out straight ahead to 1,000 feet or so (unless that path crosses a noise-sensitive area). Turns rob an aircraft of climb ability.
- Avoid prolonged run-ups, and perform them well inside the airport area, if possible, rather than at its perimeter.
- Try low-power approaches, and always avoid the low, dragged-in approach.
- If you want to practice night landings, stay away from residential airports. Do your practice at major fields where a smaller airplane's sound is less obtrusive.

Instructors—

- Teach noise abatement procedures to all students, including pilots you take up for a biennial flight review. Treat noise abatement as you would any other element of instruction.
- Know noise-sensitive areas and point them out as you come and go with students.
- Assure that your students fly at or above the recommended pattern altitude.
- Practice maneuvers over unpopulated areas and vary your practice areas so that the same locale is not constantly subjected to aircraft operations.
- During practice of ground-reference maneuvers, be particularly aware of houses or businesses in your flight path.
- Stress that high rpm prop settings are reserved for takeoff and for short final but not for the traffic pattern. Pushing the prop to high rpm results in significantly higher levels of noise.
- If your field is noise-sensitive, endorse your students' logbooks for landing at a more remote field, if available within a 25-nm range, to reduce touch-and-go activity at the home airport.

Fixed-Base Operators—

- Identify noise-sensitive areas near your airport, and work with your instructors and customers to create voluntary noise abatement procedures.
- Post any noise abatement procedures in a prominently visible area and remind pilots who rent your aircraft or fly from your airport of the importance of adhering to them.
- Include copies of noise abatement procedures with monthly hangar and tiedown invoices. Make copies available on counter space for transient pilots.
- Verify regularly that your instructors are teaching safe noise abatement techniques.
- Call for use of the least noise-sensitive runway whenever wind conditions permit.
- Try to minimize night touch-and-go training at your airport if it is in a residential area. Encourage the use of nonresidential airports for this type of training operation.
- Initiate pilot education programs to teach and explain the rationale for noise abatement procedures and positive community relations.

And For The Surrounding Community—

- Send a copy of the noise abatement pattern established for your airport, along with a brief explanation of its purpose, to the local newspaper. Let the public know PILOTS ARE CONCERNED.
- See that the pattern, approach, and departure paths are designated on official ZONING AND PLANNING MAPS so that real-estate activity is conducted in full awareness of such areas.
- Lobby for land use zoning and building codes in these areas that are compatible with airport activity and will protect neighboring residents.
- Stress, publicize, and communicate the value of the airport to the community and how its operation adds to the safety, economy, and overall worth of the area.
- Sponsor "airport days" at the airport to involve nonfliers with the business and fun of aviation and possibly attract potential new pilots. Ensure these events themselves do not generate undue noise or safety hazards.
- Encourage beautification projects at the airport. Trees and bushes around the run-up and departure areas have proven effective in absorbing ground noise from airplanes.

Source: Aircraft Owners & Pilots Association

Volatile Customer Base. The number of pilots has fluctuated over the past few decades, as has the number of aircraft. Without steady growth trends, most industries have problems planning for the future and identifying the strong markets to focus on and weak markets to avoid. Aviation is very much this way, and although there are promising trends, they may prove to be just more fluctuations. The FBO may be best able to address this problem by having reasonably diverse services, by staying very close to the underlying causes of apparent trends, and by acting quickly when analysis shows that a new market can, in fact, be relied upon to make a profit for the company. Some flight operations activities may be tried and dropped. The termination of a recently-offered service should be internally documented and analyzed to determine why the service did not succeed.

Choosing What Services to Offer

Certain types of flight operations will be in heavy demand at particular airports. For example, an island or resort community may have a strong demand for air taxi and sightseeing flights. A metropolitan airport may have a high level of business charters and rentals, whereas a rural airport may be extensively involved in aerial patrol and agricultural application. Because of the varied types of aircraft required for many of the more specialized functions, the FBO will have to be selective about what services to offer. Where several FBOs are located on the same airport, they may each specialize in specific areas and avoid direct competition. Moreover, on many airports small specialized operators engage in just one type of activity, such as a parachute center, glider service, aerial photography, agricultural application, and so on. Whether these single-service operators can be defined as FBOs is dependent on their location. While FAA guidelines do not specify a minimum number of services that must be offered by an operator in order to be considered an FBO, some individual airport standards do.[10] The choice of what services to offer depends on the clients' needs, the services offered elsewhere on the field or in the region, the owners' preferences, and the expected profit from each area. Chapter 3, Marketing, discussed the use of forecasts for various types of activities, and should be referred to here as the Flight Operations business plan is developed.

Organization

Organizing the aviation business for safe, profitable flight operations is one of the most important aspects of the entire company. Goals and objectives must be identified, structure formulated, responsibilities assigned, duties outlined and understood, rules, regulations and procedures established, and controls provided in order to ensure that flight operations goals are achieved safely and profitably.

We next review the nature and requirements of the various types of flight operations listed previously.

Air Transportation

Benefits

The use of private business aircraft as a means of transportation is increasing faster than many other segments of general aviation. Since the Airline Deregulation Act of 1978, scheduled passenger service, while increasing in quantity, has diminished in terms of available non-stop flights. One reason for this is the adoption by major and national airlines of a hub-and-spoke route structure, and another is a desire on the part of many companies, greatly increased since the events of September 11, 2001, to provide traveling executives with the reduced trip times, greater personal security, more privacy, and enhanced productivity of the work environment that a corporate aircraft can facilitate.

While some airlines, such as Delta, have been using the hub-and-spoke concept for over 60 years, for many others, deregulation triggered the growth of this concept. At the time of deregulation, the major carriers had less than 50 percent of their total domestic capacity devoted to hub flying. By the late 1980s these same carriers allocated over 80 percent of their domestic seat miles to hub-and-spoke flying, and this pattern has continued into the 21st century. These facts notwithstanding, the unprecedented impact on the airlines of the COVID-19 virus and the worldwide pandemic it spawned, may lead to changes in the hub-and-spoke model and a return to point-to-point routes for some air carrier markets.[11]

"Traditional" reasons, from the airline point of view, for hub development include:

1. Scheduling efficiency—fewer aircraft to serve the same points.
2. New market synergy—one new spoke added to a hub can add many new markets.

3. Market control—connections can be more easily made on the same carrier through the same hub point.
4. Market fragmentation—nonstop service becoming viable in a particular market previously served only by flights connecting through a hub.

In addition to hub-and-spoke implications, the initial reduction in airline fares after deregulation has generally been replaced by substantial fare increases. The cost is generally greater when the ticket is booked without much advanced notice, as is the case for many business trips.

These factors: hub development, higher airfares, reduced capacity, and security concerns, combine to make private air transportation comparatively more appealing to the busy executive.

Many feature articles that address the benefits of business flying have been written in the past few years. These benefits include:

1. Increasing the executive's effective time by as much as 100 percent.
2. Elimination of tiresome and slow airline processing since September 11, 2001.
3. Permitting enroute business discussions to be held in privacy.
4. Reducing overnight hotel and meal costs by eliminating awkward or unavailable (at the times needed) airline connections and circuitous routing.
5. Ability to access over 10,000 small communities, compared with between four and five hundred by airline, and far fewer with non-stop flights.
6. Ability to increase contact among managers through ease of travel among branch offices.
7. Ability to run an enterprise with fewer corporate staff because of mobility to all divisions.
8. No lost baggage or missed flights.
9. Enhanced security, particularly for flights involving international travel.

For many corporations, the gradual increase in use of private business aircraft follows this progression:

1. An airline trip is made involving multiple connections, take-off delays, and overnight stops to accomplish business.
2. An urgently needed business trip is required and the business traveler cannot get there in time and back, using airlines.
3. A charter flight is booked through the local FBO.

For several years, a leading aviation magazine reported studies of aviation use by the nation's top companies—the Fortune 1000. Year after year, the companies that used business aircraft, which now number about half the group, had about 80 to 90 percent of the gross sales, the profit, and the return on investment and about twice the labor productivity. The non-aircraft operating half of the group had only 10 to 15 percent of the performance. More recently, there appears to be a causal connection, as identified by the 2001 GAMA/NBAA study performed by Arthur Andersen. To the aviation business, capturing the local demand for business air travel can be a rewarding and profitable activity.

Charter and Air Taxi

While often used interchangeably, it is perhaps desirable to make a distinction between the terms "charter" and "air taxi." Charter operations are non-scheduled flights that carry passengers or cargo when the party receives the exclusive use of the aircraft. Certificated air carriers may negotiate charter agreements, as they often do for tourist flights, as may general aviation businesses.

Air taxi operations, on the other hand, describe either the semi-scheduled or nonscheduled commercial flights of general aviation businesses. Individuals purchase air transportation as they would a ride in a taxicab. Light-to-medium weight aircraft are used to carry passengers and cargo to and from small communities that do not have enough traffic for regular scheduled airline service and where such service cannot meet the needs of the customer.

Aircraft. The aircraft utilized for charter work vary from a large jet transporting a group on an intercontinental trip to a light, single-engine aircraft carrying two business professionals from one community to another. Air taxi flying is normally accomplished with light-to-medium general aviation aircraft, normally twin-engine types.

Rules and regulations. The Federal Aviation Administration specifies basic operating procedures for charter and air taxi flights under Part 135 of the Federal Aviation Regulations. However, the rules for Part 135 operations have become much closer to the Part 121 requirements that govern scheduled air carrier operations, and this trend is expected to continue.

The Beechcraft King Air is frequently used for business aviation and air taxi flights.

In charter operations, the manager is selling air transportation to meet a customer's specific need. The aircraft and pilot(s) are provided to the customer according to agreement for her/his exclusive use. For example, Allied Petro-Chemical may call to charter a helicopter to transport and return two of its sales executives on a vendor call. In complying with FAR Part 135 and other pertinent regulations, the manager must ensure the following are complied with:

1. A valid FAA air taxi/commercial operator (ATCO) operating certificate.
2. A current manual for the use and guidance of flight, ground operations, and maintenance personnel in conducting operations.
3. Procedures for locating each flight for which an FAA flight plan is not filed.
4. Exclusive use of at least one aircraft that meets the requirements of the operations specifications.
5. Qualified aircrew personnel with appropriate and current certificates and the necessary recent experience.
6. Maintenance of required records and submission of mechanical reliability reports.
7. Compliance with the operating rules prescribed in FAR Parts 91 and 135, including:
   > Airworthiness check;
   > Area limitations on operations;
   > Available operating information for pilots;
   > Passenger briefing;
   > Oxygen requirements;
   > Icing limitations;
   > Night operations;
   > Fuel requirements, VFR and IFR;
   > VFR operations; and
   > IFR operations.

These requirements reflect only a summary of the pertinent sections of FAR Part 135. The reader is referred to the Federal Aviation Regulations (FARs) for current and complete details of the requirements.[12]

Aircraft Rental

In this kind of flight operation, the manager is engaged in renting aircraft to customers who provide the pilot for the flight. All types of airplanes may be rented, although most businesses utilize light-and medium-weight general aviation aircraft. In providing and controlling rental activity, the manager is primarily concerned with:

1. Establishing rental rates that will cover all costs and provide a planned profit.
2. Having adequate hull and liability insurance.
3. Providing administrative procedures for handling rental activity.
4. Obtaining positive renter-pilot identification.
5. Securing evidence of renter-pilot qualification and competence:
   > Past experience and proficiency;

> Valid pilot certificate;
> Current medical certificate; and
> Ground review and flight checkout in the aircraft to be rented, by a certified flight instructor (CFI).

6. Applying accounting procedures to control income and expenses.

Fractional Aircraft

The most common way today for a company to have partial use of an aircraft is the fractional ownership arrangement. A number of companies, starting with NetJets, initiated this concept in the latter part of the 20th century. While a fractional ownership arrangement is somewhat like a vacation condo timeshare, with each of, say, eight owners having the right to a certain number of flight hours per year,[13] it is important to note that the term "timeshare" has a different and specific meaning with respect to the Federal Aviation Regulations. According to FAR 91.501(c)(1), a timeshare is a wet lease, with the lessee leasing both the aircraft and flight crew. A fractional ownership arrangement, on the other hand, incorporates three principles: shared ownership of the aircraft, exchange of dry leases among owners, and the use of a company to manage the process. Fractional aircraft owners decide when and where to fly, but essentially transfer aircraft management issues to a third party, allowing those owners to experience the convenience and time advantages of full aircraft ownership at a substantially reduced cost. A one-sixteenth interest in an aircraft, for example, might entitle the fractional owner to 50 occupied hours per year, whereas larger shares would provide for greater usage of the aircraft.

The economics of fractional ownership are dependent on the amount the owner uses the aircraft. It is possible for businesses to fly their fractional aircraft for a much lower cost than owning a corporate aircraft, but that is true only within a certain range of hours of aircraft utilization per year. In general, outright ownership is most economical when the aircraft is operated in excess of approximately 200 hours per year, while fractional ownership is appropriate for utilization from perhaps 50 to 200 hours per year. Customers requiring fewer than 50 flight hours per year will most likely find it more economical to charter. Those ranges are dependent on type of aircraft and operation, as well as personal preferences, and the result is that they will rarely be clearly delineated. Because of this, the typical FBO could have its charter market negatively impacted by fractional aircraft providers, and aviation service businesses need to explore how they can benefit from the fractional ownership trend. For example, if they can meet fractional company standards, they may be able to contract to provide "supplemental lift" when the fractional company is fully booked and one of their owners needs a flight. While the competitive threat of fractional companies is real because this segment of aviation has grown so quickly, the FBO should keep in mind that fractionally-owned aircraft number only a relatively small percentage of the total number of turbojet and turboprop aircraft in the nation. Given that fractional companies maintain rigid and rigorous standards, they will impose such standards on any firms they contract with, which may ultimately be to the benefit of the charter industry.

Aircrew and Ferry Services

Providing aircrew services to the customer may take various forms. For example, these services can range from an occasional repositioning flight of a customer's aircraft, to the corporate customer who owns a large twin (but does not employ pilots) and contracts for full crew (pilot, co-pilot, and flight attendant) to operate the aircraft on his or her schedule. In providing this kind of service, the aviation manager should regard it as a specific business opportunity to be approached on a sound economic basis.

Operational and safety guidelines and procedures should be developed, as well as administrative methods and controls, adequate insurance coverage, and a provision made for the routine flow of information on income and costs that can be examined regularly to ensure profitability.

In addition to providing pilot services, the business may offer to completely maintain and operate an aircraft for the customer. For the non-pilot business person, this may offer an economical way of operating her/his aircraft while at the same time enabling the aviation business to obtain greater utilization of its personnel and facilities. To assure a smoothly working arrangement, the manager should develop a contractual relationship that will clearly delineate the roles and responsibilities of both parties.

Air Cargo

Air cargo (freight and mail) is important to all levels of air transportation providers, and a number of all-cargo scheduled airlines offer overnight and other delivery options using hubs where packages from throughout their systems are gathered, sorted, and redistributed. General aviation also plays a key role in the air cargo area with activities such as subcontracting to the scheduled cargo carriers, transporting canceled checks, shipping urgently-needed factory parts to keep a production line going, express small package delivery, human tissue or radiological transport, and so on. The size of aircraft may range from a single-engine propeller aircraft, to an airplane able to accommodate containers. When FAA safety requirements can be met, small amounts of freight or mail may accompany passengers in the same compartment, or designated cargo area.

The market is likely to continue to be strong for any cargo that is high-value/low-bulk or any cargo that is time-sensitive, such as documents, and perishables such as blood, human organs awaiting transplant, and flowers. An air taxi operator may coordinate with a metropolitan courier service to provide door-to-door small package service.

Air Ambulance/Medical Evacuation

This type of flight operation has been growing rapidly, due in part to the disparity in quality of medical care between rural facilities and metropolitan hospitals, some of which are unique in the nation for their specialized research and equipment. Specially-equipped aircraft that can take stretchers and supply in-flight medical care are required. This type of operation is becoming subject to increasing scrutiny and regulation.

As medical care becomes increasingly sophisticated and specialized, and as hospital competition grows, air ambulance/medical evacuation services are a growing market area for the FBO. Not every FBO will want to enter this market as it is also dependent on volatile insurance reimbursement policies.

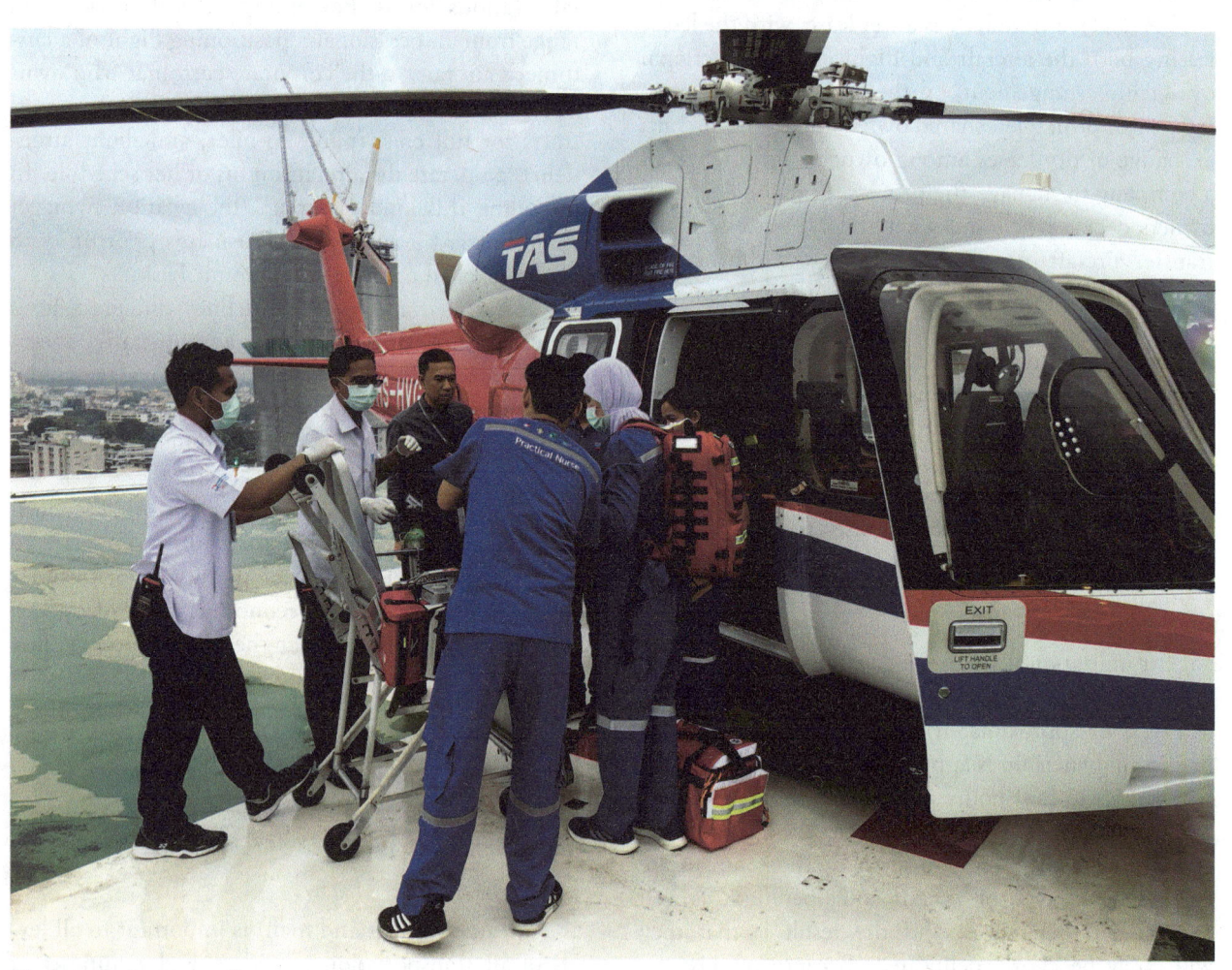

Medevac helicopter and crew

Other Commercial Flight Operations

Aerial Patrol

Flight operations for the purpose of pipeline, powerline, forest, highway, and border patrol are fairly widely used, although the actual flight activity is normally restricted geographically and seasonally. Forest patrols cover specific wooded areas and are most active in the dry season. Power and pipeline patrols cover designated lines and may be most active following heavy rainfalls, high wind events, and other inclement weather. Highway and border patrol flights are assigned specific sectors and normally have missions related to operational issues such as traffic, construction, and illegal entry. Uniformed police or border patrol personnel in aircraft owned by a federal or state agency operate many of these flights. There is, however, an opportunity for general aviation businesses to provide maintenance and other services under a contractual arrangement with enforcement agencies for the support of these aircraft.

An opportunity for conducting aerial patrol activities is the Aerial Fire Detection Service for the United States Department of Agriculture (USDA) Forest Service. The various National Forests throughout the United States develop air detection plans for their areas of responsibility. These plans establish the procedures covering air operations, including the base of operations, pilot and aircraft requirements, required flight patterns, flight frequency, the fire season, and general flight operating procedures.

The contracting officer for the particular area advertises the Aerial Fire Detection Requirements and the detailed specifications are identified. A typical contract covers the following items:

> Scope of contract;
> Descriptions of operation base;
> Flight paths;
> Flight speed;
> Observer personnel;
> Government-furnished equipment;
> Aircraft specifications;
> Pilot qualifications;
> Flight duty limitations (pilot);
> Flight time requirements;
> Inspection and approval of aircraft and pilots;
> Safety requirements;
> Flight time measurement; and
> Method of payment.

A regional air officer of the Forest Service must inspect and approve the successful bidder's aircraft and pilots. The contracting officer will authorize and regulate the standby and flight time schedule. Actual flight time will be entered on Forest Service Form 6500–122, Daily Flight Report and Invoice, and submitted by the contractor. The typical contract will contain a minimum guarantee per year in dollars, an hourly bid rate for flight time, and the rate allowed for standby time.

Powerline and pipeline patrol activities are normally conducted under contractual arrangements similar to those used by forest patrols. They are typically low-level flights providing visual inspection of cross-country pipelines or electric powerlines. Patrol activity of this type is normally accomplished in high-wing aircraft having good visual capability. Relatively slow speeds are desirable, with adequate power for emergency needs. Frequently, the Super Cub, Cessna 182, and Helio Courier are used for this type of work, although other model aircraft are being used, including helicopters.

In this situation, the aviation manager is concerned with equipment selection, pilot selection and training, operational and safety procedures, procedures for scheduling and controlling flights, insurance requirements, administrative paperwork, and the profitability of the operation.

In many cases of aerial patrol, the helicopter can be an asset because of its ability to safely handle low-altitude and low-airspeed flight, hovering, and landing in restricted areas. The decision to invest in helicopter rental or charter equipment is a major one because of the substantially higher cost of the aircraft, high insurance premiums, higher per-hour fuel consumption, and the high level of skill needed to operate it.

Aerial Application

Agricultural flying is a term that loosely describes the use of aircraft in the interest of agriculture, forestry, fishery, and public health, where that use is primarily a tool for making observations or applying product. Most agricultural flying involves aerial application: the distribution from an aircraft of agricultural chemicals or seeds. Aerial applicators spray chemicals such as insecticides, fungicides, herbicides, defoliants, and desiccants, and spread solid materials such as seeds, fertilizer, and lime. Other activities, such as minnow seeding, may also be undertaken by aerial applicators.

The aerial applicator is one of the most versatile and highly trained aviation specialists. He or she must be a businessperson as well as an experienced pilot, who is knowledgeable in areas such as chemistry, physics, agronomy, entomology, farming, engineering, meteorology, and cost accounting. A great deal of agricultural application work is done at a height of 5 to 10 feet above the target, with concern for the following factors:

1. Accurate marking or GPS navigation for successive straight swath runs.
2. Application rate, or the total quantity of material applied per acre, depending on the:
   > Output of each nozzle;
   > Number of nozzles;
   > Width of swath; and
   > Ground speed.
3. Distribution of the spray liquid on the target, determined by:
   > Droplet size;
   > Number of nozzles;
   > Flying height and speed;
   > Swath width; and
   > Meteorological conditions.
4. Operational influence of wind, rain, temperature, and humidity on the aircraft procedures and the distribution of material.

Aircraft characteristics. Aircraft used in early agricultural application work were designed for other purposes and then converted for this use. Currently, most aircraft doing aerial application work are specifically designed for their job. Those qualities desired in an agricultural aircraft include:

1. Good performance and safe operation from small, unpaved strips.
2. Safe operating speeds of 60 to 100 mph.
3. Maneuverability.
4. Docile handling characteristics.
5. Good field view from the cockpit.
6. Comfort for the pilot and protection (in the event of a crash).
7. Fire protection: fuel tanks located away from the pilot.
8. Simplicity in construction and maintenance.
9. Resistance to corrosion by agricultural chemicals.

Agricultural aircraft, used for aerial application.

The size and power required of the aircraft vary according to the type of operation and the working conditions. Most aircraft are single-engine, with 150 to 600 horsepower and a capacity to carry 600 to 3,000 pounds of material.

Rules and regulations. Agricultural application by aircraft is one of the most thoroughly and tightly controlled aviation activities in the United States. The Federal Aviation Administration, the United States Department of Agriculture, the United States Food and Drug Administration, the Environmental Protection Agency, and state departments of aviation, agriculture, and public health, have established rules, regulations, and codes. Feed-processing companies, as well as chemical and food industry groups, provide additional regulations.

Primary aviation control is provided through the Federal Aviation Regulations, Part 137—Agricultural Aircraft Operations. This regulation covers the following:

> Definition of terms involved in agricultural aircraft operation;
> Certification rules for an operator's certificate;
> Operating regulations;
> Aircraft requirements;
> Personnel requirements; and
> Required records and reports.

In addition to demonstrating knowledge of the performance capabilities and operating limitations of the aircraft, the applicant for an agricultural aircraft operator's certificate will be tested on:

1. Steps to be taken before starting operations, including survey of the area to be worked.
2. Safe storage and handling of poisons and the proper disposal of used containers and rinsate for those poisons.
3. General effects of poisons and agricultural chemicals on plants, animals, and people, and the precautions to be observed in using poisons and chemicals.
4. Primary symptoms of poisoning, the appropriate emergency measures to be taken, and the location of poison control centers.
5. Safe flight and application procedures and a demonstration of:
   > Short-field and soft-field takeoffs;
   > Approaches to working area;
   > Flare-outs;
   > Swath runs; and
   > Pull-ups and turnarounds.

Profitability. As with other flight activities, the aerial applicator is concerned with the profitability of the operation. Assuming the existence of adequate financial, accounting, and operational information, the manager must carefully determine all costs, both fixed and variable, add the desired amount of profit, and market the service to ensure a successful level of operation. The following is a detailed list of the typical expenses incurred in an aerial agricultural operation:

> Pilot wages;
> Other wages (for example, the flagger);
> Officer wages and salaries;
> Aircraft and equipment repairs;
> Shop and equipment supplies;
> Truck and automobile rent;
> Airport and office rent;
> Room rent;
> Dues and subscriptions;
> Donations;
> Travel;
> Advertising;
> Flags;
> Aircraft insurance;
> Employee insurance;
> Other insurance;
> Aircraft depreciation;
> Truck and auto depreciation;
> Radio and equipment depreciation;
> Other equipment depreciation;
> Interest (loans);
> All utilities;
> Aircraft gas and oil;
> Vehicle gas and oil;
> Office and administrative costs;
> Property taxes;
> Unemployment taxes;
> Individual insurance; and
> Licenses and other taxes.

The manager normally groups these individual expense items into categories for regular review and analysis.

The major concern to the manager is ensuring a positive margin of profit between the earnings of the aircraft and the total cost of operation. Two basic alternatives are available—to raise the price for aerial application services or to improve the efficiency of the business. The latter can be achieved through the reduction of those items that constitute the direct and indirect costs of operation and through increased aircraft utilization. Still another target of the manager is the improvement of the work rate capability of the operating aircraft. The key factors in doing so are ferry distance, payload, swath run, speed of flight, turning time, and loading time. These elements strongly influence the work rate, or the number of acres serviced per hour. By improving the efficiency of operation, the work rate can be increased and the actual cost per hour reduced. Profit will be improved by the increased acres per productive flight hour.

Aerial Advertising

Another profit potential in flight activity and another way to extend the utilization of existing aircraft is through banner towing. By making this unique advertising medium available to a wide range of businesses, additional flight-service revenue can be generated.

With a maximum recommended top speed of 80 mph (banner drag and wear increase rapidly above this speed), the aircraft used must have adequate engine cooling and positive flight control at low speeds. It is also important to ensure that all spelling is correct on the banner!

Fish Spotting

The use of small aircraft to spot fish shoals is of great importance to commercial fishing operations in certain areas, and in some cases to those involved in recreational fishing. Fish spotting is particularly useful in locations such as the Alaska coast, where the salmon season is controlled to the minute, and good timing is important. The aircraft pilot uses a radio to direct fishing boats. Although highly seasonal, this activity can be lucrative for an FBO because of the high value of the potential catch and the fact that no special equipment is needed for the aircraft.

Aerial Photography

Aerial photography has at least five important applications:

1. High-level vertical photos for photogrammetry and mapping.
2. Low-level photos for advertising, site planning, and development.
3. Low-level infrared pictures for heat-loss studies.
4. Plant disease studies.
5. Demographic studies.

Few general aviation businesses are involved in the sophisticated areas of aerial photography, because of the equipment and skills required. High-level vertical photos, for example, require a camera mounting in the floor of the aircraft, as well as oxygen, since operating altitudes may be 20,000 feet MSL (Mean Sea Level), or more. Infrared photography also requires special cameras. The likely level of involvement for most general aviation businesses will be in conducting occasional flights for realtors, land developers, news reporters, city and state planners, and engineers and commissioners who usually make their own observations and take their own pictures. The availability of software such as Google Earth provides increasingly stiff competition for these services and aerial photographers need to be ready with a good description of what their services provide that Internet-based information cannot (for example, currency of data.)

Flight Instruction

The Changing Market

The cost of flying has continued to increase over the years. Currently (2020) an hour in a Cessna 152 or Piper Warrior (representative training aircraft) may be on the order of $100-$130, with hourly instructor costs on the order of $50.00 or more. Many students are learning to fly to prepare for a career as a professional pilot, while others are learning "just for fun". Those learning just for fun may also be in the market for other discretionary pursuits, such as skiing, scuba, recreational vehicles, and tropical vacations. Thus, these activities all compete for financing with the same discretionary dollars.

Training Programs

Some flight programs involve an integrated program of ground school and in-flight instruction. This method of learning to fly is valuable for the student who wants to have some theoretical understanding of aerodynamics, navigation, regulations, and safe procedures before taking the controls. Others are best able to learn by doing and may attend formal ground school later. There are many self-study courses available to help such students prepare for the FAA written exams. In view of different needs, an FBO may offer in-flight instruction on demand, but only periodically offer ground school. It may be best to run each service as a separate profit center for this reason.

Ground school instruction comes in a myriad of forms today, including not only the traditional classroom style taught at the airport or a local college or university, but also using individualized instructional methods for which technology is the primary means of course delivery. Videotaped or DVD series can be purchased to help prepare individuals for the written and flight tests, for virtually any flight certificate or rating.

Computer-based and online training courses are additional (and increasingly popular) alternatives, with various programs available for almost any personal computer. With the introduction of computerized testing, computerized ground school has become much more attractive. Use of simulators and computer training devices can increase skills while requiring no fuel or instructor, and thus provide a much lower hourly cost to the student. The use of true simulators or flight training devices is very valuable in enabling the student to practice potentially dangerous scenarios such as an engine-out situation in a safe environment.

Instructional flight operations include the formal training that leads to private pilot, commercial pilot, instrument rating, multiengine rating, airline transport pilot (ATP), certified flight instructor, glider, and balloon certificates and ratings. Flight training also includes instruction that leads to aircraft type ratings, aerobatics, and agricultural aircraft operation.

In establishing flight-training programs, the manager is normally concerned with the requirements expected of graduates of the program, an effective and efficient syllabus, competent instructors, and the profitability of the instruction. The major source of guidance in determining pilot requirements for various certificates and ratings is the Federal Aviation Administration. The requirements for the following certificates and ratings are contained in the indicated sources:

> Recreational pilot   FAR 61.96
> Sport pilot   FAR 61.301
> Student pilot   FAR 61.81
> Private pilot   FAR 61.102
> Commercial pilot   FAR 61.121
> Instrument rating   FAR 61.65
> Instructor certificate   FAR 61.181
> Multiengine rating   FAR 61.63
> Airline transport pilot   FAR 61.151
> Glider category rating   FAR 61.63
> Lighter-than-air category rating   FAR 61.63

The sport pilot license was discussed in Chapter 3, Marketing, and has made it much easier for pilots, particularly those who do not intend to fly professionally, to enter the system. Given the dramatic rise in interest for sport and experimental aircraft, both in the past decade and predicted for the future, this seems likely to be a growth area that FBOs should consider.

Part 61 of the Federal Aviation Regulations is very important to all pilots, because it deals with pilot qualifications, privileges, and limitations. It is equally important to the manager who is responsible for ensuring that qualified personnel are piloting the aircraft.

To ensure that the pilot receiving training achieves the level of proficiency stipulated by the FAA or other sources, the manager should ensure the adequacy of the training program for the certificate or rating involved. The private pilot flight-training program is the most prevalent in the United States. There is an abundance of training programs. They range from the most fundamental topical ground and flight outline to sophisticated syllabi with associated audiovisual materials. Learning programs, which the manager can consider acquiring and installing at the business include such programs as the Cessna Pilot Center "Integrated Flight Training System" and the various Jeppesen/Sanderson programs.[14,15]

Other programs are available from various developers and publishers. In addition to complete flight and ground school programs, numerous manuals help student pilots master private pilot subjects, instrument topics, and multiengine concepts. These manuals can be obtained from a number of aviation publishers and suppliers.

Many flight-training organizations throughout the United States have developed their individual programs and syllabi for the various certificates and ratings. These schools may be of some assistance to the manager intent upon opening a training center. Of course, any aviation business that plans to open a flight school should study FAR Part 141 very carefully. This regulation prescribes the requirements for becoming a certificated pilot school and provides for the approval of pilot training courses offered by certificated schools. Federal Aviation Administration Advisory Circular 141–1B, Part 141 Pilot Schools, Application, Certification, and Compliance, sets forth guidelines to assist persons in obtaining a pilot school certificate and associated ratings under the Federal Aviation Regulations, Part 141.

Instruction Administration

The administration of a flight instruction program presents many challenges to the manager. There are procedures to develop, rules to establish and follow, schedules to create and carry out, personnel to manage, progress checks to administer, and controls to monitor. The training program may become large enough to become an independent department with the full organizational structure of a school. The administrative procedures established should provide for:

> Registration of students;
> Curriculum development and improvement;
> Scheduling of aircraft, instructors, and students;
> Identifying and coordinating aircraft maintenance;
> Conducting student progress checks;
> Obtaining FAA and state approvals of the flight school;
> Developing school operational procedures;
> Developing and maintaining an effective safety program, that incorporates SMS principles;
> Developing procedures, including visa checks, for foreign students;
> Maintaining flight training records;
> Administering student grants and loans;
> Creating and using an effective accounting system; and
> Flow of information and economic analysis of school activity.

Several organizations have recognized the need to develop a complete package for the management of a flight school, and quite a few are available today. One of the most thorough and complete is the Cessna Pilot Center System developed by Cessna Aircraft Company. This system was developed to provide "a total package with which to conduct a profitable flight-training business." The management manual for the Cessna integrated flight-training system provides a well-structured and comprehensive approach to flight training through the identification and application of the following system components:

> Integrated flight training curriculum;
> Management system for operating a flight school;
> Management training;
> Instructor curriculum training; and
> Consultation services by Center specialists.

The Cessna 172 has been widely used as a flight school trainer for over fifty years.

A review of these components clearly indicates the Cessna approach to be well-structured and comprehensive. The operational elements in the management system include:

> Flight counter procedures;
> Student enrollment forms;
> Flight training agreement;
> Appointment cards;
> Student record folders;
> Flight scheduling forms;
> Aircraft maintenance schedule;
> Instructors' work schedule;
> Instructors' scheduling cards;
> Aircraft rental agreement;
> Flight line procedures;
> Facility appearance guidelines;
> Equipment utilization;
> Accounting system;
> Financial analysis;
> Tax considerations;
> Human resources programs; and
> Reports and controls.

The total package from the Cessna Pilot Center System is designed as an integrated program with coordinated supplies, records, and procedures. The core of the system is the integrated flight and ground curriculum. Other companies have approached the development of flight instruction along the same lines. These companies include Beechcraft and the various flight instruction programs of Aero Products Research, Inc.

Flight Instructors

Flight instructors are perhaps the most critical element in the development and conduct of successful flight instruction activity. The desired image of an instructor is professionalism. He or she should be well qualified technically, possess the desired teaching skills and motivation, and demonstrate the ability to implement these skills. In many aviation businesses, a number of problems associated with flight instructor selection and utilization can create difficulties for the flight department manager; these include:

> Poor selection procedures;
> Lack of a company orientation program;
> Lack of training;
> Rapid turnover and/or lack of reliability of instructor personnel;
> Low salary scale/poor (or no) employee benefits;
> Lack of a career orientation;
> Fluctuating seasonal demands for flight instruction activity;
> Lack of marketing orientation; and
> A weak identification with the FBO organization.

Many aviation businesses do a poor or indifferent job of selecting their flight instructors. Many managers have fallen into the practice of having their office call the next name on the availability list when another instructor is needed. Too much emphasis may be placed on the mere possession of a Certified Flight Instructor (CFI) certificate, a minimal requirement, and too little attention given to other qualifications the organization should consider desirable or essential in its instructors. This approach appears to be caused by the lack of management's ability, either in time or knowledge, to engage in careful identification of personnel needs and then to exercise care in the selection of new instructors.

As discussed in Chapter 5, Human Resources, many aviation organizations, as small businesses, lack time or interest in developing adequate personnel structures. The first and obvious step is the development of a position description for an instructor position, as needed in that organization. From this description, a job specification can then be developed that identifies the qualifications the prospective employee must possess. Chapter 5 contains information on the functions and skills needed for the position of a flight instructor.

After an active recruiting program to ensure an adequate number of candidates for job openings, the next step is the utilization of an application form for all potential flight instructors. Such a form can be very effective in screening candidates and identifying those pilots best suited to the needs of the organization. Figure 5.6 in Chapter 5, Human Resources, depicts a form that is comprehensive, provides a great deal of information on the applicant, and should be extremely useful in selecting new flight instructors.

It is necessary to interview the stronger candidates in order to verify or amplify the data contained in the application form and at the same time provide the candidate with information about the position and the organization. An interview form is very useful for busy managers when conducting the search. Such a form is somewhat like a pre-takeoff checklist, in serving as a reminder to the interviewer and assisting in evaluation of the candidate.

Instructor Training

Training of flight instructors should include those elements that will enable the person to perform the total job more effectively. This means training in:

> Flight operations, safety requirements and procedures, and aircraft type specifics and characteristics;
> Effective instructional skills and techniques;
> Safety management;
> Marketing;
> Administrative duties; and
> Organizational responsibilities.

Historically, primary emphasis has been placed on flying skills, and minimal emphasis has been placed on instructional and educational skills. Frequently, due to wages being based upon the number of hours flown at an hourly rate, other responsibilities were minimized or not included, and skills or talents in other areas were never developed. Organizations have increasingly realized the advantage of hiring instructors on a full-time basis, providing training in broader organizational responsibilities, providing professional development opportunities, and realizing additional benefits from airborne and ground activities of flight instructors. There must be an adequate student load and related business activity to economically justify the individual instructor on this basis.

Freelance Instructors

Just as some FBOs must contend with "tailgate mechanics" and so-called "gypsy" operators, so too the *freelance* instructor is another possible source of unfair competition. Any pilot with a CFI certificate can legally conduct flight training, as far as the FAA is concerned. However, the owner of a based aircraft who gives lessons at someone else's airport is probably in business without a lease or operating permit. Since the owner presumably pays only hangar or tiedown rent and has little or no other overhead, such an operator can invariably undercut the FBO's prices. But he or she may not offer the student the same guarantees of high-quality instruction, the use of facilities for classroom time, or the availability of backup aircraft. Because of low-budget maintenance and lack of regulation on the field, such freelance operators can also pose a safety hazard that could cause the FBO to be liable for any damages that might be incurred, despite its lack of control over the operator.

Simulator/FTD Usage

Simulators and flight training devices (FTDs) have been designed to assist in training for basic flight maneuvers and procedures, for instrument procedures and operations, and for specific aircraft and equipment checkouts. It has been said that the worst possible place to give flight instruction is in the aircraft because of the noise, the distractions of many physical sensations, and the inability to stop the machine and talk about an instructional issue. To help overcome these problems, a wide variety of simulators and training devices have been developed.

When a flight-training device or simulator is being used as part of a training course, the extent of its use should be clearly stated in the syllabus. FAR 141.41(a) prescribes the requirements used to obtain the maximum flight training credit allowed for flight-training devices in approved courses. FAR 141.41(b) provides for the use of flight-training devices that do not meet the requirements of Section 141.41(a). The training course must clearly show that the simulator or FTD being used meets the stated goals of the syllabus and adheres to the guidelines of 141.41(a) or (b).

The simulator can be a valuable adjunct to a flight-training program by accomplishing the following:

1. Reducing the calendar time required for a given training program.
2. Providing additional utilization of flight instructors.
3. Providing a means for sustaining student interest, proficiency, and course progress during periods of inclement weather.
4. Providing an improved learning environment.
5. Facilitating the scheduling process of the organization and the student.
6. Providing a less expensive training opportunity for the student and for the organization.
7. Providing a means for the student to overcome specific learning obstacles.
8. Saving fuel.
9. Eliminating touch-and-go and other noise.

The Federal Aviation Administration has recognized the usefulness and effectiveness of the simulator in instrument training by allowing credit for instrument rating experience completed in an acceptable instrument ground trainer. Other ratings also recognize and provide for the acceptance of simulator/FTD time. Most airframe manufacturers have recognized the advantages of simulators for instructing and qualifying pilots in more sophisticated aircraft. Here, the emphasis may be on the operation of complicated systems and emergency procedures, as well as on the actual operation of the aircraft.

Simulators historically have been expensive for a small FBO, though their comparative cost-effectiveness has improved as fuel costs have increased. Today's computers have opened up a whole new era of flight training device and simulator technology. A student with a personal computer can now purchase training device software to use at home, as well as at the flight school.

Sport and Recreational Flyers

Gliders and Sailplanes

Where terrain and climate provide good soaring conditions, this is often a popular activity. It may go well with business-oriented FBO services because it is largely a weekend activity. However, the mix of slow, unpowered aircraft in the traffic pattern with other aircraft can be a hazard. Some fields are dedicated to just this type of activity. The main revenue to an FBO will come from fuel sales to the tow plane and from glider pilot tow fees. Many glider owners transport their aircraft to and from the field in special containers, so the potential for tie-down income is limited. This aspect of general aviation continues to experience modest growth in suitable locations; however, it is a very small contributor to FBO income nationally.

Parachuting

Like gliding, parachuting can present problems at an airport with a busy traffic pattern. The jump-landing zone needs to be isolated. Pilots arriving and departing in other aircraft must watch for the drop, and a NOTAM (Notice to Airmen) should be issued by the appropriate authority to warn pilots of the parachute activity and location.

However, owing to low speed and high visibility, descending parachutists do not generally present a major hazard. This, too, tends to be a weekend activity. It can generate important revenue from bystanders, for example, at an outdoor airport restaurant, as well as from participants.

Ultralights

The first ultralight fly-in was in Florida in 1974; thus, ultralights have now been around for over four decades. But the rest of the general aviation industry still seems very uncertain about how to deal with them. Ultralights require only a few hundred feet of landing area, fly at very low speeds, and (so far) the pilots are not required to be licensed or to pass any course of instruction. Since they use autogas and pilots mostly do their own repairs, they do not appear to generally offer a source of profit to FBOs and are often viewed as a hindrance to the regular customers. Whether they use separate landing areas within existing airports or completely separate fields, they may in certain locations present a new opportunity for FBO services.

The FBO owner/manager who is able to harness the enthusiasm and dedication of these avid fliers may find them an asset rather than a liability. Some of these pilots may be interested in obtaining their pilot certificate at some point and/or renting or purchasing a more traditional aircraft.

Experimental and Home-Built Aircraft

This facet of the general aviation industry appears to be growing more vigorously than personal flying in conventional aircraft. A major reason is the substantially lower cost to both operate and own such aircraft. Over half a million or more people, with close to 10,000 aircraft from all around the world, regularly attend the annual fly-in of the Experimental Aircraft Association in Oshkosh, Wisconsin. The organization also boasts a network of over 800 local chapters.

The number of experimental aircraft completed in the early part of the 1990s was almost as great as the number of general aviation factory-built aircraft in the U.S. As was indicated in Figure 1.2, the total number of experimental and "other" aircraft, that is, homebuilt aircraft, balloons, and gliders, was 27,100 in 2000, 35,200 in 2007, and 31,645 in 2018, fueled primarily by an increase in the number of light sport aircraft.

Balloons

Ballooning is very popular in some areas, but it does not need an area the size of an airport or an FBO to function, although it can add to the glamour and excitement of an air show. Conflicts with other recreational uses do, however, frequently cause ballooning to be sited at airports rather than parks.

Rotorcraft

Helicopter ratings have become more attractive in recent years. This is due, at least in part, to the availability of such machines as the Robinson R22 helicopter and the Schweizer 300CB. The R22 was designed primarily as a training helicopter, much like the Cessna 150 for fixed-wing flight training. A lightweight, two passenger aircraft, the Robinson is also a more affordable helicopter, although still more expensive than learning to fly in a small two-seat aircraft.

Helicopter flight training can occur at an airport offering fixed-wing flight training without undue burden to either operation. Typically, the helicopters fly at a pattern of 500 feet above ground level, versus 800–1,000 feet for fixed-wing aircraft. And, because helicopters are not restricted to taking off and landing on runway surfaces, they can initiate their own traffic pattern, parallel to or away from the fixed-wing traffic. Most airports with rotorcraft activity do have a marked "H" for the helipad area.

Sight-Seeing

Flying customers around the city, over sensitive scenic areas such as the Grand Canyon, or along the oceanfront so that they can view the panorama, has long been a popular part of aviation. In general, conducting sight-seeing flights requires something scenic to view and normally is seasonal work, generally peaking during the spring-summer-fall months, vacation periods, or even on sunny, pleasant weekends. With few exceptions, this type of business is conducted with existing, standard aircraft almost on a pick-up basis. Aside from the usual aircraft and pilot insurance requirements, there are some special concerns for the manager, such as ensuring safety on the ground and while airborne, pilot training, and passenger briefing, noise minimization (especially over sensitive areas), and specific flight time and route controls. With care, this type of flying can improve aircraft utilization, provide additional

Helicopter pilots may receive their training through either civilian or military flight schools.

income, and stimulate interest in aviation that may lead to business in flight instruction and aircraft sales. The notion of whether or not aircraft should be allowed to fly over U.S. national parks was a topic for discussion in the 1990s. As a result of this discussion, Congress passed the National Parks Air Tour Management Act of 2000 to regulate air tours over national parks, but the air tour management plans required by the Act have not yet been completed by the FAA and the National Park Service. In September of 2020, the FAA noted on its website, "The National Parks Air Tours Management Act (the Act) requires operators wishing to conduct commercial air tours over national parks, or over tribal lands within or abutting national parks, to apply to the FAA for authority to conduct such tours. The Act further requires the FAA, in cooperation with the National Park Service, to establish air tour management plans for parks or tribal lands for which applications are submitted. The FAA Modernization and Reform Act of 2012 amended the Act to allow FAA and the National Park Service to enter into voluntary agreements with air tour operators in lieu of developing management plans. The 2012 amendments also exempt national parks with 50 or fewer tours annually from the management plan and voluntary agreement requirements". Moreover, the FAA noted "The Federal Aviation Administration (FAA) and the National Park Service (NPS) continue working together to implement the National Parks Air Tour Management Act"[16].

For most FBOs recreational aviation of various types is not an important element of their business. At some airports, such activity is discouraged because of conflicts with faster, heavier traffic. The activity levels of the more innovative types of recreational aviation such as ultralights tend to be volatile, with many operators and manufacturers leaving the industry each year. Insurance for some aspects of sport and recreational aviation is either unavailable or a prohibitive expense. Nevertheless, for the right FBO in the right location, this industry segment may provide a previously overlooked market.

Aircraft Sales

New Aircraft

Aircraft sales are the largest item of business for some FBOs and a less significant (or zero) percentage for others. Unsurprisingly, given the low volume of aircraft being produced, NATA surveys showed that in the 1990s, the percentage of members in the aircraft sales business was shrinking. The percentage of FBOs who were franchised aircraft dealers rose steadily in the early 1980s, from 57 percent in 1981 to 84 percent in 1984. However, it decreased sharply in the late 1980s and 1990s as so few aircraft were being manufactured and sold nationwide, due to the insurance cost element of the finished new aircraft. This began to improve again during the mid-1990s.

This was particularly true for the sale of new aircraft. Selling aircraft requires the financial ability to carry an expensive inventory for unpredictable amounts of time. It also requires substantial showroom space. The average time in inventory is generally longer for new aircraft than for used, according to studies conducted by NATA.

A significant change in terms of aircraft sales relates to the number of foreign aircraft being sold in the U.S. Many companies, such as the French-based EADS, Canadian-based Diamond Aircraft, Brazil-based Embraer, and Swiss-owned companies such as Pilatus have done well in terms of domestic aircraft sales.

Used Aircraft

As with auto sales, used aircraft may be taken as trade-ins. The FBO usually refurbishes and checks out the aircraft before resale, which involves some costs that must be included in calculating the selling price. Inventory times are typically shorter for used aircraft.

Brokerage

Selling aircraft as a broker means that one does not have to carry the inventory or finance the sale, but simply act as a "matchmaker" between buyers and sellers. The commission on this activity can be substantial. Some FBOs also sell aircraft on consignment; that is, they do not pay the seller until the aircraft has a new buyer.

Demonstration Flights

Although not usually time-consuming, these flights are key to a successful sales operation, and the personnel best suited to presenting the aircraft's features and benefits should conduct them.

Flight Operations Manual

A flight operations manual has various meanings in different organizations. In many small businesses, it refers to a general guide for conducting flight operations that attempts to set the tone for the overall business and its related activities. Such a manual contains both policy and specific rules and regulations. In an effort to indoctrinate all employees in the philosophy and procedures selected by the organization, the manual becomes required reading. It also provides specific guidance to key operating personnel, including those at the front desk, on the flight line, instructors, and pilots.

In some aviation organizations the term "operations manual" refers specifically to the publication that is required in order to operate air taxi aircraft under the authority of FAR Part 135-Air Taxi Operations and Commercial Operation of Small Aircraft.

Summary

Flight operations form the core of the aviation business. Other activities, such as maintenance, aircraft storage, parts supply, and administration develop from the need to keep aircraft in airworthy condition. Flight activities may be grouped into those involved simply in transporting people or products between two points, such as charters, air taxi, business aircraft operation, and air ambulance services; and those activities involving conducting a business operation as an intrinsic part of being airborne, such as flight instruction, agricultural applications, aerial photography, fish spotting, survey, and patrol work, construction, and many others, all specialized and some requiring special aircraft equipment. Each of the various activities requires specific aircraft and flying skills. Each possesses characteristics that must be identified and understood by the FBO if they are to become profitable elements of the business. An understanding of the key elements involved in each activity, identification of the type of aircraft and other equipment needed, knowledge of the rules and regulations governing that type of flying, and awareness of the primary economic considerations are all part of the aviation manager's job. Finally, the use of internal operating procedures and guidelines to assist in achieving established goals facilitates a consistent and streamlined operation in the flight department.

DISCUSSION TOPICS

1. What requirements must an air taxi operator meet for FAR Part 135 operations?
2. What are the differences between air taxi and commuter operations?
3. What are some of the associated problems and opportunities in conducting flight instruction?
4. What flight operations techniques can be used to reduce noise and when should each be applied? What safety concerns should be addressed in conducting noise abatement arrivals and departures?
5. How does the local climate affect the market for flight operations?
6. What are some of the technical requirements facing the agricultural operator?
7. What procedures should be used to ensure the profitable operation of such activities as aerial photography, sightseeing, and banner-towing?
8. Identify five FBOs within a 100-mile radius. Which flight operations areas have they chosen to serve, and why?
9. Assume you are entering the aircraft rental business. What questions must be addressed for each rental transaction?
10. What are the differences between a simulator and a Flight Training Device (FTD)? How can the FBO that offers flight training utilize simulators and FTDs to enhance the effectiveness of flight training, and increase profitability?

Endnotes

1. GAMA 2019 Databook, pages 9-10. Retrieved October 18, 2020, from https://gama.aero/wp-content/uploads/GAMA_2019Databook_Final-2020-03-20.pdf
2. Retrieved from *http://www.nbaa.org/news/pr/2002/20020424–011.php*
3. Searles, R. A., & Parke, R. B. (1997). *NBAA's Tribute to Business Aviation*. Washington, DC: National Business Aircraft Association.
4. Retrieved November 17, 2020, from *https://nbaa.org/aircraft-operations/security/programs/large-aircraft-security-program-lasp/*
5. Arthur Andersen, Inc. (2001, Summer). *Business Aviation in Today's Economy: A Guide to the Analysis of Business Aircraft Use, Benefits and Effects on Shareholder Value* (The White Paper Series, Number 4). Retrieved from *http://www.nbaa.org/news/backgrounders/AndersenPart02.PDF*
6. NBAA Business Aviation Fact Book. Retrieved October 27, 2020, from https://nbaa.org/wp-content/uploads/2018/01/business-aviation-fact-book.pdf
7. See for example the "fly friendly" procedures adopted at Renton Municipal Airport, WA at *http:// www.ci.renton.wa.us/pw/airport/abatement.htm*.
8. See *https://nbaa.org/wp-content/uploads/2018/02/nbaa-noise-abatement-program.pdf*
9. See *http://www.aopa.org/asn/apsup04.html*.
10. U.S. Department of Transportation, Federal Aviation Administration. (2006). *Minimum Standards for Commercial Aeronautical Activities on Public Airports*. (FAA Publication No. 150/5190–7). Washington, DC: U.S. Government Printing Office.
11. Donovan, D., Forbes, March 30, 2020 "How the Airline Industry Will Transform Itself As It Comes Back From Coronavirus"., retrieved October 30, 2020, from https://www.forbes.com/sites/deandonovan/2020/03/30/how-the-airline-industry-will-transform-itself-as-it-comes-back-from-cornonavirus/?sh=7887642267b9.
12. Part 135—operating requirements: commuter and on demand operations and rules governing persons on board such aircraft; available at : https://www.ecfr.gov/cgi-bin/text-idx?SID=f9f6f8f3fdb789aecbdb3e8d35e398a2&mc=true&tpl=/ecfrbrowse/Title14/14cfr135_main_02.
13. NetJets, Inc. (2006). *The Buyer's Guide to Fractional Aircraft Ownership*. Woodbridge, NJ: Author.
14. See *http://www.cessna.com/learn-to-fly.html*
15. See *http://www.jeppesen.com*. Jeppesen Sanderson is now owned by Boeing.
16. Retrieved November 16, 2020, from https://www.faa.gov/about/office_org/headquarters_offices/arc/programs/air_tour_management_plan/

10

Aviation Maintenance

OBJECTIVES

› Identify and describe the four organizational divisions of a typical maintenance department.

› Understand the certification process for an FAA airframe and powerplant license.

› Be aware of the facilities and equipment necessary to open and operate an aviation maintenance shop.

› Understand the implications of product liability on an aviation maintenance operation.

› Recognize the advantages and disadvantages of subcontracting out all maintenance work.

› Understand the concept of flat-rate pricing.

Introduction

Goals of the Maintenance Shop

The maintenance field must be viewed in the context of four factors that do not always work in the same direction:

> FAA requirements;
> Customer satisfaction;
> Profitability; and
> Changing market conditions and technologies.

The repair and maintenance jobs must be done properly, not only from the FAA's standpoint, but from the customer's standpoint. The customer wants all work to be performed and documented in full compliance with FAA directives, and desires frequent and friendly communication. The customer seeks a sense of trust derived from quality work, understanding of tailored specifications, and reasonable prices. Profitability of the maintenance shop as an individual department is necessary and adds favorably to the bottom line of the overall business. The challenge is to cover costs and make a profit while still pricing repair and maintenance services competitively with nearby operators. The ability to recognize and address changing market conditions is essential and may even result in a decision to downsize or close the maintenance facility.

To achieve these objectives, the issues of approvals, personnel, shop facilities, shop equipment, parts

245

inventory, supply chain, quality control, subcontracting, product liability, marketing, administration, and pricing must all be addressed. In this chapter we will review all of these aspects of maintenance, emphasizing the manager's role in guiding the activity toward established objectives. The chapter addresses both mandatory and discretionary elements of maintenance operations, focusing on profitability.

Changing Issues

Fractional Aircraft Ownership. The popularity of fractional aircraft ownership is soaring because it delivers all the advantages of private jet ownership at a fraction of the cost. The first fractional ownership program was launched in 1986 by NetJets.

There appear still to be many unknowns in the fractional aircraft arena, perhaps more in relation to aircraft maintenance. While the fractional owner is usually charged a monthly maintenance fee, it is the responsibility of the fractional operating company to ensure airworthiness. An article in *Aviation Maintenance Technician* raises such questions:[1]

> "You probably have heard about fractional interests in aircraft ownership by now. It seems like they have bought out all corporate jet production for years to come. Sales are booming, but have you thought anything about the status of employees of the management arms of such companies or what the impact will be in maintenance field? These companies now employ hundreds of technicians and pilots and have a significant impact in the commercial corporate flight environment. Are they here to stay? Will their structure change? Will they be regulated?
>
> "Fractional owners don't retain their own technicians, but maybe they should. Current arrangements call for the management company to supply all maintenance. Many companies are closing their (corporate) flight departments and moving to a fractional arrangement because of the perceived dollar savings. In addition, individual aircraft manufacturers have set out on a plan to provide in-house maintenance services for most of the routine inspection process. This is the area that is of concern to some. Will these arrangements impact the average technician?"

Fractional ownership programs were subsequently regulated under 14 CFR 91 Subpart K, which became fully effective in 2005.

From *Aircraft Maintenance Technology, April 1999* by Stephen Prentice. Copyright © 1999 by Aircraft Maintenance Technology. Reprinted by permission.

14 CFR 91 Subpart K, which became fully effective in 2005, provides regulations on fractional aircraft ownership maintenance.

Technology. Computers are now an essential component of most maintenance facilities. They are used for tracking repair orders, ordering parts, tracking billable hours, and running diagnostics, among other tasks. Installing such modernizations and training all employees is necessary to stay compatible and competitive. Two major software companies, FBO Manager and TotalFBO, provide full-service aviation maintenance management packages.

Next Generation Air Transportation System. For decades, pilots have desired and discussed the idea of positive aircraft control in which the pilot, rather than the air traffic control system, makes the ultimate decisions on safe operations. With current computer technology, this dream is finally a reality. The Next Generation Air Transportation System (NextGen) is the Federal Aviation Administration's (FAA) plan to modernize the National Airspace System (NAS) with satellite-based technologies. Through NextGen, the FAA is addressing the air traffic growth by increasing the capacity and efficiency of the NAS while simultaneously improving safety and reducing environmental impacts.[2]

To operate in most of the NAS once NextGen is fully implemented, aircraft must be equipped with an Automatic Dependent Surveillance-Broadcast (ADS-B) system. ADS-B uses an ordinary Global Navigation Satellite System (GNSS) receiver to derive its position from the GNSS constellation. Position data is then combined with velocity, heading, and altitude information. This packet of data is then broadcast to other ADS-B capable aircraft as well as to Air Traffic Control (ATC) centers in real time.[3] Aircraft separation, which has traditionally been the responsibility of Air Traffic Controllers staring at blips on radar screens, will be handled automatically by the system.

The FAA mandated that all aircraft operating in Class A, B, or C airspace be installed with ADS-B capabilities by January 1, 2020. FBOs should begin familiarizing themselves with this technology now. There are a number of manufacturers who have received TSO-C166b authorization for Mode S transponder equipment. These transponders will allow equipped

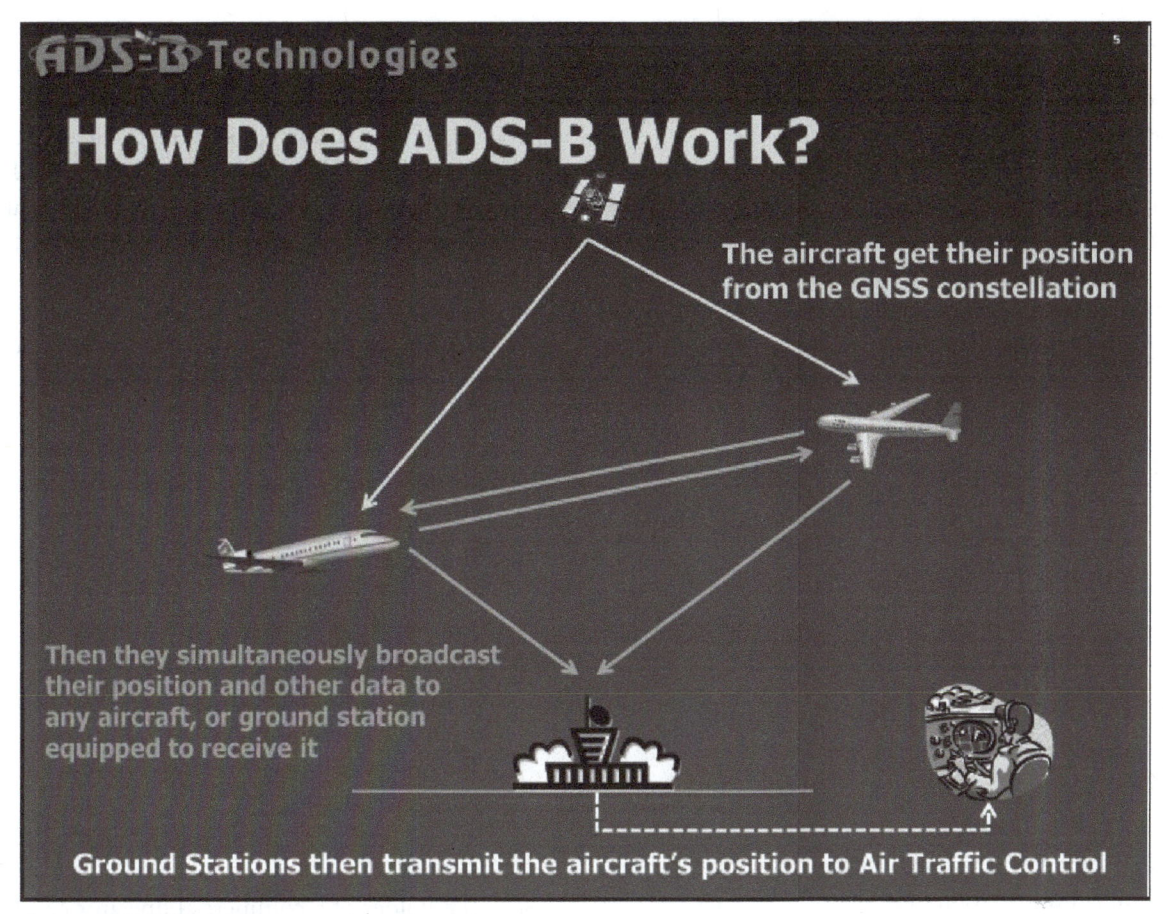

Courtesy of ADS-B Technologies

aircraft to automatically transmit position, velocity, and heading information. The sales and installation of these units has a high potential for profit.

GARA/Tort Reform Bill. Since 1994, the time period that a maintenance shop is potentially liable for faulty aircraft repairs goes back (only) 18 years. This might be assumed to mean that records must be kept for only 18 years on each aircraft that passes through the shop. However, the Tort Reform Act did not alter FAA's requirement for "cradle to grave" tracking of every part with a serial number and every action performed in aircraft repair and maintenance.

Maintenance Activity

Overview

Maintenance activities are not undertaken by all airport service businesses; however, if they are part of the service, then they are crucial to its success. Good maintenance ensures customer satisfaction and aircraft utilization, enhances the public image of the aviation business, and leads to further development of the aviation industry. Without the support of good maintenance, the reverse happens: business dries up and the industry stagnates, regardless of the products manufactured.

Maintenance activity at a typical general aviation business started when the flyer/manager found that he had to maintain his own aircraft in an airworthy status in order to stay in business. With the arrival of privately owned aircraft on the field, the manager found that he was performing maintenance on these additional aircraft. This represented a mixed blessing as, although costs are now spread over a wider base, the maintenance business adds cost and complexity in compliance with a host of FAA requirements.

With most aviation businesses today, maintenance work is done for company aircraft as well as for customers. In some larger full-service operations, where maintenance is a separate profit center, internal company requests for maintenance must compete with external customer requests. It is important that maintenance work on company aircraft is still billed at standard rates.

The aviation manager entering the maintenance business needs to understand the structure of this complex field, the pertinent rules and regulations, and the equipment and personnel needed for various levels of maintenance activity. The Federal Aviation Regulations provide the major guidance and control for the operation of aviation maintenance facilities.[4]

This guidance is contained primarily in:

1. Part 43—Maintenance, Preventive Maintenance, Rebuilding and Repair.
2. Part 65—Certification: Airmen Other Than Flight Crew Members.
3. Part 145—Repair Stations.
4. Part 147—Aviation Maintenance Technician Schools.
5. Related Advisory Circulars covering many aspects of aviation maintenance.

Maintenance activity can be categorized into the type of service performed. The following definitions are maintenance activities outlined by the Federal Aviation Administration:

1. "Maintenance" means the inspection, overhaul, repair, preservation, and replacement of parts, but excludes any preventive maintenance.
2. "Preventive maintenance" means precautionary preservation operations such as the proactive replacement of standard parts, usually not involving complex assemblies.
3. "Major repair" means repair (a) that if improperly done might appreciably affect weight, balance, structural strength, performance, power plant operation, flight characteristics, or other qualities affecting airworthiness; or (b) that is not done according to accepted practices or cannot be done by elementary operations.
4. "Major alteration" means an alteration not listed in the aircraft, aircraft engine, or propeller specifications that might appreciably affect weight, balance, structural strength, performance, power plant operation, flight characteristics, or other qualities affecting airworthiness; or an alteration that is not done according to accepted practices or cannot be done by elementary operations.

Part 43, Maintenance, Preventive Maintenance, Rebuilding and Alteration, prescribes rules governing the maintenance, preventive maintenance, repair, and alteration of the airframe, powerplant or appliances of aircraft with a U.S. airworthiness certificate. These rules identify the persons authorized to perform maintenance, preventive maintenance, repairs, alterations, and return-to-service actions for aircraft and their components. The form of maintenance records, the guidelines for performing maintenance work, and the classification of specific operations as major or minor alterations, repairs, or maintenance are all also covered.

The persons who may perform maintenance, preventive maintenance, rebuilding, and alteration are:

1. The holder of a mechanic certificate, as provided in Part 65.
2. The holder of a repairman certificate, as provided in Part 65.
3. Individuals working under the supervision of a mechanic or repairman.
4. The holder of a repair station certificate, as provided in Part 145.
5. An air carrier, as provided in Parts 121, 127, or 135.
6. The holder of a commercial operator certificate as provided in Part 42.
7. The holder of a pilot certificate issued under Part 61 who may perform preventive maintenance on any aircraft owned or operated by him that is not used in commercial air service.
8. A manufacturer operating under a type of production certificate.

Organization

Maintenance activity by an aviation business is generally accomplished in accordance with the Federal Aviation Regulations. To assist in achieving the business's objectives, most concerns are organized specifically for this purpose. The typical organizational structure of a maintenance or service department would look like that in Figure 10.1. Even if the organization is small and the personnel are few, the functions represented by this organizational chart are accomplished when the business provides service support for airframes, power plants, and associated avionics. The chart enables the business manager to identify the various maintenance functions, the relationships among them, and a framework for future

Figure 10.1 → Typical Organization of a Maintenance Department

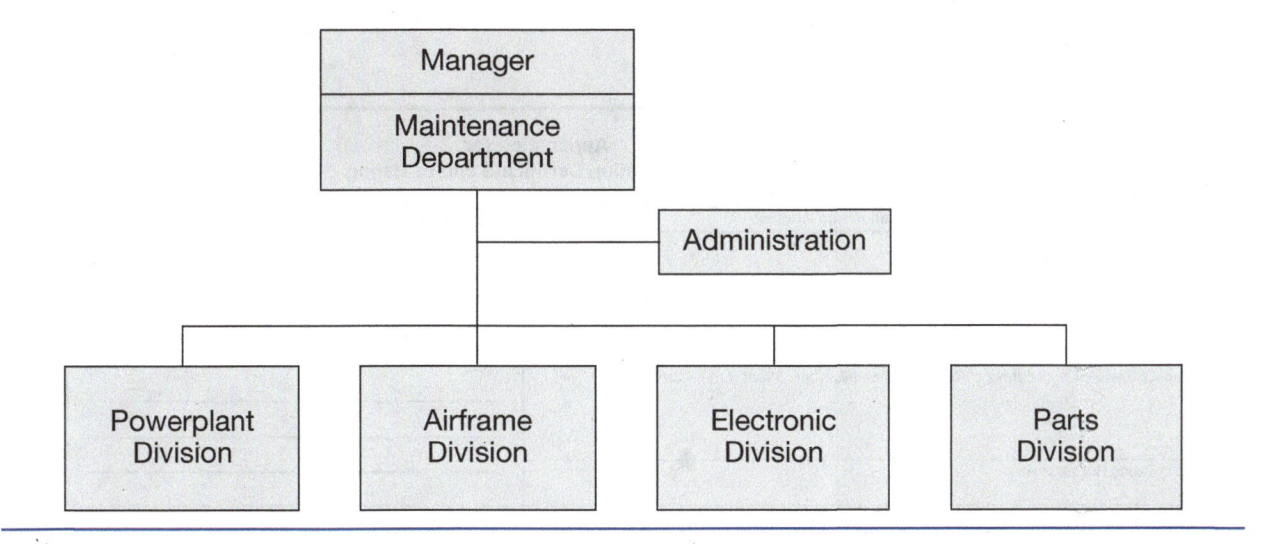

growth of maintenance activities. The components of the structure are responsible for the following:

1. Manager, maintenance department: Responsible for the overall operation of the department including profitability of service, coordination, technical competence, compliance with rules and regulations, quality of product and highest possible safety standards, and hiring and training.
2. Administration: Provides administrative support to the department by maintaining department records, maintenance manuals, Airworthiness Directives (ADs), shop orders, and required reports.
3. Powerplant division: Responsible specifically for service activity dealing with powerplants, propellers, fuel systems, and related engine accessories. This includes trouble-shooting, repair, and replacement of the equipment as authorized by the FAA.
4. Airframe division: Responsible specifically for service activity dealing with the composites, metals, and structures of the fuselage, empennage, and wings of aircraft, in accordance with FAA authorizations.
5. Electronics division: Responsible for maintenance activities dealing with communication equipment, navigational equipment, radar, and instruments. This includes accessories such as starters, magnetos, and transceiver systems.
6. Parts division: Responsible for the parts, material, and logistics support. This includes procuring stores and managing inventories at optimal levels for a most-economical maintenance operation.

Certification

Aviation businesses must be identified as facilities qualified to conduct aircraft maintenance. Part 145—Repair Stations prescribes the requirements for obtaining a certificate as a repair station that authorizes aircraft maintenance work. This part prescribes the requirements for such a certificate, the procedure for obtaining the permit, general operating rules, and the various ratings issued. An applicant for a certificate applies on a form and, in a manner prescribed by the FAA, forwards its inspection procedures manual, a list of any outsourced maintenance functions and partners, and a list of parts and accessories to be maintained according to requested ratings. An amendment to Part 145 made effective in 2014 prohibits issuance of repair station certificates to individuals whose actions resulted in the revocation of a previous certificate. Figure 10.2 illustrates the application form used for a repair station certificate. The domestic repair stations may seek the following ratings:

Airframe ratings:
> Class 1: Composite construction of small aircraft;

Figure 10.2 → Federal Aviation Administration Application for Repair Station Certificate

OMB Approved
2120-0682
April 30, 2012

If additional space is required for any item, attach additional sheets of paper.

U.S. Department of Transportation
Federal Aviation Administration

Application for Repair Station Certificate and/or Rating

1. Repair Station Name, Number, Location and Address

a. Official Name of Station   Number

b. Location where business conducted

c. Official Mailing Address of Repair Station *(Number, Street, City, State & ZIP)*

d. Doing Business As:

2. Reasons for Submission

☐ Original Application for Certificate and Rating
☐ Change in Rating
☐ Change in Location or Housing and Facilities
☐ Change in Ownership
☐ Other *(Specify)*

3. Ratings Applied for:

☐ **Airframe**
  ☐ Class 1
  ☐ Class 2
  ☐ Class 3
  ☐ Class 4

☐ **Powerplant**
  ☐ Class 1
  ☐ Class 2
  ☐ Class 3

☐ **Propeller**
  ☐ Class 1
  ☐ Class 2

☐ **Radio**
  ☐ Class 1
  ☐ Class 2
  ☐ Class 3

☐ **Instrument**
  ☐ Class 1
  ☐ Class 2
  ☐ Class 3
  ☐ Class 4

☐ **Accessories**
  ☐ Class 1
  ☐ Class 2
  ☐ Class 3

☐ **Limited**
  ☐ Airframe
  ☐ Engine
  ☐ Propeller
  ☐ Instrument

  ☐ Accessories
  ☐ Landing Gear
  ☐ Float
  ☐ Radio

  ☐ Rotor Blades
  ☐ Fabric
  ☐ Emergency Equip.
  ☐ Non-Dest. Test

Specialized Services *(specify)*

4. List of Maintenance Functions Contracted to Outside Agencies:

5. Applicant's Certification

Name of Owner *(Include name(s) of individual owner, all partners, or corporation name giving state and date of incorporation)*

I hereby certify that I have been authorize by the repair station identified in Item 1 above to make this application and that statements and attachments hereto are true and correct to the best of my knowledge.

| Date | Authorized Signature | Printed Name of Authorized Signer | Title |

Paperwork Reduction Act Statement: This form is used to apply for certification, additional ratings, or a change to a repair station in accordance with 14 CFR part 145. The FAA estimates that the average burden for this report form is 15 minutes per response. An agency may not conduct or sponsor, and a person is not required to respond to, a collection of information unless it displays a currently valid OMB control number. The OMB control number associated with this collection is 2120-0682. You may submit any comments regarding the accuracy of this burden estimate or any suggestions for reducing the burden to the Federal Aviation Administration, Aircraft Maintenance Division, AFS-300, 800 Independence Ave, SW, Washington, DC 20591, Attention FAA Form 8310-3.

FAA Form 8310-3 (12-2011) Supersedes Previous Edition   NSN: 0052-00-686-1002

continued.

Figure 10.2 ✈ Continued

Record of Action Repair Station Inspection

For FAA Use Only | For FAA Use Only

6. Remarks (*identify by item number. Include deficiencies found, ratings denied.*)

7. Findings - Recommendations

- ☐ A. Station was found to comply with requirements of FAR 145.
- ☐ B. Station was found to comply with requirements of FAR 145 except for deficiencies listed in Item 6.
- ☐ C. Recommend certificate with rating applied for on application be issued.
- ☐ D. Recommend Certificate with rating applied for on application (EXCEPT those listed in item 6) be issued.

8. Date of Inspection

9. Office	Signature(s) of Inspector(s)	Printed Name(s) of Inspector(s)

10. Supervising or Assigned Inspector

ACTION TAKEN	CERTIFICATE ISSUED Number	Inspector's Signature	
☐ APPROVED as shown on certificate issued on date shown. ☐ DISAPPROVED	Date	Inspector's Printed Name	Title

FAA Form 8310-3 (12-2011) Supersedes Previous Edition — NSN: 0052-00-686-1002

> Class 2: Composite construction of large aircraft;
> Class 3: All-metal construction of small aircraft; and
> Class 4: All-metal construction of large aircraft.

Power plant ratings:
> Class 1: Reciprocating engines of 400 horsepower or less;
> Class 2: Reciprocating engines of more than 400 horsepower; and
> Class 3: Turbine engines.

Propeller ratings:
> Class 1: All fixed-pitch and ground-adjustable propellers of wood, metal, or composite construction;
> Class 2: All other propellers by make.

Radio ratings:
> Class 1: Communication equipment;
> Class 2: Navigational equipment; and
> Class 3: Radar equipment.

Instrument ratings:
> Class 1: Mechanical instruments;
> Class 2: Electrical instruments;
> Class 3: Gyroscopic instruments; and
> Class 4: Electronic instruments.

Accessory ratings:
> Class 1: Mechanical accessories;
> Class 2: Electrical accessories; and
> Class 3: Electronic accessories.

The specific requirements for each of the ratings are established in Part 145; often, the FAA issues a limiting rating that restricts the repair station to certain makes or models. The applicant is expected to provide its own personnel, equipment, and material for safe maintenance operations. Appendix VIII, on the website for this book, contains the job function required for each rating.

Personnel

Maintenance personnel are the key to success in the aviation maintenance business. To ensure that the maintenance department is successful and contributes positively to overall business profitability, the manager must be very familiar with personnel qualifications, training procedures, certification processes, and the unique capabilities or limitations of maintenance technicians. Growing shortages of qualified aircraft maintenance personnel mean all possible steps should be taken to recruit and train people who will stay.

Qualifications

14 CFR 65 Subpart D prescribes the requirements for an aviation mechanic and associated ratings, setting forth the general operating rules for holders of the certificate. Subpart D specifies the eligibility requirements for a certified mechanic as follows:

1. At least 18 years of age.
2. Able to read, write, speak, and understand the English language.
3. Has passed all the prescribed tests within a period of 24 months.
4. Meets the requirements for the specific rating requested.

A mechanic certificate can be obtained with an airframe or power-plant rating or both, if the applicant meets all the requirements. These requirements include knowledge as demonstrated in a written test, practical experience in the rating area sought, and skill demonstrated in an oral and practical test. The knowledge requirement calls for the applicant to pass a written test covering the construction and maintenance of aircraft appropriate to the rating he or she seeks and the regulations as contained in Parts 43, 65 and 91. The experience requirements may be met through completion of a certificated aviation maintenance technician school, satisfactory evidence of constructing, maintaining, or altering airframes or powerplants for 18 months, or satisfactory evidence of concurrently performing duties appropriate to both airframe and powerplant ratings for 30 months. The oral and practical tests cover the applicant's basic skill in executing the principles covered in the rating's written test.

Federal Aviation Regulation Part 147 was modified dramatically in 1994 to reflect the changing materials used in aircraft. Of significance was a reduction in the number of hours required for learning about radial engines, fabric covering of aircraft, and woodworking. The substantial new requirements added to the curriculum include information about composite materials manufacturing and repair, avionics and electronics, as well as turbine and jet knowledge.

Training

Completion of a certificated aviation maintenance technician program usually fulfills the training or experience requirements of an aviation mechanic. A review of the requirements for such a program will provide some insight into the training given an aviation mechanic. The following are the major elements of the curriculum required by the FAA:

> Airframe—1,150 hours (400 general plus 750 airframe);
> Power plant—1,150 hours (400 general plus 750 power plant); and
> Combined airframe and power plant—1,900 hours (400 general plus 750 airframe and 750 power plant).

Coverage of the following subject areas is included:

> Aircraft covering;
> Aircraft drawings;
> Aircraft electrical systems;
> Aircraft finishes;
> Aircraft fuel systems;
> Aircraft instrument systems;
> Aircraft landing gear systems;
> Airframe curriculum;
> Airframe inspection;
> Airframe systems and components;
> Assembly and rigging;
> Basic electricity;
> Basic physics;
> Cabin atmosphere control systems;
> Cleaning and corrosion control;
> Communication and navigation systems;
> Engine cooling systems;
> Engine electrical systems;
> Engine exhaust systems;
> Engine fire protection systems;
> Engine fuel systems;
> Engine inspection;
> Engine instrument systems;
> Fire protection systems;
> Fluid lines and fittings;
> Fuel metering systems;
> General curriculum;
> Ground operations and servicing;
> Hydraulic and pneumatic power systems;
> Ice and rain control systems;
> Ignition systems;
> Induction systems;
> Lubrication systems;
> Maintenance forms and records;
> Maintenance publications;
> Materials and processes;
> Mathematics;
> Mechanic privileges and limitations;
> Position and warning systems;
> Power plant curriculum;
> Power plant systems and components;
> Power-plant theory and maintenance;
> Reciprocating engines;
> Sheet metal structures;
> Structures;
> Turbine engines;
> Weight and balance;
> Welding; and
> Wood structures.

Propellers

The curriculum must:

> Include practical projects;
> Balance theory and other instruction;
> Show a schedule of required tests; and
> Provide that at least 50 percent of the total curriculum time is in shop and laboratory instruction.

The majority of maintenance technician training is accomplished in FAA certificated schools using the recommended and approved curriculum of study. The FAA maintains a directory of all certificated schools that can be referenced in locating and selecting a school. The web-based directory can be reached at http://avinfo.faa.gov/MaintenanceSchool.asp.

Training in specific airframes, powerplants, parts, and accessories is available from manufacturers or distributors of the various product; some specialty commercial providers such as FlightSafety International also provide training.[5] This training is most beneficial in enabling the maintenance technician to meet the FAA requirements and to understand the requirements of the manufacturer and the maintenance manuals for the equipment concerned. The maintenance manager should contact the appropriate manufacturer or distributor for a schedule of schools and programs offering specialized training, and subsequently arrange attendance at the sessions required by his or her personnel.

Certification

A candidate for an aviation mechanic's certificate must apply for the certificate and associated rating on a form prescribed by the Federal Aviation Administration. After review of the application and the supplementary documents indicating successful completion of the requirements in the knowledge, experience and skill areas, a certificate and the appropriate ratings will be issued by the FAA.

Capabilities and Limitations

A certified mechanic is authorized by the FAA to perform certain functions if they meet specified requirements. They may perform or supervise the maintenance or alteration of an aircraft or appliance, or part thereof, for which they are rated. If they have an airframe rating, they may approve and return to service any airframe, or any related part of appliance, after they have performed, supervised, or inspected its maintenance or alteration—excluding major repairs and major alterations, or after they have completed the 100-hour inspection required by Part 91. If they hold a power-plant rating, they may accomplish the same for a power-plant, propeller, or any related part or appliance. An inspection authorization allows its holder to inspect and approve aircraft for return to service.

To exercise these privileges, the mechanic must meet the following requirements:

1. He or she must have satisfactorily performed the work concerned at an earlier date, showed the FAA his/her ability to do it, or accomplished the work under the supervision of a certified mechanic who has had previous experience.
2. He or she must understand the current instructions of the manufacturer and the maintenance manuals for the specific operation concerned in order to exercise the privileges of his/her certificate and rating.
3. He or she must have recent experience. The FAA has to find that either the mechanic was able to perform the work within the preceding 24 months or the mechanic has supervised other mechanics or otherwise worked in an executive capacity overseeing the maintenance or alteration of aircraft for at least 6 months.

Inspection Authorization

The next level of experience and responsibility for aviation mechanics is obtaining an inspection authorization. The holder of this authorization may inspect and approve for return to service any aircraft or related part or appliance (except those under a Part 121 or 127 continuous airworthiness program) after a major repair or major alteration. The holder of this authorization may also perform an annual inspection and perform or supervise a progressive inspection.

Candidates for an inspection authorization must meet additional requirements to qualify for this position. They must have been a certified mechanic for at least three years. For the last two years they must have been actively engaged in maintaining aircraft; have a fixed base of operations; have available the equipment, facilities and inspection data necessary to properly inspect aircraft; and pass a written test on their ability to inspect according to safety standards for returning aircraft to service after major repairs, major alterations, and annual and progressive inspections. The inspection authorization expires on March 31 of each year. To be renewed, the holder must show evidence that they (1) performed at least one annual inspection for each ninety days that he held the authority, (2) performed inspections of at least two major repairs or major alterations for each 90 days that he or she held the authority, or (3) performed or supervised and approved at least one progressive inspection in accordance with prescribed standards. Continuing education via annual FAA "refresher courses" is also required.

Repairmen

A repairman certificate may be issued to personnel employed by a certified repair station, a certified commercial operator, or a certified air carrier that is required by its operating certificate or approved operations specification to provide a continuous airworthiness maintenance program according to its maintenance manuals. These personnel must be specially qualified to perform maintenance, be employed for a specific job requiring those special qualifications, and be recommended for certification by the employer. Additional requirements include 18 months of practical experience and English language qualifications.

A certified repairman may perform or supervise the maintenance of an aircraft or its components for the job they were employed and certified, but only for their employer. They must perform these duties in accordance with the current instructions of their employer, the manufacturer of the article being maintained, and appropriate maintenance manuals.

Facilities and Equipment

Overview

A manager with the desire to operate a maintenance department must meet specific FAA requirements for facilities and equipment in order to qualify for a repair station certificate and rating. The facilities requirements include the following:

1. Housing for necessary equipment and material.
2. Space for the work to be accomplished under the rating.
3. Facilities for properly storing, segregating, and protecting materials, parts, and supplies.
4. Facilities for properly protecting parts and subassemblies during disassembly, cleaning, inspection, repair, alteration, and assembly.
5. Suitable shop space where machine tools and equipment are kept and where the largest amount of bench work is done.
6. Suitable assembly space in an enclosed structure where the largest amount of assembly work is done. The space must be adequate for the work on the largest item covered by the rating requested.
7. Suitable storage facilities used exclusively for storing standard parts, spare parts, and raw materials. This area must be separated from the shop and working space, organized so that only acceptable parts and supplies will be issued for any job, and follow standard good practices for properly protecting stored materials.
8. Adequate storage and protection for parts being assembled or disassembled or awaiting work in order to eliminate the possibility of damage.
9. Suitable ventilation for the shop, assembly area, and storage area so that the physical efficiency of workers is not impaired.
10. Adequate lighting for all work being done so that the quality of the work is not impaired.
11. Temperature and humidity of the shop and assembly area must be controlled so that the quality of the work is not impaired.
12. For an airframe rating, there must be suitable permanent housing for at least one of the heaviest aircraft within the weight class of the rating.
13. For a power-plant or accessory rating there must be suitable trays, racks, or stands for segregating complete assemblies during assembly and disassembly.
14. For a propeller rating, there must be suitable stands, racks, or other fixtures for proper storage of propellers.
15. For a radio rating, there must be suitable storage facilities to assure protection of parts and units from moisture.
16. An instrument shop must be reasonably dust-free, preferably air-conditioned.

The equipment requirements specified by the Federal Aviation Regulations for a repair station include the following:

1. The equipment and materials necessary to perform the functions of the ratings held must be present (see Appendix VIII on the book's website for a list of these functions).
2. The equipment and materials required must be those that can properly and efficiently accomplish the work. All inspection and test equipment shall be tested at regular intervals to ensure correct calibration.
3. Equipment and materials required for the various job functions must be located on the premises and under the full control of the station unless related to an authorized contract.

Managerial Concerns

The manager of a maintenance facility must be concerned with additional requirements in developing and operating the business. Included are:

> A concern for the economic efficiency of the facilities and equipment;
> The working environment for all employees;
> The image presented by the facility to the customers and the general public;
> Fire, safety, and construction laws established by city, state and federal agencies;

- Requirements of manufacturers and distributors, if operating under a franchise arrangement; and
- Special requirements that may be generated by the manager or other source.

The equipment needed to furnish a maintenance shop for business represents a large capital investment. This significant, initial investment is a major concern due to the industry's rapidly changing technology and resulting changes in aircraft construction, components, and systems. Practical problems in this area include:

- Determining the acquisition schedules and inventory levels of equipment or tools;
- Managing the control of equipment or tools;
- Ordering the replacement and repair of facilities and equipment;
- Monitoring maintenance capability and capacity;
- Ensuring full utilization of facilities; and
- Deciding whether to perform specialized work or to contract out to other FBOs.

Parts and Supplies

The inventory of maintenance supplies and parts provides another challenge to the manager, the maintenance department, and the entire organization. The goal is to maintain the proper balance between capital invested in supplies and parts, utilization of the mechanic's labor time, and customer satisfaction. An unlimited inventory of parts and supplies enables the manager to efficiently utilize maintenance labor and provide maximum customer satisfaction through rapid repair work. Such an investment level in inventory is prohibitive to any business, as cost of maintaining it far outweighs the advantages, and results in a financial drain on the organization. Figure 10.3 shows a parts sales summary from FBO Manager.

On the other hand, a minimal inventory may result in an inefficient utilization of shop personnel, customer dissatisfaction, and delay of maintenance operations. Consequently, shop personnel experience shortages, frequent waiting, and poor utilization of shop space. Customers find themselves waiting for parts to be located, ordered, and delivered. Customer dissatisfaction likely correlate directly with waiting times. All these elements can strongly influence the profitability of the overall maintenance activity.

The manager, as part of his responsibility, must consider all these factors and take the appropriate action. He must establish guidelines to be used in the operation of the parts department, identify the criteria to be used in evaluating the operation, and establish controls to be utilized in maintaining direction.

Inventory Control

A good system of control for parts and supplies will greatly improve buying and selling practices and result in a more efficient and effective service activity. Such a system will provide information on the following:

1. Optimal quantities and qualities to buy.
2. Popularity of certain items.
3. Amounts of given items regularly sold.
4. Seasonality trends in buying items.
5. Seasonality trends in selling items.
6. Types of goods wanted or needed by customers.
7. Times to display or promote items.
8. Items that are slow to turnover.
9. Items with changing demand.
10. Sources and prices of alternative suppliers.
11. Prices and details of nearby competitors.
12. Price and time sensitivities of customers.
13. Possibilities for new lines or kinds of goods.
14. Proper turnover balancing of current inventory.

A system that provides this information to the manager will eliminate many inventory problems and contribute to the success of the maintenance shop. Efficient and profitable control of the inventory of parts and supplies does not have to be difficult, complex, or expensive. Many manufacturers, agencies, and associations can furnish good inventory systems to aviation businesses. Most of these systems are economical, simple, and intuitive to operate. They are easy to install, scalable to fit the business need, and utilized in a few minutes each day by personnel with little training. Typical of the systems available to aviation businesses is the inventory control system made available by Piper Aircraft, Inc. to the organizations

Figure 10.3 ✈ Parts Sales Summary

Cornerstone Logic Inc.

Sales by Part Name for 11/01/1998 to 12/31/2001

Name	Description	Qty Sold	Taxes	Total (No Tax)
00-10-009	Narco ELT Battery	66.00	$82.94	$7,115.24
0050505725007	Paper Clips	248.16	$7.20	$1,017.64
13836	Bearing Race	88.80	$16.80	$749.65
13889CP RLR B	Timkin Tapered Roller Bearing	53.20	$27.74	$1,261.45
150-SR	Cessna 150 Seat Rails	44.00	$152.97	$3,957.80
1815179171001	Austria Book	19.00	$0.00	$246,399.33
250	250	570.00	$0.27	$33,094.00
251	251	6.00	$0.00	$0.00
252	252	4.00	$0.00	$17.00
52382	Baron Exhaust Stack	4.00	$128.99	$2,457.00
631544	Exhaust Gasket	39.00	$3.18	$72.82
631832	Upper Bracket	14.00	$3.00	$2,937.18
642917-1	642917-1 Intake Hose	38.00	$7.46	$358.19
649290	Spacers	110.00	$97.72	$2,042.23
649959	Gasket	4.00	$0.06	$1.60
652314	Lower Bracket	5.00	$6.00	$149.75
9781560273554	ASA Airframe Test Guide	15.00	$0.00	$579.15
AC Cleaning	Aircraft Cleaning	1.00	$0.00	$75.00
Aeroshell 50W	Aeroshell 50W	291.92	$47.60	$848.79
AN 4-5 A Bolt	AN 4-5 Hex Head Bolt	228.00	$0.00	$134.52
AN822-4D	90 Degree Fitting	318.00	$11.28	$18,291.36
Annual 36	Annual Inspection 36 Hours	2.00	$0.00	$3,200.00
APU	APU	1.00	$0.00	$150.00
AS 50/100	Aeroshell 50/100W	30.00	$7.02	$117.00
ASA-00-6A	Aviation Weather	21.00	$0.00	$252.00
ASA-61-21A	Flight Training Handbook	15.00	$0.00	$150.53
ASA-61-27C	Instrument Flying Handbook	17.00	$2.73	$146.00
ASA-97-FR-AM	1997 FAR/AIM	-51.00	$29.53	$3,303.54
ASA-BAG-FLT	Flight Bag	70.00	$0.00	$4,055.66
ASA-SP-30	Pilot Log Book - Black	25.00	$0.70	$173.75
CallOutFee	Call Out Fee 0-50 Gallons	3.00	$0.00	$200.00
Car Rental	Rental Car	1.00	$0.00	$78.00
Catering	Catering	7.00	$0.00	$1,010.87
CH48109	Champion Oil Filter	21.00	$0.17	$387.66
Charter Ops	Charter Operations	300.00	$0.00	$897.00
CreditCardFee	Credit Card Fees	20.00	$0.00	$5,839.41
CreditTransAc	Credit Card Transition Account	726.45	$0.00	$726.45
DeIcing	De-Icing	1.00	$0.00	$50.00

Courtesy of FBO Manager by Cornerstone Logic, Inc.

providing service to Piper aircraft. Figure 10.4 illustrates the preprinted inventory card that is part of the basic system.

There are three common methods of securing the information needed for inventory control in an aviation activity: observation, physical check, and perpetual inventory. Observation is used by the smaller aviation businesses and may be sufficient where the number of items is not large, the flow of sales is fairly constant, and the manager is in close daily contact with all the suppliers. The physical check of the inventory does become desirable, or even necessary, when the rate of sales varies on some basis or when there are very expensive items in inventory. This kind of periodic stock-taking is a physical count of the parts and supplies at a point in time, which can then be compared to the number received and sold in prior periods to estimate the amount needed to replenish the stock. The perpetual inventory record, as illustrated in Figure 10.4, allows the manager to know at all times the amount of goods that are on hand or on order.

Economic Ordering Quantity. (EOQ) In maintaining a sound inventory control, a manager must determine the most economical ordering quantity for the various materials in the aviation parts department. The most economical amount of material to purchase at one time and at a given price is that quantity where the total cost per unit is at a minimum. This low point occurs when the unit cost of preparing the purchase order for that quantity is equal to the unit cost of carrying the material in the supply room. In other words, the costs of inventory acquisition (ordering) must be balanced against inventory possession (storing). Figure 10.5 graphically presents the basic EOQ formula. The fundamental cost relationships may be expressed by the following formula:

$$Q = \sqrt{2RA/P}$$

Q = Quantity to order
R = Annual stock requirements
A = Acquisition costs per order (including shipping costs)
P = Possession costs of holding one unit of inventory for one year

When maintaining good inventory control, it is important that the manager know when to order the material as well as the correct quantity to order. In determining the correct reorder point, it is necessary that two factors be considered: (1) the consumption

Figure 10.4 ✈ Piper Aircraft Inventory Card

PART NO. ____________ ITEM ____________ MIN. ____________ LOCATION ____________
MFGR ____________ STD. PAGE ____________ MAX. ____________ LIST PRICE ____________

DATE	ORDERED	RECEIVED	SOLD	BALANCE	DATE	ORDERED	RECEIVED	SOLD	BALANCE

PART NO. ITEM

Figure 10.5 ✈ Basic EOQ Graph

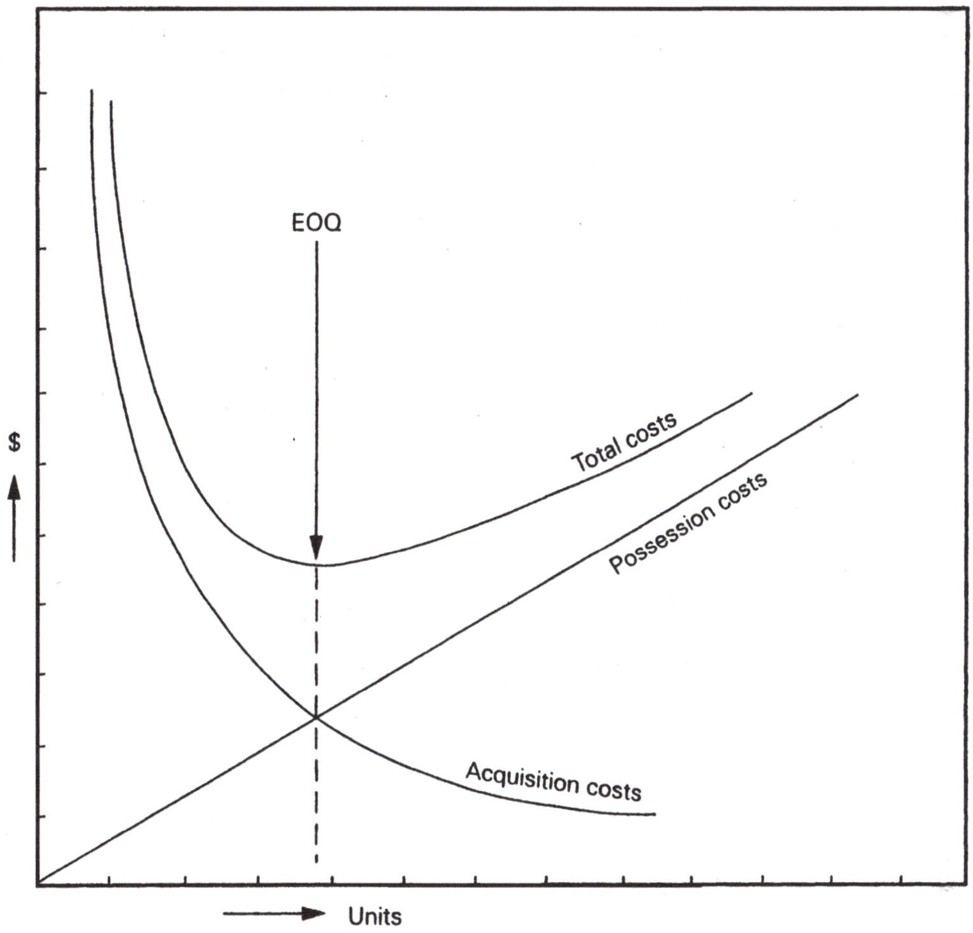
EOQ
Total costs
Possession costs
Acquisition costs
$
Units

Source: Dr. J. D. Richardson

rate in units; and (2) the time required for procurement. Normally, since an out-of-stock situation is costly, the manager adds a buffer of a minimum safe inventory level to the theoretical ordering point in order to arrive at the actual reordering point. The following example illustrates this procedure with spark plugs of a particular type:

1200	Spark plugs used annually
2	Weeks reorder time
46	Theoretical reorder point (1200 ÷ 26)
25	Average one-week safety level
46 + 25 = 71	Actual reordering point

More and more businesses are using Just-In-Time (JIT) ordering of inventory, where less is maintained in stock because required items can be obtained very fast and often overnight by air express. The items touched upon in this section represent the major elements of inventory control—the inventory records and a replenishment system. To realize the benefits listed at the beginning of this section, the manager must install the system and utilize the data from records. Careful analysis of accumulated inventory data leads to decisions that keep inventory within bounds.

Quality Control

High-quality maintenance services are important in developing and maintaining customer satisfaction. Quality control is a key method for ensuring that a maintenance department consistently produces a product that results in high customer satisfaction. Although the term "quality control" may produce an unpleasant image in some minds—akin to that of police officer or monitor—the concept actually should be much more agreeable and supportive. The best quality control is achieved when the work environment lets individual employees know that good work is desired and expected, and they strive to achieve this as a personal as well as an organizational goal. Fortunately, there are some specific steps a manager can take to foster this attitude throughout the maintenance organization:

1. Provide training in the specific procedures to be followed and requirements to be met for each maintenance operation.
2. Provide checklists and guidelines to facilitate quality work.
3. Arrange for the inspection of finished work by a different person than the person doing the work, and create an environment in which this process is understood and desired.
4. Recognize quality work and the individuals responsible for its accomplishment.

Training

Aviation mechanics who graduate from certified technical schools have a sound educational basis for their job. Throughout their training the need for quality work is emphasized. There is a need for continued training in the specific procedures to be followed and in the requirements of quality maintenance, repair, and alteration. This continued training is necessary for new procedures, new materials, new equipment, new requirements, and to refresh existing knowledge and skills. Some of this training can be accomplished on-site in the local shop, while other training must be accomplished at plants, schools, or in training seminars. The organization's emphasis on quality will develop the desired attitude toward a quality product.

Checklists

A great deal of aviation maintenance work includes lengthy procedures with many safety checks and requirements. To ensure that all the required work is accomplished in the proper sequence with the desired safety checks, guidelines must be developed and implemented. Figure 10.6 illustrates a typical list developed to ensure that all necessary items are checked on the inspection of a new aircraft after its first 50 hours. Operating and maintenance manuals frequently use checklists and procedure guides to ensure that disassembly, repair, and assembly are accomplished in proper sequence. One problem in the use of checklists is the tendency for some individuals to work from memory, and then later sign off the checklist as having been completed. Unfortunately, since this practice can lead to items being overlooked and subsequent difficulties, it is highly desirable that individuals and the organization strive to develop work patterns that utilize checklists and guides as a standard procedure, each check-off is completed as each task is performed.

Inspection

In the regulations dealing with maintenance personnel and maintenance procedures, the Federal Aviation Administration has provided for inspectors and inspection procedures. Quality of work and safety of flight are of paramount importance, and this may be accomplished in part by segregation of repair and cross-check duties. This procedure is aimed at providing a degree of objectivity and a fresh, alternative perspective that will catch items overlooked or incorrectly performed. Of course, the success of an inspection procedure will depend largely upon the qualifications and thoroughness of the inspecting personnel. The attitude of employees and the working environment must be such that the inspection procedure is universally accepted.

Recognition

One important aspect of obtaining quality work by a maintenance unit is the prompt recognition of high-quality work and of the individuals who have produced that work. This recognition can take many forms, ranging from personal praise, unit recognition, and publicity to financial incentives. Of course, the recognition must be sincere. Acknowledgement of low-quality work must also ensure that it does not become the accepted standard. Verbal reminders, corrective action, discipline, and even termination may be employed to demonstrate that low quality is not acceptable.

Balance

Quality control, like other business activities, is costly. Additional quality can be obtained with additional training, more checklists and guidelines, more inspectors, and lavish recognition programs. At some point the question must be asked: "How much quality is desired—and how much can we afford?" The

Figure 10.6 → **Aircraft Maintenance Inspection Checklist**

PIPER AIRCRAFT CORPORATION
INSPECTION REPORT
THIS FORM MEETS REQUIREMENTS OF FAR PART 43

MAKE: PIPER CHEROKEE MODEL: PA-28R-180 / PA-28R-200 Serial No. Registration No.

Circle Type of Inspection (See Note): 50 100 500 1000 Annual

Perform inspection or operation at each of the inspection intervals as indicated by a circle (○).

A. PROPELLER GROUP

1. Inspect spinner and back plate (See Note 5).
2. Inspect blades for nicks and cracks.
3. Check for grease and oil leaks.
4. Lubricate propeller per lubrication chart.
5. Check spinner mounting brackets.
6. Check propeller mounting bolts and safety (Check torque if safety is broken.).
7. Inspect hub parts for cracks and corrosion.
8. Rotate blades of constant speed propeller and check for tightness in hub pilot tube.
9. Remove constant speed propeller, remove sludge from propeller and crankshaft.
10. Inspect complete propeller and spinner assembly for security, chafing, cracks, deterioration, wear, and correct installation.
11. Overhaul propeller.

B. ENGINE GROUP

CAUTION: Ground Magneto Primary Circuit before working on engine.

1. Remove engine cowl.
2. Clean and check cowling for cracks, distortion and loose or missing fasteners.
3. Drain oil sump (See Note 2).
4. Clean suction oil strainer at oil change (Check strainer for foreign particles.).
5. Clean pressure oil strainer or change full flow (cartridge type) oil filter element (Check strainer or element for foreign particles).
6. Check oil temperature sender unit for leaks and security.
7. Check oil lines and fitting for leaks, security, chafing, dents and cracks (See Note 4).
8. Clean and check oil radiator cooling fins.
9. Remove and flush oil radiator.
10. Fill engine with oil per lubrication chart.
11. Clean engine.

CAUTION: Do not contaminate the vacuum pump with cleaning fluid. Ref: Lycoming Service Letter 1221A.

12. Check condition of spark plugs (Clean and adjust gap as required, adjust per Lycoming Service Instruction No. 1042.).
13. Check cylinder compression ref: AC43. 13-1
14. Check ignition harness and insulators (High tension leakage and continuity.).
15. Check magneto points for proper clearance (Maintain clearance at 0.016.).
16. Check magneto for oil leakage.
17. Check breaker felts for proper lubrication.
18. Check distributor block for cracks, burned areas or corrosion, and height of contact springs.

19. Check magnetos to engine timing.
20. Overhaul or replace magnetos (See Note 3).
21. Remove air filter and tap gently to remove dirt particles (Replace as required.).
22. Clean fuel injector inlet line screen (Clean injector nozzles as required.) (Clean with acetone only).
23. Check condition of injector alternate air door and box.
24. Check intake seals for leaks and clamps for tightness.
25. Inspect all air inlet duct hoses (Replace as required.).
26. Inspect condition of flexible fuel lines.
27. Replace flexible fuel lines (See Note 3).
28. Check fuel system for leaks.
29. Check fuel pumps for operation (Engine driven and electric.).
30. Overhaul or replace fuel pumps (Engine driven and electric (See Note 3).
31. Check vacuum pump and lines.
32. Overhaul or replace vacuum pump (See Note 3).
33. Check throttle, alternate air, mixture and propeller governor controls for travel and operating condition.
34. Inspect exhaust stacks, connections and gaskets (Refer to PA-28 Service Manual, Section III. Replace gaskets as required.).
35. Inspect muffler, heat exchange and baffles (Refer to PA-28 Service Manual, Section III.).
36. Check breather tube for obstructions and security.
37. Check crankcase for cracks, leaks and security of seam bolts.
38. Check engine mounts for cracks and loose mountings.
39. Check all engine baffles.
40. Check rubber engine mount bushings for deterioration (Replace as required.).
41. Check firewall seals.
42. Check condition and tension of alternator drive belt (Refer to PA-28 Service Manual.).
43. Check condition of alternator and starter.
44. Check fluid in brake reservoir (Fill as required.).
45. Lubricate all controls.
46. Overhaul or replace propeller governor (See Note 3).
47. Complete overhaul of engine or replace with factory rebuilt (See Note 3).
48. Reinstall engine cowl.

Owner 508

continued.

Figure 10.6 ✈ Continued

Circle Type of Inspection (See Note) 50 100 500 1000 Annual DESCRIPTION	50	100	500	1000	Inspector
C. CABIN GROUP					
1. Inspect cabin entrance, doors and windows for damage and operation.		○	○	○	
2. Check upholstery for tears.		○	○	○	
3. Checks seats, seat belts, security brackets and bolts.		○	○	○	
4. Check trim operation.		○	○	○	
5. Check rudder pedals.		○	○	○	
6. Check parking brake and brake handle for operation and cylinder leaks.		○	○	○	
7. Check control wheels, column, pulleys and cables.		○	○	○	
8. Check landing, navigation, cabin and instrument lights.	○	○	○	○	
9. Check instruments, lines and attachments.		○	○	○	
10. Check gyro operated instruments and electric turn and bank (Overhaul or replace as required.).		○	○	○	
11. Replace filters on gyro horizon and directional gyro or replace central air filter.		○	○	○	
12. Clean or replace vacuum regulator filter.		○	○	○	
13. Check altimeter (Calibrate altimeter system in accordance with FAR 91.170, if appropriate.).		○	○	○	
14. Check operation of fuel selector valve.		○	○	○	
15. Check condition of heater controls and ducts.		○	○	○	
16. Check condition and operaton of air vents.		○	○	○	
D. FUSELAGE AND EMPENNAGE GROUP					
1. Remove inspection plates and panels.		○	○	○	
2. Check baggage door, latch and hinges.		○	○	○	
3. Check battery, box and cables (Check at least every 30 days. Flush box as required and fill battery per instructions on box.).	○	○	○	○	
4. Check electronic installation.		○	○	○	
5. Check bulkheads and stringers for damage.		○	○	○	
6. Check antenna mounts and electric wiring.		○	○	○	
7. Check hydraulic pump fluid level (Fill as required.).	○	○	○	○	
8. Check hydraulic pump lines for damage and leaks.		○	○	○	
9. Check for obstructions and contamination in inlet of back-up landing gear extender actuator inlet head.		○	○	○	
10. Check fuel lines, valves and gauges for damage and operation.		○	○	○	
11. Check security of all lines.		○	○	○	
12. Check vertical fin and rudder surfaces for damage.		○	○	○	
13. Check rudder hinges, horn and attachments for damage and operation.		○	○	○	
14. Check vertical fin attachments.		○	○	○	
15. Check rudder hinge bolts for excess wear (Replace as required.).		○	○	○	
16. Check stabilator surfaces for damage.		○	○	○	
17. Check stabilator, tab hinges, horn and attachments for damage and operation.		○	○	○	
18. Check stabilator attachments.		○	○	○	
19. Check stabilator and tab hinge bolts and bearings for excess wear (Replace as required.).		○	○	○	
20. Check stabilator trim mechanism.		○	○	○	
21. Check aileron, rudder, stabilator, stabilator trim cables, turnbuckles, guides and pulleys for safety, damage and operation.		○	○	○	
22. Clean and lubricate stabilator trim drum screw					
23. Clean and lubricate all exterior needle bearings					

Perform inspection or operation at each of the inspection intervals as indicated by a circle (○). DESCRIPTION	50	100	500	1000	Inspector
24. Lubricate per lubrication chart.	○	○	○	○	
25. Check rotating beacon for security and operation.		○	○	○	
26. Check security of AutoPilot bridle cable clamps		○	○	○	
27. Inspect all control cables, air ducts, electrical leads, lines, radio antenna leads and attaching parts for security, routing, chafing, deterioration, wear, and correct installation.		○	○	○	
28. Reinstall inspection plates and panels.		○	○	○	
E. WING GROUP					
1. Remove inspection plates and fairings.		○	○	○	
2. Check surfaces and tips for damage, loose rivets, and condition of walk-way.		○	○	○	
3. Check aileron hinges and attachments.		○	○	○	
4. Check aileron cables, pulleys and bellcranks for damage and operation.		○	○	○	
5. Check flaps and attachments for damage and operation.		○	○	○	
6. Check condition of bolts used with hinges (Replace as required.).		○	○	○	
7. Lubricate per lubrication chart.		○	○	○	
8. Check wing attachment bolts and brackets.		○	○	○	
9. Check fuel tanks and lines for leaks and water		○	○	○	
10. Fuel tanks marked for capacity.		○	○	○	
11. Fuel tanks marked for minimum octane rating.		○	○	○	
12. Check fuel cell vents.		○	○	○	
13. Inspect all control cables, air ducts, electrical leads, lines and attaching parts of security, routing, chafing, deterioration, wear, and correct installation.		○	○	○	
14. Reinstall ispection plates and fairings.		○	○	○	
F. LANDING GEAR GROUP					
1. Check oleo struts for proper extension (N-2.75 in. /M-2.0in.) (Check fluid level as required.).		○	○	○	
2. Check nose gear steering control and travel.		○	○	○	
3. Check wheels for alignment.		○	○	○	
4. Put airplane on jacks.		○	○	○	
5. Check tires for cuts, uneven or excessive wear and slippage.		○	○	○	
6. Remove wheels, clean, check and repack bearings.		○	○	○	
7. Check wheels for cracks, corrosion and broken bolts.		○	○	○	
8. Check tire pressure (N-30 psi/M-27 psi).	○	○	○	○	
9. Check brake lining and disc.		○	○	○	
10. Check brake backing plates.		○	○	○	
11. Check brake and hydraulic lines.		○	○	○	
12. Check shimmy dampener.		○	○	○	
13. Check gear forks for damage.		○	○	○	
14. Check oleo struts for fluid leaks and scoring.		○	○	○	

continued.

Figure 10.6 ✈ Continued

Circle Type of Inspection (See Note) 50 100 500 1000 Annual DESCRIPTION	50	100	500	1000	Inspector	Perform inspection or operation at each of the inspection intervals as indicated by a circle (○). DESCRIPTION	50	100	500	1000	Inspector
15. Check gear struts, attachments, torque links, retraction links and bolts for condition and security.		○	○	○		**G. OPERATIONAL INSPECTION**					
16. Check downlock for operation and adjustment.		○	○	○		1. Check fuel pump and fuel tank selector.		○	○	○	
17. Check torque link bolts and bushings (Rebush as required.).			○	○		2. Check fuel quantity, pressure and flow readings		○	○	○	
18. Check drag and side brace link bolts (Replace as required.).			○	○		3. Check oil pressure and temperature.		○	○	○	
19. Check gear doors and attachments.		○	○	○		4. Check alternator output.		○	○	○	
20. Check warning horn and light for operation.		○	○	○		5. Check manifold pressure.		○	○	○	
21. Retract gear – check operation.		○	○	○		6. Check alternate air.		○	○	○	
22. Retract gear – check doors for clearance and operation.		○	○	○		7. Check parking brake.		○	○	○	
23. Check anti-retraction system.		○	○	○		8. Check vacuum gauge.		○	○	○	
24. Check actuating cylinders for leaks and security.		○	○	○		9. Check gyros for noise and roughness.		○	○	○	
25. Inspect all hydraulic lines, electrical leads, and attaching parts for security, routing, chafing, deterioration, wear, and correct installation.		○	○	○		10. Check cabin heater operation.		○	○	○	
						11. Check magneto switch operation.		○	○	○	
						12. Check magneto RPM variation.		○	○	○	
						13. Check throttle and mixture operation.		○	○	○	
						14. Check propeller smoothness.		○	○	○	
						15. Check propeller governor action.		○	○	○	
26. Check position indicator switch and electrical leads for security.		○	○	○		16. Check engine idle.		○	○	○	
27. Lubricate per lubrication chart.	○	○	○	○		17. Check electronic equipment operation.		○	○	○	
28. Remove airplane from jacks.		○	○	○		**H. GENERAL**					
						1. Aircraft conforms to FAA Specification.		○	○	○	
						2. All FAA Airworthiness Directives complied with.		○	○	○	
						3. All Manufacturers Service Bulletins and Letters complied with.		○	○	○	
						4. Check for proper Flight Manual.		○	○	○	
						5. Aircraft papers in proper order.		○	○	○	

NOTES:
1. Both the annual and 100 hour inspections are complete inspections of the airplane – identical in scope. Inspections must be accomplished by persons authorized by FAA.
2. Intervals between oil changes can be increased as much as 100% on engines equipped with full flow (cartridge type) oil filters – provided the element is replaced each 50 hours of operation.
3. Replace or overhaul as required or at engine overhaul. (For engine overhaul, refer to Lycoming Service Instructions No. 1009.)
4. Replace flexible oil lines as required, but no later than 1,000 hours of service.
5. Inspect in accordance with Piper Service Bulletin 309.

REMARKS:

Signature of Mechanic or Inspector	Certificate No.	Date	Total Time on Airplane

Courtesy Piper Aircraft Corp.

manager must obviously determine the desirable balance between the costs involved and the acceptable level of quality, and then strive to maintain that balance.

Competition

Nonexclusive Rights

Airports that have been federally funded (or that aspire to be) are obligated to permit all aeronautical operators who can be accommodated to use the field, provided they meet certain standards. Thus, existing FBOs are constantly threatened by the prospect of new competition at federally-funded airports. While this, in some ways, is in the best interest of the general public, it also adds difficulty to profitability at small airports.

Referrals

As a result of the competitive environment facing many FBOs, some prefer to refer their maintenance and repair work to other operators on the field or elsewhere. This referral process generally works well in both directions until there is a change in the performance of the other business. In this case not only does the FBO have very little control over its clients' maintenance needs, but also the poor service from the maintenance operator may drive those clients to another field for all their needs.

Outsourcing

Aviation managers frequently find themselves considering the feasibility of outsourcing some, or all, of their maintenance or avionics work. It may be that the light maintenance workload does not warrant having a shop or that it is not adequate to support the purchase of some special and expensive test equipment. Also, it may not be sufficient to acquire and utilize the necessary skilled avionics maintenance technicians. Any of these concerns might cause the manager to outsource maintenance work. The decision of when to outsource is normally based upon a consideration of the points just identified.

The use of financial comparisons may also be valuable in arriving at a decision on outsourcing. In each situation it becomes necessary to accumulate all the cost and income data related to the decision while being as objective as possible. Various factors that may not be easily quantified but still must be considered include available human resources, management capability, space utilization, time constraints, quality control, and customer reaction. All these elements should be considered in a rational decision-making process as described in Chapter 2, Management Functions.

When maintenance work is outsourced, the amount should be recorded, and periodically the volume of work and the overall situation should be analyzed in order to verify the decision to continue outsourcing. This analysis should be done to identify the point at which it is economically or strategically advantageous to acquire the people, equipment, space, and financing needed to perform maintenance work within the organization.

Problems. Several problem areas relate to outsourcing should be identified by the manager, considered in the decision-making process, and monitored on a day-by-day basis. Included are:

1. Negotiating an equitable contract and monitoring its operation.
2. Handling warranty questions and documentation to the satisfaction of the customer and the contractor.
3. Dealing with liability situations in the legal environment.
4. Providing for customer satisfaction and customer identification with the primary business as well as the subcontractor.
5. Ensuring that the outsourcing operation is profitable on a direct job and strategically favorable over the near-term.

Recognizing these problem areas and striving to overcome their potential difficulties becomes the responsibility of the manager and a challenge to administrative skills.

"Through-the-Fence" Operations

This type of activity refers to FBOs who are not airport tenants but own or lease adjacent land and access the airport through a real or imaginary fence at the property line. They can siphon off business from the FBO(s) on the field who are paying for the operation of the airfield, taxiways and other public areas through airport leases. In instances that a new through-the-fence proposal is made, the FAA recommends some safeguards for protecting existing tenants. But for situations that have existed for many years, as often seems to be the case, it can devastate on-airport FBO profitability. The airport owner has no real control over such an operator, except denying access until acceptable financial and performance arrangements are established. In a case where a through-the-fence operator has an existing access easement with the airport owner, it may be difficult to negate or tighten its terms. Ideally, an airport access fee should be charged.

As available land on public airports becomes scarcer, the public airport system seems likely to receive more requests for through-the-fence rights, and the semi-active through-the-fence operator may become more of a challenge also. FBOs whose airports have potential or actual sites for through-the-fence activity need to make their own interests and concerns clearly known to the public airport operator.

The FAA Modernization and Reform Act of 2012 outlines specific conditions and limitations that must be present in through-the-fence access agreements to comply with that law. This provision took effect on October 1, 2014. Airport managers should therefore make themselves aware of the law and corresponding FAA guidance material.

Tailgate, Shade Tree, and Gypsy Mechanics

As costs of repairs go up, more people are repairing their own aircraft. Provided they comply with the required inspections, this is a legitimate activity. However, commercial repair services unofficially operating in individual hangars or other locations are not legitimate. All aeronautical operators providing services for sale on a public airport must have permission and an appropriate lease. Illegal operators may be doing poor work that the legitimate FBOs may later have to correct, even to the point of liability for aircraft originally sold but never maintained by them. The "gypsy" operator conveniently disappears when trouble starts. Thus, as with through-the-fence operators, the FBO must be ever diligent in reporting such activities and insisting on enforcement of the same minimum standards for all.

Corporate and Other Self-Maintenance

Like self-fueling, this is a growing trend for economic reasons and is permitted under nonexclusive rights provisions at federally-funded airports. The FBO may still be able to capture some maintenance revenue by performing specialized work, FAA inspections, and parts sales, among others.

Administration

The administration of a maintenance department and associated paperwork is another area requiring managerial attention. Administrative functions can and should consume a fair amount of time. It cannot be over-emphasized that appropriate record-keeping in accordance with FAA requirements is essential to the long-term health of the business. The following activities are included in the term "maintenance administration":

1. Developing, understanding, and administering the overall maintenance organization and system.
2. Developing and maintaining the necessary maintenance procedures, records, archives and library.
3. Complying with the necessary records and entries required in accomplishing maintenance repair or alteration.
4. Maintaining the necessary company records associated with accounting, budgeting, work orders, and time cards.
5. Dealing with FAA regulations, procedures, and inspections.

These major activities will vary in content and meaning from business to business. The first area is primarily the job of efficiently organizing the department and then developing the administrative structure and procedures for effective operation. The second activity deals with the need to develop, maintain, and administer:

> The operating procedures and guidelines to be followed in effecting maintenance, repair and alteration,
> The necessary technical records that must be maintained, and
> The manuals and data that must be available in the department library.

The third activity area includes compliance with regulations and completion of various FAA and manufacturer records needed to for legal maintenance work. The final activity area for shop administration deals with the compilation of necessary business records within the company. Included are the records dealing with accounting, budgeting, personnel, processing shop orders, employee time cards, purchase orders, and safety reports.

Figure 10.7 shows a shop order sample from FBO Manager, and Figure 10.8 shows an example of the computer screen shop order from TotalFBO. Figure 10.9 shows an internal aircraft shop order form from TotalFBO.

Flat-Rate Pricing

Pricing determinations for the products of the maintenance department normally include all the considerations mentioned in Chapter 3, Marketing. These considerations must recognize the three cost components: material, labor, and overhead.

In many instances the traditional methods of determining prices for aviation maintenance work, assigning workloads to mechanics, and compensating individual employees have not resulted in an efficient operation. The organization might discover its maintenance department poorly utilizes labor and is thus operationally inefficient or financially ineffective. One suggested solution to this problem has been the development and application of a flat-rate price structure to aircraft maintenance. Long used in the automobile industry, flat-rate pricing structure are growing in the aviation industry. Its procedures, advantages, and problems are included here. The operating procedure when using a flat-rate billing structure is as follows:

1. A flat-rate manual covering the type of maintenance service to be offered is acquired or developed.
2. The work order to be accomplished is assessed and classified.
3. The manual is used to obtain the standard time suggested for accomplishing the job.
4. The work is accomplished by assigned personnel.
5. The individual completing the job is compensated according to the time schedule in the flat-rate manual, not how long it takes to actually complete the work.
6. If the job is returned for rework or correction, it is accomplished by the person doing the initial work, on his or her time, and without additional compensation.

When an aviation business begins operating under a flat-rate manual procedure, the following advantages are normally experienced:

1. Maintenance employees become work-oriented and time conscious.
2. The maintenance department experiences increased productivity and efficiency.
3. The quality of maintenance work improves and produces greater customer satisfaction.
4. The quality of employees is enhanced, with the marginal or less motivated employees leaving voluntarily.
5. The maintenance employees operating under a flat-rate manual experience a 10 to 20 percent increase in compensation.

The most frequent concerns regarding operating a maintenance facility under a flat-rate procedure have been:

> "My mechanics will never go for that procedure!"
> "Quality of work will go down, and we cannot tolerate that in aviation!"
> "I have no flat-rate manual for aircraft in my geographic area."

The first concern regarding mechanic reaction is best handled by a knowledgeable and determined manager who convinces his employees of the advantages of the procedure. As for the second concern, experience in several operations suggests that poor quality does not result from the system itself. Once the work force has stabilized under a flat-rate system, the quality seems to improve; perhaps the workers with higher qualifications are striving to do a better job and prevent rework and correction. Of course, the normal quality-control procedures and inspections continue in order to maintain the desired level of work.

The final problem is obtaining a flat-rate manual that adequately covers the type of aircraft serviced by a particular business in its unique geographic location. Since the various aircraft types (Beechcraft, Cessna, Piper, etc.) are constructed differently, it is necessary to have a manual for each type. The climatic differences around the country with varying ranges in temperature and working conditions influence the time involved in accomplishing repair operations. These differences must be recognized when establishing a manual for a location.

Figure 10.7 ✈ Shop Order

WORK ORDER TU-1076

Remit to:	

Bill to:	Shop Order #: TU-1076 Date Printed: 05/21/2002 Date Opened: 03/01/2002 Date Closed: 03/01/2001 Phone Number

Aircraft Information Number:

Item: 1	ACCOMPLISH INCOMING INSP. DEPARTURE INSP. LOG BOOK ENTRY AND ALL ASSOCIATED PAPERWORK. (INSP USE ONLY).
Corrective Action:	ACCOMPLISH INCOMING INSP. OF EXTERIOR AND INTERIOR A/C.

	Labor Charges -	2.50 @	$62.00	+$0.00=	$155.00
				Labor SubTotal	$155.00
				Item SubTotal	$155.00

Item: 2	ACCOMPLISH LT AND RRT ENGINE 150 HR INSPECTIONS.
Corrective Action:	COMPLIED WITH 150 HR ENGINE INSPECTION ON R/H AND L/H ENGINES AS REQUIRED. OIL FILTER REQUIRES ENGINE RUN FOR LEAK CHK. LEAK CHK GOOD - NO DISCREPANCIES NOTED AT THIS TIME.

	Labor Charges -	10.50 @	$62.00	+$0.00=	$651.00
				Labor SubTotal	$651.00
				Item SubTotal	$651.00

Item: 3	ACCOMPLISH LT AND RT 150 THRUST REVERSER INSPECTIONS.
Corrective Action:	PERFORMED A 150 HR THRUST REVERSER INSP. PER LEAK JET MM ON L&R THRUST REVERSER.

	Labor Charges -	7.50 @	$62.00	+$0.00=	$465.00
				Labor SubTotal	$465.00
				Item SubTotal	$465.00

Item: 4	REMOVE AND REPLACE ENGINE FIRE BOTTLE.
Corrective Action:	PULLED L/H FIRE EXT & REMOVED L/H FIRE EXTINGUISHER BOTTLE INSTALLED FIRE BOTTLE PER MM 26-20-01. (SAFETIED SWIVEL FITTINGS & SQUIBS, NO DISCREPANCIES NOTED AT THIS TIME.

	Labor Charges -	8.00 @	$62.00	+$0.00=	$496.00
Part #	Description	Quantity	Price	Tax	Total
30402102-1	FIRE BOTTLE	1	$1,488.00	$0.00	$1,488.00

	Shop Order #: TU-1076
Labor SubTotal	$496.00
Parts SubTotal	$1,488.00
Item SubTotal	$1,984.00

Outside Labor	$0.00
Outside Parts	$0.00
Total Parts	$1,488.00
Labor Hours	28.5
Total Labor	$1,767.00
Tax	$0.00
Total	$1,550.00

Notes: Invoice Number: TUL-0002

Courtesy of FBO Manager.

268 Essentials of Aviation Management

Figure 10.8 → Shop Order Screen

Source: Horizon Business Concepts, Inc.

continued.

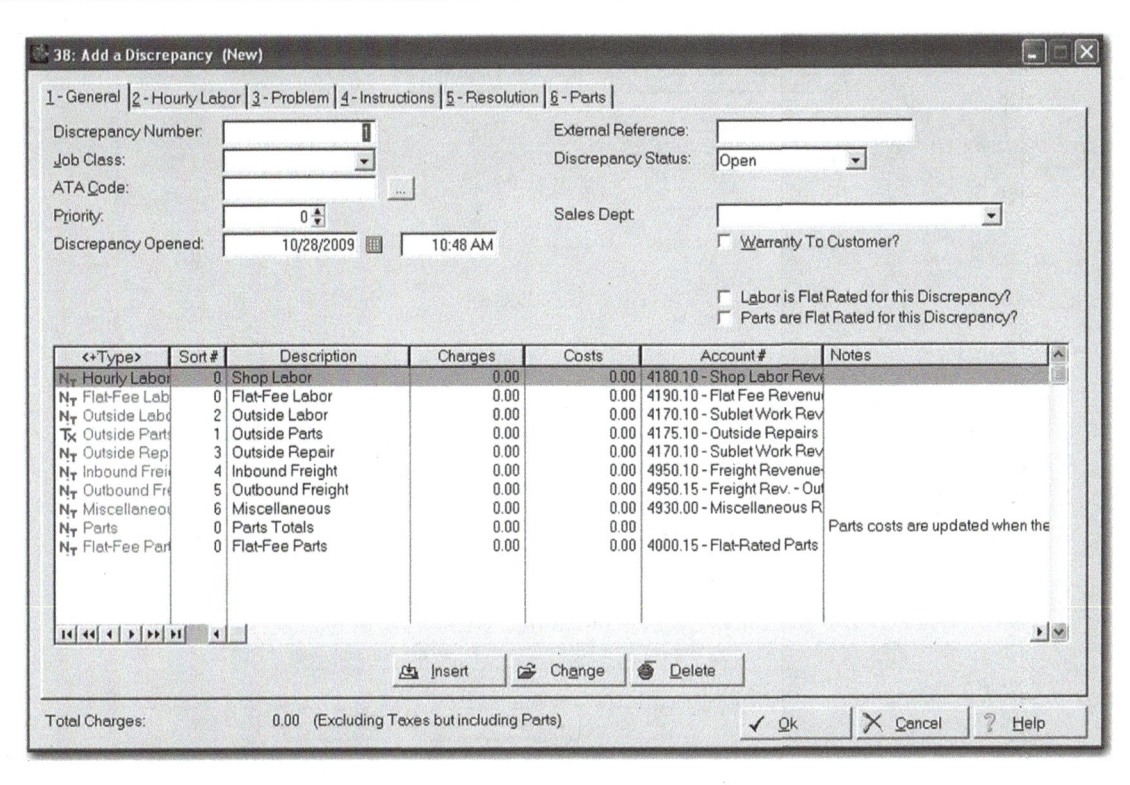

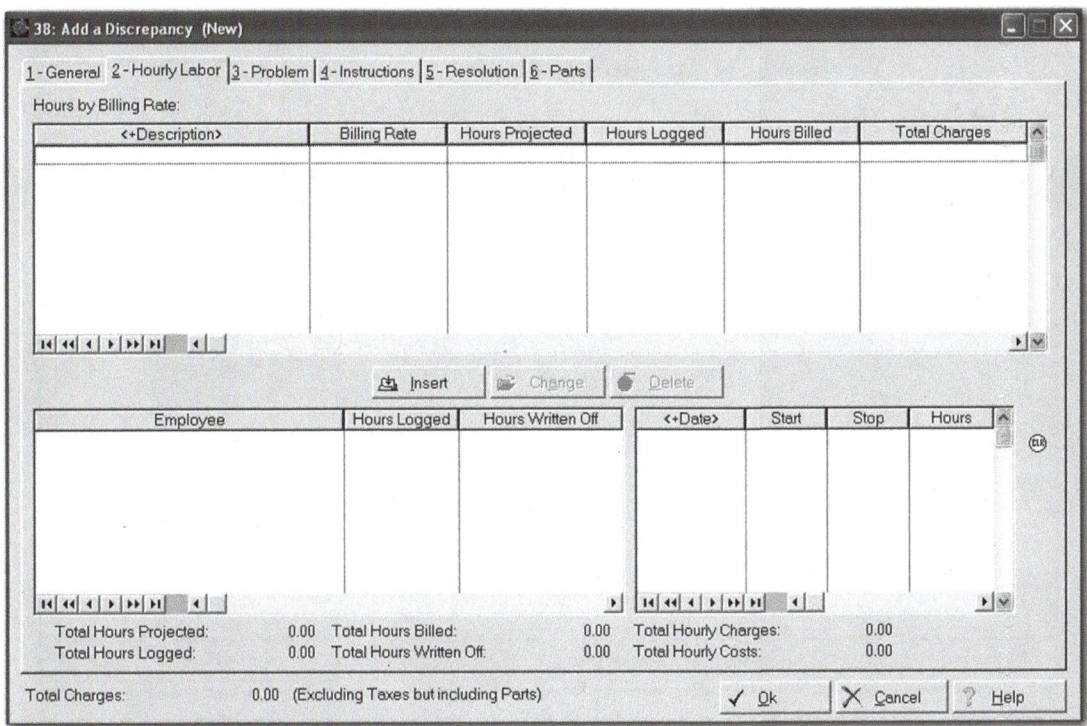

Figure 10.8 → Continued

Source: Horizon Business Concepts, Inc.

continued.

Figure 10.8 ✈ Continued

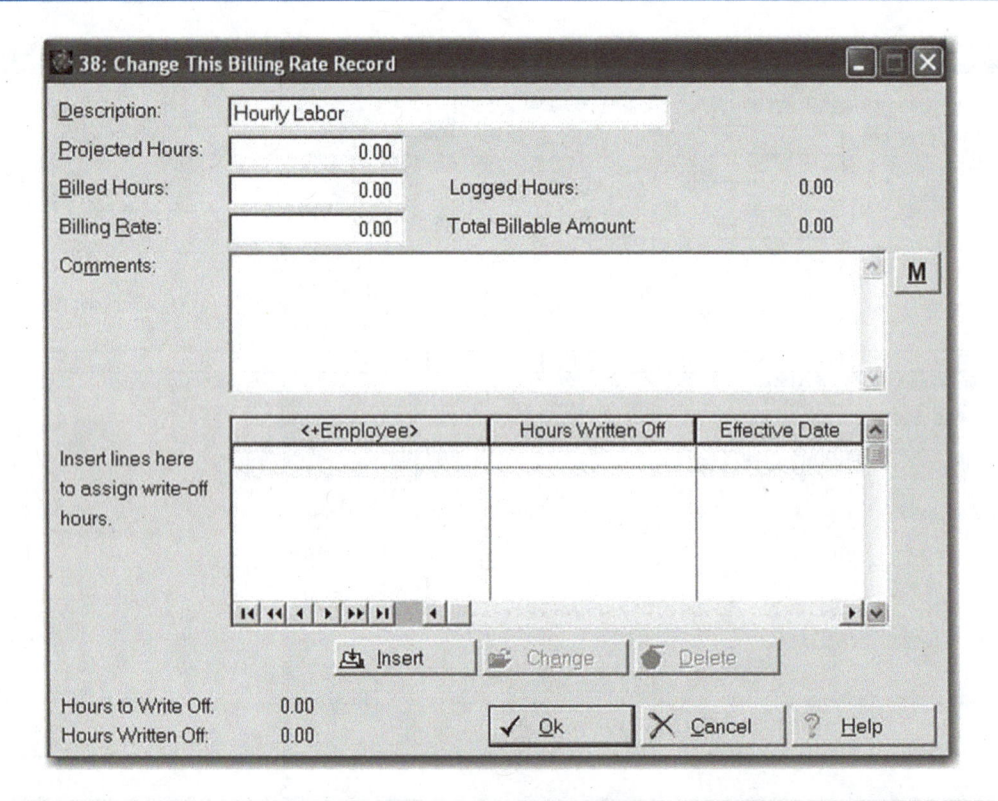

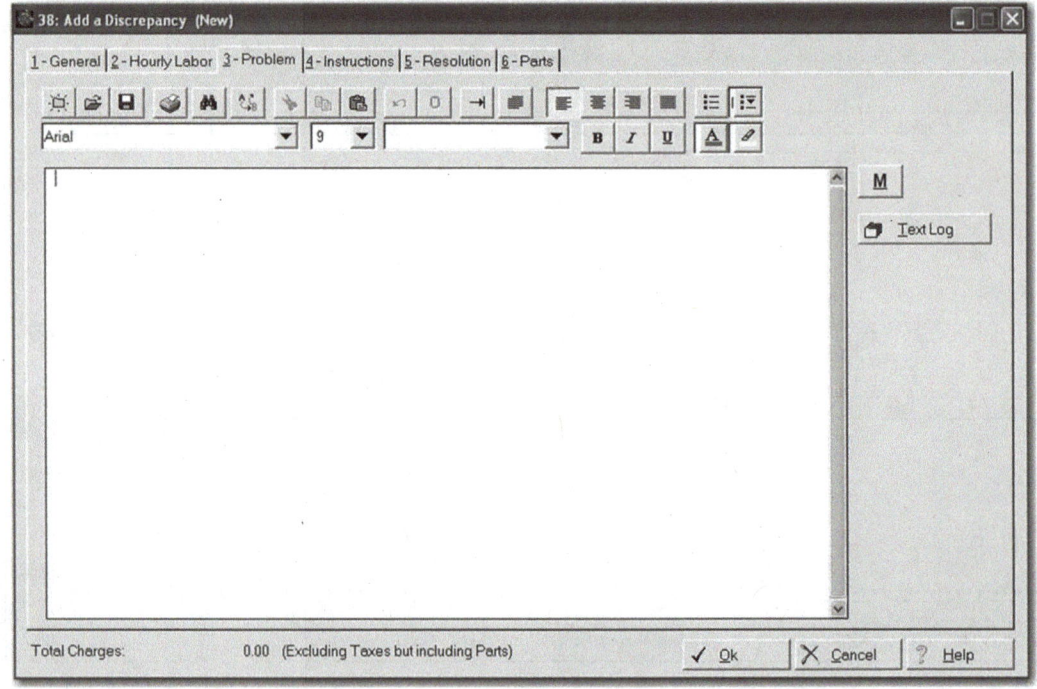

Source: Horizon Business Concepts, Inc.

Figure 10.9 ✈ Internal Shop Order Form

totalFBO®

Horizon Business Concepts, Inc.
"Where Your Dreams Take Flight"
Street Address
City, ST 99999
Phone numbers and
1 more line available

	Shop Order: 10007	Opened:	5/19/2009
		Closed:	5/21/2009

Sold To: HBC - Flight School Maintenance

Aircraft Number:	N2519	Type: C-150	S/N: 15069123

Total Time: 8,012.1   Hobbs Time: 8,012.1   Tach Time: 7,237.4   LG Cycles:

Eng#	Type	S/N	Time	Cycles	Prop Type	Prop S/N	Prop Time
1	CONTINENTAL	123456456	7,237.4				

Discrepancy: 1
Problem:
Fix squawks on lights
Action Taken:
replaced wingtip position lightbulbs

Date Completed: Completed By Technician: Accepted By Inspector:

Charges This Item: 2.75 Hours @   55.00 $ 151.25

Part Number	Description	Credit	Quantity	Units	List Price	Disc	Unit Price	Extended
93	Bulb		2.00	Each	1.910	81	0.370 $	0.74

Total For This Discrepancy: $ 151.99

Discrepancy: 2
Problem:
needs oil change
Action Taken:
drained and replaced oil

Date Completed: Completed By Technician: Accepted By Inspector:

Charges This Item: 1.25 Hours @   15.00 $ 18.75

Part Number	Description	Credit	Quantity	Units	List Price	Disc	Unit Price	Extended
AS15W50	Oil		6.00	Liter	4.950	94	0.280 $	1.68

Total For This Discrepancy: $ 20.43

Miscellaneous Charges:
                                                                   Consumables: $ 5.10

Totals:

SubTotal:	$	177.52
Sales Taxes:	$	0.60
Total Charges:	$	178.12
Amount Remaining:	$	178.12

Terms: Net 30 Days
Thank you for your business. Fly Safe!

Printed: 10/28/2009   Shop Order: 10007   Page: 1 of 1

Courtesy of TotalFBO Accounting and Business Management Software by Horizon Business Concepts, Inc.

One illustration of a successful application of the flat-rate procedure to aviation maintenance work is Southwest Air Rangers of El Paso, Texas. There, a flat-rate manual was developed for use primarily on Piper aircraft. The manual was prepared initially from the analysis of a large amount of accumulated historical data on maintenance activity. Several revisions to the manual have been made as additional experience and data have been acquired. Figure 10.10 illustrates a page from the Southwest Air Rangers flat-rate manual. Additional information on the operation of the system and copies of the complete manual may be purchased from the management in El Paso.

Computer-Assisted Maintenance

As aircraft have grown larger and become more and more sophisticated, it has become increasingly difficult for the manager to keep up with the myriad of details associated with keeping those aircraft operational on a tight schedule. Many businesses have turned to the computer for assistance in ensuring that maintenance is scheduled and completed efficiently.

Many organizations have developed completely computerized aircraft maintenance programs that are offered as a recurring service to aircraft operators. These programs include maintenance scheduling that can be customized to customer needs. Figure 10.11

Figure 10.10 ✈ Sample Page From an Aviation Flat-Rate Manual

Job Description	18	23-4	23-250	23-AT	23-LT	24	24-T	25	28-4	28-6	28-R	30+39	30+39-T	31	31-P	31-T	32-260	32-300	34	36
Absolute pressure controller (cowl off)				1																
Actuator—brake	1	1	1	1	1	1	1	1	1	1	1	1	1	1			1	1	1	
Actuator—flap (hydraulic)		3	3	3	3															
Actuator—gear door (hydraulic)		1	1	1	1									1						
Actuator—landing gear (hydraulic)		3	3	3	3						1			4						
Actuator—waste gate (cowl off)				2																
Adapter—oil filter														4						
Adjust electric trim solenoid									1½	1½	1½						1½	1½		
Aileron	1	1½	1½	1½	1½	1½	1½	1	1½	1½	1½	1½	1½	1½			1½	1½	1½	
Aileron hinge or bearing (1 aileron)		2	2	2	2	2	2		2	2	2	2	2	2			2	2	2	
Aileron hinge doubler (1 aileron)									4	4	4						4	4	4	
Air box—carburetor (cowl off)	¾	1	1	1					1	1							1			
Air filter—central (during insp)		¼	¼	¼	¼	½	½		¼	¼	¼	¼	¼				¼	¼		
Air pressure pump		1¾	1¾	1¾	1¾									1¼						
Air—propeller (service ea)		¼	¼	¼	¼							¼	¼	¼						
Align wheels (on jacks)		¾	¾	¾	¾	¾	¾					¾	¾	¾						
Alternate air cable (complete)			10	10	10	2	2		2	2	2	10	10					3	10	
Alternate air cable (core)			2	2	2	1	1		1	1	1	2	2					1	2	
Alternator (cowl on)			2¼	2¼	2¼	1½	1½	1½	1½	1½	1½	2	2	1½			1¾	1¾	3	
Alternator (cowl off)			1	1	1	1	1	1	1	1	1	1	1	1			1	1	1	
Alternator or generator belt (prop off)	¼	¼	¼	¼	¼	¼	¼	¼	¼	¼	¼	¼	¼	¼			¼	¼	¼	

By permission, Southwest Air Rangers, El Paso, Texas, 1985.

Figure 10.11 ✈ Maintenance Warning Report

10/28/2009
11:02AM

Horizon Business Concepts, Inc.
Maintenance Warning Report
Includes All Maintenance Reminders

Page: 1

Aircraft: N1234	Make: Beechcraft	Model: B200	Type: KING AIR			S/N: 44557-97864	
Eng Time	Last MOH	Prop Time	Last POHD	Last POHT	Tach Time	H/S Time	H/S MOHT	H/S Cycles	H/S MOHC	Hobbs Time
1 - 923.1	0.0	251.1	0.0	0.0	950.1	461.1	850.0	83.0	1,200.0	1,048.7
2 - 950.1	800.0				1,004.5	461.1	850.0	83.0	1,200.0	0.0

Total Time Tracks: Tach 1 Total Time Offset
Total Time: 2,200.10
Landing Cycles: 814 1,250.0

Maintenance Description (Type):	Last Done	Next Due	Hours	Days	Cycles
Engine 1 Runout			76.9		
Engine 2 Runout			849.9		
Prop 1 Runout			548.9	(3,437)	
Prop 2 Runout			(150.1)	(3,437)	
Hot Section 1 Runout			888.9		1,317
Hot Section 2 Runout			888.9		1,317
100-Hr Inspection (Date)	10/21/2008	10/21/2008			
Altimeter Certification (A 25 Hr. Inspection)	1/15/2008	1/31/2010	(1,048.7)	(372)	
Annual Inspection (Cycle)	10/21/2008	10/31/2009			0
ELT Battery Expiration (A 25 Hr. Inspection)	1/15/2009	1/15/2011	(1,048.7)		
Encoder Certification (A 25 Hr. Inspection)	3/24/2008	3/31/2010	(1,048.7)		
Static System Certification (A 25 Hr. Inspection)	6/14/2009	6/30/2011	(1,048.7)		
Transponder Certification (A 25 Hr. Inspection)	6/14/2009	6/30/2011	(1,048.7)		

Courtesy of TotalFBO Accounting and Business Management Software by Horizon Business Concepts, Inc.

shows a computerized maintenance warning report for the individual customer's aircraft. After the required data has been entered into the computer, the following reports are printed for the customer:

> Monthly aircraft status report;
> Monthly maintenance due list;
> Monthly aircraft history report;
> Annual budget performance and reliability summary; and
> Inspection and services summary.

Part of the system includes maintenance requirement cards that contain the latest acceptable maintenance procedures.

In addition to being valuable to many aircraft operators, the system is indicative of the role that the computer can play in the maintenance process.

Profitability

As mentioned earlier, the focus of an aviation business should be on maximizing profits and ensuring flight safety. By extension, the goal of the maintenance department is to contribute its share to the overall profitability of the business. The material presented in this chapter assists in the managing of the department and in achieving the desired profits. Additional assistance can be obtained through the development and use of an information system, the analysis of all relevant data, and finally the application of control techniques to assure the accomplishment of goals.

Chapter 7, Information Systems, is devoted to an in-depth examination of the development and use of information systems. Since this section is designed only to illustrate some aspects of maintenance profitability, the one area that should be clearly identified is the charge for service labor. It is important the charge to the customer cover all costs incurred by the organization in providing the service, plus a margin for profit.

Information

It is imperative that a service manager receive operational information on the various aspects of the department as timely and accurately as possible. Data on income and expenses are needed in sufficient detail to cover jobs, aircraft, individuals, and so on. Data must be provided with appropriate speed and precision to analyze and take corrective action immediately.

Analysis

The analysis of reported service data can be accomplished using several techniques. Included here are:

> Comparison with established goals;
> Comparison with previous forecasts and prior operating periods;
> Comparison with similar businesses; and
> Comparison with generally accepted industry measures.

The typical maintenance data subjected to this type of analysis could include:

> Total maintenance revenue generated,
> Revenue generated by each major working area,
> Labor utilization,
> Volume of rework,
> Net profit as a percentage of sales, and
> Indirect expenses.

Control

Control is a continuation of the analysis. It is the final stage of the comparison process that identifies the desired corrective activity and then implements that activity. For the maintenance manager, it may mean reducing personnel, increasing inventory, establishing a new inspection procedure, or engaging in a concentrated effort to promote annual inspections. The activity will be specifically related to the numerical results of analysis and may be adjusted continuously in order to achieve desired goals.

Techniques

There are numerous procedures and processes that utilize information, analysis, and control. Several practical illustrations that are useful in gaining and maintaining profitability for the maintenance department follow.

Budgeting. Various applications of budgeting may be utilized in achieving the desired level of operating profits. Sales are first estimated for each of the 12 months. Then, using the historically developed and desirable percentages for cost of sales, gross profit, expenses, and operating margin, the anticipated dollar values for each of these items is calculated for the monthly sales projections. As the operating months are completed, the actual cost of sales, expenses, gross profit, and operating margin are entered in the second line for each month, and the actual percentages for each are calculated. In this manner the budget assists in monitoring sales and maintaining the desired relationship of cost of sales, gross profit, expense and operating margin. Trends can be proactively identified, and corrective action can then be taken on a monthly basis to ensure the target operating margin is achieved.

Productivity. A second practical technique for measuring the efficiency of a shop is through the examination of its productivity. Figure 10.12 illustrates the calculation of individual and shop productivity measures. The goal is to bill out as many of the mechanics' working hours as possible. Realistically, this can seldom be 100 percent because there is administrative, training, vacation and other overhead time that should not be directly billed to the customer. In this illustration, the manager should investigate the low productivity of mechanic #4 (64 percent) and the high productivity of mechanic #10 (100 percent).

Figure 10.13 illustrates a technique for reviewing and controlling the overall shop productivity. This chart shows that the break-even point for this shop at 100 percent productivity is three mechanics. The relationship of number of personnel, their productivity, and maximum income are illustrated. By applying the shop productivity average of 89 percent for the ten employees in Figure 10.12, one can quickly assess how the department is performing. This concept can be very valuable in analyzing the overall situation and considering the need to decrease or increase the number of employees, develop additional business, or improve internal efficiency.

Ratios. Another very practical technique to use in monitoring the progress of a shop and its financial success is through ratios in monitoring the income statement. Financial ratios, or the relationship of one element of the income statement to another element, can provide a means of measuring the progress of the maintenance department and determining its degree of success or failure. A ratio can be compared with its earlier values for the same department, concurrent values for other departments or other businesses, and with standards considered generally acceptable. Some of the key ratios and standard values are compiled in Figure 10.14.

These techniques—budgeting, productivity review, and ratio analysis—are only a few of the many techniques available to the manager in maintaining the profitability of the department. However, constant surveillance of maintenance activity and decisive action are the two managerial functions most necessary to ensure a successful maintenance department.

Professional Maintenance Organizations

There are three primary maintenance-related aviation organizations. They include the Aircraft Electronics Association (AEA), Aviation Technical Education Council (ATEC), and the Professional Aviation Maintenance Association (PAMA). AEA represents nearly 1,300 FAA Part 145 Certified Repair Stations as well as most manufacturers of general aviation avionics equipment and airframes. ATEC represents aviation educators and professionals. PAMA is a national professional association of aviation maintenance technicians, with some 4,000 individual members and approximately 250 affiliated company members.

Avionics Repair Stations

Avionics repair stations located at fixed-base operations may operate independently or as an integrated part of an existing maintenance operation. As aircraft become more sophisticated and avionics become a greater part of the operating system, there will be a growing need for integrated shops and people who are competent in multiple domains.

Figure 10.12 ✈ Individual and Shop Productivity

Mechanic	Hours Paid	Hours Billed	Percent
1	184.7	160.6	87
2	176.0	171.3	97
3	169.2	164.3	97
4	193.0	123.6	64
5	169.9	145.0	85
6	198.4	165.2	83
7	175.6	157.4	90
8	184.9	175.2	95
9	160.0	152.8	96
10	160.0	160.0	100
	1,771.7	1,575.4	89

276 Essentials of Aviation Management

Figure 10.13 → **Break-Even Chart Used for Controlling Overall Service Shop Productivity**

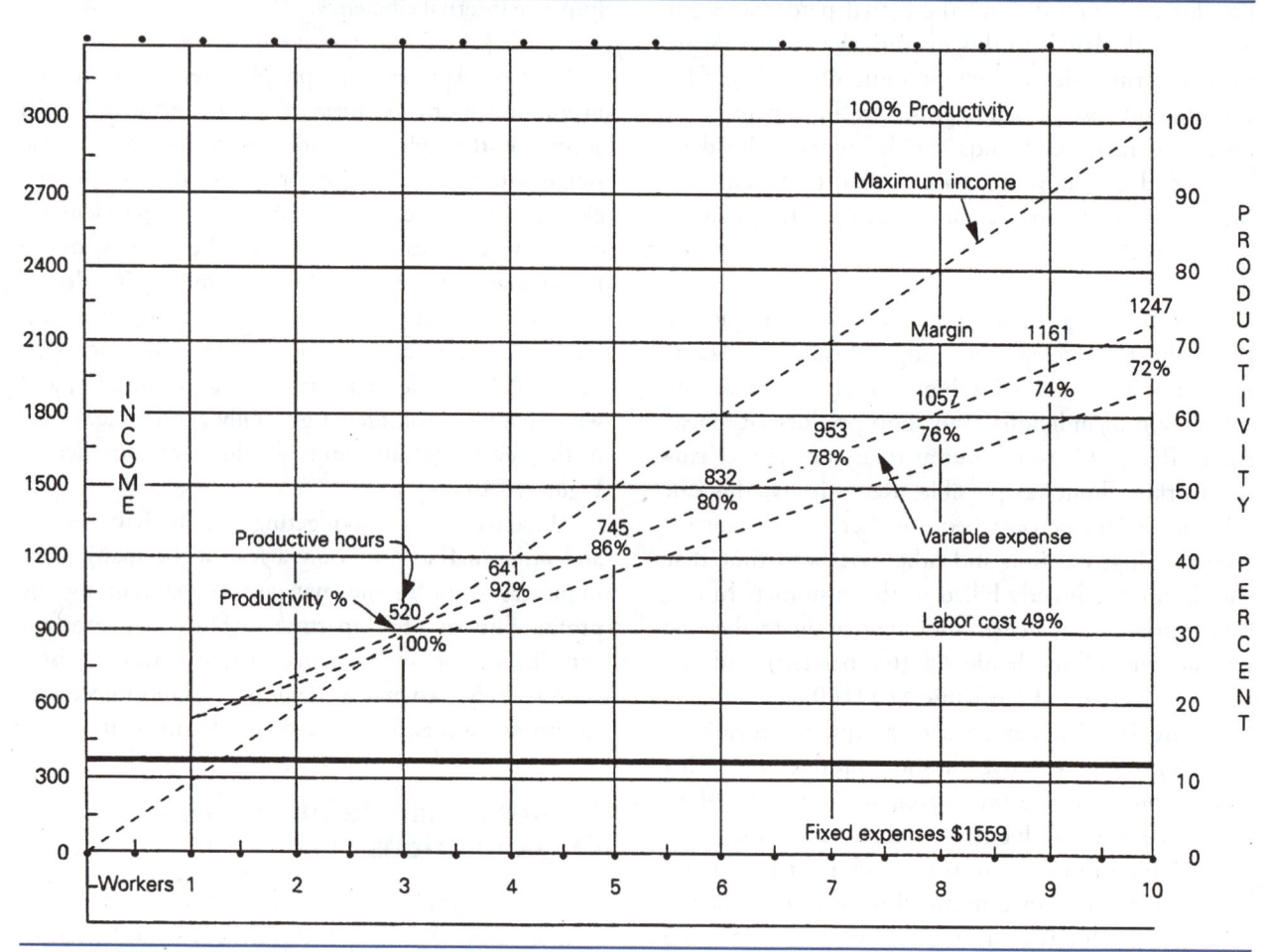

Courtesy of Robert Varner, Lane Aviation, Columbus, Ohio

Figure 10.14 → **Key Ratios and Normally Accepted Standards for Maintenance Departments**

	Type	Acceptable Range
Ratio:	*Gross profit as a percent of sales*	
	Service	45 to 47 percent
	Parts	18 to 22 percent
	Electronics	28 to 30 percent
Ratio:	*Expenses as a percent of sales*	
	Service	44 to 46 percent
	Parts	13 to 17 percent
	Electronics	26 to 28 percent
Ratio:	*Operating profit as a percent of sales*	
	Service	1 to 3 percent
	Parts	5 to 9 percent
	Electronics	2 to 4 percent

Summary

Managing an aircraft maintenance facility is a separate business or profit center requiring the same business management tools as other segments of the company. A comprehensive organizational structure is key, as is the quality of personnel. Careful attention must be paid to the selection, training, certification, and utilization of mechanics, inspectors, and repair people. Many aspects of repair shop facilities and procedures are controlled by the FAA. Identification and compliance with requirements must be accomplished carefully.

Whether as a section of the maintenance department or as a separate unit, the parts and supplies department contributes directly to the success of the maintenance shop. Keeping the optimal levels of parts inventory is a driving concern of the service manager.

Quality control depends on the individual mechanic and on the entire organization. It may be achieved through training, procedures, processes, inspection, and recognition. Other areas of concern for the service manager are outsourcing, liability, marketing, administration, and pricing. The key activity is the surveillance of departmental profitability and the harnessing of information, analysis, and controls to assure all operational and financial goals are attained.

DISCUSSION TOPICS

1. Identify four subdivisions frequently encountered in the organizational structure of a maintenance department and discuss their responsibilities.
2. How would an interested person become a certificated aircraft maintenance technician with an airframe and power plant rating?
3. What facilities and equipment are needed to open and operate an aviation maintenance shop?
4. How would one set about identifying a practical inventory level for a new aviation service shop? For a shop that has been in business for some time?
5. "Quality control expenditures must be carefully weighed against the benefits thereby obtained." Discuss.
6. What are three implications of product liability for an aviation maintenance facility?
7. What considerations weigh in favor of outsourcing all maintenance work? Against?
8. "The manager must be involved in setting and maintaining an adequate profit level for a shop." Identify and discuss these managerial activities.
9. Discuss the advantages and disadvantages of flat-rate pricing.

Endnotes

1. Prentice, Stephen P. (1999, April). *Fractional Technicians? What's At Stake?* AMT Online.
2. See *http://www.faa.gov/nextgen/*
3. See *http://www.ads-b.com/home.htm*
4. de Decker, Bill. (September 2000). *Value for Money: Choosing the Right Maintenance Program for Your Operation.* AMT Online. Retrieved from *https://www.aviationpros.com/home/article/10388546/value-for-money-choosing-the-right-maintenance-program-for-your-operations*
5. 14 C.F.R. §§ 1-200.
6. See *http://www.flightsafety.com* and *http://www.cae.com/civil-aviation.*

11

Safety, Security, and Liability

OBJECTIVES

› Understand the difference between "risk management" and "risk transfer."

› Be aware of the "deep-pocket" theory.

› Recognize the advantages and disadvantages of self-insurance.

› Be able to discuss the specialized areas of aviation insurance.

› Understand how the General Aviation Revitalization Act of 1994 changed product liability and subsequent aircraft prices sales for the aviation business owner and manager.

› Distinguish the differing roles of the Federal Aviation Administration and the National Transportation Safety Board regarding the investigation of accidents and incidents.

› Understand the mandate of the Transportation Security Administration.

Introduction

All types of business enterprises are exposed to some degree of risk. Our system of business and government rely heavily on honest individuals, be they customers, employees, or visitors. When a person has nefarious intentions, such as terrorism or theft, the aviation system may be a relatively susceptible network. The terrorist attacks of 9/11/01, and the intense concern with aviation security and safety that arose as a result, did not, for example, prevent a 15-year old student pilot in January 2002 from stealing an aircraft and crashing it into a building. This chapter discusses normal business risks as well as the special risks of aviation operations, and risk management practices of risk reduction and risk transfer. Finally, the consequences and corrective action in event of risk failure is discussed. Strong focus is placed on the evolving activities and requirements related to aviation security in a time that aircraft are used as weapons of mass destruction.

The Need for Risk Management Procedures

Risk management consists of two related areas: risk reduction, which is accomplished through careful conduct of each aspect of the business, and risk transfer, which is accomplished through purchase of insurance for particular business activities. The more effective the risk reduction, the less expensive the risk transfer. Risk reduction consists of the application of good management techniques by leaving only unforeseeable occurrences to be insured against. Insurance premiums tend to be lower with good risk reduction techniques.

An aviation business needs to address the risk management area with great care, partially because of the numerous other factors involved in airport services and facilities. There may be a lack of clarity in airport leases and operating agreements with regard to whom is responsible for which functions and areas. This can blur the distinction between aviation business liability and airport management operating liability. Reasons for this include: (1) FBOs frequently have contract responsibility to provide airport management functions, and (2) airports themselves frequently provide the same or similar functions as FBOs, and this overlap can lead to shared blame.

An adjunct to this mixed responsibility is the "deep-pockets" approach. This means litigants generally sue the wealthiest party, regardless of whether they are believed to be negligent. Some parties involved in aviation, including private airport owners, pilots, aircraft owners, and others, may not carry sufficient (or any) insurance, so the aviation business must determine whether it is adequately covered against the deficiencies of others.

Interaction of Safety, Security, and Liability

Steps to run a safe, secure airport operation with adequately trained staff are a top priority in risk management. Negligence in an area quite unrelated to a particular claim may be cited as evidence of poor attention to overall detail. A well-run airport operation not only reduces the risk of accidents, but also assists in achieving favorable insurance rates. In some areas, the safety requirements are common sense; in others, the TSA establishes them.

Risk Exposure

Normal Business Exposure

The typical business is exposed to at least these risks:

> Fire;
> Embezzlement;
> Theft;
> Vandalism;
> Trespassing;
> Data breach;
> Weather and natural disasters;
> Inadequate security of personnel, facilities, and equipment;
> Product liability;
> Third-party or non-employee liability;
> Employer's liability and workers' compensation;
> Vehicle liability;
> Injury or fatality of key persons; and
> Business interruption due to any of these events.

Aviation Risk Exposure

For FBOs, additional risk occurs because of the sensitive nature of the goods and services and the high degree of employee skills involved in their operations. In addition, several incompatible components may be found in different parts of the operation, include running engines, moving propeller blades, a mixture of automobiles and aircraft moving around on ramps, and the presence of fuel, oil, oxygen, welding materials, electrical equipment, and paint. Areas of traditional concern to an aviation business include:

> Aircraft hull damage;
> Injury, fatality, and property damage to third parties resulting from travel in the company's aircraft;
> Non-ownership liability—for occurrences in aircraft flown by other than company personnel;
> Premises and product liability, including aircraft and parts;
> Fueling safety;
> Hangar keeper's liability;
> Leakage and contamination of underground fuel-storage tanks;
> Storage and disposal of agricultural chemicals used by crop-dusters;
> Other hazardous and toxic wastes such as used motor oil and deicing fluids;
> New rules, more stringent inspections, and changing security requirements for Part 135 charter flight operations imposed by the TSA after 9/11;
> Aircraft rescue and firefighting (ARFF) response capability at air carrier airports;
> Employee exposure to hazardous products on the job; and
> Security and crime prevention.

In short, the regulatory context and insurance needs of the typical aviation business are increasing, this trend seems likely to continue as the TSA refines its proposals and as other agencies such as the Federal Bureau of Investigation (FBI), Immigration and Customs Enforcement (ICE), and Department of Homeland Security (DHS) increase involvement in aviation security.

Risk Reduction

Normal Risk Reduction

Risk management involves active risk reduction as well as risk transference through insurance. It is not simply the purchase of insurance for protection in case of loss. An adequate insurance program is only a part, and not the ultimate goal, of risk management. Actions taken by management before insurance is negotiated, during the life of a policy, and after a loss occurs are all part of risk management. They influence the premium costs to the business as well as the total loss experienced. In each risk area, specific organizational actions are available to reduce the risk and the ultimate costs involved. Additionally, in today's insurance marketplace where deductible amounts tend to be much higher than in previous decades, risk management is vital to cost control.

Fire Risk. Much can be done to reduce the risk of fire, including using fire-resistant materials, training personnel in fire prevention practices, deploying protective devices such as automatic sprinklers and fire extinguishers, and developing firefighting plans. Figure 11.1 illustrates typical fire-risk reduction procedures. If premises are remodeled to the extent that a local building permit is required, they will need to be brought up to modern standards with regard to fire risk (e.g., the installation of fire detection and extinguishing systems). However, the aviation manager may want to perform a fire risk audit even where not required by the local building department. One area vulnerable to fire and fire suppression damage, is the business' computer system. Risk reduction must involve either daily computer backups of the entire system or offsite data storage, loss of this information could create a serious business interruption.

Crime Risk. One of the most serious threats to business property today is the crime threat—acts of terrorism, burglary, vandalism, robbery, and theft. Managing these threats becomes an important aspect of risk management. Insurance is only one part of managing this risk; reduction and prevention are also extremely important. As small businesses, many aviation operations are prime targets for the burglar and robber. However, many things can be done to reduce this risk including the use of protective devices, sound operating procedures, and employee training. Protective devices include silent central station burglar alarm systems, burglar resistant locks and equipment, and indoor and outdoor security lighting. Effective operating procedures may include minimizing cash receipts through switching to credit-only transactions.

New technology is available, and constantly being improved, to assist in the area of airport security. Electronic devices that transmit a signal when unauthorized tampering occurs may be installed in aircraft. Sophisticated computerized monitoring and alarm systems are available for the premises. Lock systems for airport and premises access can be computer-controlled and codes changed frequently. Better quality locks can be installed on aircraft and hangar doors. Local police do not always realize the value of aircraft and aviation business equipment, and it's useful to actively keep them informed so that their speedy assistance is available.

A business person must deal with an occasional dishonest employee or dishonest customer. In the case of the dishonest employee, the manager's first problem is identifying the amount of the loss. Fidelity bonds can be obtained on those employees who have access to large amounts of money. Inventory shrinkage resulting from employee pilferage and other dishonest acts is substantial for many organizations. Risk management in this area includes the reduction or prevention of losses by the careful screening and selection of personnel, an effective accounting system, and varied control methods. Safeguards to be considered include

> The use of outside auditors;
> Countersignatures on all checks;
> Immediate deposit and duplication of all incoming checks;
> Bank statement reconciliation by an employee other than the one who makes the deposits;
> Joint access to safe deposit boxes;
> "Professional shopper" checks on cash register operating procedures; and
> Ensuring that all employees handling money take regular vacations.

Figure 11.1 ✈ Fire Inspection Checklist

Emergency Preparation
- # Fire organization posted
- # Fire drill held regularly
- # Fire exits well marked and unobstructed
- # Fire alarms well marked and unobstructed
- # Aisles and stairs clear
- # Evacuation procedures posted
- # Procedures for handling fuel spills posted
- # Emergency equipment well marked, in place, and ready for use

Hangars and Buildings
- # Hydrants and water supply checked and serviceable
- # Sprinkler system, checked and serviceable
- # Foam and CO_2 systems checked and serviceable
- # Fire doors checked and serviceable
- # Fire extinguishers well marked, in place, checked and tagged
- # Hose stations, checked and serviceable
- # Electrical circuits identified, enclosed, and provided with proper overload protection
- # Gas systems, checked and serviceable
- # Fuel pumping equipment, in good condition and free of leaks (extinguishing equipment adequate and available)

Maintenance Equipment
- # Spray booths, clean and properly ventilated and sprinkler heads protected from overspray
- # Power tools and accessories, wiring in good condition
- # Pressurized bottles, properly connected and secured
- # Mobile equipment extinguishers well marked, in place, and serviceable
- # Powered equipment, properly grounded
- # Test equipment, free of leaks, wiring in good condition

Management
- # General housekeeping, adequate
- # Changes (alterations, processes, methods, and procedures) first cleared with Fire Marshal
- # Floors, clean and free of flammable fluid spills
- # Storage of material, orderly and in accordance with regulations*
- # Aircraft fueling, in accordance with regulations*
- # Spray painting, in accordance with regulations*
- # Welding and other open-flame operations in accordance with regulations*
- # Ramp and grounds, clean and free of debris
- # Proper disposal of soiled shop towels and rags

*Separate checklists should be made for these operations.

From "Aviation Ground Operation Safety Handbook," copyright 2000 by National Safety Council, Chicago, IL 60611. Used with permission.

If company morale is high, petty theft tends to be minimal, but where morale is low and management is felt not to care about employees, some employees may develop a sense of entitlement. This attitude may manifest in employee behavior such as using company phones for personal calls and taking home office supplies. Thus, a prevention program related to morale and motivation may be an effective tool, in addition to benefiting the company in other ways like improved productivity and profitability.

The dishonest customer represents to the manager the additional perils of shoplifting and bad-check passing. Retail operations expect to lose a percentage of their merchandise through shoplifting. Although FBO operations do not always include retail functions, their office supplies, maintenance supplies, and inventory can be at risk. This kind of loss can be reduced by constant vigilance, special equipment, and sound operating procedures. Equipment such as two-way and convex mirrors and webcams, coupled with wide aisles, clear vision, and alert employees will help prevent the loss of merchandise. Prominently displayed warnings against shoplifting and rules that provide customer guidance will deter the would-be offender in many instances.

Today's computers and printers are so advanced that counterfeit money, fake checks, and false identification are significant risks. These white-collar crimes may be a greater threat than traditional crimes, and employees need training in how to spot them. Credit card and identity theft are growing rapidly, empowered by the world's digital transformation, and any business must take steps to reduce risk in these areas. Such steps include training all employees to check the signature line of a credit card, to ask for additional identification, or to call for manual authorization.

Bad-check losses pose a problem that can be minimized through sound procedures and well-trained employees. Proper identification should be requested before accepting checks, and checks should always reflect the exact balance due. The identification procedure should include a photograph along with a sample signature, or, alternatively, two forms of identification. Postdated, illegibly written, or two-party checks should not be accepted. Employees should be trained in following the procedures selected and in identifying potential passers of bad checks. An on-line check authorization service is an essential precautionary service.

General Emergency Risk. Airports may be affected by not only fire and theft but also by earthquakes, acts of terrorism, tornados, and similar drastic and sudden events that call for a more broad-based emergency response plan. The airport administration should be working with other units of local, state, and federal government to prepare emergency response plans and conduct table-top planning exercises and drills. The aviation business should participate if possible; regardless, the FBO manager should ascertain the status of these plans and encourage full readiness. The National Safety Council is one organization that offers a guide to emergency planning.[1]

Aviation Risk Reduction

Risk reduction comes from the application of sound management practices, as described in previous chapters. As evidenced by the many different types of aviation insurance, there are numerous areas of exposure to risk in aviation operations. Poor maintenance, bad housekeeping, nonexistent or vague guidelines and procedures, inadequately trained personnel, low quality standards, lack of emphasis on safety, and poor supervision all can lead to accidents and costly legal conflicts. The aviation business is faced with potential threats from a wide variety of consumers who may believe that they have been injured by a variety of products. Risks in this area also include careless selection of parts for installation on an aircraft, inappropriate certification of work as having been completed, or premature return of an aircraft to airworthy status when it requires maintenance. These and other similar actions should be considered carefully and steps should be taken to avoid unnecessary exposure or risk.

Because of concerns about the scale of claims for general aviation-related accidents and incidents, product liability legislation that provided some relief was successfully passed in 1994, as discussed in Chapter 1. The General Aviation Revitalization Act (GARA) provides for an 18-year time limit on product liability claims where previously no restriction was in place—a "statute of repose" for aviation manufacturers. The 18-year liability limit provides some relief from potential legal action and appears to be helping revitalize general aviation to an extent. FBOs still have additional risk exposure because of their potential involvement in the repair of engine or airframe components; they, too, may be part of a lawsuit when an accident occurs, and need to keep a good paper trail going back 18 years (or cradle to grave, from FAA's viewpoint) in case of involvement in a legal action.

Good operational risk management will:

1. Select and train qualified aviation personnel.
2. Provide adequate operational guidelines and procedures.
3. Insist on good housekeeping practices.
4. Place a high emphasis on safety.
5. Require consistently high quality standards for services and products.
6. Provide the supervision and maintenance necessary to ensure that these elements routinely take place.

Insurance companies recognize the relationship that exists between an efficient, well managed, safe aviation organization and a low accident rate and minimum risk exposure. Prior to insuring against some aviation risks, many companies will survey the organization carefully, checking all the key operational areas and developing some concept of the premium rate structure. Although overall insurance rates may be controlled more by the purchaser's clout in the marketplace than by good practices, lower premiums should be awarded to those aviation businesses that can demonstrate good organizational practices and higher premiums will be assigned to the others. The practices, procedures, and recommendations described in this book should assist the manager in identifying and mitigating some operational risk exposure, and should result in lower insurance premiums. The Operational Procedure Guide and the Operational Manual that appear as appendices should be especially useful as starting points when dealing with some of the flight operation areas, evidence that employees are required to adhere to the company's manuals should also be produced. The chapters dealing with human resources and maintenance will also provide assistance in those areas.

Aviation Safety and Security Regulations and Guidelines

A number of FAA Advisory Circulars in effect mandate safe aviation operations in such areas as fueling, aircraft handling, and the condition of premises.[2] Along with airport certification requirements and the regulations for FAA-approved maintenance, instruction, and other FBO operations, these provide a very detailed set of requirements and guidelines on how to run a sound operation.[3]

The Transportation Security Agency has taken over the FAA's former responsibilities regarding security, and is in the process of expanding them. Potential security requirements for smaller airports are somewhat unclear, as the dominant issue thus far has been air carrier airports and the need for better passenger screening. However, many general aviation businesses fly charter and air taxi flights that bring passengers to and from scheduled flights, and are thus likely to be required to enact similar security procedures. This is discussed later in this chapter.

Aviation businesses must also deal with municipal ordinances, Environmental Protection Agency (EPA) regulations and Occupational Health and Safety Administration (OSHA) requirements regarding hazards, including the handling and disposal of dangerous products. In addition, such organizations as the National Safety Council and the National Fire Protection Association provide a number of related booklets.[4] Individual airport minimum standards may provide another set of requirements for safe FBO operations in various areas.

In the words of an aviation lawyer, this is how to regard the FAA documents:

> "Rest assured that a plaintiff's attorney will acquire the Advisory Circulars and that the lawyer (and probably a jury) will fault an airport for any area in which its facility is deficient according to the Advisory Circular recommendations, even though these are not mandatory."

There are a number of FAA Advisory Circulars about safe aviation operations in such areas as fueling.

Airport Risk Audit

The attorney mentioned above recommends a self-applied risk audit whereby the airport operator, or in this case the FBO, goes over every area of activity to ensure it meets FAA, TSA, and other standards. The Management Audit shown in Appendix V is a good starting point.

Procedures Manual

The safety procedures applied by the aviation business should be written and should reference appropriate

Advisory Circulars and other standards. This document should be used for training and for recording the dates of risk audits and corrective actions.

Documentation

If a suit is ever brought against an FBO, then clear and dated documentation of standards, training, and corrective action will be an asset in presenting evidence of a risk-averse culture to investigative agencies or in court proceedings. Lack of documentation may enable the plaintiff to go unchallenged.

Inclement Weather

Weather exerts a tremendous influence on aviation operations. The level of activity and safety of personnel, equipment, and facilities are influenced by inclement weather, and they are a major management concern. Of course, the primary concern is for the welfare of personnel, equipment, and facilities. The secondary concern is for the reduction or curtailment of aviation operations. High winds, torrential rains, heavy snows, dust storms, or floods can all injure people, damage aircraft, disable equipment, and ruin physical facilities.

A well-developed weather response plan includes evacuation schedules (aircraft, vehicles, and personnel), security of utilities, protection of inventory, supplies, and records, and guidelines for getting the facility back into operation. As with other inclement weather situations, the primary concerns are knowing the situation, obtaining early warning, having a plan to follow, making the decision to act, and following through as needed.

Risk Transfer

Principles of Insurance

The principle of insurance is to pool risk with others in similar situations so that in the event of a problem some protection is obtained—an approach that assumes, sometimes incorrectly, that not all in the risk pool will be hard hit at the same time. Taking out insurance does not eliminate risk; there will still be some risks covered only by the operator (deductibles or ineligible areas), and some high-risk situations may call for such high premiums that the aviation business chooses to buy only partial coverage or to be self-insured. In some cases, the owner may have such a large and diverse risk pool that it is less expensive (or at least thought so) to self-insure. This situation occurs, for example, in some agencies that own several airports, as it does with some large corporations. However, in the case of self-insurance, funds need to be set aside to address claims. Moreover, it may be a challenge for an aviation firm to successfully self-insure for all categories of risk.

Insurance Regulations

Some insurance is optional; some is mandatory. Workmen's Compensation falls into the latter category and, in most cases, airport owners require some level of insurance for operators on their airport. This should be delineated in the airport's operating standards and also in any lease.

The U.S. Insurance Market

Industry as a whole. Since the general aviation insurance market is highly specialized and quite small compared, say, with the auto insurance market, the complaint over the years has been that there are too few providers and that premiums are too high. The industry tends to be very volatile, with new providers entering the market at times when high earnings can be achieved through the funds collected. This occurred in the early 1980s. Thus, in the 1990s, premiums were somewhat flat or decreasing. In the 21st century, small operators in particular are again having problems getting coverage, as some major providers have left the market.[5]

Much of the insurance written in the United States is reinsured. By passing on part or most of the exposure to other insurers, a company can reduce its overall risk of being insufficiently covered for a major claim from one source. The ultimate reinsurer is Lloyd's of London. It should be noted that reinsurance companies are affected by economics in a manner similar to that of the primary insurers. Companies that enter the reinsurance business because of lucrative investment opportunities often leave when claims become too frequent or onerous.

Aviation insurance. Aviation insurance is almost totally a reinsurance business; the number of sources for insurance has declined even more than for most types of coverage. There are only about a dozen companies offering aviation coverage, and not all of these offer all types of coverage. There may be only one offer for a particular need. In addition to these, there are others which have been unable to complete their treaty agreements with Lloyd's, and which may not reappear in the aviation marketplace.

Within aviation, the greatest purchasing power is held by the airlines. General aviation is perceived to have higher risks and less marketplace buying power. Within general aviation the highest risk areas are:

> Old aircraft;
> Homebuilts;
> Ultralights and other experimentals; and
> General aviation after-market modifications.

With the decline of aviation insurance availability, the lower-risk operations tend to be the only ones able to get coverage. This applies to new policies as well as renewals.

The effects are as follows:

> For small FBOs, coverage is almost nonexistent;
> Publicly-owned airports can still get coverage, but at rising cost;
> Underwriters are seeking liability limits of $100,000 per seat. Given court settlement levels, which in the United States are the highest in the world at over a million dollars, this amount is virtually useless for a severe accident.

Everyone remotely involved in a claim resulting from an accident may be sued under "joint and several liability" considerations—the aircraft manufacturer, the most recent FBO that worked on the aircraft, the maker of parts, the airport, and so on.

One effect of the insurance and claims situation has been the cost of product liability insurance. This is due to the value of many liability suit settlements of the past decades, as well as to other trends in the insurance industry. For example, Bob Martin, former general counsel of Beech Aircraft, estimated that product liability claims against Beech, Cessna, and Piper before 1994 exceeded $1 billion. This was approximately twice the net worth of these three companies combined. The costs of product liability insurance are a part of the cost of each year's production of general aviation aircraft. Liability costs have also raised FBO prices for repair and maintenance of aircraft. It would appear that aviation insurance will continue to have a negative effect on general aviation growth; although GARA's liability limit helps deflect some insurance costs, it may have the unintended consequence of transferring product liability costs from aircraft manufacturers to FBOs' maintenance businesses.

Normal Business Insurance

In general, the following types of insurance are available to the aviation manager:

1. Fire and general property insurance—covering fire losses, vandalism, hail, and wind damage.
2. Premises liability insurance—covering injury to the public, such as a customer falling on the property.
3. Product liability insurance—covering injury to the customer arising from the use of materials or service bought at the business.
4. Burglary insurance—covering forced entry and theft of merchandise, equipment, or cash.
5. Consequential loss insurance—covering loss of earnings or extra expenses in case of suspension of business due to fire or other catastrophe.
6. Fidelity bond—covering theft by an employee.
7. Fraud insurance—covering counterfeit money, bad checks, and larceny.
8. Workmen's compensation insurance—covering injury to employees at work.
9. Life insurance—covering the life of the owners, key employees, and other personnel.
10. Plate glass insurance—covering window breakage.
11. Boiler insurance—covering damage to the premises caused by boiler explosion.

These specific types of insurance can be classified into four general categories:

1. Loss or damage to property owned by the business.
2. Bodily injury and owner's property damage and liability.
3. Business interruptions and losses resulting from fire and other damages to the premises.
4. Death or disability of key executives.

The aviation business is concerned with these basic risk areas and the insurance that will serve to protect it. The list just described includes the typical operations, facilities, and resulting exposures of any business. A second area includes those exposures associated primarily with aviation businesses and their unique needs.

Loss or damage of property. The average aviation business has a considerable investment in buildings, furnishings, and inventory. These investments should be protected against fire and other perils such as smoke, windstorms and hail, riot, civil commotion, explosion, and damage by aircraft or motor vehicles. The latter form of risk insurance, or extended coverage, can be added to the basic fire insurance policy at little additional cost. Vandalism, malicious mischief, earthquake, and boiler explosion can also be added to the policy. In the beginning, the manager is concerned with the determination of insurable value and the approach to be taken. There are two basic measures: actual cash value and replacement cost value. Actual cash value is based on replacement cost and is generally considered to be replacement cost minus depreciation. Replacement cost means the cost of replacing the facility with a similar structure of a like kind and quality at present-day prices.

In planning for property insurance, the manager should consider accepting a coinsurance clause as it may result in a substantial reduction in premiums. Statistically speaking, most property loss claims are for less than 20 percent of the total property value. Therefore, mangers may agree to maintain total coverage less than the full cash or replacement value of the property—normally 80 or 90 percent. Property loss claims will be fully covered up to the agreed total coverage amount provided the insurance carried equals the insurance required. Accepting a coinsurance clause is an effective cost-saving option when there is a low probability of total property loss (e.g. a fire or tornado that completely destroys the property.)

If, at the time of loss, the insured organization has failed to maintain the required insurance, it cannot collect the full amount of its loss. It is the insured's responsibility to carry the proper amount of insurance. That is, the amount carried must equal the percentage of agreed coverage times the current property value. Payment is made under the coinsurance provision based on the following formula:

$$\frac{\text{Amount of insurance carried}}{\text{amount of insurance agreed to carry}} \times \text{amount of loss} = \text{amount paid}$$

For example, suppose Langley Aviation owns a hangar with a current full replacement value of $100,000 and has purchased a policy with a 90 percent coinsurance clause. The agreed upon amount of 90 percent means that Langley Aviation is required to carry coverage on the lesser amount of $90,000 for this hangar as opposed to paying premiums on the full amount of $100,000. As long as the full replacement value remains at $100,000, all losses will be completely covered up to $90,000. Here are three possible situations based on this information:

1. A fire causes $10,000 in damages to the hangar. Upon assessment by the insurance company, the current full replacement value of the hangar is determined to be $100,000. Langley Aviation would receive a payment of $10,000 from the insurance company ($90,000/$90,000 × $10,000 = $10,000).
2. A fire completely destroys the hangar. Upon assessment by the insurance company, the current full replacement value of the hangar is determined to be $100,000. Langley Aviation would receive $90,000 ($90,000/$90,000 × $90,000 = $90,000).
3. A year after purchasing this policy a fire causes $10,000 in damage to the hangar. The full replacement value of the hangar has increased to $125,000; however, Langley Aviation has not increased the amount of insurance carried from $90,000 to the amount of insurance now required of $112,500 (0.9 × $125,000 = $112,500). Langley Aviation would receive $8,000 ($90,000/$112,500 × $10,000 = $8,000.)

Under recent, stagnating inflationary trends, and due also to more stringent building codes, building costs have increased rapidly beyond the overall rate of inflation. Thus, premises replacement costs are much higher and have resulted in a tendency for older buildings to be underinsured as was the case in the third situation above. The manager should check frequently to see that facilities are adequately insured. Insurance companies frequently offer assistance in determining replacement value by means of an "appraisal kit," which includes multipliers to apply to the original cost, based on the age and location of the facility.

Legal liability. Legal liability is potentially the greatest risk that a general aviation manager faces. The loss associated with business property is limited to the value of the property. However, in liability exposure there is no fixed loss limit, and a judgment against the business in a personal injury or property damage suit may be a far higher amount. The size of damage suit awards has risen sharply in recent years, and today liability coverage of $1 million or more is not considered high or unreasonable. "Wrongful death" settlements, for example, range from $1 million and up in recent U.S. aviation cases. Without liability insurance, a single judgment might force an organization into bankruptcy proceedings. Consequently, liability insurance is considered essential.

There are four types of liability exposure:

> Employer's liability and workmen's compensation;
> Liability to non-employees;
> Automobile or other vehicular liability; and
> Professional liability.

Employer's liability and workmen's compensation. Under common law as well as under workmen's compensation laws, an employer is liable for injury to employees at work caused by failure to:

> Provide safe tools and working conditions;
> Hire competent fellow workers; or
> Warn employees of an existing danger;

Employee coverage and the extent of employer liability vary from state to state.

Non-employee liability. Non-employee liability, general liability or third-party liability is insurance for any kind of bodily injury to non-employees except that caused by automobiles and professional malpractice. This includes customers, pedestrians, delivery people, and the public at large. It may even extend to trespassers or other outsiders even when the manager exercised "reasonable care."

Automobile liability. Cars and trucks are a serious source of liability. Such liability is encountered primarily in vehicles owned by the business, but can be experienced under the doctrine of agency when the employee is operating his or her own or someone else's car in the course of employment. In this instance, the business could be held vicariously liable for injuries and property damage caused by the employee. If it is customary or convenient for an employee to operate his or her own car while on company business, the business is well advised to acquire non-ownership automobile liability insurance.

Professional liability. This is insurance for errors and omissions by the business in its advisory capacity and is very costly, many companies may seek to do without this element of coverage.

Business interruptions and losses. Although losses resulting from property damage may be covered by insurance, there are other losses that may be the consequence of property damage or that are indirect. For example, a fire may force the business to move to another location or to actually cease operations temporarily. Extra expenses incurred in moving and estimated profits lost during the period may also be added to the coverage limit.

Death or disability of key people. The death or disability of a key person in the organization can cause serious loss to the business. If one person is critical to the success of the company, his or her death or disablement may result in the demise of the company. Even if the key person is a non-owner employee, his or her disability can be extremely serious to the company, for the person's services may be lost, yet the obligation to pay that person's salary may continue. These risks can be minimized by acquiring life and disability insurance on the key person(s) payable to the company in amounts that will permit the business to operate and survive.

SMP-Special Multiperil Policy. This is a comprehensive policy like the homeowner's policy but for the commercial risk insurance field. Under this policy, the manager can purchase one insurance policy to cover most of the risks that normally would require separate underwriting agreements. The only ones not included in the package are workmen's compensation and automobile. By combining the policies into one package, policy writing and handling costs are reduced by creating savings reflected in reduced premium rates. This procedure can result in as much as a 25 percent savings in insurance costs, and can cause the manager to consider his or her risk and insurance problem as one, rather than several individual difficulties. Overlapping coverage should be avoided and a program that covers all important risk exposures should be developed.

Special Aviation Coverages

Aviation is a very specialized area of insurance; general aviation especially lacks buying power in the insurance marketplace. Therefore, it is vital to choose a knowledgeable agent. The following is a guide to good insurance representation.

The aviation organization, in addition to the normal exposures of business, is faced with the special exposures and problems of the aviation world. These risks must be recognized and handled in a manner similar to that of other risks. Because of the magnitude of the risks, the premium costs involved, and the potential impact on the business if adequate protection is not provided, special emphasis and attention should be given this area by the manager. The major insurance coverages in aviation include:

> Aircraft hull;
> Aircraft liability;
> Airport liability;
> Workman's compensation;
> Aviation product liability;
> Underground tank coverage; and
> Hazardous waste.

Aircraft hull coverage may normally be written to cover two basic types of coverage: "all risks" and "all risks, vehicle not in flight." The "all risks" coverage is a broader form of insurance and protects the owner against damage to, or physical loss of, his or her aircraft while on the ground or in flight. It is frequently written with a deductible clause that applies to all losses except fire, lightning, explosion, vandalism, transportation, and theft. This deductible is frequently varied for the "not in motion" exposure and the "aircraft in motion" risks. The size of the deductible has a direct bearing on the premium, and is one of the factors considered by the manager in planning his insurance coverage and risk management. The "all risks, vehicle not in flight" is a coverage that protects the physical aircraft against loss or damage while on the ground. Deductibles follow the same pattern as for "all risks" coverage.

Aircraft liability insurance covers the insured's legal liability that results from ownership, maintenance, or use of the aircraft. There are many exposures that must be considered in this area, including:

> Passenger bodily injury;
> Bodily injury excluding passengers;
> Property damage;
> Medical payments for passengers and crew;
> Voluntary settlement coverage; and
> Non-ownership liability.

In general, the liability of aircraft owners and operators for injury or damage to persons or property, respectively, conforms to the local laws governing damage suits for accidents that occurred on the land or water. The basic legal principles applied are the common law rules of negligence—that is, the burden is upon the person who has been damaged to prove fault as a proximate cause of the accident. This has been expressed as a failure to exercise the requisite degree of legal care owed to the damaged plaintiff. The coverages included in bodily injury, property damage, and medical payments are self-evident from the terms. The limits of the coverage, especially because of the catastrophic nature of aviation hazards, are a primary concern. It is extremely difficult to select adequate limits in these three areas for the exposure involved. The trend over the last few years has been toward higher and higher limits, because of larger court settlements.

Voluntary settlement or admitted liability insurance is available in conjunction with passenger legal liability. It is written on a limit-per-seat basis. Regardless of the legal liability, it offers to pay on behalf of the insured the prearranged sums for loss of life, limbs, or similar severe disabilities suffered by passengers in the aircraft. When voluntary settlement payment is offered to a passenger, a release of liability against the insured must be obtained. In the event the claimant refuses to sign a release, the offered payment is withdrawn and the passenger liability coverage applies.

Non-ownership liability arises when the individual or corporation utilizes rented, borrowed, or chartered aircraft. Generally the owner's policy does not protect the user, so additional coverage is obtained through non-ownership policies or policy clauses that cover the use of other aircraft or substitute aircraft.

Airport liability insurance is designed to protect the owner and/or operator of a private, municipal, or commercial airport against claims resulting from an injury to any member of the public or damage to property suffered while on the airport. Owners and operators are liable for all such damage caused by their failure to exercise reasonable care. This liability extends to lessees, airplane owners, passengers, and persons using the facilities of the airport as well as to spectators, visitors, and other members of the general public who may be on airport grounds. Airport operators owe a duty to a wide range of people, and litigation may arise from a wide variety of events occurring on and off the airport. The principal areas where litigation might take place can be summarized under the following headings:

Aircraft operations (liability to bailees, tenants and invitees):

> Aircraft accidents;
> Fueling;
> Hangar-keeping;
> Loading services;
> Maintenance and service; and
> Search and rescue.

Premises operations (liability to tenants and invitees) is the same as for other industries:

> Automobile parking lots;
> Elevators and escalators;
> Police and security;
> Slips and falls;
> Special events;
> Tenants and contractors; and
> Vehicles.

This list is neither inclusive nor complete. It does, however, suggest the variety of occurrences for which the airport manager has a legal duty. The hangar-keeper's legal liability endorsement provides coverage for another exposure of concern to airport owners or operators. Damage to aircraft in the care, custody, or control, but not owned by the facility operator is normally not covered in the standard airport liability policy. Many claims have been directed against airport management for aircraft loading-stand accidents, although the majority have been against air carriers and ramp-service companies. Rescue operations, if conducted negligently, may lead to legal liability for damage to persons or property. Aircraft maintenance contracts can also be the source of claims for liability and damages.

Various activities on the airport premises include have led to legal action and judgments. Among these are negligence suits concerning parking lots, stairs, elevators and escalators, police and security actions, airport special events, airport tenants and contractors, and vehicle operations. These areas are similar risk areas for many business activities and are not unique to aviation. The law follows the general rule that the operator has a legal duty to keep the premises in a reasonably safe condition for those persons who either expressly, or by implication, come to the facility by invitation.

Workman's compensation under common law, as well as the laws of the various states, considers an aviation employer liable for injury to employees at work caused by their failure to provide adequate working equipment and conditions, have competent fellow employees, and appropriately convey existing danger to employees.

Although employee coverage and the extent of the employer's liability varies from state to state and country to country, most areas require employers to pay insurance premiums either to a public fund or to private insurance companies. The funds generated in this manner are used to compensate the victims of industrial accidents or occupational illness. Premiums are based on rates that reflect the hazards involved and safety programs implemented.

Aviation product liability coverage is another area of great concern to aviation managers. The rapidly increasing number of product liability claims, the substantial costs incurred in defending these suits, and the rising costs of adverse judgments have underscored the need for sound insurance protection in this area. Typical claims have arisen from incorrect fueling, poor maintenance, and deficient design or construction of airframes, powerplants, or components.

Product liability law works in curious ways and has created a growing problem for the aviation business. The airplane manufacturer, as a larger corporation, has been a frequent target for product liability suits. In these suits, where the product is alleged to be defective, it has been easy under our judicial system to find the jury applying present-day standards in judging the safety of a product built many years ago. Coupling this with the humanitarian impulse—the feeling that someone who has been hurt must therefore be helped—juries are inclined to provide recovery from those best able to pay rather than those responsible for causing the damage. An illustration of this is a suit brought by a widow against an airframe manufacturer. She claimed the aircraft in which her husband crashed was defective. He was a VFR pilot who loaded his plane above gross takeoff weight, flew into a raging snowstorm without checking the weather, and subsequently crashed due to icing. The jury awarded $1 million, which was paid by the aircraft company's product liability insurance policy. While the fault was clearly the pilot's, a juror remarked: "Well someone was hurt, so we felt someone had to pay." This trend is further illustrated by a California jury who awarded punitive damages of $17.5 million against Beech Aircraft Company—an amount that was about 40 percent of Beech's net worth at the time. The rising cost of product liability has been reflected in the price tag placed on new aircraft. Today, the price of a new single-engine aircraft includes multiple thousands of dollars to cover the cost of product liability insurance. The consumer ultimately absorbs this cost, just as though it were an item listed on the bill of sale. The insurance cost varies with the size, purpose, and price of the plane.

Aviation businesses engaged in maintenance, fueling, sales, or similar activities have been engaged in product liability suits. One aircraft service company that was contracted to perform a 100-hour inspection was later sued because a broken valve stem caused an accident. The company was alleged to be negligent in failing to discover the defect that caused the accident.

In another case, the underlying cause of an accident was determined to be the installation of bolts and bearings that did not meet specifications, and an inadequate inspection that failed to reveal this condition. Damages in this case were awarded at $1.4 million.

Manufacturers and installers of aircraft components can be subject to legal action. There have been court cases involving fuel pumps, nose gear actuating cylinders, cylinder barrels, and propeller controls. Both the manufacturers and the aviation business using the products are involved in legal actions of this type.

The sale of used aircraft is also subject to this type of legal action, as evidenced by a case where the court held the seller liable for latent defects affecting the airworthiness of the aircraft.

Fueling activities have led to several accidents and resulting products liability suits for aviation businesses. Using deicing fluid instead of anti-detonant injection (ADI) fluid and jet fuel instead of aviation gasoline are two errors that have led to many accidents and resulting court cases.

Underground storage tanks. A key area for aviation business concern is underground fuel storage tanks. Many older tanks have corroded and begun to leak, risking contamination of nearby water supplies. The EPA has issued stringent requirements for inspection and, if necessary, removal of tanks. One aspect of these regulations is that aviation businesses and others with underground tanks must provide a $1 million or more bond as assurance of their ability to handle any tank problem. For most aviation businesses this results in a need for new insurance coverage.

Hazardous wastes. EPA regulations regarding the handling, storage, disposal and disclosure of hazardous wastes have several effects for aviation businesses with maintenance shops and especially the agricultural operator.

Aviation Tenant-Landlord Agreement

When an FBO is a tenant, it is best to obtain a written agreement with the airport owner/landlord regarding responsibilities on the airport. This should relate to area, functions, and information flows between the two. As much risk as possible should be shifted to the airport owner. This document should be reviewed, if not prepared, by a lawyer.

Selection of Aviation Insurance

Knowing what kind of insurance to carry and how much to purchase are important aspects of good risk management. Here are some guidelines for risk management and insurance selection.

1. Consider carefully:
   > The size of the potential loss;
   > Probability of loss occurring; and
   > Resources to replace the loss, should it occur.
2. How much the business can afford to lose:
   > If the loss is likely to produce serious financial impairment or bankruptcy, then the risk should not be assumed.
3. Consider the scale of the risk in relation to insurance costs:
   > A large loss may be protected by a small premium.
4. Consider the probability and the size of potential losses:
   > Repeated losses are predictable and typically small; and
   > Small losses can be assumed and budgeted as a cost of business.
5. As noted previously, the following risks can be covered by insurance:
   > Loss or damage of property;
   > Personal injury to customers, employees, and the general public;

> Loss of income resulting from interruption of business because of damage to the firm's operating assets; and
> Loss to the business from the death or disability of key employees or the owner.

Selection of Aviation Insurer

Care should be exercised in selecting a knowledgeable, reliable, and resourceful aviation insurance broker. One aviation group recommends asking these questions about the insurance provider:

1. Will the person responsible for my policy be an aviation expert with authority to bind coverage on behalf of his/her company?
2. Will I see my representative on a regular basis?
3. Will all coverage be handled in one simple policy?
4. Can I have an itemized monthly billing showing the cost of each aircraft, plus other endorsements and coverages?
5. Will premiums cover only the actual number of days I own an aircraft with no short-rate penalty or finance charge?
6. Is my policy continuous, eliminating the annual problem of filing certificates with lien holders and others?
7. Is my policy flexible enough to handle special needs the business may become involved in, such as banner towing, pipeline patrol, and so on?
8. In the event of a loss, will an outside adjuster be called, or does my representative have authority to settle claims?
9. If an aircraft is damaged or a total loss, will I have to wait weeks or months to receive payment?
10. Is my policy tailored to my exact needs and usage?

Accident Policy and Procedures

If an accident does occur on the FBO's property, to one of its passengers, or to its aircraft, there are not only insurance claims to be filed but also a number of other regulatory agencies with which to coordinate. Moreover, the following of proper procedures in handling accidents can reduce risk at the time and maintain a higher level of confidence on the part of your insurer for the next time.

Federal Reporting Requirements

The NTSB is responsible for investigating and determining probable causes of aircraft accidents, although it delegates some investigative work to the FAA as noted below. A description of each agency's role and requirements follows. If an act of terrorism or vandalism is suspected, the TSA, FBI and Justice Department may also have roles.

The National Transportation Safety Board (NTSB) is the federal agency responsible for determining the probable cause of all U.S. civil transportation accidents. This responsibility is vested solely in the Safety Board and cannot be delegated to any other agency. If, during its investigation of accidents, the NTSB discovers facts, conditions, and circumstances require corrective action to preserve public safety, it may issue recommendations calling for remedial changes in any phase of civil aviation. The knowledge gained from accident investigation is used to prevent additional accidents. The Board also generates safety recommendations from the findings of special studies. The NTSB is separate from the FAA as the NTSB may fault the FAA for short-comings in the creation or enforcement of laws.

In carrying out its responsibility to determine the cause of all U.S. civil aviation accidents, the Board has issued United States Safety Investigation Regulations (SIR). Part 830 of the Regulations specifies rules pertaining to aircraft accidents, incidents, overdue aircraft, and safety investigations. It is important for managers and pilots to be familiar with, and to comply with, the provisions of this regulation. Important sections of Part 830 follow.

"An accident. The National Transportation Safety Board has defined an 'aircraft accident' as an occurrence associated with the operation of an aircraft that takes place between the time any person boards the aircraft with the intention of flight until such time as all such persons have disembarked, in which any person suffers death or serious injury as a result of being in or upon the aircraft or by direct contact with the aircraft or anything attached thereto, or the aircraft receives substantial damage."

Serious injury means any injury that:

1. Requires hospitalization for more than 48 hours, commencing within seven days from the date the injury was received.
2. Results in a fracture of any bone (except simple fractures of fingers, toes, or nose).
3. Involves lacerations that cause severe hemorrhages, nerve, muscle, or tendon damage.
4. Involves injury to any internal organ.
5. Involves second- or third-degree burns or any burns affecting more than 5 percent of the body surface."

NTSB rules involving accident notification responsibilities include that the definition of a "fatal injury" includes any injury that results in death within 30 days of an accident. The definition of "incident," is an "occurrence other than an accident, associated with the operation of an aircraft that affects or could affect the safety of operations."

Substantial damage is harm that adversely affects the strength, structural integrity, aerodynamic performance, or flight characteristics of the aircraft and that would normally require major repair, alteration, or replacement of the affected component. The following describes the NTSB's requirements for action in the event of accidents, incidents, overdue aircraft, and safety investigations.

Immediate Notification. The operator of an aircraft shall immediately, and by the most expeditious means available, notify the nearest National Transportation Safety Board, Bureau of Aviation Safety Field Office when:

1. An aircraft accident or any of the following listed serious incidents occur:
    a. Flight control system malfunction or failure;
    b. Inability of any required flight crewmember to perform normal flight duties as a result of injury or illness;
    c. Failure of any internal turbine engine component that results in the escape of debris other than out the exhaust path;
    d. In-flight fire;
    e. Aircraft collision in flight;
    f. Damage to property, other than the aircraft, estimated to exceed $25,000 for repair (including materials and labor) or fair market value in the event of total loss, whichever is less.
    g. For large multiengine aircraft (more than 12,500 pounds maximum certificated take-off weight):
        i. In-flight failure of electrical systems which requires the sustained use of an emergency bus powered by a back-up source such as a battery, auxiliary power unit, or air-driven generator to retain flight control or essential instruments;
        ii. In-flight failure of hydraulic systems that results in sustained reliance on the sole remaining hydraulic or mechanical system for movement of flight control surfaces;
        iii. Sustained loss of the power or thrust produced by two or more engines; and
        iv. An evacuation of an aircraft in which an emergency egress system is utilized.
    h. Release of all or a portion of a propeller blade from an aircraft, excluding release caused solely by ground contact;
    i. A complete loss of information, excluding flickering, from more than 50 percent of an aircraft's cockpit displays known as:
        i. Electronic Flight Instrument System (EFIS) displays;
        ii. Engine Indication and Crew Alerting System (EICAS) displays;
        iii. Electronic Centralized Aircraft Monitor (ECAM) displays; or
        iv. Other displays of this type, which generally include a primary flight display (PFD), primary navigation display (PND), and other integrated displays;
    j. Airborne Collision and Avoidance System (ACAS) resolution advisories issued either:
        i. When an aircraft is being operated on an instrument flight rules flight plan and compliance with the advisory is necessary to avert a substantial risk of collision between two or more aircraft; or
        ii. To an aircraft operating in class A airspace.
    k. Damage to helicopter tail or main rotor blades, including ground damage, that requires major repair or replacement of the blade(s);
    l. Any event in which an aircraft operated by an air carrier:
        i. Lands or departs on a taxiway, incorrect runway, or other area not designed as a runway; or
        ii. Experiences a runway incursion that requires the operator or the crew of another aircraft or vehicle to take immediate corrective action to avoid a collision.
2. An aircraft is overdue and is believed to have been involved in an accident.

The notification shall contain the following information, if available:

1. Type, nationality, and registration marks of the aircraft.
2. Name of owner and operator of the aircraft.
3. Name of the pilot-in-command.
4. Date and time of the accident.
5. Last point of departure and point of intended landing of the aircraft.

6. Most recent position of the aircraft relative to some easily defined geographical point.
7. Number of persons aboard, number killed, and number injured, as is known.
8. Nature of the accident or incident, the weather, and the extent of damage to the aircraft or surroundings, as is known.
9. A description of any explosives, radioactive materials, or other dangerous and sensitive articles carried.

Manner of notification. The most expeditious method of notification to the NTSB by the operator will be determined by the unique circumstances, geography, and time of the event. The NTSB has advised that either of the following are considered examples of acceptable notification:

> Direct telephone notification;
> Notification to the Federal Aviation Administration, who would, in turn, notify the NTSB by direct communication—that is, dispatch or telephone.

Reports. The operator must file a report on NTSB Form 6120.1 or 6120.2, available from the NTSB Field Offices or the NTSB headquarters in Washington, D.C.:

> Within ten days after an accident;
> When, after seven days, an overdue aircraft is still missing; or
> On an incident for which immediate notification is required by §830.5(a), only as requested by an authorized representative of the Board.

If physically able at the time the report is submitted, each crewmember shall attach thereto a statement setting forth the facts, conditions, and circumstances relating to the accident or occurrence as they appear to him or her to the best of his or her knowledge and belief. If the crewmember is incapacitated, he or she shall submit the statement as soon as physically able. Statements filed with the NTSB or FAA by individuals may be used in court or subjected to perjury charges.

Where to File the Reports. The operator of an aircraft must file any report required by this section with the Field Office of the NTSB nearest to the accident or incident location.

The Safety Board is a relatively small organization, and has delegated to the Federal Aviation Administration the task of investigating nonfatal minor crashes involving light aircraft with maximum gross weights less than 12,500 pounds, with the exception of air-taxi aircraft and helicopters. The Safety Board still determines the probable cause of these minor crashes after evaluating the FAA's investigation findings.

In these instances, the FAA is tasked with the investigative role as well as other related safety functions. A statement of the FAA's investigative role has been given as follows:

"FAA's responsibility in accident investigation is of a two-fold purpose: to assist the National Transportation Safety Board in carrying out its prime investigative task and to determine whether there may have been a breakdown in any of the following areas of responsibility charged to the FAA under the Federal Aviation Act of 1958:

> Violations of the Federal Aviation Regulations;
> Operation and performance of air navigational facilities;
> Airworthiness/crashworthiness of FAA-certified aircraft; or
> Competency of FAA-certified airmen, air agencies (such as repair stations and flight schools), commercial operators or air carriers."

State and Local Reporting Requirements

State roles in accident investigation vary. Some states have staff that work with FAA (and NTSB, when applicable) to locate downed aircraft, keep onlookers away, and assemble evidence. Some states issue their own reports of probable cause, generally much faster than NTSB. Traditionally, larger states uniquely suited to air transportation such as Alaska employ aviation accident investigation personnel. Local police and municipal chief executives may also be required by law to be informed, regardless, this is a standard courtesy in the interest of public relations. The FBO should be fully aware of all local requirements.

Aircraft Rescue and Firefighting (ARFF) Procedures

Airports operating under FAR Part 139 are required to have ARFF equipment and procedures. Extensive debate and study have taken place in the last few years about how small an airport needs to be for exemption from ARFF rules. General aviation airports do not have any federal requirements, but good management suggests that the FBOs and the airport owner should have a joint plan with local fire and rescue departments and subsequently hold periodic drills.

Airports other than certificated airports that maintain fire-fighting and rescue services will find the guidelines contained in FAR Part 139 and AC 139.49–1 very useful. Airports that do not maintain fire-fighting services might benefit from these sources as well as from the bibliography contained in the appendix of this circular. In addition, Advisory Circular 150/5200–15 is a valuable document. Availability of the International Fire Service Training Association's (IFSTA) Aircraft Fire Protection and Rescue Procedures Manual and the report "Minimum Needs for Airport Fire Fighting and Rescue Services"[6] should be of value and interest to the aviation manager concerned with providing adequate facilities. Firefighting training may also be required of FBO employees and employers.

Search and Rescue

Search and rescue functions are handled slightly differently by each state. The state police, National Guard, or other group may lead and organize joint efforts. The FBO may be actively involved by taking actions such as loaning aircraft, coordinating Civil Air Patrol (CAP) spotters, or providing a communications base; the FBO may also choose to limit the company's involvement.

Aviation Security

Flight Security

Risk management and risk transfer through insurance may apply to other risk areas such as aircraft theft, drug trafficking, hijacking, and so on. Perhaps unlikely events for the typical FBO to witness, these could be very serious threats to life if they occur.

Widespread criminal and terrorist activities directed against the aviation community have increased in both likelihood and intensity in recent years; the aviation community's focus on this issue was heightened by the terrorist attacks of 9/11. As a result, all FAA security functions were transferred on February 17, 2002 to the newly-established TSA under the DHS.

The main impact of airport and aircraft security regulations so far has occurred at air carrier airports. However, some of this impact is felt at feeder airports where scheduled air-taxi operators have customers who are "through" passengers, connecting with Part 121 carriers. In order for the ATCOs (Air Taxi Commercial Operator) to discharge their passengers into secure concourses, they are required to develop and maintain approved security programs that meet the minimum acceptable standards. The TSA works diligently to conform to mandates from Congress, and to react to new threats to security detected by the intelligence community. As a result, security requirements can change quickly, and every aviation professional must be vigilant in keeping abreast of current requirements and restrictions.

New rules for flight training implemented by the TSA seek to address the fact that several of the 9/11/01 terrorists were in the United States on expired or inappropriate visas and were learning to fly at FBOs that might have noticed suspicious aspects if they had been alert. These rules apply to flight schools or flight instructors who wish to provide flight training to aliens or non-U.S. citizens in aircraft weighing less than 12,500 pounds. Flight school employees who have direct contact with flight students (regardless of citizenship or nationality) and flight instructors must complete an initial security awareness training course. When first implemented, this training was required to have been completed by January 18, 2005; however, it is still required for active instructors who have yet to complete it. New employees or instructors must complete the initial training within 60 days of being hired or certificated. Each flight school employee or independent instructor must also receive recurrent security awareness training every 12 months from the month of their initial training. Both flight training providers and flight schools are required to document (via certificate of completion) the fulfillment of initial and recurrent security awareness training in accordance with TSA regulations; the documentation must be produced upon request from TSA. Additionally, flight school and independent instructors must verify, document, and record the citizenship of all flight students prior to providing flight training. Non-U.S. citizens and aliens seeking initial flight training inside or outside the United States for a U.S. airman certificate under 14 CFR are required to participate in the Alien Flight Student Program (AFSP) and undergo a security threat assessment. An alien is not required to participate in the AFSP and undergo a security threat assessment if;

> He/she is seeking recurrent training, such as a flight review, instrument proficiency check, or flight training listed under 14 CFR 61.31; or
> He/she is seeking ground training; or
> He/she is participating in a discovery or demonstration flight for marketing purposes; or
> The Department of Defense or U.S. Coast Guard (or a contractor with either) is providing his/her training.

Complete program guidelines for aliens or non-U.S. citizens seeking flight training in aircraft with a maximum certified takeoff weight less than 12,500 pounds are provided by AOPA and can be viewed online at https://www.aopa.org/advocacy/pilots/alien-flight-training-program.[7]

Prior to a non-U.S. citizen or alien obtaining training in the operation of aircraft with a maximum certificated takeoff weight greater than 12,500 pounds, the flight school must notify TSA and the candidate must submit identification in the form of passport and visa information as well as his/her birth and citizenship history. Applications will either be processed via Category 1 Regular Processing or Category 2 Expedited Processing. Both processing methods require candidates to register with TSA online at https://www.flightschoolcandidates.gov, submit fingerprints and pay a processing fee. Applicants applying for expedited processing must also submit information establishing that they are eligible for expedited processing. Once TSA notifies the flight school that a candidate is not a threat to aviation or national security, flight training must be initiated within 180 days.[8]

Other elements of the TSA legislation include a GA security study, and improved perimeter access security.[9]

Thus, new federal agency requirements to meet the ever-changing security concerns that affect aviation operations occur on a frequent basis. This adds significantly to the regulatory burden and complexity of maintaining compliance by managers in every facet of aviation operations. However, the ramifications of failing to comply with security regulations are so severe that the manager must use every source of information and guidance available to ensure the operations for which he or she is responsible are in good standing.

NATA suggests the sample mission statement in Figure 11.2 on security to its members:

Figure 11.2 → Sample Security Mission Statement

ABC AVIATION
SECURITY MISSION STATEMENT

ABC Aviation is committed to the safety and security of our customers, co-workers, and community.

To ensure the highest level of protection for you and your aircraft, and in support of national efforts to increase aviation security across the country, please adhere to the following procedures:

— Positive ID required for ramp access. Please see the front desk.

— All baggage must remain under your control prior to boarding aircraft.

— Maintenance hangars are limited to employees only.

— Flight crew must identify all passengers and baggage prior to boarding aircraft as a group.

— Immediately report any suspicious activities or individuals to the front desk.

Thank you for your patience and cooperation.

Provided As a Member Service of
NATA
The Voice of Aviation Business

Source: National Air Transportation Association

Summary

Aviation safety, security, and liability issues are similar in many cases to those arising in any business. The four general categories of insurance are:

1. Loss or damage to property.
2. Bodily injury and property damage liability.
3. Business interruptions and losses resulting from fire and other damage to the premises.
4. Death or disability of key executives are as much applicable to an aviation business as any other.

Due to the nature of operations, there are other insurance needs stemming from the unique risks faced in aviation. The discussion is divided into three sections: risk exposure and how to reduce it; risk transfer and how to optimize I; and the requirements to comply with aviation accident and incident policies in a most expeditious and fair manner.

Endnotes

1. See *https://www.osha.gov/harwoodgrants/grantmaterials/bytopic/*.
2. Federal Aviation Administration (2012). Advisory Circular 150/5230-4B, Aircraft fuel storage, handling, training, and dispensing on airports. Washington, DC: FAA; also, Federal Aviation Administration (1974). Advisory Circular 00-34A, Aircraft ground handling and servicing. Washington, DC: FAA.
3. Federal Aviation Regulations, 14 CFR 141, 147 (2014).
4. National Safety Council. (1982). General aviation ground operation safety handbook. Chicago, IL: The Council; National Fire Protection Association (2012). NFPA 407 - Standard for aircraft fuel servicing. Quincy, MA: NFPA.
5. Casualty Actuarial Society (2003). Pricing issues in aviation insurance and reinsurance. Retrieved from *https://www.casact.org/education/specsem/sp2003/papers/lane.pdf*
6. AOPA (n.d.). AOPA's guide to TSA's Alien Flight Training/Citizenship Validation Rule. Retrieved from *https://www.aopa.org/advocacy/pilots/alien-flight-training-program/aopas-guide-to-tsas-alien-flight-training-citizenship-validation-rule*
7. AOPA (n.d.). Alien Flight Training Program. Retrieved from *https://www.aopa.org/advocacy/pilots/alien-flight-training-program*.

DISCUSSION TOPICS

1. Distinguish between risk management and risk transfer. What would be the drawbacks of a risk management plan that had only the first of these elements? Only the second?
2. Explain the "deep-pocket" theory.
3. What are the pros and cons of self-insurance?
4. What are the problems associated with reducing insurance costs by using a high deductible?
5. What are the four main types of risk exposure and which is the hardest to obtain adequate coverage against?
6. What are five types of insurance coverage particular to aviation?
7. Discuss trends in product liability suits and their effect on the typical FBO.
8. Discuss the top three aviation security needs of an aviation service business and of a general aviation airport and how they can best be met.

12

Physical Facilities

OBJECTIVES

› Convey the role of the airport business owner/manager in using, protecting, and promoting the airport.

› Describe the major parts of an airport master plan.

› Describe the major issues in negotiations for financing of developments on leased airport land.

› Discuss major environmental issues relevant to airport properties and businesses.

› Understand noise reduction techniques utilized around airports, and describe the possible impacts of noise pollution on an FBO.

› Understand why and how general aviation airports are threatened, and describe ways an FBO can help address this situation.

Introduction

The closing of domestic airports in the U.S. is widely acknowledged in the industry to be one of the greatest threats facing FBOs. Airport losses have been mainly among the mid-size general aviation airports, where FBOs make a good deal of their income, as well as among the privately-owned public-use or wholly private fields that previously would have often been absorbed into the public-use system. This chapter addresses physical facilities, starting with the national airport system and moving to the local airport environment, the airport itself, and the FBO facilities on that airport. Each of these four areas is described, and then two final sections discuss problems and opportunities. Obviously, problems and opportunities that face the national system can very quickly impact the individual local airport, so even if an airport business is not currently facing a problem, this chapter should help identify what may occur in the future and what can be done about it.

The Four Levels of Airport Service Business Involvement in Physical Facilities

Airport businesses are dependent, for the operation of their companies, on the provision of a national airport and airway system, and on continued availability and maintenance of the particular airport(s) where they are located. Few other businesses are so dependent on facilities over which they have so little control. In recent years, the national airport system has been losing, on average, one airport per week, and replacing this inventory is extremely difficult. Debates about siting new airports, expanding current ones, and preventing noise, air, or light pollution restrictions can go on for years before being resolved; far from all issues are resolved favorably for aviation stakeholders. Thus, businesses operating on airports need to be concerned with four levels of airport physical facilities. These are:

1. The national and international airport and airway systems.
2. The community affected economically and environmentally by the operation of the FBO's local airport.
3. The airport itself: runways, taxiways, ramps, terminal buildings, parking lots, and so on.
4. The FBO's own facilities.

Each of these four levels must be well-planned and safely-operated for the individual FBO to run an efficient, profitable business. Yet only the last, the FBO's own facilities, are under FBO control. At publicly-owned airports, even the use of premises is subject to many obligations and restrictions. An important part of running a successful airport business involves regularly monitoring and interacting with the other private or public entities responsible for operating the four levels of physical facilities.

The National Airport Hierarchy

The Airport System

Chapter 1 discusses the national airport hierarchy and its operation. Figure 12.1 (source: U.S. Department of Transportation, Federal Aviation Administration. (2020). *Report to Congress: National Plan of Integrated*

Figure 12.1 → **National Hierarchy of NPIAS Airports**

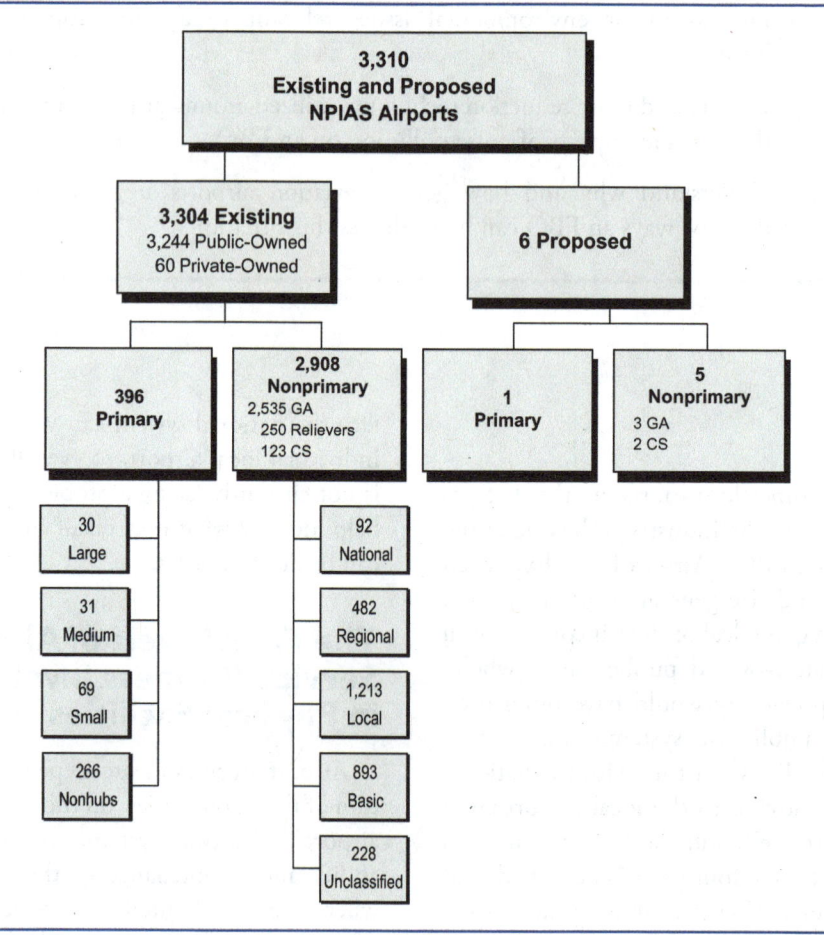

Source: FAA; NPIAS 2021–2025

Airport Systems (NPIAS) (2021–2025). Washington, DC: U.S. Government Printing Office. Available online at http://www.faa.gov/airports/planning_capacity/npias/current/) shows the numbers and types of NPIAS airports in the U.S. As noted in the National Plan of Integrated Airport Systems (NPIAS) report for 2021–2025, there are 19,636 airports in the United States.[1] Of those, 14,556 are privately-owned and restricted to private-use. Another 5,080 airports are open to public use. The NPIAS report also notes "More than 664,500 pilots, 7,628 commercial aircraft, and 211,749 general aviation aircraft perform about 99 million annual operations at more than 19,600 landing areas in the United States." The NPIAS contains 3,310 airports, which includes 3,304 existing and 6 proposed airports that are anticipated to open within the 5-year period, as covered by the 2021–2025 NPIAS report.[2] Airports are situated to make air travel convenient and to give rural communities access to rapid transportation services such as air ambulances. According to the 2021–2025 NPIAS report, census data shows that 95% of the American population lives within 30 miles of a primary airport and 99.7% lives within 30 miles of a NPIAS airport. Most NPIAS airports, however, are located in metropolitan or micropolitan statistical areas.

All 3,304 NPIAS airports are eligible for federal grants for planning and construction. Figure 12.2 shows an example of the distribution of activity levels in the system, based on 2018 data included in the 2021–2025 NPIAS report.

Publicly-owned airports have government funding opportunities not always available to privately-owned fields. The private fields usually compete for the same aviation consumers as public airports and need to understand their relative position in the marketplace. Among the nation's NPIAS airports, facility planning includes the use of federal Airport Improvement Program (AIP) funds collected from passenger ticket taxes, fuel taxes, and a variety of GA taxes. In addition, commercial service airports over a certain volume are eligible to apply Passenger Facility Charges (PFCs), the revenue from which does not go into a national fund but directly back to the airport levying the PFC.

Primary commercial-service airports are defined as those with over 10,000 annual passenger enplanements; primary airports are further classified into four categories: large hub, medium hub, small hub, and non-hub airports. The classification is based on the amount of civilian air traffic generated by the aviation community. A large hub services at least 1.0 percent of the nation's total annual enplaned passengers, a medium hub 0.25 to 0.99 percent, a small hub 0.05 to 0.24 percent, and a non-hub less than 0.05 percent. Non-primary commercial service airports are those having more than 2,500 annual passenger enplanements but fewer than 10,000. Other types of airports

Figure 12.2 → Distribution of Activity

Number of Airports	Airport Type	Percentage of Total Enplanements[1]	Percentage of All Based Aircraft[2]	Percentage of Total Operations[3]	Percentage of NPIAS Cost
30	Large-Hub Primary	71.39	0.7	13.3	29.4
31	Medium-Hub Primary	16.65	1.7	5.1	10.7
69	Small-Hub Primary	8.46	4.4	6.8	11.7
266	Nonhub Primary	3.43	11.0	12.0	14.2
396	**Total Primary**	**99.93**	**17.8**	**37.1**	**66.0**
92	National Nonprimary		9.7	8.9	4.6
482	Regional Nonprimary		20.4	23.3	9.6
1,213	Local Nonprimary		18.7	22.2	12.7
893	Basic Nonprimary		3.6	6.4	6.6
228	Unclassified Nonprimary		1.0	2.1	0.0
2,908	**Total Nonprimary**	**0.07**	**53.4**	**62.9**	**33.5**
3,304	**Total NPIAS**	**100**	**71.2**	**100**	**99.4**

[1] Based on 2018 enplanements data.
[2] Based on active general aviation fleet of 211,749 aircraft in 2018.
[3] Based on approximately 99 million operations in 2018.
Source: FAA; NPIAS 2021–2025

include general aviation airports and general aviation reliever airports.

Reliever airports are some of the most important in the system; the following excerpt sums up their role:[3]

> "General aviation, basically everything other than scheduled passenger transportation, does not always fit cohesively at large commercial airports. Even though large corporate aircraft fit easily into the commercial carrier environment, their flexible schedules may cause perturbations to commercial airport operations. Add in the full range of general aviation aircraft, single- and multiengine piston aircraft, single- and multiengine turboprop aircraft, corporate aircraft, and rotorcraft, and the scene becomes more complicated and far more difficult to manage effectively. When further consideration is given to the myriad of services afforded by general aviation operators, including flight instruction; banner towing; aerial photography; sky diving; air evacuation; corporate/executive transportation; air taxi; and charter, the need to separate commercial carriers and general aviation is clear. Reliever airports were intended to resolve this incompatible mix of aircraft and operations. The United States Congress defines a Reliever Airport as an airport that relieves congestion at a commercial airport and provides general aviation access to the community."

The Airspace System

The national airspace system, which was described briefly in Chapters 1 and 9, of course is also a vital element to the airport business. General aviation activity was severely restricted in the aftermath of the September 11, 2001 terrorist attacks and unrestricted access has not been completely restored. The FAA has made many changes to how it operates the airway system, including an increase in the past two decades in the number of FAA air traffic control towers operated under contract instead of directly.

Public Airport Organizational Structure

An airport's organizational structure can vary greatly depending on its size and ownership. In a large metropolitan area served by several airports, one authority may have jurisdiction over all aviation operations or even an entire transportation system. For example, the Massachusetts Port Authority operates both Logan International Airport and Hanscom Field, a suburban airport. Some airports have small port authorities that govern them, and may also run ports and marinas. Other airports are run by city departments, county boards, or within public works, parks, or utilities agencies. Airports run by counties instead of cities are afforded a degree of separation from elected officials, both because the jurisdiction may be larger and because county functions are quite diverse. When the airport has little direct governance by an elected body devoted just to this function, the public authority frequently sets up an airport advisory committee with considerable power of recommendation. If such an arrangement exists, airport tenants need to make sure that they are fully represented.

The Airport's Wider Environment
Overview

Airports are but one component of the national and local transportation system. The aviation environment consists of the airport, airspace around an airport, the approach zones, the flight patterns, and the noise impact areas; the aviation community consists of airport users, be they recreational or business fliers, administrators, taxpayers, and neighbors.

There is little that can be said about this wider environment that is not problematic. As noted earlier, airport business owners and pilots often do not live in the local community, and local voters who are satisfied with the airport are often less vocal to elected officials than those who are dissatisfied. The biggest problem is generally noise that impacts the community, for three reasons that have grown in importance over the past five decades:

1. Residential and other development has been allowed to get closer and closer to existing airports.
2. Air traffic has substantially grown; airports that saw just a few operations annually in the past years may now have 200,000 or more operations annually.
3. The fleet mix now includes jets, helicopters, and intensive touch-and-go training at the smaller number of towered airports in the system, as well as the relatively unobtrusive single-engine or twin-engine aircraft conducting cross-country flights. General aviation turbine aircraft have also grown in popularity.

How FAA Handles Aviation Noise

As was noted in Chapter 9, Flight Operations, aircraft noise lends itself to three types of abatement and control: source control, operating controls, and land use controls.

Source control means quieter aircraft technology by controlling and reducing the amount of noise emitted. Major steps have been taken in this direction; however, it is largely outside the control of FBOs. In recent decades, much was done to develop quieter "Stage 2" and "Stage 3" aircraft, defined by 14 CFR 36. As of December 31, 1999, all Stage 2 and noisier aircraft over 75,000 pounds were banned from conducting domestic flight operations. Stage 3 aircraft may be new, re-engined, or hush-kitted. In Europe there is considerable debate about whether re-engining and hush kits are acceptable, and further joint efforts seem likely to set standards for aircraft making international flights.

For some years, the conventional wisdom was that a Stage 4 (even quieter) aircraft was not possible, because of the noise the aircraft airframe makes in motion, however, recent chevron exhaust and blended wing-body designs have improved noise.

Operational controls mean various flight techniques to reduce noise around airports. These were discussed more fully in Chapter 9, Flight Operations.

Land use controls consist of various mechanisms to reduce or stabilize the number of noise-sensitive activities around the airport. In fully-developed areas this approach has sometimes involved the purchase of homes and schools and their relocation or demolition.[4]

Other tools include soundproofing of existing buildings and the construction of berms, noise walls, and ground run-up enclosures to contain noise on the airport. The largest soundproofing program to date is at Chicago-O'Hare airport, with 11,500 homes and 124 schools retrofitted at a combined cost of $691 million since 1982. In some cases, it is possible to purchase an aviation easement to the title of a house so that the owners and their successors forgo their right to complain about noise in exchange for a fee.[5]

Where land is not yet developed, there is more opportunity to prevent problems. However, this is often accompanied by less urgency, until one day someone realizes that a new nursing home or elementary school is under construction off the end of the main runway, close to the airport. Available land use controls for undeveloped areas include:

1. Transfer or sale of development rights to the airport owner.
2. Aviation easements.
3. Soundproofing as part of the airport area building code.
4. Zoning restrictions; for example, in an "overlay" airport zone that adds more stringent conditions to existing zoning, or a complete rezone that eliminates incompatible land uses entirely from the airport area zoning. However, where this is perceived as "down-zoning" to a less commercially-desirable use, there can be a question of "taking" private property.
5. Full disclosure of the airport's proximity in leases of new buildings.

The Oregon Aeronautics Division provides sample legal language for these and other techniques and several FAA publications are also available.[6]

FAR Parts 150 and 161

These parts of the Federal Aviation Regulations deal with aviation noise. They provide for a percentage of each year's Airport Improvement Program funding to be allocated to noise studies and to land acquisition for noise-abatement purposes.

The noise study requirements include the preparation of a noise contour map showing present or future noise sensitive areas. Once this map is accepted, the next step of a Part 150 study is an abatement plan that, working through a participatory process, evaluates all possible operational and land use strategies and presents a plan that will achieve the most effective results for that particular airport area. If appropriate, federal grant funds are available to purchase land for noise-abatement purposes. Smaller airports may be unable to procure funding for noise activities, even though they may have noise-sensitive neighbors, because their 65 decibels (dB) DNL (day-night average sound level) noise contours are contained within the airport property line. However, a more flexible recent approach by the FAA may improve this situation.

The FAA has for many years funded the removal of homes in highly noise-sensitive areas such as under the arrival and departure paths utilized at a major jet airport. In other locations, considerable funding has been made available for soundproofing of residences—at a cost of over $20,000 per home. The FAA is now addressing schools and multi-family housing. In addition, it has funded noise and land use studies to help local policy-makers.

Facilities on the Airport

Introduction

Some FBOs are owners of the airport and control all the facilities. Others may manage a private or public airport on contract or may simply be one of several tenants. The state of facilities on the field requires constant monitoring and feedback, especially in those functions for which the FBO does not have direct control.

Facilities on the airport affecting the FBO's operations may include the following:

> Runways and taxiways;
> Air traffic control tower;
> Terminal buildings;
> Lights, radio, etc.;
> Navigational aids such as NDB, ILS, and VASI;
> Utilities;
> Maintenance;
> Auto access and parking; and
> Space leased to other FBOs.

The Airport Master Plan and ALP

Airports in the NPIAS are eligible for federal planning and construction grants under the Airport Improvement Program (AIP). Eligible airports are encouraged to prepare a full Master Plan about every five years and to keep track of intermittent changes to facilities and activities. Smaller airports may not do a full Master Plan but rather an Airport Layout Plan (ALP) that shows the ultimate layout of the airport as it expands over the next 20 years. The ALP is also a key end product of the full Master Planning process. Master Plans are supported by textual and numerical analysis explaining how and why the ALP was selected. Smaller airports with fewer complexities may not need all this documentation.

The cost of installing adequately designed fuel tanks is a key consideration for an FBO.

A full Master Plan involves the following steps:

1. Inventory—of facilities, of activity and of financial performance.
2. Forecasts of demand—usually over 5-, 10-, and 20-year time-frames. Forecasts are generally prepared for based aircraft, operations and passengers, and broken down by aircraft type, year, season, peak day, and peak hour for noise analysis.
3. Demand and Capacity analysis—evaluation of the difference between existing facilities and needed facilities in the future based on peak demands. Traditionally, this topic has been examined in terms of runway or gate capacity; however, at many general aviation airports, the constraints on aircraft-basing areas will be reached long before runway saturation so that basing capacity may become the limiting factor. This analysis includes safety requirements.
4. Development alternatives—evaluation of different layouts and changes to accommodate the forecasted demand. These can include land acquisition and runway expansion, though that is not necessary.
5. Financial plan—an examination of capital and operating accounts and revenue sources to examine how to pay for the needed capital improvements. Increasingly, too, this task will examine how the operating account can be improved, for example, by new revenue sources.
6. Noise and environmental analysis—assessment of whether growth and expansion will be within acceptable environmental limits. Certain topics may require a full Environmental Impact Statement (EIS) for federal or state agencies.
7. Implementation program—how to stage the desired improvements, usually including a phased capital improvement program for the 5-, 10-, and 20-year periods.
8. Public participation program—solicitation of feedback on airport plans, usually at two levels. A technical committee will receive advance copies of the study and review them in a public process. Certain proposals, such as runway extensions, also require formal public hearings with legal notice periods, transcripts, and a formal schedule. In today's world, all interested parties must be consulted for opinions on draft concepts. There should be no surprises on either side at a public hearing.

Participation by Airport Businesses in Airport Policy and Planning

The FBO, air-taxi operator, commuter, or other aeronautical services provider at a public airport should actively seek participation in the technical-level process of a master plan. What is done to the public portions of the airport and the decisions that are made about how to allocate still-vacant land at the airport can vastly affect the FBO's prosperity. An FBO may have prospective clients who are being held back by lack of another 500 feet of runway, by lack of availability of the right fuel, or other deficiencies that the federally-supported planning and construction process could remedy. There may be inadequacies in the condition of existing facilities such as the runway needing pavement maintenance, the taxiway needing widening, snow removal needing additional plows, or a host of other possibilities for which the master plan will set budgets and priorities. The master plan will address what role the airport is to play. Will it serve primarily single-engine trainers? Does the airport owner see a major corporate twin-engine aircraft and jet market? Are capacity limits looming that will inhibit growth for the FBO's business?

The FBO should seek a full role in the study. Airports are unique in the national transportation system because they rely on teamwork between public and private sectors to keep them viable. The master plan is a key opportunity to audit and alter this symbiosis.

Private Airports

Since the AIP legislation of 1982, certain privately-owned airports have become eligible for AIP funding. In order to do so, they must meet the following criteria:

1. Be a privately-owned reliever airport, or
2. Be a privately-owned airport that is determined by the Secretary of Transportation to enplane 2,500 or more passengers annually and receive scheduled passenger service of aircraft, which is used or to be used for public purposes.

The following conditions also apply:

1. The airport must continue to function as a public-use airport during the economic life of the federally-funded facilities (at least ten years).
2. The airport must comply with the other grant conditions made to all sponsors, summarized here in Figure 12.3.

Figure 12.3 → Grant Conditions Under Airport Improvement Program

1. The airport will be available for public use on fair and reasonable terms and without unjust discrimination (all like users shall have like rates and terms).
2. No exclusive right shall be granted for the use of the airport by any person providing aeronautical services to the public (the providing of services by a single FBO is not considered as an exclusive right if it should be unreasonably costly, burdensome or impractical for more than one FBO to provide such services, and if allowing more than one FBO would require reduction of space leased to an existing FBO.
3. The airport shall be suitably operated and maintained, with due regard to climatic and flood conditions.
4. Aerial approaches to the airport must be adequately cleared and protected.
5. Appropriate action, including the adoption of zoning laws, has been or will be taken, to the extent reasonable, to restrict the use of land adjacent to or in the immediate vicinity of the airport to activities and purposes compatible with normal airport operations.
6. Facilities will be available for the use of United States aircraft.
7. The airport owner will furnish to the government, without cost, any land needed for air traffic control, air navigation, or weather services.
8. Project accounts and records will be kept in accordance with prescribed systems.
9. The fee structure will permit the airport to be as self-sustaining as possible.
10. The airport owner will submit financial reports as requested.
11. The airport and its records will be available for inspection.
12. Revenues at public airports will go to aviation purposes.
13. Land acquired for noise control with federal funds will, when resold, retain restrictions making development on the land compatible with the airport.

Source: Public Law 97-248, September 3, 1982, paraphrase of Section 511. For precise conditions see P.L. 97-248 itself.

Airport Revenue Planning

One component of the airport master plan is a financial plan. It typically examines needed construction costs compared with available federal, state, and local revenue sources. It also examines future operating revenues and rates to see if reserves can be set aside to cover the construction plan. Some airports have general funds or bonding powers available. Funds are often sought directly from airport users, manifested in FBO leases and concession fees. The FBO, therefore, needs to be very close to policy-making in this area and be prepared to explain the costs and revenues of the business, especially if he or she believes that higher rates will be self-defeating because of lost demand.

On the other hand, as discussed in Chapter 3, Marketing, the FBO and airport owner may overlook many profitable areas for new aviation and non-aviation business. The FBO's knowledge of these areas and their break-even levels may help to refocus the search for more money away from raising rates and toward provision of new services.

The FBO's Own Facilities

An FBO on a public airport is not only a tenant renting certain physical space, but is also involved in an operating agreement that provides certain rights and obligations relating to how the business is run. FBO facility management and lease development is thus somewhat more complex than the usual commercial lease.

The FBO in a full-service operation normally leases or owns the following:

> Aircraft parking;
> Tie-down areas;
> Hangars;
> Fueling facilities;
> Administrative space, front desk, and waiting area; and
> Maintenance shop.

Data Collection

The FBO will have accumulated considerable data about current facilities and activities, and must keep this information in a current and usable format. Data on market trends, new opportunities, and demand growth for existing services, are discussed in Chapter 3, Marketing.

Data on existing activities should include:

> Number of based aircraft, by corporate and personal ownership, by equipment type, and amount of use;
> Identity of frequent transients (see Chapter 8, Flight Line and Front Desk, for a discussion of a useful transient log or guest book);

- Seasonality and time-of-day peaking trends;
- Noise complaints and responses given;
- Revenue and profit by area; and
- Activity of any concessions or any ancillary operators, such as car rental, restaurant, vending machines, and industrial parks.

Data on existing facilities should include maps, plans and condition of:

- Navigational aids under FBO control;
- Buildings;
- Ramps;
- Tie-downs; and
- Hangars.

The FBO may find it useful to review its database with the airport owner to see if other information would assist in the overall airport planning process.

Planning a New Airport

A study conducted in the late 1970s suggested that the newest airports entering the system were airports on private land, and this observation is still valid. Such facilities may be as simple as a farmer rotating which field is used for crop-dusting flights, or they may be planned, public-use facilities. Any new landing area requires FAA approval and, in most cases, either approval from or registration with the state aviation agency. In addition, since airports are often not permitted land uses under municipal or county zoning, there may have to be zoning alterations or approvals. There could also be requirements for compliance with state or local environmental regulations and ordinances.

Facility Expansion

The decision to build a second maintenance hangar will depend on break-even analysis, as discussed in Chapter 4, Profits and Cash Flow. In addition, it will almost certainly require approval from the landlord, even if the site has already been leased.

Facility replacement may likewise require approvals. Design control may be involved, and certainly compliance with all FAA requirements on obstructions and setbacks will be necessary, as will compliance with local building codes and possibly airport requirements beyond the municipal code. Building codes have become much more stringent over the last three decades, increasing expansion costs; even a modest remodel may require the presumed unaffected parts of the premises to be brought up to code. Many states also have regulations affecting airports that will need to be considered. For its own purposes, the FBO will, of course, have to develop a staging plan for continuation of business during construction, in order to minimize disruption.

Preventive Maintenance

Preventive maintenance—repairing and maintaining physical facilities prior to immediate need—is a key to good facility planning. Proactive versus reactive repairs is a trade-off between suffering unexpected crises that disrupt service and incurring frequent expenses. One aviation expert suggests a split of the maintenance budget into 70 percent preventive maintenance and 30 percent crisis troubleshooting. The 70 percent may have a proportionally larger labor element because it involves such things as a planned schedule for replacing light bulbs, while the 30 percent may involve the sudden purchase of major equipment to replace equipment that has unexpectedly failed, such as the furnace or the air-conditioning system.

Scheduled Replacement of Plant and Equipment

The expected life of all well-maintained equipment provides the basis for a planned replacement schedule that may extend over as much as 20 years. Each year various major items are replaced. This must be part of the facilities budget. The replacement schedule will only be adjusted when the unexpected premature collapse of a major item occurs. Even then, the short-term preventive maintenance budget should ideally be able to handle such replacement costs rather than the long-term equipment budget. The deferral of major replacement tasks is to be avoided, if possible, because every delay drastically increases the risk of several, concurrent collapses that would seriously disrupt the smooth operation of the business. Regular inspection of facilities will help identify and implement any appropriate changes in priorities over time.

As with many other areas of the economy, there is an increasing interest at airports in encouraging the private sector to provide facilities that were previously built by the public airport authority. For example, as mentioned, there is a growing number of air traffic control towers run on contract under FAA regulation by private companies. As discussed in Chapter 9, Flight Operations, the reduction in flight service stations has triggered a new service area: commercial weather data providers whose services are subscribed to by FBOs and provided to customers. Potential profits for FBOs and/or other private investors are in such areas as the development of business offices on the airfield, restaurants, and hangars. Hangars have often been built by public airport operators, but recent years have seen increasing questioning of why scarce public funds should be used to build facilities that (1) benefit only certain airport users; and (2) can be profitable for the private sector. Certain types of portable hangars can be particularly profitable, as well as versatile, in deployment.

FBOs wishing to expand their base of operations should be constantly aware that the lines between public and private sector investments at airports are shifting and blurring; this change may present new opportunities for profit.

Zoning and Other Local Controls

Zoning regulations are varied and may not necessarily be aimed at fostering compatible uses around the airport. If existing zoning restricts something an FBO manager would like to do, he or she has the choice of either seeking permission for one special case or seeking a rezoning of all pertinent parcels. The latter is usually a more major undertaking than an FBO might want to attempt. However, if a rezone has already been initiated, the FBO manager will want to have input into the decision-making process. The FBO manager may also want to have input to any zoning changes in the immediate airport neighborhood since approval of some types of land-use, namely residential, will simply favor new noise complaints.

In addition to zoning, developments the FBO may want to undertake will be restricted by local fire and building codes. Any state requirements, FAA rules, and runway protection zones are other considerations for the FBO manager.

Environmental Compliance

Environmental compliance is another area for which codes and requirements have become much more stringent in the last several decades. Aircraft wash pads, deicing areas, fueling areas, paint shops, and other basic facilities come under intense scrutiny and often require corrective actions and expensive remediation. Some airports, for example Dayton International in Ohio, have completely banned the use of ethylene glycol for deicing, and others are capturing and recycling it on-airport.[7] Underground storage tanks must usually be removed and replaced above-ground.

There is a strong push for airports, along with all other parts of the economy, to be more conscious of sustainability through recycling, re-use and more frugal consumption of everything from oil to paper. In 1999, then-President Clinton adopted an Executive Order, *Greening the Government Through Efficient Energy Management,* which puts a fairly forceful approach to environmental consciousness on the radar of federal agencies.[8] This order spearheaded an emphasis on environmentally-sustainable construction. Leadership in Energy and Environmental Design (LEED) is a suite of standards for such construction, developed by the U.S. Green Building Council. These standards have begun to influence airport design and construction projects across the country and world.

Leases

The right to operate an aviation business at a particular airport may be obtained through legal usage of land on airport property. The FBO may receive these rights by being the airport owner or, more commonly, signing a lease arrangement with a public or private airport owner. Providing the basic direction of business operations and establishing a framework for success or failure, the manager must fully understand the lease process and obtain as favorable an arrangement as possible. Of course, from the onset the manager should recognize that it is necessary to obtain the services of an attorney to assist in molding the contractual arrangement, it is also recommended to find an attorney who is familiar with the aviation business or to be prepared to spend time acquainting an attorney with the technical implications and operational difficulties of the subject issues.

Initial considerations. Prior to dealing with the contents of a lease and the typical framework of the contract, one should clearly identify the following considerations:

1. The aviation lease is a composite agreement, a combination of a real estate lease as it is normally understood by lawyers and non-professionals, and an operating agreement that sets forth the

obligations, duties, and restrictions that apply to the manner in which the aviation business shall be conducted on the leased premises. As normally recommended by lawyers, specialists, and practitioners, or required by many public transportation authorities, the operating agreement may be made as a separate document that is incorporated into the lease by numerous references.

2. The pre-lease situation is normally one in which the lessor and lessee (land owner and aviation manager) are in a bargaining situation. It is not a precast, fixed situation. Guidelines are available, but each lease is different, representing the local situation and various local and state laws.

3. The contractual relationship between the fixed-base operator and the owners of the airport has an enormous impact upon both parties and, more importantly, upon the community they both serve. Because of the economic impact upon the community, the lease should attract a competent aviation organization and provide the opportunity for an adequate margin of profit on the required investment of time, money, and experience. A lease should cover a period of time that is sufficient to procure and sustain the financing required to establish and operate the business.

4. Developing a lease is not a one-time or periodic activity. A lease is a living instrument that controls a constantly-changing relationship and as such should be under constant review. This is reflected in the comment "start working on your second lease the day after signing your first one."

5. Developing a lease is in reality developing a plan. As such, the plan should include those elements that will ensure success for both parties. A basic concern in creating the plan is the need to set the terms and length of the lease with full consideration of the requirements of the potential lenders and the sources of potential investments. In the eyes of the lender, the amount loaned will be limited to an amount that can be adequately amortized from the funds expected to be generated by the business during the term of the lease. Therefore, a 25-year or longer lease will be needed for many financing purposes.

6. To meet the requirements of aviation business flexibility and the needs of the lending institution, there is a basic interest in the assignability provisions of the lease. Normally the lessee should be empowered to assign or sell the lease for financing purposes upon written notice to the lessor, with approval thereof not unreasonably withheld. The lessee should typically have the right to sublease part of the space covered by the lease, provided the sub-lessee is subject to the same conditions and obligations as those in the basic lease. Furthermore, it is desirable that the lease state the lessee has the right to sell without restriction to any corporation formed by, consolidated or merged with it, provided, however, that the purpose of the surviving organization is to perform under the lease.

7. An understanding of local and national economic aviation trends is important in projecting FBO sales over the life of the lease as well as anticipating increases in lease payments.

Preliminary Planning. Prior to any actual negotiations over the terms of the lease, considerable planning needs to be accomplished. Achieving a successful lease and ultimately a profitable business will depend to a major degree on the thoroughness of the planning efforts. In this plan you should:

1. Become acquainted with the airport owner, manager, board, businesses and users.
2. Review and study the lessor's previous agreements with parties operating similar businesses.
3. Review and study the lessor's agreements with any airlines serving the airport; such contracts may have direct and indirect influences upon proposed general aviation businesses.
4. Develop a prospectus for the business and make financial projections for the future. This information is vitally necessary to the lessee for making decisions on the lease costs that are acceptable without compromising profitability objectives. The airport board, in acting on the bid for a lease, has the right to probe into many personal and business financial issues to satisfy itself of the FBO's ability to perform.
5. Determine exactly which lease terms are acceptable to the lending institutions expected to provide financing.

6. Obtain the advice of the intended fuel supplier on the lease agreements dealing with fuel-storage facilities, fuel handling, and the related fee structure, such as the fuel flowage fees.
7. Determine, through a title search, that the lessor has clear title to the property being considered and is legally empowered to offer the desired lease.
8. Determine that the individuals negotiating the lease as lessors have the legal capacity and authority to represent the community or agency having actual title to the property.
9. Review the master plan for the airport and determine whether the future projections are in harmony with the proposed lease and the included business and financial projections of the aviation business. Ensure that the areas planned for lease will not be harmed by future facility developments elsewhere on the field.
10. In a new lease, review such issues as the underground storage tank conditions on the leased property; the airport noise situation and community relations. Also ensure that the airport is in compliance with FAA requirements, OSHA regulations, local fire codes, and other pertinent constraints; review the procedures and processes needed to keep compliance.
11. Review existing and future liability and insurance needs for the leased property, and examine coverage held by the airport operator to determine that there are no nebulous areas of liability between lessor and lessee.

Procedural steps. The normal procedural steps followed in negotiating and awarding a lease are:

1. The development and release of an invitation to bid by the lessor. The invitation is aimed at soliciting responses from all interested and qualified parties and contains the basic information necessary to identify the property, the desired services, and the basic leasing arrangement. In order to ensure coverage of desired items and an element of standardization among prospective lessees, a sample lease bid is usually included in the bid announcement, along with any applicable airport minimum standards that serve to screen out unqualified bidders. Bonding capability is a likely requirement.
2. The interested lessees then prepare and submit lease bids to the lessor. These are usually closed bids with simultaneous bid opening for all respondents.
3. All proposals received by the owners are evaluated fully through a detailed analysis of the major elements in each proposal. An airport board may establish a lease committee for this purpose.
4. Negotiations are conducted with the bid respondents submitting the most acceptable proposals in order to assure complete understanding of the proposals submitted and to reach any modifications deemed necessary.
5. The bidder offering the most favorable final proposal is identified, selected, and notified.
6. The detailed lease and operating agreement will be completed and agreed upon.
7. The final lease proposal is agreed upon, and the unsuccessful bidders advised and released from their offers.

Major lease components. Most aviation leases utilize a format that covers certain basic concerns and common elements. The major components in a lease include:

> Site location;
> Terms of the lease;
> Options;
> Terms of termination before expiration;
> Rights after termination (reversion);
> Lease release;
> Lease disputes and how they will be handled;
> Rights and obligations; and
> Rent amounts and conditions under which it may be raised.

We will examine these components, defining each and identifying elements of concern.

Site location. The premises being considered should be described clearly, fully, and accurately the official plot plan and survey drawing of the properties should be attached and incorporated by text reference.

First right of refusal for adjacent areas to the site should be accorded to the lessee.

The privilege to use all general public airport facilities and improvements, such as landing areas, runways, taxiways, parking areas, aprons, ramps, navigation facilities, and terminal facilities, should be identified.

Term of the lease. The lease term should be established by setting the dates for the beginning and end of the lease period.

Provisions should be included for extension of the term of the lease.

The term of the basic lease should be long enough to permit the amortization of loans made for physical improvements on the property and the erection of hangars and other installations. Financiers will be reluctant to make loans for longer periods than the basic lease, even if renewals are available.

Options. Provision for extending the lease for an additional term should specify the length of the term, the maximum amounts by which rents, fees, and payments can be increased during the option period; the basis on which rent fee increases are to be calculated, such as a local cost-of-living index, should be specified.

Allowance should be included for appropriate extension of the term of the lease in the event of interruptions due to causes beyond the control of the lessee. In lieu of such extension, the lease should provide for a moratorium on rent payments under such circumstances.

Termination before expiration. The rights of both parties must be clearly set forth in the lease to cover the contingency that the lease may be terminated before its stated expiration date.

The lessor can usually terminate for one or more of the following reasons:

1. Substantial non-performance or breach of contract.
2. Failure of the lessee to observe and conform to the terms of the lease and his or her continuing failure to bring his operations into compliance within 30 days after receiving notice from lessor to do so.
3. Failure of the lessee to pay the rent when it is due. The lease should set a period of time (usually 30–60 days) during which the lessee may remedy the fault.
4. If the lessee becomes bankrupt.
5. If the lessee makes a general assignment of the lease for the benefit of creditors.
6. If the lessee abandons the premises. Under this circumstance, the lessor generally has the right to remove the lessee's abandoned effects without being liable for damages.

The lessee may terminate the lease prematurely for the following reasons:

1. If the lessor fails to perform substantially under the terms of the lease. Termination may require written notice to the lessor and the lapse of a stated period of time (usually 30 to 60 days) before the actual termination.
2. If the lessor commits any act or engages in any activity that prevents the lessee from doing business for a period of time (normally specified in the lease).
3. The lease might provide that the lessee can terminate the lease in event of civil commotion, acts of the military, acts of God, damage to runways, court orders restraining the use of the airport, or similar events that may interrupt normal business for a specified period of time. In lieu of premature termination of the lease, there may be a provision for a moratorium on rent payments during such interruptions and/or the extension of the primary term of the lease during such time period.
4. If the lessee is hindered by the lessor, or unreasonably prevented from doing business in accordance with the terms of the lease, he or she has recourse to the courts and may receive compensatory damages.

Rights after termination. The lease should contain provisions covering the rights of both parties after termination. Some leases provide that the lessor has a purchase option on improvements to the real estate at a depreciated value. Depreciation schedules should be spelled out in the lease and provisions included for handling property under early termination conditions.

Reversion. When an airport tenant builds facilities on public airport land, the most common form of agreement is that after an amortization period of 25 or 30 years, the facilities' title will revert to the airport owner. The reversion issue was originally used in leases because the typical airport building was not designed to last more than 30 years and usually needed to be torn down at the end of the lease period. In the past few decades, building codes have gradually become more stringent, so that premises may have a much longer life and owners often seek lease extensions to reflect that life.

Usually, the tenant will have rights of first refusal to become the renter of the facilities and at a rate to be negotiated. If the FBO in question has not been maintaining a quality operation, the airport operator may, not unreasonably, seek to use the end of the lease period as a means of getting a better operator into the premises. Affected FBOs must address whether (1) they are operating in such a way as to ensure their rights of first refusal are in good standing and (2) they are receiving a reasonable rent for the older buildings that they now must start to lease. The monthly cost may be higher than the mortgage, given original construction prices and interest rates. Some FBOs have torn down their hangars rather than rent them from the airport, and the reversion issue is often contentious.

One innovative solution that has tax benefits is for the building to revert to the airport as soon as it is completed. The FBO then becomes a tenant from the outset. Instead of paying property taxes on the full value of the premises, he or she pays an occupancy or leasehold tax related only to the value of that year's lease payments. Over a 25 or 30-year period, this cuts taxes by a factor of ten or more. The landlord and tenant must provide for maintenance, repairs, utilities, and insurance on the premises within the lease.

Lease release. The lease may also be terminated before the date of expiration in a mutually-acceptable release.

By a subsequent written agreement, both parties can agree to terminate the original agreement. For example, both parties may agree to terminate the lease if performance becomes impossible or impractical, due to causes beyond the control of either party.

The lease can be written so that all obligations shall be held in abeyance during the period of interruption, and when operations under the lease are resumed, the term of the lease is extended for a period equal to the period of interruption.

Lease disputes. Even the most thoroughly prepared lease may not cover all potential problems, and disputes may arise. The lease should include a means for settling such disputes. One method provides for a three-person arbitration committee, with one arbitrator picked by each of the parties to the lease and the two selected arbitrators picking a third disinterested party. The selected manner of dispute resolution should be agreed upon and specified in the lease.

Rights and obligations. Both parties to the lease must clearly understand their respective rights and obligations. A "right" is what the lease says *may* be done; an "obligation" or "duty" is what *must* be done. This section of the lease is actually the basic operating contract because it specifies what the lessee can do and what functions must be performed to satisfy the requirements of the lessor. Likewise, the rights and obligations of the lessor should be identified. It is extremely important that the terms of the lease be designed to give the lessee sufficient latitude to operate the business profitably. The rights and obligations of each party, if the other fails to perform, should also be spelled out. The following items are frequently included in an aviation lease under rights and obligations:

Covenant not to compete. Assuming the lease covers a business operation, there should be a provision that the lessor will not compete with the lessee; or if there is competition, its extent and character should be specifically identified. An example of something to avoid in this area would be an airport manager (lessor) employed by the city yet also engaged in a general aviation business in competition with the FBO (lessee). Another example would be both the lessor and the lessee having the right to sell fuel.

Operation. The lease must give the lessees and their customers the right to ingress, egress, and to have free access to the premises, as well as "peaceful possession and quiet enjoyment" thereof. There should also be an assurance that the lessor will continue to operate the airport as a public airport for the duration of the lease, consistent with government regulations, and that there will be no restrictions to normal operations or contingent restrictions that might apply to the proposed leasehold operation during the term of the lease.

It is usual for a lessor to require indemnity insurance holding the public airport harmless from all claims, risks, accidents, or injuries caused by lessees or their employees acting in the airport's behalf in the operation of the leasehold business. The amount of such insurance coverage should be negotiated after competent consideration of all factors.

Duty to make improvements. The lease frequently requires the lessee to provide physical improvements and installations on the premises. Examples of this requirement could include the erection of hangars, shop facilities, office facilities, landscaping, paving, parking areas for motor vehicles and aircraft, taxi strips, ramps, aprons, and advertising displays and signs. The lessee might also be required to install fixtures, decorations, and equipment, as well as to construct specific facilities for the use of the public. The lessee should know the requirements of the lease before negotiating or awarding any construction contracts.

Right of prior approval. A lease usually gives the lessor the right to approve or disapprove all architectural plans and designs for required improvements. Since a lease will likely require the lessor's prior approval of the contractor selected to perform the improvements, it is recommended that the lessee take every precaution to prevent the development of arbitrary or capricious demands by the lessor. It is far better to clearly and carefully establish required improvements in the proposed plan and consider the design or implementation of requirements during the negotiations.

Minimum criteria. Many leases specify that the lessee comply with certain minimum criteria in meeting the requirements for improvements. Some specify a minimum fiscal expenditure, some describe the requirements in terms of space (for example, 200,000 square feet for a hangar), and others identify that the lessee must abide by the regulations promulgated by the authorized officials of federal, state, county, and city governments, and by airport officials. The lessee should carefully review the implications of these types of requirements and ensure his or her capability and interests. As in other areas, the advice of an attorney is beneficial, if not essential.

Deadlines. Deadlines for required improvements should be clear and feasible. There is a possibility that the lessor will want a deadline for any construction and improvements required in the leasehold. Clauses requiring completion "within 120 days of signing the lease" or stating that "construction must be complete within six months of design approval" are typical of this type of provision. They may be extremely difficult for the lessee, especially when non-completion penalties are involved. In negotiating these kinds of lease clauses, it is desirable that the lessee obtain as much flexibility as possible in schedules for completing hangars, buildings, and other physical improvements. It is also desirable that the lessee be protected where there are delays caused by such events as fire, earthquake, flood, military action, civil strife, strikes, picketing, or other issues beyond the control of the lessee.

Equipment and fixtures. It is desirable that the lessee have the right to install (at his or her own expense) the equipment and fixtures required to perform the functions of the business and the right to remove them at any time (also at his or her own expense). In some instances, the lessor's prior approval is required before removal of equipment; in these cases, provision should be made to ensure that it would not be unreasonably withheld. Under removal circumstances, the lessee is normally expected to restore the premises to a condition satisfactory to the lessor.

Maintenance and repair. The lease should clearly specify which party is to be responsible for repair and maintenance of each element of the leased facilities. In cases where the tenant obtained financing and built a building, the tenant would be fully responsible for its maintenance. At issue may be determining who is responsible, landlord or tenant, for maintaining the leased apron area of the leased premises. Where the building has reverted to the landlord and is being rented back, then the lessor normally pays for structural repairs and specific major items, and the lessee pays for maintenance needed because of ordinary wear and tear. Under these circumstances, if the lessee does not perform the necessary maintenance and repair, the lessor is normally given the right (after a specified time delay) to enter the premises and perform the necessary work at the lessee's expense.

Fire loss. Most leases require the lessee to replace buildings or facilities destroyed by fire and to return them to the pre-damaged condition so that the replacement is equivalent in value to the original facilities. The following provisions are normally made in this area:

1. 75-80 percent of fire, extended coverage and hangarkeeper's liability coverage are usually required, with the lessor approving the insurance company.

2. The lessor and any mortgage holders are normally named on insurance policies as additional insured.
3. The abatement of rent should be sought while facilities are not in use due to the fire.

Ownership after termination. The normal provision for ownership after termination of the lease indicates that title to the improvement shall remain with the lessee during the term of the lease, but passes to the lessor at the end of the lease. This would allow the lessee to have depreciated the property and liquidated the debt incurred with any financial supporters. The lease, and both parties, should ascertain the rights of lenders in the event the lease is prematurely terminated.

As noted previously, some leases provide that the property title must pass to the lessor as soon as construction of the improvements agreed to in the lease are completed. These arrangements have advantages and disadvantages that should be considered when developing the lease.

Removal at termination. Lease negotiations should cover the question of removing tenant improvements, fixtures, or equipment from the site at the termination of the lease. Normally, a "reasonable" time is given to the lessee to exercise his or her option to remain. If the option to renew is not exercised and the lease is thereby terminated, the lessee should be given the authority to remove certain clearly-defined fixtures or equipment within a stated time (say, sixty days) or otherwise the title of these will revert to the lessor.

Relocation of site. In some situations, lessors determine that they need the leased premises for the expansion or further development of the airport. The lease should include provisions for this contingency. One possible arrangement indicates that the lessor has the right (say, on six months' notice) to relocate or replace at the lessor's expense the lessee's facilities in substantially similar form at another generally-comparable location on the airport. If this is done, the lessee's loss of income during the transition period must be considered, and all contingencies covered, such as abating the rent and/or extending the period of the lease. If not covered in the lease, it should be remembered in most cases that the public authority owning the airport has the right of eminent domain and has the inherent right to condemn the leasehold and improvements for appropriate compensation in the usual manner provided by law.

Performance bonds. The lease frequently includes the provision requiring the lessee to furnish bonds for the "full performance" under the terms of the lease for such items as facility construction, guarantee of wages, and payment of contracts.

Rent. The remaining lease component, rent, will be covered fully in a later section of this chapter dealing with lease payments.

The operating agreement. A major portion of the lease constitutes an operating agreement between the lessor and the lessee. The earlier portions of this material are common to most real estate leases, but this area is unique to the lease dealing with aviation businesses. Since people outside the aviation industry may not understand many of the following topics, care should be exercised in selecting legal advice and in negotiating items in this portion of the lease. The typical aviation business lease covers the following topics in an operating agreement:

> Permitted sales;
> Permitted flight operations;
> Permitted line service;
> Signs and advertising;
> Service charges;
> Lessor inspection;
> Security;
> Snow removal;
> Uniforms;
> Fuel sales and charges;
> New functions;
> Collection of any fees on behalf of the lessor;
> Motor vehicle parking;
> Maintenance, repair, and overhaul;
> Subleases;
> Taxes or payments in lieu;
> Towing disabled aircraft;
> Business practices;
> Exclusive rights;
> Collection of landing and parking fees; and
> Vending and game machines.

A review of each of these operating topics with special emphasis on its application to the aviation business is desirable.

Insurance. Due to the complexity, cost, and rising deductibles in the ever-changing and volatile aviation insurance market, an important consideration in the FBO's operating agreement with the airport owner is responsibility for specific liabilities in the event of a problem. This requires the attention of an expert in aviation insurance, as does the premises insurance.

Sales. The lease should specifically permit the lessee to sell (retail and wholesale) new and used aircraft, new and used avionics, communication, or navigation equipment, aircraft parts, radios, and pilot supplies. In addition to the sale of aircraft, equipment, parts, and accessories, the lessee should have the right to finance such equipment, to insure aircraft and contents, or to act as agent for another party for these purposes.

Flight operations. The lessee should be specifically given the right to engage in flight operations that may include:

> Demonstration of aircraft for sale;
> Charter flights;
> Air taxi activity;
> Commuter airline operations;
> Flight training (primary and advanced);
> Aircraft leasing;
> Aircraft rental;
> Test flights;
> Sight-seeing flights; and
> Aerial application.

In some instances the lessor may feel that certain flight activities, such as primary flight training, should not be conducted at the airport. Here, the lessee should not be prohibited from conducting such flight activity away from the airport. For example, the lessee should not be prohibited from transporting the primary student by air to an outlying airport or practice area. Normally it should be acceptable for the lessee to have the right to conduct advanced, recurrent, or periodic flight training of licensed pilots at the airport. Some activities, such as aerial application, will entail special requirements on the part of the lessor.

Maintenance, repair and overhaul. The lessee in an aviation business should have the right to maintain, repair and overhaul all types of aircraft, engines, instruments, radio and electronic gear and to remove, install or re-install such equipment in aircraft in his or her care, custody, and control. Depending upon the circumstances, it may be desirable to indicate in the lease that the lessee also has the right to maintain, repair, and overhaul motor vehicles used in his business.

Line service. It is highly desirable that the lessee have the right to conduct aircraft fueling and other line-service activities. In many instances, this should include the right to service the large aircraft operated by scheduled carriers who do not maintain their own servicing equipment on the airport, and the servicing of military aircraft under government contracts. The location of fueling and line service operations should be clearly defined in the terms of the lease. The lessee should have the right to load and unload passengers and cargo and to transport passengers from transient aircraft parking areas to the terminal, auto parking areas, and other areas of the airport.

Service charges. The lessee on an airport must have the right to assess charges and fees to customers for services rendered. The lessor may periodically be given the right to review the schedule of fees and charges. Some leases provide that the lessee must set charges at levels that are reasonably competitive with those in the surrounding area.

Towing disabled aircraft. The lessee should be given the *right* to tow disabled aircraft from or about the airport. The stipulation that this also be an obligation of the lessee should be considered carefully due to equipment requirements. If necessary, it may be wise to limit the size of the aircraft to less than 12,500 pounds.

Security. The lease will probably require the lessee to prevent unauthorized persons from transiting the facilities or entering into restricted flight or loading areas. This will likely mean that the lessee has the additional responsibility of providing fencing of a quantity and quality acceptable to the lessor, and personnel escort practices that will ensure control of visitors and customers. The terrorist attacks of 9/11 and their aftermath have raised awareness for the need of greater general aviation security, regardless of airport size. Depending on whether the airport is governed by FAR Part 139, the lessor may ask the lessee to commit, as part of the lease language, to security requirements such as employee background checks that are a result of new Transportation Security Administration regulations.

Snow removal. TThe lease should be clear regarding the duties and obligations of both parties to remove snow and similar hazards, including a specific description of the various areas of responsibility. An FBO tenant may have the opportunity to contract with the airport to remove the snow for other airport users.

Uniforms. Some leases stipulate that the lessee's employees are to wear uniforms at the lessee's expense; this sometimes requires the prior approval of the lessor for uniform designs to conform to airport standards.

Motor vehicle parking. The lease must also provide for the identification of adequate space for the parking of motor vehicles. Provisions must be made for vehicles operated by lessee, lessor, customers, employees and the public at large. This should include vehicles operated on the normal vehicle roadway, as well as the airport ramps and taxiways.

Vending and game machines. The lessee should be given the right to operate vending and game machines on leased premises. If the lessee does not own such machines, their installation may be subject to the approval of the lessor and, if the lease is on a proportional rental basis [i.e. lessee pays a percentage of the gross revenue in rent] their income may be included in the calculation of the rent for the facilities.

Signs and advertising. The lease should specifically provide the lessee with the right to install signs and advertisements promoting the business name and the brand names of any aircraft, fuel, and other products or services. If the lease provides that the lessor's consent be required prior to sign installation, it should also specify that such consent will not be unreasonably withheld or that unreasonable criteria will not be established. Municipalities usually have local sign ordinances that may be applicable to the airport, and this must be taken into consideration. Guidelines should be established that provide for the removal of signs after the termination of the lease or end of affiliation with the manufacturer or supplier of products.

Subleases. When desirable, the lease should authorize the lessee to operate ancillary businesses on the leased premises, such as rental car agencies, gift shops, restaurants, barbershops, newsstands, and so on. It should also authorize the subleasing of space to provide related services such as electronic repair-related businesses, corporate offices, or aircrew spaces.

Taxes/payments in lieu. The income tax burden of a lessee is the responsibility of the business, not the responsibility of the lessor. In some states, leaseholds are exempt from property tax per se but are taxed instead under a leasehold tax. The potential lessor should ensure clarity about the tax requirements of the specific jurisdiction and budget accordingly for the necessary taxes.

Lessor inspection. The lessor may desire provisions in the lease that provide rights to inspect the leased premises with reasonable advanced notice, and to review and audit the records and accounts of the lessee's business. If these rights are included in the lease, it is recommended that the lease specify that such rights be exercised reasonably, that they not be used improperly to harass the lessee, and that the expenses of such reviews, inspections, and audits be borne by the lessor. Sometimes, this is written up to say that if all is found to be in order, the costs will be borne by the lessor, but if not, then the lessee must pay. Extreme caution should be taken in this area to ensure that any lessor rights resulting in an unfavorable business situation for the lessee are not created. If the lessor or a subsidiary business competes with the lessee, the lease should be designed to prevent an unfair advantage through intimate knowledge of the lessee's operations, finances, and records.

Business practices. The lease should provide the lessee with broad latitude in the exercise of judgment in areas of normal business activity. Lease terms requiring the lessor's approval of prices and discounts should be avoided. The lessee should have broad latitude in establishing charges and in the inventory to be maintained and sold. He or she should avoid lease terminology that forces an investment in slow-moving items of merchandise, including some grades of aviation fuel. The lease should also provide that the lessee may engage in other aviation—or aviation-related—businesses not specifically covered in the lease, by making a request in writing to the lessor, and that approval will not be unreasonably withheld. It will be important to know whether such approvals can be made administratively, or must be made by a decision of the elected body such as an airport board, commission, city department or county council.

Exclusive rights. An airport is a limited geographical area, and investors providing funds for a business in this space would normally want protection against fierce competition. Frequently, a clause is requested for the lease that would prevent additional aviation businesses from entering the airport for a specified period of time. Such clauses will not normally be permitted at airports that have received federal funds, as indicated in the FAA's Advisory Circular on Non-Exclusive Rights. However, the FAA's Advisory Circular on Minimum Standards can help an FBO combat competition from unqualified operators such as "tail-gate mechanics" and free-lance flight instructors.

Minimum standards. Any FBO on the field may have to comply with minimum standards set by the airport owner. The FAA provides guidelines on these. High minimum standards may help to keep out under-financed operators who would likely "skim the cream" of aviation business for a short time and then perhaps move on. The manager of the FBO already on the field can encourage stringent standards for his/her own protection but must, of course, be prepared to also comply with them.

Collection of landing and parking fees. Leases vary, and frequently the lessee is required to collect landing and parking fees as an agent of the lessor. Naturally, it would be desirable if the lessee were compensated for this service. The question of parking fees for aircraft awaiting or undergoing maintenance should be anticipated and normal business latitude be provided the lessee in such matters.

Lease payments. In exchange for the privilege of operating an aviation business and receiving the benefits of such activity, the lessee normally agrees to some form of lease payment to the lessor. The calculation of this payment and the rate of payment vary widely from lease-to-lease, primarily according to local conditions, the value of the lease, and the bargaining power of the two parties. The typical aviation operation is potentially a complex business operating in a volatile and sometimes erratic environment. The goal of airport owners, developers, and managers is to achieve the best possible aviation facilities. The economic balance of a lease is very delicate, with a need to consider two major factors: (1) adequate economic incentive to support and motivate the lessee in providing satisfactory aviation services and (2) minimum cost to the lessor in meeting the obligation to provide these services. Because of wide differences in the economic potential of airports, there have been some leases that paid the lessee to manage the airport, while others have provided substantial returns to the lessor. It is important to recognize the existing situation and develop a lease that has the correct balance. The lessee must be able to operate at a reasonable profit level and not be forced out of business. The lessor, normally a public agency, must meet the need for providing adequate public aviation services and must do this as economically as possible.

Lease payments may be made in one of three different ways, or any combination of them:

> Rent on real estate by cents-per-square foot;
> Fees based on the lessee's unit sales, such as cents-per-gallon of fuel sold; and
> Percentage of gross sales.

Frequently, the lease provides for all three sources: rents, fees, and revenue percentage payment. While rents and fees are fairly simple, the percentage of revenue may result in an excessive burden to the lessee if not kept within prudent bounds. The areas of concession fees and revenue percentages are full of pitfalls for both landlord and tenant, and should be explored carefully. The tendency of the lessor to extract heavy payment from the lessee must be guarded against in order to have a dynamic business that meets the needs of the growing aviation public. All three forms of payment may be combined into one rate per square foot, which then allows comparison with FBO rates on similar fields.

Real estate rent. Rent is the fee charged for the use of real property where the business operates. Real property is normally identified in the lease as unimproved property, improved property, buildings, office space, terminal building space, aprons, and ramps. Establishment of the rental rates is a bargaining process, and the lessee must determine whether the proposed rates are competitive and within the business potential for the location. Rates are constantly changing because of the economic pressures being exerted upon the airport operators. The trend over time is toward higher rental rates; however, since such rates are competitive, there is considerable variation among airports.

Comparable figures should be obtained from other airports, as well as from other leases on the lessor airport. These figures provide at least some guidance and a starting point in calculating the existing competitive situation. Also, when examining a given rate structure, a lessee should consider the entire lease because it is possible that a low rate in one area will be offset by a high rate in another area. Individual rates can be raised by the lessor using different lessees to achieve a higher overall rate. Frequently, a sliding scale of rentals is developed for the benefit of a new aviation business. In the early years of operation the rate is low, and it increases over time as the business grows. This arrangement encourages the growth of a stable business that can provide necessary services rather than maximizing the immediate cash flow for the lessor. It is realistic for both parties to anticipate that as the business grows, the lease may include provisions for increases in future rentals.

Percentage of the gross. Many aviation business leases provide for rent schedules based upon a "percentage of the gross," usually with fixed minimum rental payments. This manner of rental payment calculation can be beneficial to an aviation business because if sales go down, rent goes down. On the other hand, it makes it much harder to budget expenses as rent is now a variable cost instead of a fixed cost. Also, using a percentage of the gross rather than net income means that if there are some high costs involved in the business, the tenant will be paying unduly high rent that is not due to high income. If the percentage is set too high, it may establish a basis for rent payments so burdensome that the lessee cannot operate at a profit.

If the lessor insists upon using gross sales revenues as the selected formula for establishing rents, the lessee may maintain some predictability and fairness by striving for a reduced schedule for the fixed real estate rental and for the use of an "adjusted gross formula," which is defined as the gross income, less taxes, less bad debts, less fuel flowage fees, and less aircraft sales.

Aircraft sales. Some lessors strive to obtain payments on a percentage of the gross or net revenue from aircraft sales. For the typical aviation business, however, it is felt that aircraft sales should be encouraged by not taking a direct percentage on any sales (wholesale or retail), presuming that a long-range benefit will result from more aircraft operating hours, more maintenance, storage, fuel, and other related businesses. Some states apply a sales tax that is returned in part to the local jurisdiction where the sale took place.

Fuel sales. Payments on the sale of fuel through the imposition of flowage fees should receive careful consideration by both parties to the lease. Fuel is part of the line service, which is expected and provided in the support of transient aircraft. The income from fuel sales often provides the revenue base that contributes to the development of a stable organization. It tends to support a business when sales or other revenues may not be forthcoming. Where the public body (for example: city, county, airport authority) retains fueling rights without providing other services, transients are frequently discouraged by the arrangement, service is not as responsive, and the aviation businesses tend to be weak, with higher failure rates. Fuel flowage fees range from one-half cent to a few cents per gallon on fuel pumped by the lessee. Flowage rates above this level create a heavy economic burden on the lessee and may cause additional burdens as fuel prices increase, controls become heavier, and customer reaction becomes stronger.

If fuel flowage fees are included in the lease, it is desirable that the following practical guidelines be followed:

1. The lessee has the right to fuel the business's own aircraft without flowage charges.
2. The lessee has the right to fuel aircraft operated by the local, state, or federal government without charge.
3. Air carrier aircraft with "turn-around" fuel contracts should not incur a charge.
4. Fuel fees should be based upon fuel sold and not on volume delivered into the storage facilities.
5. The fuel fee should be based upon a specified sum (for example, two cents per gallon) only for fuel sold at retail to transient aircraft.
6. The lessee should have the absolute right to select the fuel supplier.
7. Clear provisions should be included in the lease to cover bulk fuel storage agreements so as not to restrict the lessee or to proliferate the number of storage units.

Landing and parking fees. The lease may have a requirement that the lessor collect airport landing fees from its customers and transient visitors on behalf of the airport administration. Landing fees have long been a problem at airports, especially those open to the public. Many operators have found that the administrative costs of collecting landing fees for

Fuel is part of the line service, which is expected and provided in the support of transient aircraft.

small aircraft and small airports exceed the revenues. In addition, these funds are so small that they contribute very little income to the public agency. Many operators have found that landing fees act as a deterrent to itinerant pilots, especially to general aviation aircraft with alternative landing options. In going elsewhere, the aircraft take their additional, and perhaps major, business activity with them. If landing fees are required, it is suggested that they are collected only from commercial aircraft. Those aircraft involved in commercial passenger or freight operations, such as certificated carriers, commuter airlines, charter, and air taxi, are normally charged landing fees.

A partnership. The lease for the operation of an aviation business is more than a simple contractual agreement. Both the lessor and the lessee are partners who work for the benefit of the entire community. The agreement is more than a lease; it is a lease, an operating agreement, a partnership agreement, and a community service plan that establishes a mutually-beneficial relationship between the two parties and the public they serve. This kind of relationship is necessary if the airport's service and support facilities in a given community are to grow and to provide a stimulus to the growth of the aviation sector and the overall economy of that community. For aircraft to be useful in business and personal transportation throughout the network of almost 20,000 U.S. landing facilities, there must be a sustained improvement in the quality of the aircraft-support facilities across the country. The number of aircraft is growing, and more and more individuals and organizations are using aircraft for personal as well as business transportation. The partnership element of the aviation business lease will do a great deal to promote the necessary growth of aircraft support activities. This agreement cannot be one-sided; rather, it must be equitable and represent the interests of both parties.

Competition with Other FBOs

FAA non-exclusive rights policy. Any airport that has received federal funds from the Aviation Trust Fund must by law be available to all users. This includes being available to all FBOs that are interested and can meet required standards. This means that even airports that hope to be funded must comply with FAA regulations. Yet the rules also recognize that if new competition would cause neither the old nor the new FBO to have enough business to be viable, some protection for the existing FBO may be added. In many cases, the airport minimum standards permit self-fueling and self-maintenance by qualified operators. In instances that the airport standards prevent a certain type of newcomer due to a lack of service variety, area coverage, or capitalization, the FBO needs to be vigilant that the standards are being enforced and applied fairly.

Since an existing FBO at a publicly-funded field has limited protection against the granting of leases to new competitors, the best protection may be to provide a full-service, high-quality operation so that all aviation community needs are satisfactorily met.

Through the fence. A "through-the-fence" operation is one on land not owned by the airport owner but with legal or long-standing rights of access through an actual gateway or across an imaginary property line onto the airport. Some airports actively encourage such through-the-fence activities as, for example, it is common for corporate aircraft hangars to be connected to industrial parks surrounding airports. In other cases the through-the-fence situation may involve only one operator who may have an established aviation business. The problem for the airport operator is that there may not be safety, legal, or financial control over operations off an airport, meaning no revenue can be collected.

The problem for other FBOs is that the through-the-fence operator may not be paying the true cost of airport facilities. Businesses on the field may be subsidizing the through-the-fence operator. FBOs already on an airport will generally find that it is not in their favor if through-the-fence activities are permitted. However, FBOs seeking to enter operations at a busy field may find that there is no available airport land and that a through-the-fence easement to adjacent land may be their best alternative. The FAA discourages it, but local airport administrations may be persuaded to permit it, as may private airport owners. As existing airports use up their available land, this type of situation may become more common, and FBOs need to fully understand both sides of the question.

Threats to the General Aviation Physical System

Lack of Appreciation of GA's Role

The role of the national general aviation airport system is little understood, except by those directly involved. While most people in the U.S. have at flown on an airline and appreciate its availability, most do not hold the same expectation for general aviation. However inaccurate, General Aviation's image of frivolity dates from the early days of the barnstormer and includes jet-set playboys buzzing around on weekends, business executives traveling to Washington on corporate jets to request bailout funds from the federal government, and so forth. The general aviation airport may be seen by some members of the public as "a marina for planes" at best, or a public nuisance at worst. General aviation fulfills many important functions, both in providing a major element of the national transportation system and in enabling many revenue-generating and protective activities to be performed from the air. However, the average member of the public has very little knowledge or understanding of the general aviation industry's contribution; if impacted, though, the general public is much more likely to be vocal about its nuisances than about the benefits.

As was discussed in Chapter 9, Flight Operations, corporations find general aviation to be of great value to their efficient operations. Yet, business aviation operators may exacerbate general aviation's poor image in several ways:

> Business or corporate aircraft may be seen as overly luxurious;
> Some companies allow employees to use their aircraft for private purposes. Reinforcing the idea that general aviation is an elitist perk, this is perceived as taking the tax write-offs without appropriate employee reimbursement; and
> Some corporations are secretive about their reliance on general aviation travel, not only for fear of offending stockholders but also due to concern over the security of their executives or the environmental sustainability of their operations.

Recent marketing campaigns sponsored by industry organizations, such as AOPA's "General Aviation Serves America" and NBAA's "No Plane, No Gain" programs, have contributed greatly to the overall effort by the general aviation industry to improve general aviation's image among the non aviation-oriented public. It is important that those involved in the industry realize the necessity of working to counteract the commonly unfavorable image of general aviation.

Airspace Restrictions

Poor images of general aviation, from either safety or social standpoints, may translate to unfavorable treatment when priorities must be set. Especially in the allocation of the national airspace and airport system, Figure 12.2 showed that reliever and other general aviation airports handle almost 70 percent of based aircraft in the U.S., and thus substantially decongest the larger airports of small aircraft traffic.

Noise and Operating Restrictions

Many small airports were sensibly built outside of towns. But many of those towns have spread in all directions and now often surround the airport. At the same time, the airport traffic has often grown heavier and more frequent; and if jets or frequent touch-and-go training flights are involved, the airport environment is often significantly noisier.

The early phase of airport master and system planning in the 1970s often led to confusion in the public mind. For example, the term "reliever airport" was defined by legislation as general aviation airports relieving general aviation traffic from airline airports yet is still often misconstrued by the public to mean that their little suburban airport may start serving jumbo jets. It does not matter that there is not room for the necessary runways and that airline logistics preclude split operations except at the biggest hubs. Such misinterpretation over the years has led to substantial community fears about airport growth, noise, and safety. However, residents' fears about increased airport noise are not a myth but a reality. Such noise concerns must be addressed by all airport businesses if they are to survive and thrive.

Before 1990, many airport communities acted independently to introduce night curfews, and restrictions on particular classes of aircraft and types of aircraft operations. In 1990, Congress enacted the Aircraft Noise and Capacity Act (ANCA), which seeks to protect the entire national airport system by not allowing operating restrictions unless it can be proven (through a very onerous process) that the economic benefits outweigh the costs. This is now codified as FAR Part 161. These regulations were written with such stringent requirements that it is extremely difficult to get a Part 161-based airport operating restriction in place; in fact, no community has thus far been successful, although a few have proceeded with local restrictions. For example, Naples, FL has been strenuous in its pursuit of operating restrictions and, given other preoccupations, the industry appears to not be battling this particular case at present.

In lieu of getting a Part 161 or local restriction established, the community's only recourse is to close the airport completely; unfortunately, this can happen all too easily.

Airport Closures

Public airports are at risk for closure. In theory, the fact that a public airport has accepted federal funds commits it, through the grant assurances, to stay open another 20 years. A landmark case that reached legal resolution in 2002 suggests otherwise. Kansas City, Missouri wanted to give up the runway portion of Richards-Gebaur airport and use the area instead for a freight rail/truck cargo transfer point. AOPA filed suit to prevent FAA from releasing the city from its grant obligations. A second suit by friends of the airport was joined with the AOPA suit, but the 8th Circuit Court ruled in favor of FAA in 2000 and in 2002 the U.S. Supreme Court declined to hear the appeal. So the airport was closed, establishing an unfortunate legal precedent.[9] NATA created a list of the nation's 100 "most needed" airports, which are in fact the 100 most vulnerable airports. Many key airports in the system are on this list, for it is the busier fields that generate the most resident concern.[10]

Disappearance of Private Airports and Back Country Strips

A study in the late 1970s showed that the private airport system is the main source of new airport facilities. While there have been some results in recent years in the effort to expand publicly-owned airports, it is still true that the pipeline of new fields into the system depends heavily on private actions. A private-use field may open, may eventually become a public-use field, and then eventually offer a full array of aviation services and facilities. Yet the privately-owned airports, which provide needed airport system growth, are some of the most threatened airports in the country for such reasons as:

1. There is no guaranteed succession when the owner dies or retires. Today, World War II, Korean era, and Vietnam War era pilots who opened FBOs are retiring or dying; the next generation may see much greater value from selling the land for development and much less reward in taking over a stressful business.

2. As private facilities, such airports are often required to pay property taxes from which their publicly-owned counterparts are exempt,

making it harder to compete financially. They may be taxed at highest and best use, rather than actual use, which can be financially punitive and often forces development to generate revenue to pay the taxes.
3. They must finance all capital improvements, even those for which fees cannot be directly charged, from their own funds. Public airports, on the other hand, can be subsidized by 90 percent. Since 1982, private airports in the NPIAS can apply for federal funding; however, there are many hoops to jump through and strings attached in applying, causing the process to be of marginal value.
4. As private businesses, these airports may not get the support of local decision-makers for needed land use and airspace obstruction controls.

In 2001, GAMA and others were active in supporting a bill to protect general aviation access to airstrips on federal land and the airspace over it.[11] The bill was aimed at ensuring that the state aviation agency and the FAA would have an opportunity to determine the necessity of an airstrip before other organizations are allowed to restrict access or permanently close it. It is ironic to note that, while a major concern of the FAA is to maintain and promote the nation's airport system, other federal agencies may be operating indirectly against this objective.

Siting New and Expanded Facilities

At the same time that the privately-owned airports are coming under such pressure, the public system has its own problems. Because of fears about noise, safety, and unlimited expansionism, it is becoming increasingly difficult to obtain all the necessary approvals for a new publicly-owned airport. As Figure 12.1 shows, the NPIAS anticipates building only 11 new GA airports in the next five years. From inception to ribbon cutting, a new public airport can take 10 to 15 years and cost millions of dollars; thus, it is unlikely that all projected airports will actually be constructed. It can be very difficult to site a new airport or expand an existing one. Encroachment by housing and commercial development may make it costly to acquire needed land for larger runway protection zones and runway lengthening. The site requirements of even a modest new airport are extremely challenging because in metropolitan areas, conveniently-located flat land, suitable for a runway, is the first to be developed for other purposes since this is cheaper than building on a slope. While new GA airports do not require vast acreage, new air carrier airports have been getting larger and larger; Denver International Airport, which opened in 1995, occupies 33,000 acres of land, while Dallas-Fort Worth International Airport, opened in 1974, lies on over 18,000 acres. The older Seattle-Tacoma International Airport, by comparison, rested on just over 3,000 acres prior to a runway addition in 2008.

For a period after the new international airport at Denver was opened, experts were heard to comment that this might be the last new airport of any reasonable size to be created in the U.S. In 2001 and early 2002, action began regarding the new Chicago-area "South Suburban" airport at Peotone that would relieve congestion at nearby O'Hare and Midway airports. Even though the first land parcels have been purchased, opponents are still raising questions about whether the need could be handled in another way.[12] Similarly, plans to expand Fort Lauderdale's airport, in development for more than a decade, have run into heavy community opposition. A public meeting in 2002 to review runway extension proposals drew 650 opponents.[13]

Lack of Suitable Land Use Controls

Poor land use planning and zoning to protect small airports are still a problem. Incompatible developments, such as housing, are still often permitted adjacent to airports. Some of the most avidly anti-airport organizations are in communities still permitting such developments next to the runway or under the flight path. Communities and local elected officials seem unwilling to protect airports from encroachment through land use controls. This is partly because, in the face of the numerous competing local priorities, the value of the airport does not appear high enough to take special steps against popular, city-wide housing and economic development policies. In addition, voters tend to complain to elected officials about noise more than airport users and proponents voice airport benefits. A typical example was reported in the Atlanta Citizen: an area of land just south of Falcon Field, one of Atlanta's airports, was zoned for industry but, unknown to the airport authority, the land-owner and the city entered conversations about rezoning it for a 121-unit subdivision directly under the flight path.[14]

The FAA's own procedures help to create the problems. The FAA deals only with aircraft in the air but leaves it to local communities to address land use, development, and zoning. Yet the FAA has operated until very recently from a strictly-interpreted guideline that airport noise is only a community impact if the annualized noise contour is above DNL 65 dB. This noise metric has a long history; one of its rationales is that, in urban areas, it is difficult to distinguish airport noise from ambient city noise due to traffic and people at levels below DNL 65 dB. However, the EPA suggests new housing should not be built on sites exposed to above DNL 55 dB. The FAA does appear to be getting more flexible about enabling a lower noise metric to set the standard.

Even the FAA has not completely succeeded in restricting encroachment by those who wish to use land near airports for unsuitable purposes. The Wendell Ford Aviation Investment and Reform Act (AIR-21) of 2000 further extended the prohibitions on the construction of new municipal solid waste landfills found in the previous FAA Reauthorization Act of 1996. AIR-21 prohibits new landfill construction within a 6-mile radius of public airports that have received AIP funding and serve scheduled commercial aircraft with fewer than 60 passenger seats.[15] While this includes many airports, it is clear that general aviation airports without commercial service are exempt from this regulation.

Our airport system is finite; it therefore needs preserving. Even a small new airport requires at least one mile of flat land with road and utility access as well as unobstructed approaches. Such prime real estate, if it can be found, cannot serve major metropolitan and urban centers unless it lies close to the developed areas. An inherent conflict exists between airports and communities.

Bird Strike Threat

In another example of inadequate use of land, local officials often allow such incompatible facilities as landfills near airports; as it was shown in the previous section that many general aviation airports are not regulated in this respect. Since landfills almost always attract large numbers of birds, their presence near airports of any kind creates a bird strike hazard. Birds are capable of causing aircraft accidents, as was seen in 2009 with the landing of an airliner on the Hudson River due to multiple bird strike-induced engine failures. A protected waterway may encourage wildfowl too near the airport. Bird control is anathema to animal rights activists. Airport administrators use falcons, firework-type explosions, shotguns, and the planting of unappealing grass as ways to minimize the presence of birds on and around the airport, however, the control of birds in the air around airports continues to be a challenge.

Revenue Diversion

In the past fifteen years or so, revenue diversion has become a major aviation issue. Communities with profitable airports have sought to use—and have actually used—the surplus to benefit other community programs. The FAA and the U.S. DOT Inspector General have been assiduous in seeking to ensure that these funds are retained for aviation use, since national airport and airway system funding needs are much greater than available resources. FAA Reauthorization Acts beginning in 1982 provided progressively clearer guidelines regarding what is and what is not revenue diversion.

This is not an area that an FBO manager can do much about unless he or she is also the publicly-owned airport operator—in which case she or he will be provided with detailed guidelines about how to proceed. Despite the recent publicity the issue has received, it appears likely that communities facing budgetary constraints will continue to look to profitable airports as a source of local government finance.

Inadequate General Aviation Funding

The Airport and Airway Trust Fund (AATF), which was established in 1970, funds airport planning, improvement and construction projects at NPIAS airports. From the beginning, the larger airports have generated the majority of trust fund revenue through airline ticket taxes. They have also been allocated most of the grants through AIP, the mechanism for disbursing funds from the AATF. The current legislative program distributes most of the funds to primary commercial service airports, which are those with more than 10,000 annual passenger enplanements.

At GA airports the total available funding is relatively small, due to AIP funding formulas. An FBO desiring to identify funds slated for its NPIAS airport may find the data on the web; Figures 12.4 and 12.5 show how the $43.6 billion of infrastructure development proposed in the 2020 NPIAS report is allocated by airport category or purpose of development. Small airports, even though a key component of the national airport system, must therefore compete against one another for limited funding. This is compared with primary commercial service airports, which are entitled to PFCs, other unique fees, and a certain amount of funding based upon passenger enplanement levels.

Figure 12.4 ✈ **NPIAS Development Cost Estimates by Airport Category**

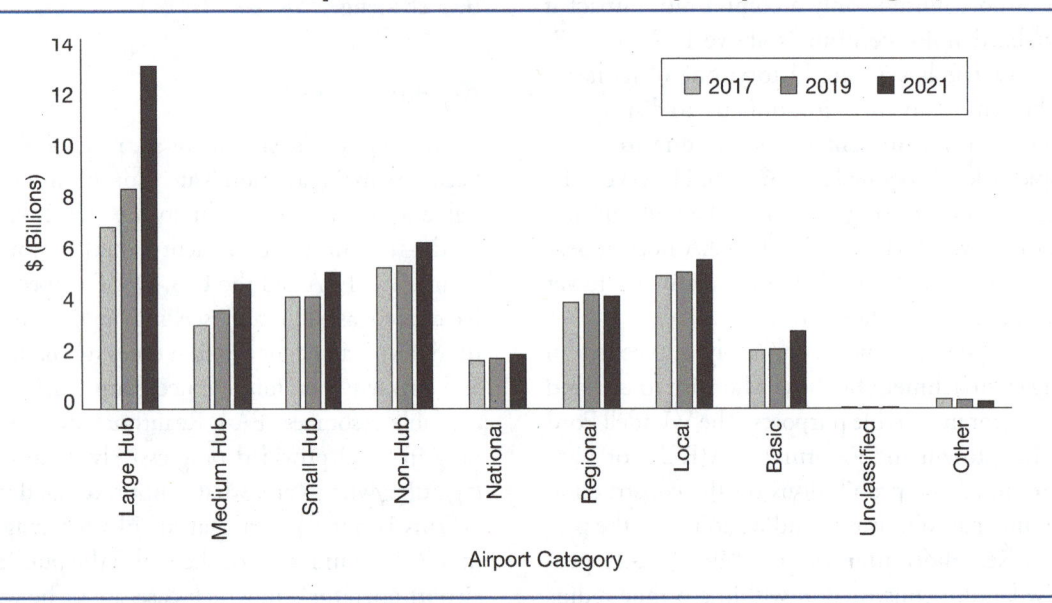

Source: FAA

Figure 12.5 ✈ **Total NPIAS Cost ($43.6 Billion) Estimates by Type of Development**

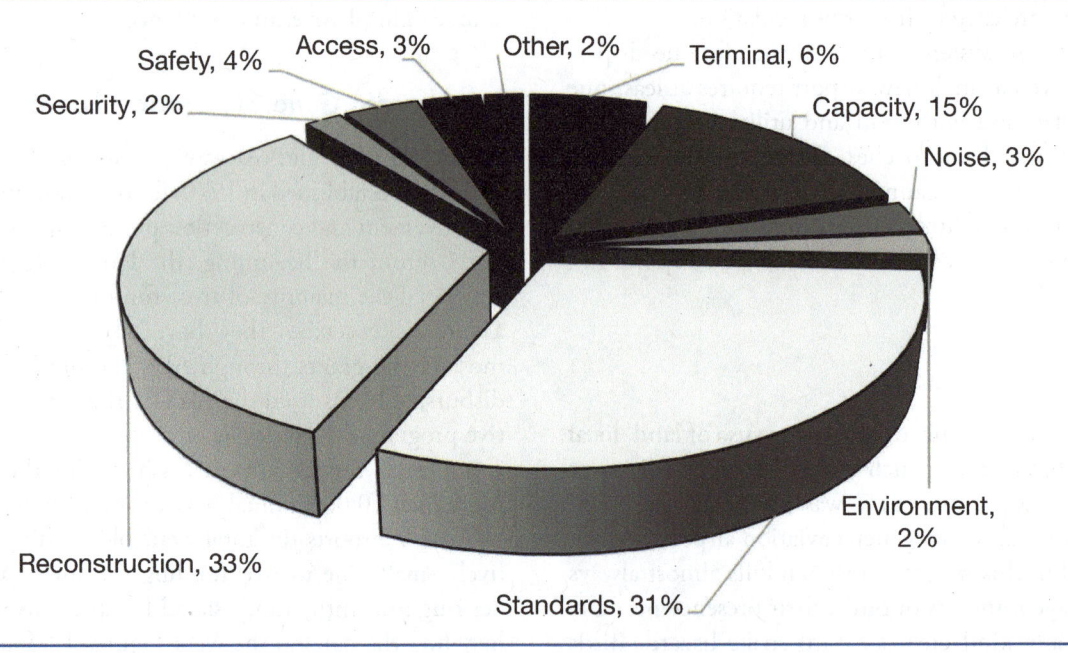

Source: FAA

Beginning in the 1980s, the available funds for airport improvements accumulated more rapidly in the AATF than they were expended. Pressure is mounting for a higher level of funding, especially for larger airports. If this does not occur, the passenger ticket tax will be cut back to reduce the balance in the trust fund, even though national airport needs far exceed available funds. Tenants and operators at small and large airports alike need to ensure that these airport improvement funds continue to be available at appropriate levels.

The airport system nationwide can demonstrate a much greater need for capital improvements than revenue projections would appear to support. Continued growth in passenger travel and inflight operations means that existing airports all around the country are approaching capacity in their airspace, terminals, aircraft parking, maintenance areas, and runway capacity. The problem has been greatly exacerbated by the unwillingness of Congress to appropriate all funds accruing in the AATF. As noted by the FAA in the 2020 NPIAS report, the 2,908 nonprimary (commercial service, general aviation, and reliever) airports make up 88 percent of the airports yet account for only 34 percent of the total development. General aviation and reliever airports, however, generate only approximately 7 percent of the tax receipts, primarily through fuel taxes, that flow into the AATF.[16] Thus, scheduled passenger service strongly cross-subsidizes the general aviation system. This is even more apparent when one considers that general aviation aircraft comprise around 96 percent of the civil aircraft fleet and generate around 63 percent of flight hours, thereby using a larger share of the airway system than they contribute through AATF revenues. The General Aviation industry needs to maintain or increase the level of cross-subsidy, but for this to happen it needs to vigilantly and persistently educate the public about the role and value of the GA system.

New Initiatives and Opportunities

The list of issues above means it is becoming increasingly important to the health of the general aviation system—and that includes FBOs—to preserve and enhance the viability of the airport system that already exists. To accomplish this requires an intensive two-pronged nationwide strategy, addressing noise and economic benefits.

Noise Abatement Programs

It appears that the FAA is becoming more flexible about measuring and addressing airport noise, and more willing to adapt national noise metrics to local conditions. In a strongly-worded article in *Airport Magazine*, it is suggested that airports adopt their own measures for incompatibility—if an airport is surrounded by hills for example, the higher homes will experience much more noise than if the land were flat, even though the FAA Integrated Noise Model may not generate noise contours above DNL 65 dB. Thus, if people are being disturbed, DNL 65 dB is too loose a measurement for this community.[17]

According to this article:

"Minneapolis and Cleveland have recently taken steps to formally establish DNL 60 dB as their local threshold for compatible land use. Both announced programs to expand their Part 150 residential sound insulation programs to the DNL 60 dB contour line. But will the FAA approve the use of Federal funds for sound insulation programs outside of DNL 65 dB noise contours? Cleveland's Part 150 update contains a measure to sound-insulate residences within or contiguous to the 60 DNL band of the . . . noise contours. FAA approved the measure in August, 2000 on the basis that the airport operator has adopted the DNL 60 dB noise contour as the designation of non-compatible land use, thus making the measure fully eligible for AIP or PFC funding."

In FAA's noise policy update draft published for comment in 2000, the FAA promised to support efforts of establishing local noise standards and to recognize those standards in Part 150 noise compatibility programs. However, NATA, in an October, 2000 statement to FAA about its noise policy, opposed the lowering of the 65 DNL contour. This is an ironic move, since this is a tool that could help communities do more about appropriate land use around airports, especially if it made them eligible for federal funds to soundproof homes in a broader area. If implemented, it would to some extent reduce the pressure for operating restrictions.

Continuous Noise Abatement Strategy

A noise study will not, in one step, eliminate noise and the threat it can pose to an airport's future. Indeed, in many cases the opportunity to be heard encourages airport neighbors to mobilize in opposition to the airport. Prevention, rather than cure, is the best strategy. Prevention means setting up good operational techniques before they are demanded. It means making sure that chronic complainers are invited to see the positive side of the airport in every way possible. It means keeping the business community and town leaders aware of the way airport users depend on FBO services in order to contribute to the local economy.

Economic Benefits Studies

The two prongs of an approach to airport preservation and enhancement are (1) noise abatement and

(2) convincing the neutral or hostile airport neighbor of the economic importance of that local airport to his or her community. This section discusses communicating the economic benefits of the airport. It is a difficult challenge that requires continuous effort.

In the past, there have been two basic approaches to documenting the economic benefits of general aviation airports. One is heavily statistical and seeks to quantify many impacts generated directly or indirectly by the airport. This includes airport operator, FBO and airline salaries and taxes, income from airport users, spending in the community by air travelers, and "multiplier" effects as revenue generated by the airport is respent in the local or regional economy.

The other approach is more qualitative or journalistic and presents profiles of key airport users and what they produce, showing descriptively how the airport is vital to their business.

The former approach can cause those examining the situation to become inundated with data, making it difficult to see the big picture. The latter, on the other hand, may not provide enough hard facts to convince skeptics. A blended approach may therefore be best. Regardless of the airport owner's activities in pursuing such studies, it is advantageous for marketing and public relations if the FBO makes its own ongoing effort in this regard. The FBO should know who its aircraft tenants are, how often they use the airport, how their businesses are changing, and how their businesses affect the community. Employment and payroll figures should be readily available for the FBO itself and for its customers. The FBO should keep a good transient log and, periodically, when a new company signs in, should call to find out if there will be regular visits, and whom they are coming to see. The FBO should be active in at least one local business group such as Kiwanis, Lions, Jaycees, or Rotary Clubs or the Chamber of Commerce so that the facts about the FBO operation, and the role of the airport, reach ears other than those in the aviation community. FBOs should become involved in airport economic studies sponsored by others in order to maximize their own favorable publicity.

Some airports do much more than this. For example, the Cheyenne Regional Airport in Wyoming has conducted multiple economic impact studies that demonstrate the airport's increasing contribution to the economy.[18] The airport manager makes frequent presentations to various stakeholder groups regarding the merits of the airport. Many other airports are undertaking similar research and communications efforts. Airport business owners can press for such studies where they are not being done, contribute to them when they are, and work with the airport administration to help spread the word about the airport's benefits to the community and region.

NATA offers a Community Relations Toolkit, and AOPA also has a kit.[19]

Airport Privatization Program

Worldwide, there is a growing trend toward selling public airports to private companies, which then continue to operate them for a profit. For example, BAA Limited (formerly the British Airports Authority) was formed as a private entity in 1986 to own and operate seven British airports, including London Heathrow. This privatization trend has also become popular in the railroad industry and in other areas of public infrastructure operation, such as the long-term leasing of toll roads to private operators. In the U.S., the trend toward airport privatization is much less prevalent. Regardless, it is important to consider the implications of the transition from public to private operation, since, for the fixed base operator, this could mean a change in landlords from a governmental entity to a private company. It is noted that the type of ownership may also determine the likelihood of competitors entering the market.

The FAA Reauthorization Act of 1996 first acknowledged the concept of domestic airport privatization by authorizing a privatization pilot program.[20] Its goal was to encourage privatization at a minimum of five locations. The revenue diversion issue, already mentioned, is an obstacle when a private company wants to take a percentage of airport revenue as profit in return for assuming the risk of operating the airport. In spite of this deterrent, however, some examples of domestic airport privatization have occurred. In 1994, prior to the 1996 Reauthorization Act, the Indianapolis Airport Authority handed over management of Indianapolis International Airport and several reliever airports to BAA on a ten-year contract that was subsequently extended for two additional years. Stewart International Airport in Newburgh, New York, an early pilot program applicant, became private when another British group acquired a 99-year lease on the airport.[21] Several other airports explored privatization possibilities in the early years of the new century, but then the privatization trend in the U.S. began to reverse. Stewart International once again became a public airport with the acquiring in 2007 of the remaining 93 years of its lease by The Port Authority of New York and New Jersey. Similarly, the Indianapolis Airport Authority resumed operation of Indianapolis International and its relievers in December of 2007. Finally, a $2.5 billion deal to privatize Midway International Airport in Chicago collapsed in early 2009 due to the global credit crisis. This effectively ended all current efforts at domestic airport privatization.

How any further attempts at airport privatization will affect airport tenants remains to unfold. Privatization efforts seem unlikely to grow quickly for various reasons, including revenue diversion for business profit. Another reason is that those current managers assigned to investigate the merits of privatization might stand to lose their jobs if a transition to private management were to occur. An additional stumbling block is the program's provision that airlines at commercial service airports may have veto power over airport projects and related fee increases.[22] Barriers such as these will most likely have a negative effect on future privatization efforts in the U.S.

NASA Small Aircraft Program

NASA has become increasingly active in slower and lower domains of flight, including general aviation and advanced air mobility. Between 2001 and 2006, NASA and the FAA performed a joint research venture named the Small Aircraft Transportation System (SATS). Involving a fleet of automobile-sized, high-tech, on-demand, jet-powered, aerial taxis and accompanying aerospace enhancements, SATS studied the expansion of point-to-point general aviation travel. Results of this program may encourage aviation-dependent business development at many communities nationwide that have smaller, less congested airports.[23] Early demonstrations of such air transportation oper-ations using very light jets (VLJs) modeled on SATS have seen limited success. DayJet, an operator of the Eclipse 500 VLJ, operated as many as 28 aircraft before scaling back and finally suspending its operations in 2008 due to financial concerns. Eclipse Aviation itself ran into financial problems, partially due to the loss of many orders from DayJet, and filed for bankruptcy in 2009.

Airport and Aviation System Planning

The single airport has little value itself; it must be part of a system. The FBO that recognizes this and is aware of changing markets, demographics, and roles of competing airports, is poised to take advantage of new opportunities. Periodically, planning agencies perform regional and national aviation studies. The FBO should be aware of these opportunities to present data express opinions on aviation concerns beyond the individual airport's boundaries.

Summary

While the continuous improvement of airport facilities is necessary for aviation businesses, FBO managers—provided they do not own the airport—find that they have relatively little direct control. In addition, the health of their business depends on having access to a viable national and international airport network. The attitudes and concerns of airport neighbors can also have a major impact. Sound physical-facility planning, therefore, takes into account the airport's wider environment and community attitudes. Airport physical facilities may be grouped into two categories: (1) the public-use facilities which may be shared by other FBOs, such as runways, taxiways, terminals and auto parking, and (2) the private-use facilities that the FBO uses through an operating agreement and lease. These facilities may include hangars, ramps, maintenance facilities, and office buildings. Public planning for an airport usually starts with a master plan that follows FAA guidelines. Funding may come from several public sources for the public portions of the airport; the FBO will have to raise private funding. FBOs can benefit greatly—as well as be adversely affected by—the planning and publicity created by an airport's public sponsors. Therefore, it is advisable to stay closely involved with all public actions affecting the field, to understand the planning and funding processes, and to communicate the economic value of the individual FBO operation.

Endnotes

1. U.S. Department of Transportation, Federal Aviation Administration. (2020). *Report to Congress: National Plan of Integrated Airport Systems (NPIAS) (2021–2025)*. Washington, DC: U.S. Government Printing Office. Available online at *https://www.faa.gov/airports/planning_capacity/npias/*

DISCUSSION TOPICS

1. How is an FBO affected by the operation of those parts of the airport not covered in its lease?
2. Why should an FBO be concerned with promoting the airport as a whole to the business community and airport neighbors?
3. What are six steps involved in an airport master plan?
4. What might be some of the considerations to be resolved when an FBO seeks financing for developments on leased airport land?
5. What is the most significant environmental issue at most airports and why? What can an FBO do about mitigating it?
6. What is the value of preventive maintenance and how would you set about developing such a program for a combination of maintenance hangar and office?
7. Identify five operational techniques that can be used to reduce noise around airports. What are their pros and cons from (1) the community's viewpoint; (2) the FBO's viewpoint?

2. Ibid.
3. NewMeyer, D. A., Hamman, J. A., Worrells, D. S.; & Zimmer, J. R. (1998). Needs Assessment of a Major Metropolitan Reliever Airport. *Journal of Air Transportation World Wide, 3*(2), 49–63.
4. For example, Boston-Logan International Airport, Boston, MA; also, Sea-Tac Communities Plan, Seattle-Tacoma International Airport, Seattle, WA.
5. For an excellent summary of noise control efforts around the country see Bremer, K. (1999). Aircraft Noise: Cooperation, Not Confrontation. *Airport Magazine, 11*(4).
6. See for example the latest editions of the following:
Oregon Department of Transportation, Aeronautics Division. (1981). *Airport Compatibility Guidelines* (Oregon Aviation System Plan, Volume VI). Salem, OR: Author.
U.S. Department of Transportation, Federal Aviation Administration. (1978). *Airport Landscaping for Noise Control Purposes.* (FAA Publication No. 150/5320–14). Washington, DC: U.S. Government Printing Office.
U.S. Department of Transportation, Federal Aviation Administration. (1983). *Noise Control and Compatibility Planning for Airports.* (FAA Publication No. 150/5020–1). Washington, DC: U.S. Government Printing Office.
Airport Noise Compatibility Planning, 14 C.F.R. 150 (2009). Notice and Approval of Airport Noise and Access Restrictions, 14 C. F.R. 161 (2009).
U.S. Department of Transportation, Federal Aviation Administration. (2007). *Airport Master Plans.* (FAA Publication No. 150/5070–6B). Washington, DC: U.S. Government Printing Office.
U.S. Department of Transportation, Federal Aviation Administration. (2006). *Minimum Standards for Commercial Aeronautical Activities.* (FAA Publication No. 150/5190–7). Washington, DC: U.S. Government Printing Office.
U.S. Department of Transportation, Federal Aviation Administration. (1978). *Potential Closure of Airports.* Washington, DC: U.S. Government Printing Office.
U.S. Department of Transportation, Federal Aviation Administration. (1987). *A Model Zoning Ordinance to Limit Height of Objects Around Airports.* (FAA Publication No. 150/5190–4A). Washington, DC: U.S. Government Printing Office.
7. See for example Bremer, K. (1998). The Three R's: Reduce, Recycle, Recover. *Airport Magazine, 10*(2).
8. See *http://energy.gov/nepa/downloads/eo-13123-greening-government-through-efficient-energy-management*
9. See *https://caselaw.findlaw.com/us-8th-circuit/1306288.html*
10. National Air Transportation Association. (2001). *America's 100 Most Needed Airports.* Alexandria, VA: Author. Retrieved from *https://www.nata.aero/data/files/gia/airport_misc/2007ga_airportsinitiativewhitepaper.doc*
11. General Aviation Manufacturers' Association. (2001, June 14). Press Release. Washington, DC: Author.
12. (2002, April 16). *Suburban Chicago News.*
13. (2002, April 16). *South Florida Sun Sentinel.*
14. (2002, April 14). *The Atlanta Citizen News.*
15. U.S. Department of Transportation, Federal Aviation Administration. (2006). *Construction or Establishment of Landfills Near Public Airports.* (FAA Publication No. 150/5200–34A). Washington, DC: U.S. Government Printing Office.

16. See *https://www.faa.gov/about/budget/aatf/*

17. Albee, W. (2001). The Compatible Land Use—Noise Challenge. *Airport Magazine, 13*(2).

18. Cook, B. (1995). Small Airports—Front Door or Doormat? *Airport Magazine, 8*(1).

19. See *https://www.nata.org/sites/default/files/nata-communications-toolkit.pdf*.

20. Federal Aviation Administration Reauthorization Act of 1996, Pub. L. No. 104–264, 110 Stat. 3213 (1996). Retrieved from *http://www.gpo.gov/fdsys/pkg/PLAW-104publ264/content-detail.html*.

21. Arthur, H. (1998). Airport Privatization—A Reality Check. *Airport Magazine, 10*(5).

22. Arthur, H. (2000). FAA's Airport Privatization Pilot Program—Are We There Yet? *Airport Magazine, 12*(5).

23. Croft, J. (2002). Small Airports: To Be or Not To Be? *Aviation Week and Space Technology, 156*(April 15), 58–61.

Index

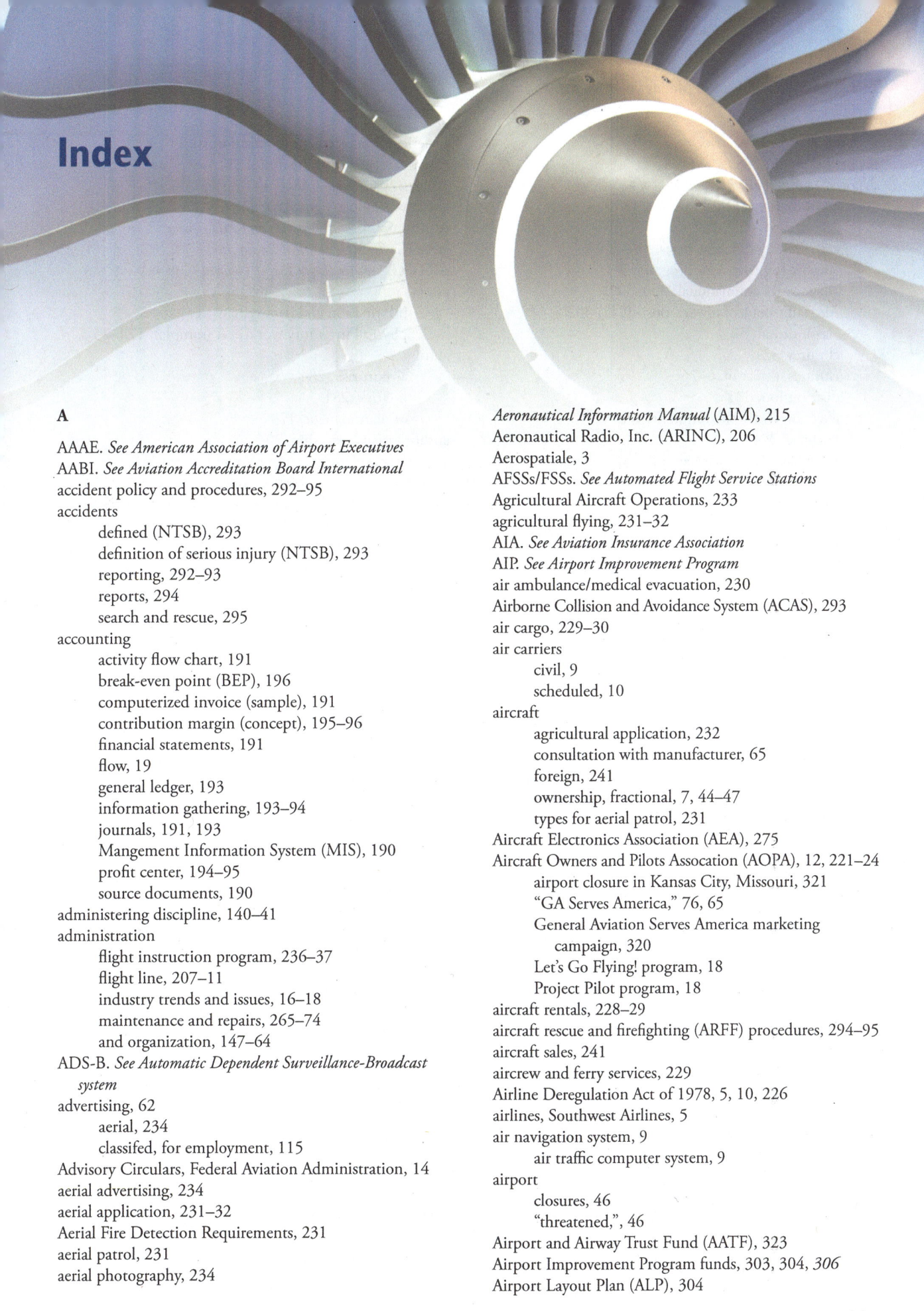

A

AAAE. *See American Association of Airport Executives*
AABI. *See Aviation Accreditation Board International*
accident policy and procedures, 292–95
accidents
    defined (NTSB), 293
    definition of serious injury (NTSB), 293
    reporting, 292–93
    reports, 294
    search and rescue, 295
accounting
    activity flow chart, 191
    break-even point (BEP), 196
    computerized invoice (sample), 191
    contribution margin (concept), 195–96
    financial statements, 191
    flow, 19
    general ledger, 193
    information gathering, 193–94
    journals, 191, 193
    Mangement Information System (MIS), 190
    profit center, 194–95
    source documents, 190
administering discipline, 140–41
administration
    flight instruction program, 236–37
    flight line, 207–11
    industry trends and issues, 16–18
    maintenance and repairs, 265–74
    and organization, 147–64
ADS-B. *See Automatic Dependent Surveillance-Broadcast system*
advertising, 62
    aerial, 234
    classifed, for employment, 115
Advisory Circulars, Federal Aviation Administration, 14
aerial advertising, 234
aerial application, 231–32
Aerial Fire Detection Requirements, 231
aerial patrol, 231
aerial photography, 234

Aeronautical Information Manual (AIM), 215
Aeronautical Radio, Inc. (ARINC), 206
Aerospatiale, 3
AFSSs/FSSs. *See Automated Flight Service Stations*
Agricultural Aircraft Operations, 233
agricultural flying, 231–32
AIA. *See Aviation Insurance Association*
AIP. *See Airport Improvement Program*
air ambulance/medical evacuation, 230
Airborne Collision and Avoidance System (ACAS), 293
air cargo, 229–30
air carriers
    civil, 9
    scheduled, 10
aircraft
    agricultural application, 232
    consultation with manufacturer, 65
    foreign, 241
    ownership, fractional, 7, 44–47
    types for aerial patrol, 231
Aircraft Electronics Association (AEA), 275
Aircraft Owners and Pilots Assocation (AOPA), 12, 221–24
    airport closure in Kansas City, Missouri, 321
    "GA Serves America," 76, 65
    General Aviation Serves America marketing campaign, 320
    Let's Go Flying! program, 18
    Project Pilot program, 18
aircraft rentals, 228–29
aircraft rescue and firefighting (ARFF) procedures, 294–95
aircraft sales, 241
aircrew and ferry services, 229
Airline Deregulation Act of 1978, 5, 10, 226
airlines, Southwest Airlines, 5
air navigation system, 9
    air traffic computer system, 9
airport
    closures, 46
    "threatened,", 46
Airport and Airway Trust Fund (AATF), 323
Airport Improvement Program funds, 303, 304, *306*
Airport Layout Plan (ALP), 304

331

airport noise. *See noise abatement*
airport risk audit, 284
airports. *See also noise abatement*
    accessibility to general public, 18
    access, post-9/11, 220
    activities, by different types, 301–2
    Airport Improvement Program, 9
    airspace system, national, 300–302
    Air traffic control towers, post- 9/11, 302
    classification, 301
    closures, 321
    closures, threat of, 6–7
    distribution, 301
    effect on wider environment, 302–4
    facilities, 304–6
    FBO facilities, described, 307
    FBO facilities on airport premises, 307
    four levels of physical facilities, 299–300
    hierarchy, national, 300–302
    landfills and bird strike hazard, 323
    land use, development, and zoning, 67, 307, 322–23
    losses of, 299
    marketing the facility, 67–68
    NATA list of "most needed" airport, 321
    National Plan of Integrated Airport Systems, 9
    new initiatives and opportunities, 325–27
    non-aviation sources of revenue, 92
    nonexclusive rights and federal funding, 264, 319–20
    operating agreement with FBO, 314
    operating restrictions, 321
    participation by FBOs in planning, 305
    privately owned, and back country strips, 305
    privatization program, 326–27
    public, 305
    Reliever Airport, 321
    reliever airport, 302, 305
    revenue planning, 306
    system planning, 327
    threatened, 321
    threats to continued operations, 320–25
airport system, 9, 300–302
Air Route Traffic Control Centers, 9
airspace system, 302
Air Taxi Commercial Operator (ATCOs), 295
Air Taxi Operations and Commercial Operation of Small Aircraft, 242
Air Traffic Control (ATC) centers, 277, 246
Air Traffic Controllers, 246
air traffic control towers, post- 9/11
    Federal Aviation Administration (FAA), 300
air transportation, benefits, 226–27
Alaska Air Carriers Association, 12
Alien Flight Student Program (AFSP), 295
American Association of Airport Executives, 12

American Management Association, 128
American Petroleum Institute, 212
"analysis paralysis," 148
anticipate industry trends, failure to, 22–23
AOPA. *See Aircraft Owners and Pilots Association*
ARINC. *See Aeronautical Radio, Inc.*
ARTCC. *See Air Route Traffic Control Centers*
assets
    information and data systems, 169
    physical inventory worksheet (sample), 174–75
audit
    airport risk, 284
    fire risk, 281
    of management, 179
authoritarian leadership, 30
autocratic leaders. *See authoritarian leadership*
Automated Flight Service Stations (AFSSs/FSSs), 215
Automatic Dependent Surveillance-Broadcast (ADS-B) system, 246
automobile liability, 288
Aviation Accreditation Board International (AABI), 115
aviation forecasts, 47–53
    budget development, 81
    expenses, 76
    Master Plan and Airport Layout Plan (ALP), 304
aviation, general (GA), 10–11
    and public perceptions, 17
aviation hours flown, 12
aviation industry groups, 11–14
aviation industry, growth, *versus* insurance costs, 286
aviation insurance, 285
Aviation Insurance Association, 12
Aviation Maintenance Technician (AMT), annual salary survey, 139
aviation manager, practical decision tools for, *34*
aviation product liability coverage, 290
aviation publications, list of, *54*
aviation risk exposure, 280–81
aviation risk reduction, 283–84
aviation safety, 284
aviation security, 295–96
aviation tenant-landlord agreement, 291–92
Aviation Trust Fund, 7
Aviation Weather Center (AWC), 216
avionics technology, 7
    new devices, 7
AWC. *See Aviation Weather Center*

B

balance sheet
    budgeting, 85
    operating account, 81
    sample, *82–83*

balance sheet (sample), *82–83*
balloons, 239–40
billing, invoice form (sample), 207
body language, communication skills, 130
Boeing, 10
Boeing Aircraft Company, 3
borrowing, cash flow problems, 72
branding, 65
    institutional promotion, 65
    issue of, 27
Brandt, Steven, 107
break-even point (BEP) analysis, 78–79
British Aircraft Corporation, 3
budgeting
    annual sales (worksheet), 85
    assembly and approval, 88, 91
    balance sheet, 85
    balance sheet (sample), *82–83, 90*
    cash budget worksheet, 92
    chief difficulties in, 83
    expense, 85, 89
    expense worksheet (sample), *89*
    flexible or variable, 91
    income statement, 85
    income statement (sample), 81
    information systems, 180
    introduction, 81
    maintenance facilities, 274–75
    monthly report (sample), *86–87*
    operation and control, 91
    preparation, 84
    purchases, 84–85, 88
    purchases budget worksheet (sample), *88*
    sales, 84
    timetable, 84
    timetable for development, 84
    worksheet (sample), 90
business. *See also organization structures*
    organization and administration, 147–64
business activities, analyzing, 177–79
Business Aviation Software Engine (BASE) software, 199–200
business ethics, 37–38
business plan, *26*
    company history and background, 27
    financial information, 28
    functional schedules, 28–29
    growth strategy, 28
    human resource needs, 106–7
    industry overview, 27
    legal requirements, 27
    management, 28
    marketing analysis, 27
    marketing plan, 27
    market niche and goal, 26
    mission statement, 25, 152
    need for, 24–25
    operations; production plan, 27
    personnel, 27
    products, 27
    strategy and objectives, 28
    values, 25
business practices, unsound, 16
business security
    combating losses, 201
    confidentiality and control of information, 200
    types of losses, 200–201
business travel
    advantages of aircraft owning, 55
    effects of 9/11 terrorist attacks, 44
    "No Plane, No Gain,", 65

C

CAA. *See Civil Aeronautics Authority*
career track in aviation
    advancing within the business, 114
    hiring and recruiting, 115
    job descriptions, 107, *117*
    pipeline concept, 101–3
    students, 106
    Women in Aviation, International, 14
cash discount, 99
cash discount value, 99
cash flow
    data needs, 169
    forecast, 91
    methods for improving, 77
    month- by- month analysis, 76–77
    monthly chart (sample), *77*
    planning for positive, 75–77
    profits and, 71–72
cash *versus* credit, 75, 99
Center Weather Advisories (CWAs), 215
certification and licensing
    aviation mechanic's certificate, 254
    inspection authorization, 254
    training for maintenance workers, *110*
Cessna Pilot Center System, 237
channel pricing policy, 61
charter and air taxi operations, 227–28
charter operators, 45
Civil Aeronautics Authority (CAA), 1
Civil Air Patrol (CAP), 295
Clinton, William J., 308
closing a sale, 65–66
collection of payment, 66, 97
colleges and trade schools, 115

communication
- barriers to, 130
- dyslexia, prevalence of, 130
- feedback, 131
- improving, 131
- internal organization, 157–58
- verbal and non-verbal, 130

communication skills, poor, 23

communications systems
- analysis of, 189–90
- company correspondence, 184–85

community relations
- economic effect of airport on region, 326
- landfills and bird strike hazard, 323
- membership in business groups, 326

Community Relations Toolkit, 326

comparable worth, 113–14

competition, 57
- federal funding and airport use, 264
- flexible or variable, 92

complaints system, 67

computers
- -assisted maintenance, 272–74
- company correspondence, 184–85
- desktop models, 197
- record keeping, 187
- selecting, 198
- service bureaus, 200
- software choices for management, 198–200

Congress
- Aircraft Noise and Capacity Act (ANCA), 321
- attitude toward aviation, 11
- funding through Airport and Airway Trust Fund, 323–25
- mandates for security measures, 295
- National Transportation Safety Board (NTSB), 15

consulting, 65

contribution analysis of sales, 66

contribution margin (concept), 195–96

corporation, 151
- mission statement, 152
- philosophy, 152

correspondence, land use and zoning, 67

Corridor Family software, 199

cost-based pricing, 57–58

cost containment, 74

cost control, 74

cost of sales, 180

credit check form, 98

credit, extending, 93

credit, line of, secured, 93

credit management
- four Cs of credit, 94–95
- functions of, 94
- policy (sample), 95
- policy creation, 94
- process, 94–98

credit period, 99

crime risk, 281

culture of organization, 152–53

current ratio, 177

customer, feedback, 65

customer orientation, 42
- flight line personnel and, 211

customers. *See also flight line; front desk; marketing; promotion*
- cash *versus* credit, 75, 99
- charter operators and FBOs, 45
- credit applicants, 96
- effects of negativity, 204
- flight line employees, 211
- fulfilling a want, 56
- funds on account, 93
- growth plan, 55
- segmentation of markets, 43

CWAs. *See Center Weather Advisories*

D

data systems, flow of information, 170, 172

Davis, Ralph, 154

debts, bad, 75

debt to equity ratio, 178

decision-making process, 32–35
- decide something, 35
- tools, *34*, 34–35

Defense Science and Technology Organization of Australia, 3

delegation, motivation of employees, 134

delegative leadership, 31

demand and supply curves, market planning, 58

demand-based pricing, 58

democratic leaders. *See participative leadership*

Denver International Airport, 322

department activities, analyzing, 179–83

Department of Homeland Security (DHS), 281

depreciation, 75

direct mail, 62, 65

dirigibles, 3

discrimination in employment, 111–12

distribution systems, 61–62

Domestic Reduced Vertical Separation Minima, 221

Drucker, Peter, 22

dyslexia, prevalence of, 130

E

EAA. *See Experimental Aircraft Association*

EBITDA, 178

economy
    benefits studies, 325–26
    benefits of aviation, 221–22
    cash *versus* credit, 71–72, 75
    effect of airport on region, 326
    effect on labor market, 103–4
    studies of benefits of airports, 325–26
    training to keep abreast of changes, 128–29
education
    general public, 18
    non- U.S. residents as students, guidelines, 295–96
    promoting the airport, 67
education [ADD flight training, mechanics, ETC] non- U.S. residents as students, guidelines, 295–96
ELS Fuel Management System, 199
employee evaluation, 134–38
employees. *See also human resources*
    application form, 116, *119–22*
    fair information practices, 114
    hiring and recruiting, 115
    role in marketing, 42
    screening, *118*
employment. *See human resources*
    lifetime, 133
employment agencies, 115
employment laws and regulations, 109, 111–14
Entrepreneuring: The Ten Commandments for Building a Growth Company (Brandt), 107
environmental compliance, 308
environmental concerns
    Greening the Government through Efficient Energy Management (Executive Order), 308
    Leadership in Energy and Environmental Design (LEED), 308
Environmental Protection Agency (EPA), 284
environmental requirements, underground storage tanks, 291, 308
expansion, profit maximization, 72
expenses
    budget development, 85, 89
    forecasting, 76
Experimental Aircraft Association, 12, 18, 239
    Young Eagles program, 18

F

FAA. *See Federal Aviation Administration*
FAA Intengrated Noise Model, 325
FAA Reauthorization Act of 1996, 326
facilities, supplied by FBOs, 216
FAR. *See Federal Aviation Regulations*
FBO. *See Fixed Base Operator*
FBO facilities on airport premises. *See also FBO operations*
FBO Manager software, 167
FBO operations
    aircraft sales, 318
    airport system planning, 327
    competition with other FBOs, 319–20
    data collection, 306–7
    equipment, planned replacements, 307–8
    fuel sales, 318
    landing and parking fees, 317
    leases with airport facilities, 308–19
    nonexclusive rights and federal funding, 264
    operating agreement, 314
    participation in airport planning, 305
    planning new airport, 307
    preventive maintenance, 307
    Small Aircraft Transportation System (SATS), 327
    zoning regulations, 308
FBOperational Fuel Management System software, 199
Federal Aviation Administration (FAA), 1
    accident investigations, 292
    Advisory Circulars, 14, 211
    airport noise and noise abatement, 325
    air traffic control towers, post- 9/11, 307
    application for repair station certificate, 250–51
    charter and air taxi flights, 227
    charter and air taxi regulations, 227
    Domestic Reduced Vertical Separation Minima, 221
    fractional aircraft ownership, 44–47
    functions, 14
    general, through 2040, *50*
    mandated aircraft positioning (ADS- B), 246
    private flight instructors, 2
    repair station certification, 250–51
    revenue diversion issue, 6, 326
    safety regulations, 279
    simulator training, 238–39
    U.S. civil airspace, 9
    weather data, 218
Federal Aviation Regulations (FAR), 229
    aviation mechanic's certification, 252–53
    concerning air carriers or air taxi operators, 45
    equipment requirements for maintenance, 255–56
    FBO line employees and fueling, 212
Federal Bureau of Investigation (FBI), 281
feedback
    communication skills, 131
    customer, 65
file control, 188
financial information, 28
financial records, 180
financial survival, 167–68
financing
    sources of money, 93
    types of money, 93
finger or pier layout of flight line, 206
fire inspection checklist, 282
fish spotting, 234

Fixed Base Operators (FBO)
    business practices, unsound, 16
    early history, 1–3
    industry trends and issues, 16–18, 44–47
    services supplied by, 15–16
    tasks and skills required, *108–9*
    tasks required, *108*
fixed *versus* flexible pricing policy, 60
flight hours, 2020, *49*
flight instruction
    administration, 236–37
    changing market, 234
    instructors, 237
    programs, 234–36
    training of instructors, 238
flight instructors, 237
    freelance, 238
    training, 238
flight line
    administration, 207–11
    aircraft handling, 207
    baggage and cargo handling, 207
    checklist for periodic review, 208
    customer service, 203–4
    fueling, 212
    functions of, 203–4
    layout, 206
    operations, 206–7
    radio services, 206
    Safety Management System (SMS), 211
    security, post–9/11, 211
    "send-off,", 2007
    service array and profitability, 212
    service request form (sample), 208, 209
flight operations
    access to major airports, 220–21
    aerial advertising, 234
    aerial patrol, 231
    aerial photography, 234
    agricultural flying, 231–34
    air ambulance/medical evacuation, 230
    air cargo, 229–30
    aircraft rentals, 228–29
    aircraft sales, 241
    aircrew and ferry services, 229
    air transportation, benefits, 226–27
    choosing services to offer, 226
    fish spotting, 234
    flight instructions training programs, 234–35
    fractional ownership, 229
    manual, 242
    market trends, 220
    organization, 227
    post–9/11, 220
    sport and recreational flyers, 239–41
    system issues, 220
    types of flights, 219–20
Flight Options, 45
flight planning and services, 212
flight plans, 9
    pilot services, 215
flight school
    administration, 236
    break-even analysis (sample), 78–79
    Cessna Pilot Center System, 237
    non-U.S. residents as students, guidelines, 295–96
    operations, 237
    required activities, 162
flight security, 295–96
Florida Aviation Trades Association, 12
flying lessons, break-even analysis (sample), *78*
"follow-me" vehicle, 66
"Forecasting Aviation Activity" (FAA guide), 48
four Cs of credit, 94–95
four Ps (product, price, place, and promotion), 42
fractional aircraft ownership, 222
    benefits of, 45
    trend toward, 44–47
free-on-board (FOB) pricing policy, 61
frontal or linear layout of flight line, 205
front desk
    appearance, 205
    car and hotel reservations, 215
    coordination of services and administration, 207
    log book for transient pilots, 215
    procedures, 214–15
    up-to-date tools, 205
    voice mail, disadvantages, 214
fuel
    autogas, trend in, 213
    cost-based pricing (example), 57–58
    cost increases, 5
    demand curve (example), 58
    ELS Fuel Management System, 199
    equilibrium of supply and demand (example), *60*
    extension of credit, 99
    FBOperational Fuel Management System software, 199
    Federal Aviation Administration requirements, 207
    sales, 318
    self-fueling, trend toward, 213
    supply curve (example), *60*
    systems for servicing aircraft, 206
    underground storage tanks, 291, 310
Fuller, Reed, 2
funding
    Airport Improvement Program (AIP), 301, 303, 304, *306*

cash flow problems, 72
federal, and airport use, 264
private airports, 321–22
public federal funds, 67
public, organizational structure, 302
future developments, scramjet, 3

G

GAMA. *See General Aviation Manufacturers Association*
GARA. *See General Aviation Revitalization Act of 1994*
"GA Serves America" (AOPA), 65
General Aviation industry
funding, 325
threats to physical system, 320–25
general aviation industry
flexible schedules *versus* commercial carriers, 302
General Aviation Manufacturers Association (GAMA), 10, 12, 220
General Aviation Revitalization Act of 1994 (GARA), 5–6, 44, *48*, 247, 283
general emergency risk, 283
Geographic Information Systems (GIS), 27
geographic pricing policy, 61
gliders and sailplanes, 239
globalization and airline industry, effect on labor force, 105
globalization of market, 5
Global Navigation Satellite System (GNSS), 246
Global Positioning Systems, 7
government. *See also Congress*
corporation regulations, 151
emergency response, 283
fire, safety, and construction laws, 255
land use, development, and zoning, 67
local, and noise abatement, 223
local, and "threatened" airports, 46
local, support for airport, 67
promoting the airport, 67–68
requirements and effect on business, 162
GPS. *See Global Positioning Systems*
Greening the Government through Efficient Energy Management (Executive Order), 308
gross profit, 180
growth of aviation business, market research, 44–45

H

handling credit transactions, 97
helicopter. *See rotorcraft*
Helicopter Association International, 14
Herzberg, Frederick, 152
Hindenburg, 3

hobby *versus* business, 73
hotel reservations, offered by FBOs, 215
hub-and-spoke route structure, 226
human resources. *See also personnel*
alcohol and drug testing, 123
background check of job candidate, 123, 125
benefits, 139
business planning, 106–7
communication with employees, 129–31
comparable worth, 113–14
compensation systems, 139–40
credit checks on job candidates, 125
disciplinary problems, 140–42
dishonest employees, 281
employee organizations, 144–45
employment outlook for mechanics, 105
employment outlook for pilots, 103–4
evaluating employees, 134–38
fair information practices, 114
fidelity bonds on employees, 281
hiring and recruiting, 115
information systems, 172, 173
interchangeability of most staff, 102
interviewing job candidates, 116, 123
job application form (sample), *119–22*
job candidate's suitability, 125
job description format (sample), *117*
job descriptions and specifications, 107, *108, 109, 117*
job offer, 125, 127
labor market trends, 103–4
laws against discrimination in the workplace, 111–12
laws and regulations, 109, 111–14
medical conditions, employees, 123
morale, reduction of petty thievery, 282
motivating employees, 132–34
Motivation- Hygiene theory, 152
orientation and training of new employees, 127–29
payroll taxes and deductions, 114
permanent *versus* temporary help, 107
personality testing, 125
personnel policy manual, 142, 144
physical examination, 125
pipeline analogy, *102*
pipeline concept, 101–3
promotion of employees, 139
recognition of high- quality work, 260
reference checking, 125, *126*
screening guide (sample), *118*
selecting employees, 116–27
separation, 142, *143*
skills required, 107
special activities, 107, 109
training, 107, *110*

338 Essentials of Aviation Management

    training requried for maintenance positions, *110*
    turnover, 101
    white collar crime, 283
    workers compensation, 113, 285
    workman's compensation, 113, 285, 288, 290
    workplace safety, 112
Human Side of Enterprise, The (McGregor), 132
hypersonic transports, 3

I

IFR flight plans, 215
IMG (consulting firm), 45
Immigration and Customs Enforcement (ICE), 281
income statement (sample), budgeting, *80, 85*
industry contacts, for employment, 114
industry norms, 162
industry overview, 27
industry trends, 44–47
    information systems, 176
information exchange, five phases of, 129
information systems, 28–29, 74
    accounting, 193, 214
    analysis of business activities, 177–79
    analysis of departments, 179–83
    analyzing business as a whole, 177–79
    cycle of information, 184
    human resources, 172, 173
    performance monitoring, 169
    processes, 170–72
    purposes of, 168–70
    record keeping, 183–90
    special reports, 169–70
    taxes, 176
    tools, 196–200
inspection checklist (sample), 261–62
insurance
    aviation insurance, 285–86
    aviation, kinds of, 291–92
    business interruptions and losses, 286
    costs, *versus* growth of aviation industry, 285
    death or disability, 286
    effect on aviation in 1980s, 106
    effect on cost of aircraft, 53
    leases, 315
    normal business insurance, 286–88
    personal liability, 288
    reduction *versus* costs, 281
    regulations, 285
    selection of broker, 292
    special aviation coverages, 289–91
    special multiperil policy, 288
    tort reform, 47

"Integrated Flight Training System" (Cessna), 235
integrated marketing, 67
International Brotherhood of Teamsters, 144
International Fire Service Training Association's (IFSTA), 295
Internet
    use by FBOs, 4
    web- based advertising, for employees, 115
Internet-based technology, 22
Internet-based teleconferencing, 22
inventories, 178
inventory. *See also parts and supplies*
    assets and liabilities, 75
    just- in- time (JIT) ordering, 259
    minimum *versus* unlimited, 256
    sales summary (sample), 257

J

jet engines
    development of, 3
    early history, 3
job application form (sample), *119–22*
job descriptions and specifications, 107, *108, 109, 117*
job evaluation, 139
job satisfaction
    enlargement of job description, 156
    evaluating training program, 129
    Motivation- Hygiene theory, 152
    other than financial compensation, 133

L

laissezfaire. *See* delegative leadership
land use and zoning, 67, 307, 322–23
land use, development, and zoning, 67, 322–23
Langley, Samuel, 2
Large Aircraft Security Program (LASP), 221
Leadership in Energy and Environmental Design (LEED), 308
leadership styles, 30–31
    group leader in informal organization, 161
leases, 308–9
    aircraft sales, 318
    business practices, 316
    components, 310
    deadlines, 313
    disputes, 312
    duty, 313
    equipment and fixtures, 313
    exclusive rights, 317
    fire loss, 313–14
    flight operations, 315

fuel sales, 318
insurance, 315
landing and parking fees, 318–19
landing and parking fees, collection of, 317
lessor inspection, 316
line service, 315
maintenance and repair, 313
maintenance, repair and overhaul, 315
minimum criteria, 313
minimum standards, 317
motor vehicle parking, 316
operating agreement, 314
operation, 312
options, 311
ownership after termination, 314
partnership, 319
payments, 317
percentage of the gross, 318
performance bonds, 314
preliminary planning, 309–10
procedural steps, 310
real estate rent, 317–18
release, 312
relocation of site, 314
removal at termination, 314
rent, 314
reversion, 311–12
right of prior approval, 313
rights after termination, 311
rights and obligations, 312
sales, 315
security, 315
service charges, 315
signs and advertising, 316
site location, 310–11
snow removal, 316
subleases, 316
taxes/payments in lieu, 316
termination before expiration, 311
term of, 311
towing disabled aircraft, 315
uniforms, 316
vending and game machines, 316
Legal Information Institute at Cornell University, 111
legal liability, 287–88
legal structures. *See also organization structures*
corporation, 151
partnership, 150–51
sole proprietorship, 149–50
leisure time *versus* stress, 74
leverage ratios, 178
liabilities
employer, 288
list of areas, 290

lifelong learning, 128
light-sport aircraft, 4
Likert, Rensis, 155
liquidity ratios, 177
location, critical element in FBO's success, 61
lumped-order terms, 99

M

mail-service control, 187
maintenance and repairs. *See also mechanics*
activities, 247–52
administration, 265–74
application for repair station certificate, 250–51
certification, 249–52
checklists, 260
competition, 264–65
facilities and equipment, 255–56
Federal Aviation Regulations (FAR), 248, 252
"gypsy,", 265
inspection, 260
inspection authorization, 254
inspection checklist (sample), 261–62
internal shop order form (sample), 271
leases, 313
liabilities in providing, 290
organizational structure, typical, 248
parts and supplies, 256–59
personnel, 252–55
pricing, flat- rate, 266, 272
professional organizations, 275
quality control, 259–64
record- keeping, 265
record-keeping, maintenance and repairs, 265
referrals, 264
repair ratings (FAA), 255
repair station certification, 250–51
required activities, 162
versus self- maintenance, 265
self- repairers, 169, 265
service data, analysis, 274
shop order (example), 267
shop order on computer screen (example), 268–70
"through- the- fence" operations, 264–65, 320
warning report (sample), 273
Maintenance Technician Magazine, 151
management, 14
analysis and action, 183
analyzing business as a whole, 177–79
audit, 179
bad-check losses, 283
definition of, 21
impact on, 144

    risk reduction practices, 281
    by walking around, 154
    white collar crime, 283
management audit, 179
management functions
    acknowledge and accommodate people's workstyles, failure to, 24
    anticipate industry trends, failure to, 22–23
    areas to delegate, 32
    business ethics, 37–38
    business plan, need for, 24–25
    decision-making process, 32–35
    develop, train, and acknowledge people, failure to, 23–24
    dos and don'ts of delegation, 32, *33*
    establish and adhere to standards, failure to, 24
    focus on profit, failure to, 24
    indecisiveness and lack of systems, 23
    lack of personal accountability and ethics, 23
    lack of priorities, 23
    leadership styles, 30–31
    managerial control, 31–32
    managing, *versus* doing, *29*, 29–30
    objectives of delegation, 31
    planning and organizing, 24–29
    poor communication skills, 23
    poor time management, 23
    problem-solving and delegation, style of, 30
    recognize needs people fulfill by working, failure to, 24
    support company policy in public, failure to, 24
    time management, 35–37
    traditional, 21–22
Management Information System (MIS), 190
managers
    analysis and action, 183
    budget development, 81
    budget operation and control, 91
    credit, extending, 93
    cycle of information, 184
    effects on profits, 75
    financing, 93
    goals of organization, determining, 147–48
    "Three-legged stool" analogy, 153, 163
    leadership of employees, 134
    management by result, 154
    management by walking around, 154
    personnel (*See human resources*)
    risk reduction practices, 281
    routines, percentage of work day, 149
    security issues, 200
    span of control, 154–55
    workload and delegation, 153

manuals
    Aircraft Fire Protection and Rescue Procedures, 295
    flight operations, 242
    maintenance, flat-rate, 266
    "Minimum Needs for Airport Fire Fighting and Rescue Services," 295
    organizational, 164
    safety procedures, 284–85
    Southwest Air Rangers (example), 272
manufacturers
    aviation, 10
    consulting, 65
    distribution systems, 61–62
    General Aviation Manufacturers Association (GAMA), 10, 12
    training workshops, 128
market analysis, budget development, 84
marketing. *See also promotion*
    airport facilities, 67–68
    analysis, 27
    budgeting, 67
    contribution analysis, 66
    contribution analysis of sales, 66
    customer orientation, 42
    customers' needs, 55–56
    definition of, 42–43
    effect of location, 53
    four Ps (product, price, place, and promotion), 42
    "General Aviation Serves America," 320
    industry trends, 44–47
    information systems, 67
    integrated, 67
    marketing plan, 66
    and market research, 44–45
    nature of aviation market research, 45
    need for, 41–42
    "No Plane, No Gain" plan (NBAA), 65, 320
    orientation, 74
    performance evaluation, 66–67
    plan, 27
    product and service definition, 56
    promotional activities, 62–65
    prospective customers, 55–56
    quality control, 67
    segmentation of markets, 43
    typical department, 42
marketing and promotion, merchandising, 62
marketing plan, 66
market-level pricing policy, 61
market niche, 57
    cash flow problems, 72
market projections. *See aviation forecasts*
market share, forecasting techniques, 53–55

Martin, Bob, 286
Master Plan and Airport Layout Plan (ALP), 304–5
Master Plan and forecasting, 305
McGregor, Douglas, 132
mechanics. *See also maintenance and repairs*
    aviation mechanic certification, 254
    employment outlook, 105
    inspection authorization, 254
    productivity, measuring, 275
    repairman, 254–55
    tailgate, 238
    "tailgate," 238
    training, 253
    wages and benefits, 106
meeting management, *37*
meetings
    face-to-face *versus* teleconferencing, 44
    "real-time," 157
merchandising, 62
military, pilots, 8
mission statement, security (sample), 296
motivation
    employment, lifetime, in Japan, 133
    Maslow's hierarchy of needs, 132
    Motivation-Hygiene theory, 152
    rewards for work, 133
    teamwork, 133
    workplace as social center, 133
MyFBO.com, 167

N

NAAA. *See National Agricultural Aviation Association*
NAS. *See National Airspace System*
NASA. *See National Aeronautics and Space Administration*
NASAO. *See National Association of State Aviation Officials*
NATA. *See National Air Transportation Association*
National Aeronautics and Space Administration (NASA), 7, 327
National Agricultural Aviation Association, 14
National Airspace System (NAS), 211
National Air Transportation Association (NATA), 14
    annual salary survey, 139
    Community Relations Toolkit, 326
    list of "most needed" airports, 321
    noise abatement programs, 325
    survey of corporate self-fuelers, 213
National Association of State Aviation Officials, 14
National Business Aviation Association (NBAA), 14, 128
    fractional aircraft ownership, 46
    No Plane, No Gain, 320
    "No Plane, No Gain" plan (NBAA), 65
National Fire Protection Association (NFPA), 188, 284

National Parks Air Tour Management Act of 2000, 241
National Plan of Integrated Airport Systems (NPIAS), 9, 48, 301, 322, 323, *324*
National Records Management Council, 189
National Safety Council, 283, 284
National Transportation Safety Board (NTSB), 15
    accident reports, 293, 294
    reporting accidents, 292
National Weather Service (NWS), 216
navigation, technology advancements, 47
NBAA. *See National Business Aircraft Association*
NBAA noise abatement techniques, 223
needs assessment, sales planning, 65
neighbors, marketing the airport to residents, 67
NetJets, 45, 229
net working capital, 178
noise abatement
    Aircraft Noise and Capacity Act (ANCA), 321
    AOPA Guidelines for noise abatement, 224
    community fears over airport growth, 321
    complaints system, 67
    continuous noise abatement strategy, 325
    FAR Parts 150 and 161, 303–4
    NBAA noise abatement techniques, 223
    soundproofing, 303
    stage 2 and Stage 3 aircraft, 303
    "threatened" airports, 46
non-employee liability, 288
non-ownership liability, 289
"No Plane, No Gain" plan (NBAA), 65
NPIAS. *See National Plan of Integrated Airport Systems*
NPIAS ? what is this?, 9
NTSB. *See National Transportation Safety Board*
NWS. *See National Weather Service*

O

Occupational Health and Safety Administration (OSHA), 284
Occupational Outlook Handbook, 2008–2009, 103–4
O'Hare International Airport, 322
One Minute Manager, The (LOOK UP), 131
open-ramp or transporter layout of flight line, 205–6
operations, aviation (sample list), 176
opportunity cost, 37
ordering, economic ordering quantity (EOQ), 258–59
organizational culture, 152–53
organization structures. *See also legal structures*
    choice of structure, 162–64
    communications technology, 157–58
    culture of organization, 152–53
    decentralization and consensus, 153–54
    external pressures, choices of structure, 162–64

formal, 158
formal *versus* informal, 158–61
functional or matrix, 158–59
government and regulatory influences, 162
human factors, 156
industry norms, 162
integrated business activities, 163
internal, four major considerations, 158
Japanese productivity, studies of, 152
line organization, 159
line organization (chart), 159
line- staff organization (chart), 160
management by walking around, 154
manual, 164
Motivation- Hygiene theory, 152
new approaches, 157
practical applications guidelines, 162–63
rational model *versus* new approaches, 152
"rumor board," 161
span of control, 154–55
specialization, 153
specialization, pros and cons, 153
staff organization, 159
staff support, 155–56
teams and work groups, 155
orientation of new employees, checklist for, 127–29
Ovington, Earle L., 2
ownership of FBO, 93

P

Packard, David, 154
PAMA. *See Professional Aviation Maintenance Association*
parachuting, 239
Parkinson's First Law, 35
participative leadership, 30–31
partnership, 150–51
parts and supplies. *See also* inventory; inventory
economic ordering quantity (EOQ), 258–59
inventory card (sample), 258
sales summary (sample), 257
Passenger Facility Charges (PFCs), 301
payroll taxes and deductions, 114
personnel, 8. *See also mechanics; pilots*
budgeting procedures, 91
compensation, 139
policy manual, 142, 144
pilots
active, by type of certificate, *51*
advanced training, 46
Certified Flight Instructor (CFI) certificate, 237
employment outlook, 103–4
FAA medical certificate, 125
"gypsy," 238
instructor training, 238
log book for transient pilots, 215
military- trained *versus* civilian, 8
noise abatement techniques, 222–24
regional differences in airports, 67
simulator training, 238–39
Sport Pilot certificate, 4, 8, 46, *47*
students, 46
ultralight and homebuilt aircraft, 4
weather data, 215
women, 46
pollution, 17
ethylene glycol (for deicing), 308
hazardous wastes, 291
underground storage tanks, 291, 308
poor communication skills, 23
post-analysis action, 183
prepayment terms, 99
price-based pricing, 58–59
pricing
cost-based, 57–58
demand-based, 58
distribution systems, 61–62
policy, 60–61
policy, seven areas of, 60–61
price- based, 58–59
true pricing theory, 59
product and service definition, 56
productivity
break- even chart, service shop (example), 276
Industrial Revolution, 153
Japanese workers, 152
maintenance, flat- rate pricing, 266
maintenance, repairs, and service, 274
mechanics, 275
product liabilities in providing, 290–91
product liability law, 290
product life-cycle pricing policy, 61
product line pricing policy, 61
products
classification, 56–57
competition, 57
variations in profitability, 79–80
Professional Aviation Maintenance Association (PAMA), 14
professional liability, 288
profitability ratio, 178
profit and loss statement (P&L), 28, 80–81
profits, 71–72
agricultural application, 233
array of services, 212
balance sheet (sample), *82–83*
break- even analysis, 78–79
definitions of, 72
flight line, 212

forecasting sales and revenues, 76
fueling, 212
historic context, 71–72
income statement (sample), 80
levels, 73–74
maximization, 72–73
monthly cash flow chart (sample), 77
non-monetary profits, 73
objectives, 72–73
pricing, flat-rate, for maintenance, 266
profit and loss statement, 80–81
ratios, in maintenance department, 275
versus security, 4
service and maintenance, 274
"supernormal,", 71
techniques for generating, 74
variations among product lines, 79–80
profit to sales ratio, 72
promotion. *See also marketing*
consulting, 65
effectiveness, evaluating, 66–67
"GA Serves America" (AOPA), 65
institutional, 65
"No Plane, No Gain" plan (NBAA), 65
planning chart (example), 63
word-of-mouth, 65
promotional activities
advertising, 62
direct mail, 62, 65
methods, 62–65
referrals, 65
sales, 65–66
promotional pricing, 61
promotional pricing policy, 61
promotion of employees, 139
property taxes, 57
public awareness, 17–18
pollution, 17
public ownership of an FBO, 93
purchases budget worksheet (sample), 88
purchasing
budget development, 84–85, 88
budget worksheet (sample), 88

Q

quality control, 67, 259–64

R

radio systems, 206
record keeping, 75
computers, 187
control of procedures, 186–87
design of records, 184
documentation of risk-prevention activities, 285
essential *versus* nonessential, 189–90
mail service, 187
management program, 186
National Fire Protection Association guidelines, 188
National Records Management Council, 188
retirement *versus* retention of records, 188
success of organization, 183–84
types of records, 186–88
records, financial, 180
Recreational Pilot certificate, 4
recruiters, or "head hunters" for employment, 114–15
"Red-Hot-Stove" Rule, 141
referrals, 65
regulations
agricultural application by aircraft, 231–33
charter and air taxi, 227
reliever airports, 302, 321, 325
reports
income report (sample), 181–82
maintenance warning (sample), 273
monthly budget (sample), 86–87
service and maintenance, 274
special, and information systems, 169–70
state and local requirements, for accidents, 294
restaurant, including at airport, 56
retained earnings, 92
return on investment (ROI), 72
revenue diversion, 6, 323, 326
revenue, loss since 9/11, 44
revenue sources, 92
rewards for business, 72
rewards for effort, profit motive, 72
risks of entrepreneurship, 72
rotorcraft, 3–4

S

safety. *See also accidents; security*
accident policy and procedures, 292–95
accident reporting, 292–94
communities adjacent of FBOs, 17
landfills and bird strike hazard, 322
National Transportation Safety Board, 294
risk exposure, described, 281–82
risk management procedures, 279–80
risk reduction, 281–85
search and rescue, 295
technical issues, 18
Safety Board, 294
Safety Investigation Regulations (SIR), 292
Safety Management System (SMS), 211
sales
annual budget, 84

collection and analysis of data, 66
collection of payment, 66
contribution analysis, 66
cost of, 180
total, 180
sales planning
    alternatives, 65
    closing, 65–66
    collection of payment, 66
    needs assessment, 65
    techniques to determine market share, 55
Sampling theory, 34
satellite system layout of flight line, 206
SBA. *See Small Business Administration*
SBIC. *See Small Business Investment Companies*
scramjet, 4
search and rescue, 295
search engine optimization, 4–5
Seattle-Tacoma International Airport (Sea-Tac), 17, 322
security. *See also accidents;* insurance; safety
    air carrier *versus* air feeder airports, 295
    airport access, post–9/11, 220
    background checks, 125
    electronic alarms, 281
    embezzlement, 200
    flight, 295–96
    information systems, 200
    post–9/11, 295
    versus profitability, 4
    since 9/11, 44
    Transportation Security Administration, 15
    Twelve Five Standard Security Program (TFSSP), 221
Security Directive 1542-04-8F (SD-08F), 4
self-interest, enlightened, 73
"send-off,", 207
separation, 142
September 11, 2001 terrorist attacks, 44
sightseeing, 240–41
skills, transferrable, 107
Skype (teleconferencing software), 44
Small Aircraft Transportation System (SATS), 7, 327
Small Business Administration (SBA), 30, 93, 128
Small Business Investment Companies (SBIC), 93
Smith, Adam, 152, 153
SMP-special multiperil policy, 288
SMS. *See Safety Management System*
social change
    training to keep abreast of, 128
    workplace as center of social life, 133
social responsibility, 73
sociometric techniques, informal organization, 161
software choices, 198–200
sole proprietorship, 149–50

sound planning process, 24
Southwest Airlines, 5
span of control, 154–55
special- purpose ratios, 179
sport and recreational flyers
    balloons, 239–40
    experimental and home- build aircraft, 239
    gliders and sailplanes, 239
    parachuting, 239
    rotorcraft, 240
    sightseeing, 240–41
    sport pilot license, 235
    ultralights, 239
Sport Pilot certificate, 4, 8, 46, *47*
SR-71 Blackbird, 3
storage
    of inventory, 75
    record keeping, 189
subsonic jet, 3
surveys
    advantages of aircraft owning, 55
    aircraft sales and inventory, 241
    Cessna survey on new pilots, 46
    corporate self-fuelers, 213
    market research, 44–45

T

tangible net worth, 178
taxes
    corporation, 151
    depreciation, 75
    information systems, 170
    payroll taxes and deductions, 114
    planning, 91
Taylor, Frederick, 152
teamwork, 133
technology
    alarm systems, 281
    avionics, 7
    information concerning, 176
    information systems and software, 74
    up-to-date tools, 205
    World Wide Web, 4–5
Terminal Area Forecasts, 48, *52*
terrorism, and security, 4
tiltrotor, 4
time management, 35–37
    poor, 23
    sources of, problems, 35
    strategies, 35–37
tort reform, 47, 247
Total FBO software, 167, 198, 199, 207

total sales, 180
training
- aircraft parts covered by training, 253
- communication skills, 129–31
- communications systems, 189–90
- customer orientation, 211
- developing supervisors and managers, 128
- dyslexia, prevalence of, 13
- five phases of information exchange, 129
- flight line personnel, 211
- on the job, 127
- keeping abreast of social and economic change, 128–29
- motivation, 132–34
- new employees, 127
- security procedures, 211
- value orientation, 134

training requried for maintenance positions, *110*
transportation, car reservations offered by FBOs, 215
transportation, national system of, role of aviation, 3
Transportation Security Administration (TSA), 4, 15, 211
- airspace usage, 221
- background checks of employees, 123
- origins, 15
- reference checks on job candidates, 125, *126*
- safety regulations, 281
- security functions previously covered by FAA, 295
- U.S. civil airspace, 9

trends. *See also aviation forecasts*
- advanced certificates and ratings for pilots, 46
- factors affecting future, *51*
- flight operations as key profit center, 220
- fractional aircraft ownership, 7, 14, 44–47, 222, 229
- general, through 2040, *50*
- historical data, 55
- inflationary, and cost of insurance, 287
- labor market, 103–4
- market, in flight operations, 220
- privatization of airports, 326–27
- self- fueling, 213
- self- repairers, 170, 265

troubled worker, 141–42
true pricing theory, 59
TSA. *See Transportation Security Administration*
Tupolev TU-144 (supersonic transport), 3
Twelve Five Standard Security Program (TFSSP), 221

U

ultralights, 4, 239
UNICOM (radio link), 206

U.S. Air Force, 3
U.S. Bureau of Labor Statistics (BLS), annual salary survey, 139
U.S. Department of Agriculture (USDA) Forest Service, aerial patrol, 221
U.S. Green Building Council, 308
U.S. insurance market, 285–86

V

venture capital, 93
vertical takeoff and landing, 4
very light jets (VLJs), 327
VFR flight plans, 216

W

wages and benefits in general aviation fields, 104–5
- compensation systems, 139

WAI. *See Women in Aviation, International*
weather
- insurance, 285
- services, 216

Welch, Jack, 152
Wendell Ford Aviation Investment and Reform Act (AIR-21), 323
Women in Aviation, International, 14
Woomera test range (Australia), 3
word-of-mouth promotion, 65
work force. *See also human resources*
- quality circles, 161
- social activities, 162
- task forces, 161–62

workman's compensation, 113
workplace safety, 112
worksheets
- annual sales, 84
- budgeting worksheet (sample), 90
- cash budget, 92
- expense worksheet (sample), *89*
- monthly budget (sample), *86–87*
- purchases budget worksheet (sample), *88*

World War II, 3
World Wide Web, 4–5
Wright brothers, the (Orville and Wilbur), 3, 4

Y

Young Eagles program, 18